History of Analytical Chemistry

Ferenc Szabadváry
Technical University of Budapest

The author presents a history of analytical chemistry from ancient times up to the present day. The development of analytical chemistry is treated chronologically up to the end of the eighteenth century; the first part of the book is concerned with the analytical works of ancient and mediaeval times and covers the period when iatrochemistry was practised; the phlogiston age; the era of Lavoisier to Berzelius; the first section concludes with a biographical chapter on Berzelius. In the following chapters the development of various branches of analytical chemistry is described separately. Other chapters deal with: the further development of qualitative and gravimetric analysis; titrimetic analysis; organic analysis; electrogravimetry; optical methods; electrometric analysis and other methods; the development of theoretical analytical chemistry. The material has been collected mainly from original contemporary papers and more than one thousand references are featured. One hundred figures mostly facsimiles and a number of portraits of old and new analytical chemists complete the work. This book will be invaluable for both analytical chemists and researchers of scientific history.

INTERNATIONAL SERIES OF MONOGRAPHS IN

ANALYTICAL CHEMISTRY

General Editors: R. Belcher and L. Gordon

Volume 26

HISTORY OF ANALYTICAL CHEMISTRY

HISTORY OF ANALYTICAL CHEMISTRY

BY

FERENC SZABADVÁRY

Technical University of Budapest

Translated by

GYULA SVEHLA

Technical University of Budapest

PERGAMON PRESS

OXFORD · LONDON · EDINBURGH · NEW YORK
TORONTO · PARIS · FRANKFURT

Pergamon Press Ltd., Headington Hill Hall, Oxford
4 & 5 Fitzroy Square, London W. 1
Pergamon Press (Scotland) Ltd., 2 & 3 Teviot Place, Edinburgh 1
Pergamon Press Inc., 44—01 21st Street, Long Island City, New York 11101
Pergamon of Canada, Ltd., 6 Adelaide Street East, Toronto, Ontario
Pergamon Press S.A.R.L., 24 rue des Écoles, Paris 5^e
Pergamon Press G.m.b.H, Kaiserstrasse 75, Frankfurt-am-Main

© Copyright 1966
Akadémiai Kiadó

First English edition 1966

This is a translation of the original Hungarian book
Az analitikai kémia módszereinek kialakulása
published in 1960 by Akadémiai Kiadó, Budapest

Library of Congress Catalog Card No. 64—23711

2046/66

CONTENTS

PREFACE	vii
I ANALYTICAL CHEMISTRY IN ANTIQUITY	1
1. Ancient science	1
2. The origin of chemistry	3
3. The earliest knowledge of analysis	5
Notes and references	8
II KNOWLEDGE OF ANALYTICAL CHEMISTRY IN THE MIDDLE AGES	9
1. Alchemy	9
2. Knowledge of analysis	13
Notes and references	20
III ANALYTICAL KNOWLEDGE DURING THE PERIOD OF IATRO-CHEMISTRY	21
1. The clarification of chemical concepts	21
2. The beginning of analysis in aqueous solutions	26
3. Robert Boyle	35
Notes and references	40
IV THE DEVELOPMENT OF ANALYTICAL CHEMISTRY DURING THE PHLOGISTON PERIOD	42
1. The phlogiston theory	42
2. The blow-pipe	50
3. Further light on reactions in solution	55
4. The beginnings of gas analysis	62
5. Torbern Bergman	71
Notes and references	81
V THE ESTABLISHMENT OF THE FUNDAMENTAL LAWS OF CHEMISTRY	85
1. Quantitative analysis before Lavoisier	85
2. The principle of the indestructibility of matter	90
3. Stoichiometry	97
Notes and references	111
VI THE PERIOD OF BERZELIUS	114
1. The recognition of the composition of minerals	114
2. The life and personality of Berzelius	125
3. The establishment of atomic weights	139
4. The state of analytical chemistry in the age of Berzelius	144
5. The first analytical textbooks	150
Notes and references	156
VII FURTHER DEVELOPMENTS IN QUALITATIVE AND GRAVIMETRIC ANALYSIS	161
1. Introduction of systematic tests for the ions	161
2. Gravimetric analysis	174
3. Microanalysis	185
Notes and references	192
VIII VOLUMETRIC ANALYSIS	197
1. Ancient history of titrimetry	197
2. From Descroizilles to Gay-Lussac	208

3. From Gay-Lussac to Mohr 227
4. Friedrich Mohr 237
5. The development of titrimetric analysis before the introduction of synthetic indicators 250
6. Development of indicators 257
7. The development of titrimetric analysis up to the present day 265
   Notes and references 271

IX ELEMENTARY ORGANIC ANALYSIS 284
1. From Lavoisier to Liebig 284
2. From Liebig to Pregl 290
   Notes and references 305

X ELECTROGRAVIMETRY 309
   Notes and references 316

XI OPTICAL METHODS 318
1. Preliminaries of spectroscopy 318
2. Kirchhoff and Bunsen 324
3. Further development of spectrum analysis 333
4. Colorimetry 337
   Notes and references 344

XII THE DEVELOPMENT OF THE THEORY OF ANALYTICAL CHEMISTRY 349
1. The development of physical chemistry 349
2. Wilhelm Ostwald 353
3. Development of the concept of pH 361
4. Theory of titration 365
   Notes and references 371

XIII ELECTROMETRIC ANALYSIS 375
1. The measurement of pH 375
2. Potentiometric titration 378
3. Conductometric titrations 383
4. Polarography 384
   Notes and references 387

XIV OTHER METHODS OF ANALYSIS 390
1. Radiochemical analysis 390
2. Chromatography 393
3. Ion exchange 397
   Notes and references 399

AUTHOR INDEX 403
SUBJECT INDEX 411
OTHER TITLES IN THE SERIES 419

PREFACE

As far as I know this work is the first attempt in the literature to describe the historical development of analytical chemistry in a systematic, continuous way.

Analytical chemistry is the oldest and one of the most important branches of chemistry. Modern scientific chemistry has been developed with the aid of analytical chemistry. Even later, almost all important new results in chemistry were preceded by the suitable development of analytical methods.

It is easy to prove this statement. The basic laws of general chemistry, like the laws of simple and multiple weight rates, could be established only when the chemists already had sufficient knowledge and skill in the analysis of minerals and inorganic compounds, and so could determine the chemical composition of a number of compounds. Stoichiometric calculations were made possible by the determination of atomic weights using analytical methods. Development of organic chemistry could begin only after the introduction of reliable methods of elementary organic analysis, by which the new preparations could be identified. Biochemistry had to develop in the "dark" until microanalysis could produce suitable methods for the identification of very small amounts of substances. But examples can be taken from other sciences also, although more could be mentioned from chemistry too. Beyond the boundaries of chemistry we can see many highly valuable services rendered by analytical chemistry. Everybody knows, for example, that clinical chemical analysis is an important part of medical diagnostics. Our century is known as the atomic age; this probably began in 1938 with the discovery of uranium fission. The latter was also proved by the methods of analytical chemistry. Nuclear physicists made uranium fissions even before then, but, since they did not analyse their products, could not recognize the process.

It is, I think, unnecessary to mention that industrial production is impossible without analytical chemistry. There are few factories today, whatever they produce, which do not have well equipped chemical control laboratories.

This important and widespread use of analytical chemistry provides the reason for examining and describing the so far neglected history of this science.

This history covers about four thousand years, and starting with the earliest data available to us lasts up to the present. The most reliable and objective judge of a discovery or scientific result is time. It is advisable therefore to complete a book on science history with one human era before the time of writing. From this historical aspect it is relatively easier to select the really important, lasting results, which have to be mentioned. When starting this book, I wished to do this, too.

Development however is very rapid. Had I finished the treatment of the history of analytical chemistry at the period of the early nineteen hundreds, this work would be very incomplete. There are so many new important discoveries in this field which originate from the present century, that I felt it impossible to omit them completely. Therefore, I made up my mind to give at least the (probably highly incomplete) outlines of the most recent developments.

For this an author must have much courage, because he cannot rely on the best judge, time, but on his own imperfect and subjective knowledge. It is more than probable, therefore, that a historian of a later age will find a number of deficiencies in that part of this book connected with the recent past.

In my work I have tried to use the original literary sources whenever possible, but sometimes these were unattainable. This is indicated clearly in the corresponding references.

The structure of the whole book was carefully considered. In historical books a chronological presentation is commonly used. This, however, would cause some trouble to the reader, who, being presumably a chemist, is accustomed mainly to a thematical treatment, and therefore would be less interested in a chronological one. Therefore, the author chose a mixed composition. From the beginning while separate methods were indistinguishable, i.e. until the middle of the 18th century, the history of analytical chemistry can be handled uniformly and chronologically; later, however, a thematical system is used. This has the inevitable disadvantage that sometimes a recession in time over several centuries becomes necessary.

A great number of the scientists who have helped to develop the methods of analytical chemistry are mentioned in this book. A name in itself means little to the reader except where it is that of a world-famous scientist. Therefore, I wanted to give brief biographies of the people mentioned in order to show the person behind the name. This has not been completely possible in spite of the work and time spent on it by the author. There are papers from early periods about whose authors no biographical information could be found, and there are also people from the recent past and even from modern times about whom no information was available.

I want to express my sincere thanks to Professor László Erdey for encouraging me in my work, and to Professor Ronald Belcher for the careful examination of the English text and for his valuable remarks. I am indebted to my friend and colleague Dr. Gyula Svehla for the tedious work of translating the Hungarian text into English and to Mr. Bernard Fleet for the editing work done on the English text.

Finally I am grateful for valuable data, suggestions and information to the following persons: Professor I. P. Alimarin (Moscow), Professor E. Asmus (Berlin), Dr. É. Bányai (Budapest), Dr. J. Barnard (Phillipsburg), Dr. W. C. Broad (Phillipsburg), Professor K. Fajans (Ann Arbor), Professor R. Flatt (Lausanne), Dr. E. Gegus (Budapest), Dr. É. Gere—Buzágh (Budapest), Professor W. Gerlach (Munich), Professor G. Hevesy (Stockholm), Professor J. Heyrovsky (Prague), Dr. E. H. Huntress (Cambridge), Dr. J. Inczédy (Budapest), the late Professor G. Jander (Berlin), Professor A. T. Jensen (Copenhagen), Dr. L. Káplár (Budapest), Professor I. M. Kolthoff (Minneapolis), Professor P. Mesnard (Bordeaux), Dr. I. Meisel (Leipzig), Professor I. Obinata (Sendai), Dr. F. Pilch (Graz), Professor J. Proszt (Budapest), Professor E. Rancke-Madsen (Copenhagen), Dr. H. Richter (Berlin), Dr. H. Roth (Limburgerhof), Dr. Schellhas (Freiberg), the late Professor E. Schulek (Budapest), Professor L. G. Silen (Stockholm), Dr. I. Tachi (Kyoto), Professor J. Teindl (Ostrava), Professor H. H. Willard (Ann Arbor), Professor A. Zeller (Vienna).

As a pioneer work, my book is presumably incomplete in many topics. I should therefore appreciate the comments and criticism of the reader.

Budapest

Ferenc Szabadváry

CHAPTER I

ANALYTICAL CHEMISTRY IN ANTIQUITY

1. ANCIENT SCIENCE

The so-called historical age of mankind only began 5000—6000 years ago. At that time men already possessed considerable knowledge. They worked in agriculture and in other branches of industry and they could build and sail ships. To fulfil these achievements they needed knowledge of many natural phenomena. Science was born with the classification and use of this knowledge for man's own needs, as well as by establishing the internal relations between them. During the further development of human society, science and production were always in close contact; the development of the one resulting in the development of the other. This correlation became more marked as the structure of human society became more elaborate, so that in modern times science and technology are indistinguishable.

The extent of scientific knowledge in ancient times is known mainly from Greek sources, therefore the history of all branches of science originates from the ancient Greeks. In Classical Greek times (that is, up to the death of Alexander the Great) Greek science was primarily human science, almost entirely philosophy.

The origins of numerous rules and ideas of the modern natural sciences are to be found in the teachings and natural philosophical views of the different Greek schools of philosophy. These views, however, did not come as the results of experiments, but mostly only by speculation. Greeks did not indulge in the analysis of the natural sciences, they examined nature as a whole and studied its pattern. It should be noted that in this field the Greeks were very far advanced. The various philosophical schools of thought had different views on the nature of materials and elements, and most books dealing with the history of chemistry usually give a brief outline of them.

Before the advent of the Greek civilisation, however, there existed people possessing a culture several thousand years old. These people, Sumerians, Egyptians, Babylonians, Phoenicians and Jews lived on the shores of the Mediterranean Sea. It is obvious from their architectural achievements that they must have had a considerable knowledge of mathematics, and also of physics. This knowledge must have had some influence on the early Greek philosophers. Recent studies have shown that numerous ideas, which had hitherto been thought to be of Greek origin, were known at a much earlier date. For example, Aristotle's theory of the five elements of the universe, fire, water, earth, air and ether is recorded in a very similar manner in one of the earliest Hindu works. The historian Diodorus of Sicily (1—2 B.C.) refers to this oriental influence as follows,

> Orpheus, Musaios, Homer, Lykurgos, Solon, Plato, Pythagoras, Eudoxos and Democritus of Abdera travelled to Egypt, and surely it was here that they learned much of that which was to make them famous later[1].

In all these oriental countries, however, the people lived in absolute monarchies governed by a strong central power, where state and religion were inseparable. In Egypt, as in Babylon, the priests were very powerful and the control of scientific knowledge was in their hands. Presumably, most of the existing scientific knowledge was known only to the temples, and this knowledge was passed from one generation of priests to the next and was a closely guarded secret. In contrast to this the city-states of the ancient Greeks were mainly democratically ruled communities in which neither strict dogmatic state-religion, nor a powerful religious order was allowed to develop. They wanted to express and discuss their knowledge and opinions quite openly. Greek philosophy could only develop from free discussion and its influence is to be felt even up to the present day.

Later, Philip II of Macedonia conquered the Greek city-states. When the conquered is much more cultured than the conqueror, then in most cases the latter adopts its culture. So it was that the Macedonian Empire adopted the Greek culture. Alexander the Great, son of Philip, was for a time tutored by Aristotle the greatest philosopher of ancient times. Afterwards, when Alexander conquered Persia, Egypt, and a part of India, Greek influence spread over the whole of Western Asia and this influence remained even after the untimely death of the conqueror. Although Alexander's empire later disintegrated, the enormous broadening of the sphere of influence of Greek science was to have a far-reaching effect.

This resulted in the second flowering of Greek science, the so-called Hellenistic age, the cultural centre of which was Alexandria. The knowledge of this age has a different character from that of the preceding classical ages.

The Hellenistic age saw the origin of the natural sciences of today. Although no great philosophical treatises similar to those which were written in the classical ages were produced, the discussion and explanation of the great philosophers' ideas was begun. This study was continued throughout the Hellenistic age and lasted even until the 16th and 17th centuries. In addition to this critical study of philosophy, the recently originated natural sciences underwent an enormous expansion. It is very probable that this sudden surge in the growth of natural sciences was due to the fact that the scientific knowledge of thousands of years, which had hitherto been a closed secret in the hands of the Egyptian and Babylonian priesthood was now open to investigation by the free-thinking Greeks.

Archimedes (287—212 B.C.) and Heron (2nd century B.C.) the first great names in the field of mechanics, lived at the beginning of Hellenistic times; Euclid (3rd century B.C.) was the founder of geometry; Hipparchos (2nd century B.C.) and Eratosthenes (276—195 B.C.) were the first great masters of astronomy.

This age, especially on the oriental shores of the Mediterranean Sea, lasted until the Arabs conquered Syria and Egypt in the 8th century.

Chemistry first appeared at the end of the second part of this age, and the early form of this science in particular stemmed from the marriage of Greek philosophy and the mystic, secret sciences of the Eastern peoples.

2. THE ORIGIN OF CHEMISTRY

Chemistry as a Science did not exist in ancient times although mankind possessed knowledge of many subjects, some of which now belong to chemistry. Even at the beginning of recorded history men were able to dye fabrics, tan leather and make vessels of pottery and of glass. He could also make soap and preserve foods. Women used a variety of cosmetics, the manufacture of which required the ability to work up fatty oils and to distil volatile liquids. What is much more important, however, man could make metals and alloys from ores and knew how to work them. In ancient times seven metals were known, namely gold, silver, copper, tin, lead, iron and mercury. In addition, most of the oxides of these metals were also known. Acetic acid was the only acid known to the ancients, and of the alkalis only the lime and alkali metal hydroxides, prepared by the caustification of soda or potash. They also used several naturally occurring salts, soda, saltpetre and rock salt. Potash was also prepared from vegetable ashes. They also knew the vitriols from which copper and iron sulphates were prepared. Together with a few metal acetates, prepared by dissolving metals in acetic acid, this was the extent of the chemical compounds known in pre-Hellenistic times.

Several common chemical operations, distillation, crystallization, evaporation and filtration were also well known. However, all these substances and operations were only used by craftsmen in the pursuit of their various trades. Accordingly, chemistry as such did not exist at this stage, but rather chemical technology.

Between the second and third century A.D. an aspiration was born; this was to produce gold and silver from base substances. To achieve this purpose conscious experiments were carried out, these experiments being supported by theories. This aspiration of "science" was called alchemy. The Roman Emperor Diocletian, after the suppression of a minor rebellion in Egypt, caused a diligent search to be made for all the ancient books which treated of the admirable art of making gold and silver, and without pity committed them to the flames, apprehensive as we are assured, lest the opulence of the Egyptians should inspire them with confidence to rebel against the Empire. This is the first recorded event in the history of chemistry, although Diocletian himself does not appear to have had much faith in this vain science [2]. The Sicilian astronomer and mathematician Julius Maternus Firmicus, who lived at the time of Constantine the Great, discusses in his book (336 A.D) the influence of the stars on the lives of men. In it he mentions that the study of chemistry is recommended to those who were born under the sign of Saturn [3]. This is the first known use of the word chemistry.

There are many theories as to the origin of this word, the most probable being that it is derived from the Egyptian word for Egypt, *Khemi*, and so literally it means Egyptian art. To this word the Arabs attached their article, *al*, so that it became, *alchemia*, and this word even in present day usage means the science of making gold. Others maintained that it was derived from the Greek words λυμος meaning fluid and κεω meaning to pour.

By the beginning of the 18th century the evolution of scientific thought had finally brought about the collapse of alchemy as a serious science, but it is interesting to note that in the light of modern knowledge on the structure of matter the basic assumption of the alchemists, namely that the difference between chemical elements is caused by variation in the amounts of the same basic components is fundamentally correct.

This basic assumption of the alchemists can be traced to the theory of Aristotle, which stated that all substances in the worldar e composed of four primary elements, namely water, earth, fire and air; a fifth element known as the ether having a rather spiritual character. The fact that substances are different is due to the fact that they contain differing proportions of the primary elements. Metals are formed in the depth of the earth from water and earth, over a long period under different conditions. The more earth a metal contains, the lower is its resistance to fire. In addition to earth, mercury also contains air.

If, however, all metals are formed from the same components then it should be possible to change one metal into another simply by altering the proportion of the various components, and although in nature the rate of this process is very slow, it was hoped that the choice of suitable laboratory conditions would increase the rate.

Numerous examples of changes in material were available; liquids were changed into air (or vapour) by the effect of fire (heat) and an earth-like residue remained. When iron was added to a solution of copper sulphate it disappeared and copper was formed. When mercury was melted with sulphur, a black substance was formed which on heating changed its colour to red. As they had no knowledge of compounds, elements or of solutions, or even of allotropic forms, the only conclusion that could be drawn from these phenomena was that new substances had been formed.

Hellenistic science was born from the marriage of Greek philosophy and the Babylonian – Egyptian occult sciences, and the effect of this double origin is to be seen most clearly in alchemy. Alchemy inherited besides the clear Greek theoretical foundation also the mysticism associated with the oriental religions and was never able to rid itself of it.

The age of alchemy lasted for more than a thousand years, through the whole of the middle ages, and it was active even until the early 19th century.

Although it did not achieve its aim, a large amount of knowledge and experience was gained on which the science of modern chemistry was built.

3. THE EARLIEST KNOWLEDGE OF ANALYSIS

Analytical chemistry is the mother of modern chemistry. Substances must always be first examined to find their composition and only then is it possible to use them for a definite purpose. Therefore, without analysis there would be no synthesis, without analytical chemistry there would be no chemistry. Consequently, the development of analytical chemistry was always ahead of general chemistry and only after analytical chemistry had reached a certain stage was it possible to formulate new chemical rules.

It has already been mentioned that in pre-Hellenistic times chemistry did not exist as a science. However, it was necessary in everyday life to have some form of control on the quality of various goods and merchandise. Methods were in existence at that time for this purpose which can be regarded as analytical procedures. Probably far more methods were in existence than we have record of today, because the general tendency in writing in those days was towards the philosophical rather than the practical.

Control of the purity of gold and silver, and the prevention of counterfeiting was always of primary importance to the administrators of the early communities. It is not surprising, therefore, that methods for analysing gold and silver were developed. The earliest known procedure which is still in use today, is known as the fire-assay or cupellation. The fire assay is a quantitative procedure and depends on weighing the substance to be tested before and after the heating. The balance has been known since very early times and is so old that the peoples of the ancient civilizations attributed its origin to the Gods.

In 2600 B.C. Gudea was king of Babylon and there remain some weights which were used in those times. These weights were made of stone and were made in the shape of a geometric or animal figure. They were marked with the size of the weight and they also had sacerdotal seals. Dungi was a king of the town of Ur in Sumeria (2300 B.C.). During his reign an institute was formed for the testing of measures, this institute being presided over by the priests. In the Babylonian system of weights and measures, which was similar to the metric system, there was a relationship between length, weight and volume measures [4]. Although it may be accidental it is worth while to note that the double ell, the unit of Babylonian measure of length, and the basis of this system of measures, is equal to the length of the second pendulum on the parallel adequate of Babylon (992·5 mm) [5]. In ancient Egypt also, the balance was known and its picture is to be seen on a great many paintings, where it is supposed to represent judgement over the dead. Walden has established interesting relationships between the measures of ancient times and European measures before the metric system, and as a consequence he inferred that the latter were derived from the former. A few examples are mentioned here. The Babylonian unit of weight, the *leight mina* was equal to 491·29 g (the Athenian and Ptolemaios units of weight were similar) compared with the weight of the old French pound which was 489·5 g and that of the Dutch pound 492·17 g, while that of Hannover was 489·6 g. Another example is that the volume of the English gallon, 4·5435 l. is approximately equal to 10 ancient Egyptian hins, 10×0.4548 l., etc. [6].

The methods for testing gold and also the early methods of metal and ore testing were simply scaled down versions of the manufacturing process. There were no alternative methods because of the lack of mineral acids, hence only dry methods could be employed.

The fire-assay has been used since the beginning of recorded history of analysis. In the Old Testament there are several references to its use. For example: "And I will bring the third part through the fire, and will refine them as silver is refined, and will try them as gold is tried" [7]. And "Son of man, the house of Israel is to me become dross: all they are brass and tin, and iron, and lead, in the midst of the furnace; they are even the dross of silver . . . As they gather silver and brass, and iron, and tin, into the midst of the furnace, to blow the fire upon it, to melt it . . . As silver is melted in the midst of the furnace, so shall ye be melted in the midst thereof" [8]. And also, "And I will turn my hand upon thee, and purely purge away thy dross, and take away all thy tin . . ." [9]. References to the fire assay have also been found on the Tel al Amarna tablets, cuneiform tablets in the Babylonian language found in the Nile Valley. The King of Babylon complains to Amenophis the IVth Egyptian pharaoh (1375–1350 B.C.) "Your Majesty did not look at the gold which was sent to me last time, they were sealed only by a clerk, therefore after putting them into the furnace, this gold was less than its weight" [10].

The most detailed description of refining is left to us by the Greek traveller Agatharchides of Cnidos:

> Metal founders, after getting a certain weight and measure of gold ore, put it into an earthen vessel, they add to it lead, salt, a little bit of tin and barley bran, then they cover the vessel with a lid, which is carefully plastered. Afterwards the vessel is heated in a furnace for five days and five nights continuously. After this the vessel is allowed to cool, and then pure gold is to be seen, free from impurities.

During this process the metal lost some of its weight [11]. This is undoubtedly a cupellation procedure, even if the reason for the addition of the tin and barley bran is a bit obscure, and also why the vessel was plastered over. Pliny described the metals which were used for counterfeiting and records that a law was passed to prevent it, ordering the examination of coins *(ars denarios probandi)*. This law resulted in such universal satisfaction, that a statue was erected in honour of Marius Gratidianus, who introduced the bill [12].

The examination of gold by the procedure of cupellation was called *obrussa* and hence the examined gold, *aurum obrisum* [13]. Suetonius records that the emperor Nero ordered that taxes must be paid on such tested gold *(Exigit aurum ad obrussam)*. Subsequent Roman emperors gradually reduced the noble metal content of coins, so that their examination was no longer a primary concern. The Emperor Alexander Severus wanted to restore the value of money so he ordered that all the coins in circulation be melted down and new money be minted. He then appointed coin examiners *(artifices)* part of whose task in the examination of coinage was to separate different alloys [14]. The other method which was in common usage for testing the purity of gold was by scraping it on a touchstone and examining the colour and extent of the yellow streak produced. Theophrastus (372 (?) — 287 B.C.) described this technique.

> Also the effect of a touchstone on gold is admirable, being similar to that of fire in that it betrays its content. Whilst fire changes the colour of the gold, with the touchstone conclusions are drawn from scratches. From these one can establish whether the gold is pure or if it contains silver or copper, and also the amount of these metals present in one *stater* (weight measure). The best touchstones are to be found in the River Tmolos [15].

The separation of gold and silver from one another was a much more difficult task. According to legend, Hieron, the King of Syracuse entrusted Archimedes with the task of finding out if the goldsmiths, who had made him a new crown, had cheated him. Archimedes solved this problem by a completely new physical method, measuring the specific weight of the crown. 250 years later Pliny refers to a chemical method:

> Into an earth vessel together with the gold must be placed two parts of salt and three parts of *misy*, and covered with a mixture of two parts of salt, and one part *schiste*. The whole must then be placed in the fire, when the mixture takes up everything that is not gold, and pure gold remains [16].

According to Strunz *misy* is iron sulphate, salt, of course, means sodium chloride, and *schiste* is brick powder. When the mixture is heated silver chloride is formed which then melts and is adsorbed on the brick powder [17]. This procedure is called *cementation* and was in use for hundreds of years.

Two ancient Egyptian manuscripts deal with chemistry and pharmacology, and are known as the Stockholm and Leyden papyri respectively, the names being derived from the libraries where they are now to be found. These manuscripts refer to the "qualitative" examination of gold and tin. If gold undergoes no change when heated it is considered pure; if it becomes harder then it contains copper, and if it becomes whiter it contains silver. Tin is first melted and then poured on to papyrus. If the papyrus burns to ashes then the tin is pure, but if the papyrus does not burn it is an indication that lead is present [18]. Evidently the explanation of this phenomenon is that the melting point of pure tin is higher than that of tin alloys.

Copper sulphate was quite often adulterated with iron sulphate. This was tested for in the following way: the copper sulphate was placed on a red hot iron and if any red spots appeared on the latter (evidently Fe_2O_3) then this would indicate that it contained iron sulphate [19].

There are also a few references to tests in solution; these were similarly used for the prevention of adulteration. According to Pliny the adulteration of copper sulphate was apparently rather profitable and he first recorded the use of a chemical reagent to detect adulteration by iron sulphate. A strip of papyrus was soaked in the extract of gall-nuts, and was then treated with the solution under examination. If this solution contained iron sulphate *(atramentum sutorium)* then the papyrus became black. *(Deprehenditur et papyro galla prius macerato; nigrescit enim statim aerugine illita* [20].)

The medical men of ancient times were concerned with the examination of water. Pliny makes reference to sulphuric, acidic, saline and iron-containing waters. Turbid waters were filtered and cleaned with white of egg. Water was also distilled, and it was noted that sea-water became drinkable after evaporating it, and condensing the vapour [21]. According to Hippocrates, the great Greek physician, the water which boils quicker is much better and cleaner than that which is slower to boil [22]. Though this is theoretically true, it is difficult to imagine that the slight delay caused by the presence of the salt in solution could have been observed. A much more accurate method is due to Vitruvius (lst century A.D.) who determined the purity of water by weighing the residue after distillation [23].

These few examples show how limited is our knowledge of the ancient methods for the examination of substances.

NOTES AND REFERENCES

1. BERGMANN, T.: *Geschichte des Wachstums der Chemie.* Berlin (1792) 66
2. GIBBON, E.: *Der Untergang des römischen Weltreichs.* Stuttgart 330
3. JULIUS MATERNUS FIRMICUS: *Math.* 3 15. Reference Kopp, H.: *Beiträge zur Geschichte der Chemie.* Braunschweig (1876) 43
4. WALDEN, P.: *Mass, Zahl und Gewicht in der Chemie der Vergangenheit.* Stuttgart (1931) 13
5. LEHMANN, Z.: *Z. f. Ethnologie* (1889) 21 322
6. WALDEN, P.: *Mass, Zahl und Gewicht in der Chemie der Vergangenheit.* Stuttgart (1931) 20
7. ZACHARIAH 13 9
8. EZEKIEL 22 18 22
9. ISAIAH 1 25
10. KNUTSON, J. A.: *Die el-Amarnatafeln.* Leipzig (1915) 93
11. MÜLLER, C.: *Geographi Graeci Minores.* Paris (1855) 1 125
12. PLINIUS, C. S.: *Naturalis historiae libri.* 33 45
13. SENECA, L. A.: *Epist.* 13 1 2
14. HOEFER, F. *Histoire de la chimie.* Paris (1866) 1 126
15. THEOPHRASTUS: *Opera quae supersunt omnia.* Paris (1866) 346
16. PLINIUS, C. S.: *Naturalis historiae libri.* 33 24
17. BERTHELOT, M.: *Die Chemie im Altertum und Mittelalter.* Wien (1909) 15
18. BERTHELOT, M. *Collection des anciens alchimistes grecs.* Paris (1887) 36
19. PLINIUS, C. S.: *Naturalis historiae libri.* 24 11
20. PLINIUS, C. S.: *Naturalis historiae libri.* 34 26
21. APHRODIUS, A.: *Meteorologia.* 2 19
22. DAREMBERG, M.: *Oeuvres choisies d'Hippocrate.* Paris (1855) 378
23. VITRUVIUS, P. M.: *De architectura.* 8 5

CHAPTER II

KNOWLEDGE OF ANALYTICAL CHEMISTRY IN THE MIDDLE AGES

1. ALCHEMY

With the collapse of the Western Roman Empire, political and economic stability no longer existed in Western Europe. The storms of the great invasions dashed on the cultured areas of Europe. Under circumstances such as these science could not develop and for hundreds of years there are only faint traces of its presence. However, in the eastern part of the empire, which was ruled by the Byzantine emperors, a stable government ensured a peaceful existence. Industry and trade were carried on in the rich cities, libraries flourished, scientific research prospered, and classical literature was taught in the schools. This period of calm temporarily ceased in the 8th century, when from the depth of the Arabian desert a fanatical, barbarous race broke forth, to spread Allah's flag throughout the world. Within less than 100 years most of the ancient areas of human culture, Mesopotamia, Syria and Egypt, were ruled by the Arabs. The ancient culture of these countries, however, managed to survive and once again the phenomenon of an uncultured conquering nation being profoundly influenced by the culture of its vanquished enemy is observed. The Arabs continued the study of the classics with great eagerness and translated many works from the original Greek. For example Al Mamum, Caliph of Bagdad (9th century) the son of Harun al Rasid of Arabian Nights fame, established a translating institute in his capital, and when a peace treaty was concluded with the Byzantine emperor, one of the clauses ensured that the emperor must give the Caliph a copy of all the books which are to be found in the Byzantine libraries [1]. All branches of science flourished under the Arabs, and around all the big mosques schools were formed. These early temple schools, which can be regarded as the ancestors of our modern universities, were found in Damascus, Bagdad, Samarkand, Cairo and in Cordova in Spain. The last was to exert a very great influence on Christian Europe.

Meanwhile European civilisation was beginning to consolidate. The German tribes settled and intermingled with the Latin ancestral population. The feudal society was comprised of small land-owners or tenants under the protection of a liege lord. Instead of a unified empire, many small principalities came into being and with the Church of Rome as the only co-ordinating factor. The authority of the Church in all these countries was felt equally, and its power continually increased. The Crusades, initiated by the Church, acquainted the descendants of the Barbarians with the culture of the East. Apart from the Crusades there were other sources of contact between East and West. Both in Sicily and in Spain there was either warlike or peaceful contact between them, lasting for hundreds of years. Arabian science began to have an influence on Europe, the philosophy of Aristotle particularly, the interpretation taught by the Arabic philosopher Ibn Rusd (Averroes) being introduced to the West. In A.D. 1100 Salerno opened a medical academy, followed in 1150 by Montpellier, although before this time many European physicians had been taught at Arabic schools. The academies of Salerno and Montpellier became the citadels of Averroesian philosophy.

This philosophy proclaimed the thesis of double truth which proclaims that religious and scientific truths are quite separate. The spread of these views was responsible for the Church revising its own philosophy. This work was done mainly by Saint Thomas of Aquinas in the 13th century, and involved a great deal of the teachings of Aristotle. This scholastic philosophy was later enforced by the Church who prevented its further development, causing it to become dogma. In order to teach its own philosophical views the Church supported the foundation of academies and universities throughout Europe. Most of the ancient European universities, Bologna, Paris, Coimbra, Salamanca, Oxford were founded in the 13th century, and in the next hundred years universities were founded in nearly every country. In these universities the study of theology played the leading role, but soon the study of the natural sciences, especially in the medical faculties, was revived. Although scholastic philosophy dominated the controversy the nominalist and realist tendencies awakened the spirit of polemics [2]. Discussion widened the horizon of knowledge and prepared the way for a new era.

It is possible to divide the progress of science in the Middle Ages into three periods. These are the periods of Byzantine, Arabian and Christian influence. The development of alchemy can be similarly subdivided.

Before the end of the 19th century, when Berthelot and his co-workers translated many Greek alchemical manuscripts [3], little or nothing was known of Greek alchemy; indeed the origins of alchemy had previously been thought to be Arabic. The translation and explanation of these works is often arbitrary and artificial, yet it can be clearly seen that a great deal of alchemical knowledge is due to the Greeks, and that the Arabs merely acquired it from them. The Greeks developed many chemical operations, particularly distillation, and their manuscripts contain pictures of many types of distillation apparatus. Very little is known of the individual Greek alchemists, the most notable among them were Zosimos, whose work *Chemical instruments and furnaces* is lost, and remains only in the second extract, and Olympiodoros, Pelagios, Pseudo-Democritus and Synesios. Presumably they all worked between the 4th and 5th centuries A.D. The language of their manuscripts is, however, so mystic and symbolic, that the exact chemical meaning is unsolvable. An extract from the work of Zosimos shows this; he writes,

> Prepare, my dear, a church of one lead white or alabaster stone, in the building of which there is neither beginning nor end, but in the inside there is a spring with clean

water, and brightness which glows like the sun. Watch, where is the entrance of the church, take your sword in your hands, and look for the entrance. Because the place is narrow, where the inside opening is, and a dragon lies beside the door, guarding the church. Sacrifice this first.

The author of the book from which this extract is taken says that it is not difficult to realize that the passage is a description of a distillation apparatus [4] (Fig. 1). Olympiodoros refers to a certain *"nitronoil"*, which dissolves metals [5]. It is conceivable that this was nitric acid. Greek alchemists make frequent references to a woman named Mary the Jewess, and a manuscript in the Parisian Bibliothèque Nationale describes a water bath which is attributed to her. [6]. In connection with this it is interesting to note that the French name for a water bath is *bain-marie*

The state of development of Arab alchemy is known mostly from Latin translations. The Arab alchemists although agreeing in principle with Aristotle's theory, modified it slightly to include two further primary elements, namely mercury and sulphur which of their known elements appeared to represent the qualities of "moist" and "dry" vapours most closely.

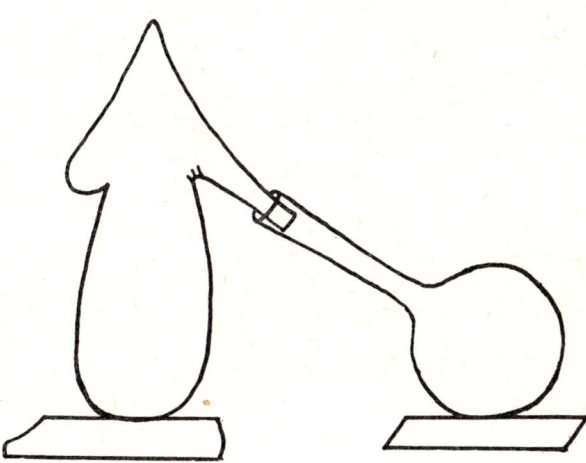

FIG. 1. Distillation apparatus from the alchemistic manuscript of Zosimos. (From Hoefer: *Histoire de la chimie* [1842])

Whether one of the most decisive discoveries in the progress of chemistry, the discovery of nitric and sulphuric acids and of *aqua regia*, is due to the Arabs or to the later Christian alchemists hinges on the identity of the author of the book *Summa perfectionis magisterii*. This book was published in Europe in the 13th century, and the author's name was given as Geber. In actual fact, Geber is the distorted, latinized version of Jâbir Ibn Hayyân, an Arab physician and alchemist, who lived in the 8th or 9th century and about whom very little is known. However his reputation was very notable and this was sustained by the *Summa* for hundreds of years. Whether Jâbir Ibn Hayyân did write the *Summa* has been the subject of considerable doubt during the last 60 years, many authorities believing that the book is the work of a Christian alchemist of the 13th century. This scepticism is quite understandable; ancient alchemists had the habit of ascribing their works to other authors, partly to gain greater importance, and partly in order to remain anonymous as the Church actively opposed alchemy.

In spite of the fact that many alchemists were priests, Pope John the XXII in 1317 had forbidden the study of alchemy.

There are several reasons for doubting that Jâbir Ibn Hayyân wrote the *Summa*, the main objection being that no reference to this work is found before the 13th century. Albertus Magnus and Roger Bacon, who lived in this century, make

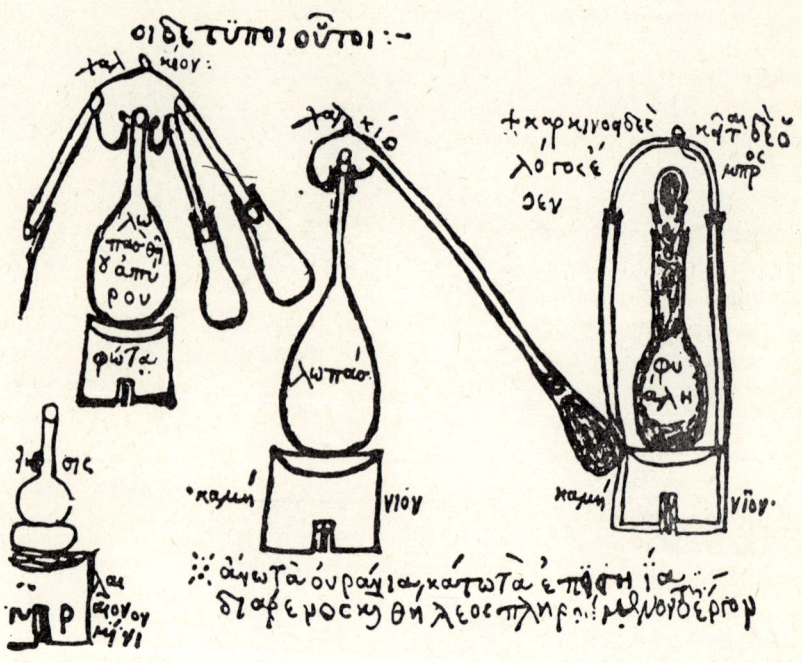

Fig. 2. Chemical apparatus of Greek alchemists. Taken from a manuscript in Bibliothèque Nationale. (From Bugge: *Buch der grossen Chemiker*)

no mention of it; also its dry and objective style does not resemble Jâbir's eloquent declarations, although the original version of this work is not known. In addition, it is difficult to imagine that this excellent work was written at an early stage in the development of Arab alchemy. Alternatively, it can be argued that before the 13th century there was no western alchemy worth mentioning, and also that the translation of the *Summa* could have changed its style.

It should also be remarked upon that there is still a large number of Arabic manuscripts awaiting translation. Ruska [7] claims that it is difficult to imagine that so much knowledge was available in the 8th century but he also says of some authentic Arabic texts that he is amazed at the extent of Jâbir's medical knowledge.

However, the preparation of nitric acid is described by Geber as follows:

Let us distil a pound of cyprian vitriol (iron or copper sulphate), one and half pounds of saltpetre, a quarter of a pound of alum, and obtain the water (acid). This water dissolves metals very well. Its effect will be even greater by adding a quarter of a pound of salmiac to it *(aqua regia)* [8].

Geber also mentions sulphuric acid which he obtained by distilling alum [9]. This procedure is referred to in the work of Rhases (Abu Bekr Al Rasi), an Arabian alchemist of the 10th century. In that medieval Latin translation of his work the name for sulphuric acid is *oleum*, a name which is still used [10].

In the subsequent period of western alchemy zinc, antimony and bismuth were all discovered, and this initiated the use of salt solutions and processes which occur in solutions. It was only later in the 16th century that alchemy adopted its dubious character the form of which it was to maintain until it finally collapsed.

FIG. 3. Arabian distillation apparatus from the 13th century. (From Schelenz: *Geschichte der pharmazeutisch chemischen Destilliergeräte*)

2. KNOWLEDGE OF ANALYSIS

The idea of specific gravity has been known since antiquity, or at least since the time of Archimedes, the first known reference to the use of a hydrometer being of Greek origin. Hippocrates had previously observed the variation in weight of the same volume of different liquids, and Galenus, the great Roman physician, checked the density of salt waters recommended as a medicine [11] with eggs. Synesios, a Greek alchemist of the 4th — 5th century had introduced the use of a hydrometer for this purpose which he described as follows:

> This is a cylinder-tube, on which horizontal lines are marked to measure the depth it submerges into the liquid. In order to keep it in a vertical position in the liquid, a small weight is attached to the lower end [12].

This instrument he called a *hydroscopium*. From a poem of the 6th century entitled *De ponderibus et mensuris* (concerning weights and measures) of uncertain authorship which describes the hydrometer, the following extract is taken:

> With this apparatus the weight of all liquids can be measured. In less dense liquids the cylinder submerges more, whereas in more dense liquids more marks are to be seen above the liquid. Taking the same volume of different liquids, the most dense will be the heaviest; taking the same weight of them, the volume of the one that is least dense will be greatest. If one of two liquids one covers twenty, the other twenty-four marks, you will know of it, that the former is the heavier, by one drachma.

The discovery of the hydrometer was later forgotten and it was not until a thousand years later that Boyle rediscovered it and claimed the achievement for himself.

As we can see, the text mentions weight instead of specific weight. Arabian authors also mix these two ideas. Albiruni, an Arab scientist (973—1048), determined the specific weight of numerous substances, and there also remain many data due to Alchazini (12th century) who was the author of the book *About the balance and the wiseness*. We know that he determined specific weight with the help of a balance. An examination of his results shows that the measurements carried out a thousand years ago were astonishingly accurate in comparison with present-day values, the only difference being that the value for milk was considerably denser [13].

TABLE 1

Matter	Density	
	According to Albiruni or Alchazini	Today
Rainwater	1·000	1·000
Seawater	1·041	1·028—1·042
Wine	1·022	0·992—1·038
Honey	1·406	1·410—1·445
Olive oil	0·915	0·914
Milk	1·110	1·028—1·034
Vinegar	1·027	1·013—1·083
Rocksalt	2·19	2·161
Tin	7·32	7·28
Wrought iron	7·74	7·86 —7·76
Copper	8·66	8·30 —8·92
Lead	11·32	11·34
Mercury	13·560	13·546
Gold	19·05	19·21

Specific weight (weight) is also referred to by Geber as a characteristic criterion of metals. Metals are similarly characterized by their melting point, colour, lustre and ductility. For example here is the description of a few metals taken from Geber's *Summa perfectionis magisterii* [14].

Gold. Metallic substance, its colour is lemon-yellow, it is very heavy, lustrous, ductile on hammering, is resistant to the test *(cineritium)*, roasting and burning with coal. It is soluble, giving a red solution and renovates the body, it can be alloyed easily with lead and mercury. Only with the greatest effort can it be fixed with the spirits (dissolved in acids?); this is the great secret of this art, which slips out of the memory of those who have a hard head.

Silver. It is a white, clinging ductile metal and melts very easily; it resists the fire test. If it is combined with gold, they cannot be separated by fire. It gives a wonderful blue colour on treating with acid and salmiac vapours, *(super fumum autem acutum, sicut aceti, salis armoniaci et agrestae, fit caelestinus color mirabilis).* Its ores are not as pure as the ores of gold and generally it is mixed with many strange substances.

Lead. It is a dun-coloured or dull white, heavy, not clinging metal. It is easily malleable, ductile with hammering and has a low melting point. It forms *cerussa* (lead white) with acetic acid, and minium when it is roasted. Though it is not at all like silver, with the help of our art we can change it to silver! (The high silver content of lead ores might have been the cause of this positive but erroneous declaration). Its weight does not remain constant during this change. (It concludes by stating that "Lead is used for fire assays.")

Tin. It is a metallic, dirty-white, somewhat clinging, metal. It is easily malleable, ductile and also melts very easily giving a curious sound *(sonans parum stridorem).* It does not resist the fire test. The *tutia* (tin ore?) easily fuses with copper, and in doing so becomes yellow. It is easily affected by air or by acid.

Iron. It is a greyish white, high melting metal, very clinging, not very ductile and with difficulty malleable. It is very dense and is difficult to work because of its high melting point. None of the metals which are difficult to melt are suitable for *transmutatio* (transformation of elements).

These descriptions are extremely clear and completely different from the language of the earlier Greek alchemists. Geber also gives a clear description of the determination of noble metals by cupellation:

Let us take sifted ash or lime *(cinis cribellatus aut calx)* or the powder of animal bones consumed by fire, or a mixture of these, or something similar. This must be moistened with water and moulded until it is compact; in the middle of this we must make a round hole and put into it a certain quantity of glass powder. This must be dried, and then the substance to be examined is placed into the hole *(ponatur illud, de cuius intentione sit tolerare examen),* A good coal fire is then made and the flames blown on to the examined substance until it melts. An excess of lead is added to it *(Saturni* [lead] *partem post partem projiciamus)* and the mixture reheated strongly. If the substance shows any change then this is an indication that it is impure. The procedure must be continued until the whole of the lead disappears *(donec totum plumbum evanescat).* If when this is complete the substance still shows some change then this indicates that the cleaning was not perfect. In this case more lead must be added, and the mixture reheated, and this should be repeated until the mass becomes calm with the pure gold glittering on its surface[15].

Geber notes that only gold and silver are resistant to this test.

Albertus Magnus also describes the cupellation procedure but not in such great detail; he also describes the cementation process for the separation of gold and silver [16].

The cupellation procedure was already in general use in the 14th century, and many records show that at this time it was in statutory use for the examination of gold. Charles I of Hungary in 1342 published a decree which ordered that in

FIG. 4. Gold testing in the 15th century. (From Biringuccio: *Pirotechnia* [1540])

all mining towns a regal house *(domus regalis)* should be established; here all the ores and metal destined for melting must be produced so that the royal *camerarius* could establish the noble metal content [22].

A more detailed description of the examination is given in a statute of King Philip the Sixth of France in 1343. This can be regarded as the earliest example of a standard method of analysis.

The cupells are described as small, flat, recessed and well washed vessels, made of vine-shoot ash and burnt leg-bone of sheep. To free it from any substances which would interfere with the test, it must be well polished. Afterwards the cavity must be moistened with a liquid which is a suspension of powdered deer-antlers in water. This causes a white glaze to be formed so that the examined substance can occupy a better position and is easier to remove afterwards. The decree also prescribes that the lead used for the test should be of good quality and must not contain either gold, silver of copper or any other impurities. It also mentions the precision required in weighing.

> The balance used for the test should be of good construction, precise and should not pull to either side. The test should be carried out in a place where there is neither wind, nor cold, and whoever carries out the test must take care not to burden the balance by breathing upon it [17].

The balances employed for these tests must have been quite sensitive for breathing to have any effect on them.

The fire test is often described in later works. Paolo de Canotanto, an alchemist probably born in Taranto in the 15th century, states that one-sixth part of lead must be added to the molten metal *(et metalle injiciatur ibi plumbi pars sexta)*; he describes also a special furnace used for the cupellation *(ad faciendam cupellam)* [18].

Antimony was discovered in the 15th century, and from this discovery a new method of separation of gold and silver was developed, also using the fire process. On melting a mixture of the metal with antimony, the antimony takes up the copper and silver, leaving the gold at the bottom of the melting pot as a regulus [19].

The scratch test was also improved during the Middle Ages. In the 15th century the byelaws of Nürnberg describes the series of needles used for the examination of gold, and by 1518 a comparative series of 24 needles was employed; 23 of these were gold–copper alloys, with a gold content of 1–23 carats, and the remaining one being pure gold. The scratch from these needles on a touchstone was compared with the scratch obtained from the examined object. The scratch was also treated with nitric acid, which would not affect the scratch resulting from pure gold [20].

With the use of mineral acids, metals could easily be dissolved and it then became possible to work with the solutions obtained. This point marks the beginning of analytical chemistry as we know it today, but for a long time this process of solubility was regarded as a thing of mystery, as if in some way the dissolved substance disappeared. This idea was responsible for keeping alive the old alchemical ideas on the changing of substances, and during the whole of the 17th century only van Helmont stated definitely that the dissolved substance does not disappear. In spite of this a large number of procedures was carried out using solutions, and as a result of this many reactions of the metals were discovered.

Mineral acids were first used for the testing of gold and silver, and it was discovered that although *aqua regia* dissolves gold, silver is precipitated by it. It was also found that silver can be dissolved with nitric acid, and this gave a simple method for the separation of gold and silver. It is interesting to note in this respect that the original name of nitric acid was *separating water*. Albertus Magnus gives an account of the preparation of nitric acid and describes some of its properties. He mentions that it will dissolve silver, thus enabling it to be separated from gold, and that mercury and iron are converted by it to the oxide [21] *(Liquor est dissolutiva lunae, aurum ab argento separat, mercurium et martem calcinat, convertit in calces)*. In addition, he recommends the use of glass or stone vessels for carrying out these reactions, because of the effect of acid on metal apparatus. He also notes that a solution of silver nitrate will blacken skin.

Albertus Magnus (1193–1280), a Dominican monk who later became a bishop, was a professor at the Paris University and was also the teacher of Thomas Aquinas. He was one of the outstanding scientists of the Middle Ages and a pioneer of the natural sciences.

At the beginning of the 15th century, nitric acid was adopted for the industrial separation of gold and silver. At about this time it was also discovered that the separation is more effective if the ratio of gold to silver is 1 to 3, and hence ores of a low silver content had silver added to them before the separation. This

FIG. 5. Albertus Magnus (1193—1280). Painting by Fra Giovanni da Fiesola, to be found in the Dominican Abbey in Florence. (From Bugge: *Das Buch der grossen Chemiker*)

resulted in the name *inquartation* being given to the process. The earliest records of this process are from Selmecbánya (Schemnitz) in the upper Hungarian mining district (now in Czechoslovakia). This was one of the most important noble metal producing areas in Europe in the Middle Ages and presumably the inquartation process was discovered here. The process was also used for the testing of gold, and there are numerous references to its use. In 1499 the citizens of

Selmecbánya complained of the unfairness of the test with nitric acid carried out by an official tester [22]. By the 16th century, the separation of gold and silver with nitric acid was widely used throughout Europe [23].

At this time it was customary in the Hungarian mining towns to employ, in addition to the official analyst, a tester elected by the citizens. These acted as a check upon one another, and there are several records of their disputes over differing results of analyses. There was considerable official interference with the work of the town analyst, whereas the official analyst was sponsored by the authority and had to take an oath which for the sake of interest is reproduced below:

> Swear to the most majestic and most powerful sovereign Maximilian, Roman Emperor, King of Hungary and Bohemia, Archduke of Austria, our most merciful master, that you will be obedient and faithful to His Excellency and to his Head Chamber Count, and that you will behave honestly in your position and that you will test everybody's gold and silver honestly, so that the tax due to his Excellency can be honestly exacted. Also promise to test all samples either ore or ore-stone in a similar manner and for to treat all people both rich and poor in the same way, honestly and to disregard either friendship or enmity, or pay, or present or poverty and to avoid being misled or deceived [22].

Paracelsus tells us that after the separation of gold and silver with nitric acid, the silver can be recovered from the solution by the addition of copper wire [24], addition of iron wire to the solution will cause copper to be precipitated [25]. Raymond Lull, the Spanish alchemist who led an adventurous life (1235—1315?) and who was highly esteemed by the succeeding generation because of his prediction that if the sea were made of mercury then he would be able to change it to gold. In his description of the preparation of nitric acid, he mentions that the water used must first be cleaned with silver, in order to remove the *pinguedo salis nitri*. Probably by this salt he meant chloride, so that he freed the water from chlorides [26]. It should also be mentioned that in one of his works Lull gives a drawing of a refrigeration apparatus which is very similar to the ball refrigerator [27] (Fig. 6). It is very probable that apart from the few chemical reactions described, the alchemists of the Middle Ages knew of many similar to them, and although they had no idea of the nature of mechanism of these reactions, they still form the basis of the knowledge on which modern chemistry and analytical chemistry has been built. The beginning of this development marks the start of the period of medical chemistry, the next important period in the history of chemistry.

FIG. 6. Cooling tube of Raymund Lull from the 13th century. (From Manget: *Bibliotheca chemica* [1705])

NOTES AND REFERENCES

1. DANNEMANN, F.: *Naturwissenschaften in ihrer Entwicklung und in ihrem Zusammenhange.* Leipzig (1922) **1** 302
2. In the Middle Ages two opposing schools of thought existed, the realists and the nominalists. The former group, also called the Platonic realists, supported the views of Aristotle who taught that general concepts are the true realities with which natural science should be concerned. The nominalists on the other hand considered that true reality lay in concrete individual objects and that general concepts were mere names.
3. BERTHELOT, M.: *Collection des anciens alchimistes grecs.* Paris (1887)
4. BUGGE, G.: *Das Buch der grossen Chemiker.* Berlin (1929) **1** 10
5. HOEFER, F.: *Histoire de la chimie.* Paris (1866) **1** 275
6. Manuscript No **2249**. fol. **103**. Quotation HOEFER, F.: *Histoire de la chimie.* Paris (1866) **1** 285
7. BUGGE, G.: *Das Buch der grossen Chemiker.* Berlin (1929) **1** 18, 60
8. *Artis chemicae principes Avicenna atque Geber.* Basle (1572) 734
9. KOPP, H.: *Geschichte der Chemie.* Braunschweig (1845) **3** 303
10. *Raxis Liber, pui dicitur lumen luminum magnum.* Quotation: HOEFER, F. *Histoire de la chimie.* Paris (1866) **1** 341
11. GALENUS, C. L.: *De simplic. med. facultate.* Leyden (1550) 61
12. SYNESIOS: *Epistolae.* **15**. Quotation from Hoefer. loc. cit. **1** 281
13. KHANIKOFF, N. J. *J. Amer. Oriental. Soc.* (1859) **6** 1; WALDEN, P.: *Mass Zahl, und Gewicht in der Chemie der Vergangenheit.* Stuttgart (1931) 29
14. *Gebri Arabis Chimia sive Traditio Summae Pefectionis* etc. Leyden (1668). Quotation: KOPP, H.: *Geschichte der Chemie.* Braunschweig (1845) **2** 40
15. DARMSTAEDTER, E.: *Die Alchemie des Geber.* Berlin (1922) 88
16. *Beati Alberti magni episcopi Ratisbonensis Opera omnia.* Leyden (1651). *De rebus metallicis et mineralibus libre.* **2** 265
17. HOEFER, F.: *Histoire de la chimie.* Paris (1866) **1** 499–500
18. CANOTANTO, P.: *De Theoria ultra aestimationem peroptima ad cognitionem totius alkimiae veritatis.* Bibliothèque Nationale. Paris. Hand-archives: 7159. Quotation: HOEFER, F.: *Histoire de la chimie.* Paris (1866) **1** 468
19. DARMSTAEDTER, E.: *Berg-, Probier- und Kunstbüchlein.* München (1926) 14 BIRINGUCCIO.: *Pirotechnia.* Venice (1540) 71
20. SUDHOFF, K.: *Arch. Gesch. Naturw. Technik* (1909) **1** 84
21. MANGOT: *Theatrum chimicum.* (1703) **4** 937 (ALBERTUS MAGNUS: *Compositum de compositis*)
22. SZABADVÁRY, F.: *Magy. Kém. Lapja* (1965) **20** 138
23. DARMSTAEDTER, E.: *Berg-, Probier und Künstbüchlein.* München (1926) 54 BIRINGUCCIO, V.: *Pirotechnia.* Venice (1540) 63
24. PARACELSUS, P. T. B.: *Bücher und Schriften usw.* Basle (1589) 159
25. SAVOT: *Traité de metallurgie* 74
26. KOPP, H.: *Geschichte der Chemie.* Braunschweig (1845) **4** 202
27. RAYMUNDUS, LULLUS: *Ars auriefera.* 3. Quotation: MANGET: *Bibl. chimica.* Geneva (1702) **1** 866

CHAPTER III

ANALYTICAL KNOWLEDGE DURING THE PERIOD OF IATROCHEMISTRY

1. THE CLARIFICATION OF CHEMICAL CONCEPTS

The development of iatrochemistry commenced in that remarkable period of European history known as the Renaissance. This was a period of the rebirth of the natural sciences and marks the start of their very rapid progress. The discovery of printing simplified the method of making books, and also made them much cheaper than before. The works of ancient classics were multiplied by thousands, and were far more widely read. This resulted in a great thirst for knowledge as people became more erudite. Meanwhile the area of the known world was suddenly expanded because of the many new geographic discoveries. The problem of safety in navigation turned the attention of many scientists first towards astronomy, and then to mathematics and physics. Practice and theory came closer to one another than ever before. It was soon discovered that it was not possible to solve the scientific problems which arose on the grounds of existing knowledge, but that by a suitable examination of the factors involved in each problem it was often found possible to arrive at a solution. The role of experience thus became very important in the formulation of ideas. The origin of this new method of research is generally attributed to Francis Bacon (1561–1626) the English philosopher and statesman who pointed out the importance of the experimental method. Bacon gives a remarkable account of the inductive theory of research into the natural sciences. The next citation illustrates this:

> As in also everyday life natural endowments, and the secret characteristics and emotions of human soul manifest above all in trouble, also the secret phenomena of nature manifest sooner by the effect of mechanical interventions, than if it freely runs on its way [1].

Bacon, however, was not the discoverer of this new method, it had already been developed long before his time. His only contribution was to clarify the scientific methods in use at that time; but the brilliant manner in which he expressed his ideas on the importance of science and the scientific method was to prove a very great stimulus to scientific thought. The importance of experience had been stressed long before Bacon. Leonardo da Vinci for example wrote:

> The interpreter of the wonders of nature is experience. It never misleads us, only our grasp can do it with us. Until we can establish a general rule, we must accept the help

of experience. Although nature begins with the cause, and with the experiment, we must do it inversely, we must discover the cause with experiments [2].

The role of experience is also symbolised by one of the works of Bernard Palissy (1499–1589), written in the form of a dialogue between two people, the vain, empty Theorique (theory) and the clever Practique (practice), the topic of their conversation explaining the inductive scientific method [3].

The progress and gradual acceptance of this new scientific spirit naturally required considerable time. One of its precursors, Roger Bacon, lived in the 13th century, but it was not until the 15th century that it became widespread. The 16th century, which was to see a collision between old and new in so many fields of experience, was also to see the clash between the new natural sciences and scholastic ideology. The former had triumphed by the 17th century, and this was to result in the establishment of academies of science [4].

This also started a new period in the progress of chemistry. From the fruitless experiments of a thousand years the futility of the attempts to make gold was realised. The interest of chemists now turned to a new field and the making of medicines became prominent. The first and most conspicuous propagator of this new tendency was Paracelsus, whose full name was Aureolus Philippus Theophrastus Bombastus Paracelsus von Hohenheim (1493–1541). He was a true Renaissance character, lawless, passionate, unscrupulous and was simultaneously admired as the greatest scientist or was considered a charlatan by his contemporaries. His dual character is reflected in his varied life. From a vagabond of the roads he became a professor at the University of Basel, but was eventually compelled to leave because of his extreme views, so he once again became a vagrant charlatan and eventually died in a hospital for the poor in Salzburg. He fervently attacked the ancient medical authorities, Hippocrates, Galenus and Avicenna, who he said were "unworthy to untie his sandal-straps". He proclaims:

> And you, who have learned from these believe that you know everything although you know nothing. You want to prescribe medicines and yet cannot make them: Chemistry solves for us the secrets of therapy, physiology and pathology; without chemistry we are trudging in darkness [5].

This marks the path of the new chemistry. After Paracelsus the development of chemistry and medicine proceeded together and the study of chemistry was undertaken by physicians. Their aim was to explain the processes going on in the living organisms and because solutions played an important part in the mechanism of these organisms, they turned their attention towards the solution process. The discovery of the nature of the processes occurring in solution led to the development of qualitative chemical analysis. In this time the clarification of many chemical concepts was started, and many of the principles established, which are valid even today, will shortly be discussed in more detail as they are of fundamental importance to the development of analytical chemistry.

Up to this time the ancient alchemical concepts of the nature of substances had not changed essentially. But Paracelsus introduced a third element, the "salt". According to Paracelsus the three elements sulphur, mercury and salt compose not only the metals, but all the organic and inorganic substances such as "iron, steel, lead, emerald, sapphire and quartz". He also stated that "the human body is similarly composed of these" [6]. However, until this view existed, the possibility of converting elements had to remain open — at least theoretically.

FIG. 7. Paracelsus (1493—1541). Painting by an unknown master. (From Bugge: *Das Buch der grossen Chemiker*)

And the great chemists of this time did not doubt it, but did not deal with the question.

Van Helmont also modified the alchemical conception of elements and, of the four elements of Aristotle, considered only water as a primary element. He proved this view with experiment, namely with a "quantitative analytical" experiment. Into a pot he weighed 200 lb of dried soil and in this he planted a willow branch which weighed 5 lb. A lid was used to cover this and the branch was watered with rainwater daily. This was continued for five years. After drying and reweighing the soil he found that the mass was almost unchanged, a mere difference of 2 oz less. But the willow branch by this time had become a tree and weighed 164 lb. Van Helmont argued that as only rainwater had been added then the tree had been formed from water. Then the tree was burnt; gases were evolved and only soil (ash) remained, therefore both of these had also been formed from water [7].

This experiment may appear to us to be somewhat naive but when one considers the knowledge available at the time it is quite remarkable. When one considers the inductive nature of van Helmont's experiment and also the careful avoidance of experimental errors it is even more remarkable.

Much later Robert Boyle gave a definition of elements which is extremely close to the modern definition:

> I call elements those primary and simple bodies, which do not contain anything, cannot be made from one another, form the complex bodies, and into which complex bodies can be reduced [8].

He supposed that there were many elements but did not enumerate them. He still believed, however, in the transmutation of elements.

With his works *The Sceptical Chymist* (1661) and *The Origin of Forms of chemical elements* (1667) although he made a difference between these and the "metaphysical elements" by which he meant primary elements and the philosophical problems concerned with them. Boyle considered that the primary elements are intangible, but that this is of no importance to a chemist, who in his work is concerned only with the actual elements. Although the relationship between atoms and molecules is not very clear Boyle makes a clear distinction between compounds and mixtures. Boyle revived the concept that elements consist of minute, indivisible particles which because of their affinity for one another combine to form compounds [9]. This theory was simply a restatement of the theory of atomism originally proposed by Democritus in 400 B.C. This theory of Democritus was soon displaced by Aristotle's theory of the continuity of matter.

Attempts had been made even before Boyle to revive the atomist theory and traces of this are to be found in the works of the scientists of the Renaissance. One of these was Pierre Gassendi the French mathematician and astrologer (1592–1655) and many other philosophers and scientists were to follow him [10].

During the period of iatrochemistry the first division of compounds into three classes, acids, alkalis and salts, was made. Everything which was soluble was regarded as salt and Palissy records, "everywhere is salt", saltpetre, alum, borax, tartar, even sugar was regarded as salt [11]. It was soon discovered that salts were formed from the combination of acids and alkalis. Angelus Sala (1639) established that ammonium chloride was formed from ammonia and hydrochloric acid [12] and Tachenius made the very general statement *Omnia salsa in duas dividuntur substantia, in alcali nimirum et acidum*, namely that every salt can be decomposed into acid and alkali [13]. It is obvious that by this time acid and alkali were collective terms. Boyle defined acids on the basis of their possessing certain properties, and giving similar reactions with certain compounds. Thus acids had the power to dissolve metals to varying degrees, to precipitate sulphur and other substances dissolved in alkalis, and also to change the colour of certain vegetable extracts to red. All these properties were lost, however, when they were brought into contact with alkalis [14]. The definition of alkalis was rather more vague. Boyle described alkalis as those substances which give blue or green colourations with certain vegetable extracts [15]. Alkali carbonates were difficult to classify but were considered as alkalis, as a further criterion of alkaline character was that the effect of acids caused effervescence.

Lémery (1645–1715) writes that

> Alkalis can be recognised by the effervescence which occurs when acids are poured on to them. This ensues almost at once and continues until the acid finds something to dissolve [16].

Effervescence was later to become the first indicator to be used in volumetric analysis. Boyle had noted that this effervescence did not occur with all alkalis and states:

> I found it on the occasion of my experiences, that if ammonia is mixed with sulphuric acid of medium strength, great heat developed and it did not fizzle at all or only a little, although in other cases it behaved like alkalis [17].

It was not until a century later that this difference was explained by Black, and it was only in the 19th century that the anomaly of bases was finally solved.

Speculation as to the nature of the solution process also continued. Norton in 1477 had stated that substances remain unchanged when dissolved in acids [18] but a number of erroneous ideas persisted for a long time. The most popular idea was that substances on solution were completely destroyed. Van Helmont opposed this idea and considered that dissolved substances were not destroyed and could be recovered from the solution entirely although the solution had the appearance of water: *Licet argentum in chrysulca dissolutum, periisse, quatenus aquae formae, videatur permanet tamen in pristina sui essentia; prout sal in aqua solutum, sal est, manet et inde reperitur, sine salis mutatione* [19]. By distillation he was able to establish that the substance remained unchanged, although he discovered this practically by distilling lead.

The first traces of the concepts of chemical affinity and chemical relationship are also to be observed about this time. Although the word affinity had been used by Albertus Magnus, Glauber was the first to note its significance, namely that the tendency of a substance to combine with other substances varies greatly, and depends on the nature of the second substance. He shows that when sal-ammoniac is treated with potassium hydroxide, lime, or zinc oxide, ammonia is driven off, while other oxides are incapable of doing this, because the nature of the former is such that they have a great affinity for all acids, they like them very much and are also liked by them very much so they fasten on to them; by this the salmiac part remains without a pair and it can be distilled.

He also observed that sulphuric acid drives out hydrochloric and nitric acids[20]. Glauber noted that nitric acid dissolves metals with increasing ease in the following order: gold, mercury, copper, tin, iron, lead *(aber doch immer das eine Metall lieber als das andere, nach denne es ihme in seiner Natur verwand* [chemical relative] *und zuguthen ist* [21]*)*. The first recorded observation of double decomposition was also made by Glauber. He examined the reaction between gold and potassium silicate, both dissolved in *aqua regia* and comments

> *Aqua regia* by its acidity killed the *sal tartari* (alkaline part) and because of this it had to set free its silicic acid; at the same time the *sal tartari* killed the acidity of the *aqua regia* so that it cannot keep the gold which it had taken up, consequently gold and silicic acid are set free from the solution *(zugleich Gold und Kiessling von ihrem Solvente erlediget wird)* [22].

The discovery of gases also occurred during this period. The evolution of gases during different reactions had previously been noted but had been considered to be air. Van Helmont realised the existence of gases other than air. He says for example

> In consequence of burning coal, *spiritus sylvestris* comes into being. 62 lb of oak-coal gives one lb of ash, therefore it provides 61 lb of *spiritus sylvestris*. This *spiritus*, which was formerly unknown and cannot be kept in vessels, and cannot be converted into a visible form, I call by the new name *gas*...(...*Hunc spiritum incognitum hactenus, novo nomine gas voco* ...) [23].

The examination of gases led, in the next century, to the establishment of modern chemistry.

It can be seen that this period of chemistry was responsible for the clarification of many chemical concepts. This clarification was a direct result of the development of analytical chemistry. The next section considers this development in more detail.

2. THE BEGINNING OF ANALYSIS IN AQUEOUS SOLUTIONS

The examination of noble metals by dry processes was well established at a very early stage in history and has been described previously. The description of the scratch test found in the works of Georgius Agricola is essentially the same as the procedure in use today and it would still be possible to use the docimastic method of Agricola for the analysis of gold. Agricola, however, was mistaken in thinking that he was the first to write about these methods of analysis [24]. in this field there was no further development, and subsequent authors were limited to repeating previous work. The blow-pipe which was the most important instrument of analysis by dry processes only appeared at the end of this period and was confined mainly to glass manufacture. This instrument was later used for the qualitative analysis of a variety of minerals.

FIG. 8. Georg Agricola (1494–1555). (From his *Twelve books on Mining and Metallurgy*)

Georgius Agricola (1494–1555) was born near Meissen in Saxony. He studied medicine in Leipzig and at various Italian universities and then settled in Joachimstahl and later in Chemnitz. His interests did not lie in medicine and he is chiefly remembered for his contributions to mining and metallurgy. The books he wrote on these subjects were very important both to his own and to later generations, and give a remarkable picture of the state of development in this field in the Middle Ages.

Of all the processes occurring in solutions, only the separation of gold and silver with nitric acid was used for the purpose of analysis during the Middle

Ages. The best description of this process is given by Agricola: gold was hammered into a thin sheet and was then rolled up and treated with nitric acid [25]. Most of the chemical reactions on which the classical system of qualitative analysis is based were only discovered during the period of iatrochemistry. The most important of these discoveries was that substances can be divided into groups on the basis of these reactions and the single substances can then be identified.

One of the first people to stress the importance of processes occurring in aqueous solution was Basilius Valentinus who stated that:

> To remove the silver from copper and restore its original colour is a great art which is not known to metallurgists, but belongs to chemistry and the laboratory [26].

He established further,

> that philosophers know two procedures, the aqueous process which was the one that I used, and also the dry process [27].

In this same work he also mentions that Hungarian iron is brittle because it contains copper and that Hungarian silver contains gold. Consequently this iron can only be used after refining [28]. He makes no mention, however, as to how these results were obtained, i.e. his analytical methods.

The use of the words "precipitation" and "precipitate" are also found in this text; he precipitates gold from *aqua regia* with potassium carbonate, and mentions that the precipitate *(praecipitatum)* must be air-dried, because it explodes on direct heating [29]. Basilius Valentinus carried out experiments on the separation of metals with solutions of different acids and alkalies. This is referred to in his work *Offenbarung der verborgenen Handgriffe: Vitriol schlägt nieder das Mercurium vivum, und Sal tartari das ☉, ♀ und gemein Salz das ☽, ♂ die ♀, eine Lauge von Buchenaschen den Vitriol, Essig den gemeinen Schwefel, ♂ tartarum, und Salpater den Antimonium* (☉ = Sun = Au, ♀ = Venus = Cu, ☽ = Luna = Ag, ♂ = Mars = Fe; these are the ancient alchemical symbols for the metals, and refer to the mystic connection of astrology and alchemy and from them are derived the chemical symbols in use today. These symbols went out of use in the 16th century. *Sal tartari* is potassium carbonate and vitriol refers in some cases to iron sulphate and in others to copper sulphate).

Further works appeared under the name of Basilius Valentinus in the year 1604. Their author was supposedly a deceased Benedictine monk who had lived in the second half of the 15th century. These books contained many new facts and contain references to hydrochloric acid, bismuth, antimony and to the spirit-lamp. These represented a considerable advance from the alchemists and also a step forward in the analytical field. The theoretical views of Basilius Valentinus are very similar to the views of Paracelsus. Paracelsus was later accused, after his death, with knowing of the works of Basilius Valentinus and of drawing information from them. If these accusations are true then it would seem rather strange that Paracelsus alone knew of their existence.

As soon as these works were published suspicion was aroused as to the identity of the author, and an active investigation was carried out. In 1675 a researcher named Gudenus claimed that he had evidence to show that in 1413 a monk of the same name lived in Erfurt [30]. This date subsequently proved to be wrong. It is now reasonably certain that the works of Basilius Valentinus were written at a later date than those of Paracelsus, and that the author was one of the most talented chemists of this period. It is possible that he used a false name in an attempt to spoil the reputation of Paracelsus. According to Felix Fritz [31] these works were written by N. Soleas, who also wrote other books under this name. Other researchers, however, claim that Soleas also is a pseudonym. For further information on this question reference must be made to the studies previously mentioned.

During the period of iatrochemistry a great deal of work was carried out on the examination of water. First of all information was required concerning the medicinal effect of various mineral waters. This examination of water contributed greatly to the development of analysis in aqueous solution. One of the most detailed reports dealing with the examination of different waters to which we can refer is the work of a student of Paracelsus, Leonhard Thurneysser. This work, *Pison oder von kaltem, warmen minerischen und metallischen Wassern*, was published in 1572.

Thurneysser (1530–1596) was a physician who led an adventurous life. He travelled about the world from England to Arabia. After a rather promising introduction the book becomes rather confused. According to his method water is poured into the *mensura*, which is a vessel divided into 24 parts. It is first of all weighed, filled with rainwater and then with the water under examination. From the difference in the weights it is possible to deduce the quantity of the dissolved substance. The water must be distilled off and the residue weighed; the residue is then redissolved and allowed to crystallize. The crystals are separated and then heated and if they ignite then they are composed of *nitrium* (nitrate), if they are easily soluble and on heating become red then they contain vitriol, and if soluble only with difficulty they contain lead. The distillate must be redistilled and if the residue becomes blue on heating then copper or gold is present, whereas if it evaporates then it contains mercury. If it turns brown copper is present, if it remains white then the water contains tin. He also drew conclusions from the colour of a flame when a portion of the residue was ignited in it [32]. The method of Thurneysser, although containing several correct assumptions and observations is for the most part rather confusing.

The account of Libavius written in 1597 is rather more concise. He draws attention to the fact that the examination of mineral waters must take place near their springs in order that the gaseous components *(spiritus)* are not lost during transportation. He puts the water in a retort which has a tight fitting end. It is then gently warmed while the lid is kept cool. The volatile gases are then trapped in the lid *(segregatio spiritum)*. The next step *(segregatio aquositatis et contentorum)* is the examination of the dissolved substances. His method of determining the quantity of dissolved salt is to dip a piece of linen of a known weight into the solution and it is then dried and reweighed. The difference in the two weights gives the quantity of dissolved substances. In order to establish the nature of the dissolved substances he distils the water until it becomes as dense as a syrup. Into this he hangs a straw or a thread on which salts crystallize out. From the shapes of the crystals he establishes whether the water contains alum, vitriol or saltpetre. The extract of gall nuts is used to identify iron. The ammonia-content of the water can be determined by observing the blue colour formed when copper solution is added [33].

Andreas Libavius (1540—1616) was born in Halle and was a teacher in a secondary school in Jena, later becoming a school-director in Coburg. He was the author of many progressive chemical works, and was a follower of the principles of Paracelsus in that he made a careful classification of minerals and natural substances, but did not agree with some of Paracelsus' more radical views.

The work of Boyle concerning the examination of mineral waters, *Memoirs of a natural history of mineral waters*, published in 1685, also contained many original contributions to existing knowledge. In the introduction he states that the geological environment of the water should be noted, and discusses the conclusions to be drawn from this [34]. Then follows a description of the test. The temperature and the specific weight (using a hydrostatic balance) of the water must be measured, and the transparency, colour, smell and touch of the water must be observed. A drop of water must be put under a microscope and examined to see how many moving minute particles there are in it. The presence of any sediment must be noted, and also if any solid deposit is formed on standing in the air, or by boiling or freezing the water. The viscosity is also measured. After this it is possible to carry out the chemical examination of both the water and distillation residue. Boyle also uses the juice of gall nuts to test for iron (black colouration) and copper (red colouration or precipitate) but states that the amount of gall-nut powder added must be carefully measured as the strength of the colour then gives the amount of metal present. He also suggests other colour reagents, for example rose-extract, the juice of pomegranate, brazil wood extract, etc., but says that these are not very reliable as metal reagents because there are numerous metals with which they do not give any colour. Therefore, he describes a new reagent which he calls *volatile sulphureous spirit*, which with lead gives a black precipitate [35]. He tried this reagent with other metal solutions such as gold, mercury and tin. He dissolved gold in *aqua regia*, mercury in nitric acid, and tin, because it does not dissolve in nitric acid, "in a peculiar solvent, that readily acts upon it and keeps it permanently dissolved". Gold and mercury gave a black precipitate and tin gave a yellowish-brown precipitate [36]. He made the reagent by melting flowers of sulphur and to this he gradually added a similar quantity of potassium and distilled the mixture with an aqueous solution of ammonium chloride. Only the last part of the distillate must be absorbed in water, because this is the only part suitable for the purpose of the reagent. The author has repeated this experiment and has found that hydrogen sulphide is formed [37]. Consequently Boyle was using hydrogen sulphide as a qualitative reagent. It is strange that it did not occur to him to use hydrochloric acid instead of ammonium chloride for the discharge of the gas, when he would have obtained a much more effective reagent. Hydrogen sulphide was not used in analytical practice after this until the end of the following century.

Later on Boyle spent a great deal of time attempting to prepare a reagent to test for the poisonous arsenic. With hydrogen sulphide he did not get a precipitate because he had diluted his stock solution with basic mineral water. So in the end he recommended the use of sublimate ($HgCl_2$) to indicate arsenic.

Boyle measured the chemical reaction of water with an extract of violets. If the water was alkaline then the violet solution became green. He also mentions that alkaline waters can be recognized by the fact that they give a precipitate with mercury chloride and effervesce when treated with acids [38].

His next step was to distil the water and to do a fractional crystallization on the residue, and to draw conclusions from the shape of the crystals. After redissolving them he carried out tests with the above mentioned reagents.

Another important book which deals with the examination of waters is the work of F. Hoffmann: *Methodus examinandi aquas salubres*, published in 1703.

Friedrich Hoffmann (1660—1743) came of a medical family, and he himself studied medicine at Jena. Later he travelled widely and while on a visit to England he became acquainted with Boyle. In 1685 he commenced medical practice and his reputation grew quickly. In 1693 he was appointed the first Professor of Medicine at the recently established University of Halle and was one of the most famous physicians of his time. He worked at the University for 48 years and during this time students came to him from all over the world. For three years (1709—1712) he was a court physician to the King of Prussia, but preferred a scientific life and returned to his professorship. The Hoffmann drops preserve his name to the present day. He also contributed to the study of chemistry, but the medical side of his work is more important than his chemical activities.

In his *Methodus* Hoffmann describes the examination of numerous mineral waters. For example, he records that a red deposit indicates the presence of iron, which after heating to incandescence can be magnetized. Alternatively, he uses tannic acid or even the juice of pomegranate or oak-extract to detect iron. He also used a mixture of lime and shell-powder with which iron can be separated. He considered the question of the nature of the dissolved iron and concluded that the *gas silvestre* (carbon dioxide) keeps it in solution because after boiling the solution the iron separates out. In some cases it is present as vitriol (iron sulphate) [39].

Hoffmann precipitated copper with iron. He was of the opinion that copper is seldom to be found in water, although the mineral water in Besztercebánya (Hungary) contains copper [40]. Rock salt can be indicated with *lunar caustic* (silver nitrate). In the residue after evaporation the presence of alkali can be identified with vegetable extracts (by indicator) or with *salmiac* as ammonia is formed in the presence of alkali. An evil smelling red substance, hepar, is formed if the residue from alkaline waters is mixed with sulphur and melted [41]. Oil of

Fig. 9. Van Helmont (1577—1644). (From his *Aufgang der Artzneikunst* [1683])

vitriol (sulphuric acid) was used to precipitate lime. Hoffmann also discovered the presence of magnesium salts in water and distinguished them from calcium salts:

> Numerous springs contain a neutral salt, which is as yet unknown... This salt is similar to potassium sulphate *(arcanum duplicatum)*, it is bitter and gives a cold sensation when placed on the tongue, and does not effervesce either with acids or with alkalis, neither does it give a precipitate with vitriol [42].

Sulphuric waters can be recognised by their smell and also by the fact that they stain silver black. Hoffmann was of the opinion that water was a complex substance containing both an etheric and a salt component.

This abbreviated list of water-analyses clearly shows how the use of reagents assisted the development of wet analytical methods during this period. However, in the earlier literature, we can find many references to reactions used for the identification of various substances. Some of these are mentioned below.

Helmont had obtained data on the behaviour of silicic acid from a study of water-glass manufacture. The silicic acid was dissolved with excess alkali, and nitric acid added to neutralize the excess or, as he describes it, that all the alkali should become saturated *(quae saturando alcali sufficit)*. He observed that the silicious earth separates out entirely unchanged in both form and weight *(eodem pondere)* [43].

Jean Babtist van Helmont, the first great chemist of modern times was a practising physician. He was born in Brussels in 1577 into an old and illustrious family. He chose his profession despite parental opposition and left to study at the University of Louvain. He was at first a follower of the old school but was later influenced by the ideas of Paracelsus. He was also aware of the mistakes that Paracelsus made, and his cultured upbringing protested against his vulgar style. He was later to modify the chemical view of Paracelsus to such an extent that a new theory emerged, particularly with regard to the nature of the elements. He was offered the post of Imperial Court Physician at Vienna, but refused it preferring to devote the rest of his life to research at his estate near Brussels, where he died in 1644. Most of his works were not published during his lifetime but were collected and edited by his son in 1648.

Van Helmont is a puzzling character, sometimes he shows himself to be a clear thinking natural philosopher, drawing conclusions from experiments, and on other occasions he reveals his alchemical leanings believing in the possibility of making gold and also that mice are formed from a dirty shirt dusted with wheaten meal.

Of the several extant analytical notes of Tachenius, one of the most interesting is concerned with the systematic examination of the effect of gall-nut extract on metal salt solutions. He observed the colour formed with solutions of many metals [44]. He also established that iron is not removed from the human body via the urine as had previously been supposed as no colour was obtained when urine was treated with gall-nut extract [45]. This is the first example of analysis being applied to biochemistry. He also examined *silicic marl* (SiO_2) and proved that it was an acid as it formed a salt with potassium carbonate. Its acidic character is confirmed by the fact that even the strongest acid, nitric acid, does not attack it [46]. Tachenius devoted a great deal of time to the study of mercury sublimate. He found that it gives a yellow precipitate with caustic soda and a white precipitate with pure soda [47]. He also gives many examples of applied analysis. In Venice there was a flourishing trade in scent, particularly with rose-water which was used to prevent ascarides. It was found that some solutions of rose-water taken for this purpose caused the patient to vomit. Tachenius examined this problem and concluded

The cause of this is not the rose-water but is due to copper particles which have come from the copper flask used for the extraction. As a proof of this, if a few drops of alkali are added to the rose-water a green precipitate immediately forms. After filtering it is found that it will not cause any further vomiting. If the green precipitate is fused with borax it shows the presence of copper [48].

Tachenius also made potassium sulphate by adding potassium carbonate to a solution of iron sulphate until precipitation was complete, then after filtering he obtained the potassium sulphate by evaporating the filtrate [49].

Otto Tachenius was born in Westphalia at the beginning of the 17th century. He was a pharmacist and a physician. Little is known about his life, most of which was spent in Venice. He was a passionate and provocative man and during his lifetime was involved in many scientific polemics, and an equal number of disputes with authority. His works were published from 1655 onwards in Venice, where he was still living in 1669. Although he is a somewhat neglected figure in the history of chemistry, a reappraisal of his work would ensure him of much more recognition. It is easy to appreciate his sharp observation and that he was an expert in the technique of analysis. His technological descriptions are equally important. Discussing soaps, for example, he notices that fats contain a "hidden acid" [50]. Another fact worth mentioning is that he observed the variation in strength of acids and that stronger acids will displace weaker acids from their compounds [51].

The Romans had initiated the practice of storing wine in lead vessels to prevent it from going sour. This practice was continued in many parts of Europe and resulted in wine being contaminated with lead. A method for the determination of lead in wine was proposed by Eberhard Gockel (1636—1703) and consisted simply of adding sulphuric acid to the wine, when a white turbidity occurs if lead is present [52].

Numerous analytical reactions are also to be found in the works of Johann Rudolf Glauber.

Glauber, one of the great figures in chemical technology, was also a very skilled chemist as well as being a successful business man. He employed journeymen in the production of his commercial chemicals. He consistently strove to improve quality and reduce prices. His new method of making hydrochloric and nitric acids from rock salt and saltpetre respectively, he kept secret and only towards the end of his life when he had retired from business did he disclose them.

Glauber (1604—1670) was born in Germany the son of a hairdresser and in his youth travelled widely over the whole of Europe. Later he lived and worked mainly in Amsterdam which was almost untouched by the storms of the Thirty Years War. Towards the end of his life he was ill for a long time. His disease was probably acute mercury poisoning, if one judges from the noted symptoms. Glauber's interest in economics is shown in his book *Teutschlands Wohlfahrt* which gives advice to assist the recovery of the German economy which was at that time in a state of depression. The proposals he made were based on the increase of exports, at the same time decreasing the amount of imports by making more and cheaper goods at home. Glauber is a typical example of this type of early capitalism.

Glauber's analytical knowledge was also considerable and several discoveries emerge from his work. He found that silver chloride would dissolve in ammonia solution and that silver gives a precipitate with alkali and with carbonates [53]. He also observed that lead chloride was only slightly soluble in water, and he noted the reactions of cobalt blue *"bereitet von flüssiger Sand-Pott-Asche und Kobolet"* [54]. Another interesting observation he made was that the colour of a cochineal extract became scarlet when nitric acid was added to it, but he did not pursue the possibility of using this as an acid–base indicator, his only use for it being as a paint for hair and nails.

Kunckel, who is referred to again later, examined the reaction of dissolved lime with *salmiac* and showed that it liberates ammonia.

3. ROBERT BOYLE

After discussing the rather sparse analytical knowledge of the 16th and 17th centuries we must turn to the work of Robert Boyle. Many authorities consider that the work of Boyle marks the beginning of chemistry as a Science. Although reluctant to underestimate the value of Boyle's discoveries I myself consider this view to be a slight exaggeration; but without doubt, Boyle helped chemistry to dissociate itself from the medical science to which it had become bound and gave it a new direction.

> It is, in my opinion, one of the principal impediments to the advancement of natural philosophy that men have been so forward to write systems of it. When an author, after having cultivated only a particular branch of physics, sets himself down to give a complete body of them, he finds himself, by the nature of his undertaking and the laws of method, engaged to write of many things wherewith he is unacquainted. I saw that several chymists had by a laudable diligence, obtain'd various productions, and hit upon many more phenomena considerable in their kind, than could well be expected from their narrow principles; but finding the generality of those addicted to chymistry, to have had scarce any view but to the preparation of medicines or the improving of metals, I was tempted to consider the art not as a physician or an alchymist but a philosopher. And, with this view, I at once drew up a scheme for a chymical philosophy; which I should be glad that any experiments or observations of mine might any way contribute to complete.
>
> And, truly, if men were willing to regard the advancement of philosophy, more than their own reputations, it were easy to make them sensible that one of the most considerable services they could do the world is to set themselves diligently to make experiments and collect observations without attempting to establish theories upon them before they have taken notice of all the phenomena that are to be solved [55].

This elegantly composed introduction was the basis of the belief that Boyle had worked out a new chemical theory, although from his own words it can be seen that he was not in favour of advancing theories; rather he conducted experiments and from them drew certain inferences. He followed this principle — which was in effect the English empiricist philosophy of the time — throughout the course of his work.

Once having chosen the pursuit of science Boyle was fortunate enough to have sufficient wealth to enable him to do so. Boyle's contributions to the fields of mechanics, optics and chemistry are of equally great importance.

It is probably an exaggeration to say that Boyle was the creator of analytical chemistry, although he contributed greatly to its development, but in the progress of any science even the greatest experts are dependent on the experience of their predecessors. The previous account of the state of analytical chemistry in the Middle Ages shows that Boyle had a great deal of knowledge to draw on. In one respect, however, Boyle was the creator of this branch of science, for he originated the use of the word analysis in the sense that we know it today. This word often occurs in his works and also appears in the title of one of them: *The chymical analysis of seed pearls* [56], which describes the tests necessary to determine their composition and the inferences to be drawn from the results.

Robert Boyle was born in 1627 the 14th child of an aristocratic family. The family estate was wealthy enough to provide for all these children. He was given the best possible education and afterwards made the "Grand Tour" of the continent. He returned to England on hearing of the death of his father in the Civil War, and lived in Oxford from 1654 until 1668, and after this in London, where he died in 1691. Boyle was in poor health throughout his life and never married. He was a founder member of the Royal Society and took an active part in its affairs and later became President. His first book was published in 1660 and the rest followed almost unceasingly, about forty in all. His works are concerned with many problems in physics, chemistry and the medical sciences, and were always based on his experiments. In a dry and non-commital way he describes more than a hundred experiments and discusses the conclusions that can be drawn from them.

The most striking feature about Boyle's studies in chemistry lies in his use of reactions to identify various substances. He introduced many new reagents and he also used hydrogen sulphide which was referred to earlier. He initiated the use of certain vegetable and animal extracts (the extracts of violets and cornflowers, cochineal, and litmus solution) for testing for acidity and alkalinity, and he used these in a systematic manner.

Boyle also studied the process of precipitation and his definition of this phenomena is extremely clear and would still apply today.

> By precipitation I here mean such an agitation or motion of a heterogenous liquor as makes the parts of it subside in the form of a powder or other consistent body. And as chymists call the substance, which thus falls to the bottom of the liquor a precipitate; so we shall term the body that is put into the liquor the precipitant; that which is to be struck down, the precipitable substance or matter; and the liquor, wherein it swims before the separation the menstruum or solvent [57].

He then goes on to discuss the causes of precipitation. He mentions earlier views on this subject, the opinion of the disciples of Aristotle who considered that precipitation occurred when two substances had an aversion for one another. The precipitation of acidic substances with alkalis was quoted as confirmation of this idea. Boyle could not accept this theory because it implied that substances had a certain "occult" property [58]. He showed that it was not only alkalis (potassium carbonate) which would precipitate dissolved substances from acids as, for example, silver nitrate precipitates hydrochloric acid from *aqua regia*, and in addition, sulphuric acid precipitates coral and pearl from an acetic acid solution, although all of these are acidic substances. There are examples of precipitations where the weight of the precipitated substances is equal to the weight of the dissolved substance (an example of this is where metallic gold is separated by reduction), but it is more common as Boyle states:

> That in many precipitations, a coalition is made between the small parts of the precipitant, and those of the dissolved metal, appears by the weight of the precipitate, which tho' carefully washed and dried, often exceeds that of the dissolved metal; for if, having dissolved silver in good *aqua fortis* (HNO_3) you precipitate it with a solution of sea-salt (NaCl), in pure water and from the very white precipitate wash the salts; the remaining powder, being dried and slowly melted, will look much less like a metalline body, than a piece of horn, whence it takes its name of *Luna cornea* (AgCl); so considerable is the addition of the saline to the metalline particles ... [59].

From this it is seen that the washing and drying of precipitates before weighing them was customary at this time.

Boyle's search for analytical reagents was extremely systematic, and to illustrate his methods more fully his account of the examination of mineral waters for arsenic is reproduced below.

He first states his reasons for carrying out the study, "arsenic is a pernicious drug and yet has been suspected to be clandestinely mixed with mineral waters". First of all he prepared an aqueous solution of arsenic. This required considerable skill and although recording that it was difficult, Boyle did not include any details. He diluted an aliquot of this solution with mineral water from Spa, and then added ammonia to the solution when a white precipitate slowly separated out. Another aliquot was tested with potassium carbonate, when a heavy white cloudy precipitate descended. With sulphuric acid he could not observe any precipitate. From this

FIG. 10. Robert Boyle (1627—1691). (From his *Opera varia* [1680])

EXPERIMENTA ET CONSIDERATIONES DE COLORIBUS,

Primùm ex occasione, inter alias quasdam Diatribas, ad Amicum scripta, nunc verò in lucem prodire passa, ceu

INITIUM HISTORIÆ EXPERIMENTALIS DE COLORIBUS.

A ROBERTO BOYLE

Nobili Anglo, & Societatis Regiæ Membro.

Non fingendum, aut excogitandum, sed inveniendum, quid Natura faciat aut ferat. Bacon.

GENEVÆ
Apud SAMUELEM DE TOURNES.
M. DC. LXXX.

Fig. 11. Title page of Robert Boyle's book: *Experimenta et considerationes de coloribus* (1680)

he concluded that arsenic was an acidic substance. He then added a few drops of violet extract to another portion of the solution, when the colour changed to green instead of the expected red. (Probably due to the alkalinity of the mineral water.) A further portion was tested with hydrogen sulphide *(volatile sulphurous spirit*) but no precipitate was formed (too alkaline). Boyle also applied a more sensitive test to determine if arsenic is a veritable acid, but does not give a detailed description of it. However, even with this method acidity could not be revealed in the solution, rather an ammonical (urinous) or alcalic (lixiviative) effect.

When a portion of the arsenic solution was treated with a strong solution of sublimate a thick white precipitate was formed, which is quite natural in alkaline solution. However, the precipitate was white, similar to that formed with ammonia and not brick-red, as is usually formed with alkalis [60].

He concluded, therefore, that a solution of sublimate is a good reagent for arsenic.

Boyle made an attempt to measure the "limit of detection" of several reactions. One of these was the detection of chloride ions with silver nitrate, he writes

I took above a thousand grains of distill'd water, and instead of corporeal salt, put to it one single drop of moderately strong spirit of salt, and having shook it into the water, I let fall into a portion of this unequally compos'd mixture, some drops of our solution, of silver, which presently began to precipitate in a whitish form; so that, for aught appeared to the eye, this tryal succeeded better than if the water had been impregnated with but a thousandth part of corporeal salt. *(Prior to this he had carried out a similar test with rock salt.)* And further, diffusing one drop of spirit of salt, into two thousand grains of distilled rain water; and letting fall some drops of our precipitant into it, the success well answer'd our expectation. But to urge the trial still further, I added as much of the same distill'd rain water, as by conjecture, amounted to at least half as much more. So that one grain of spirit of salt had a manifest operation, tho' not quite so conspicuous as the former, upon above three thousand grains of saltless water, and possibly, if the vial could have contained more, and would not have been, when fill'd too heavy for our tender balance, the discolouration of the mixture would have been discernable, tho' but one grain of spirit of salt had been put to four or even five thousand grains of water ... [61].

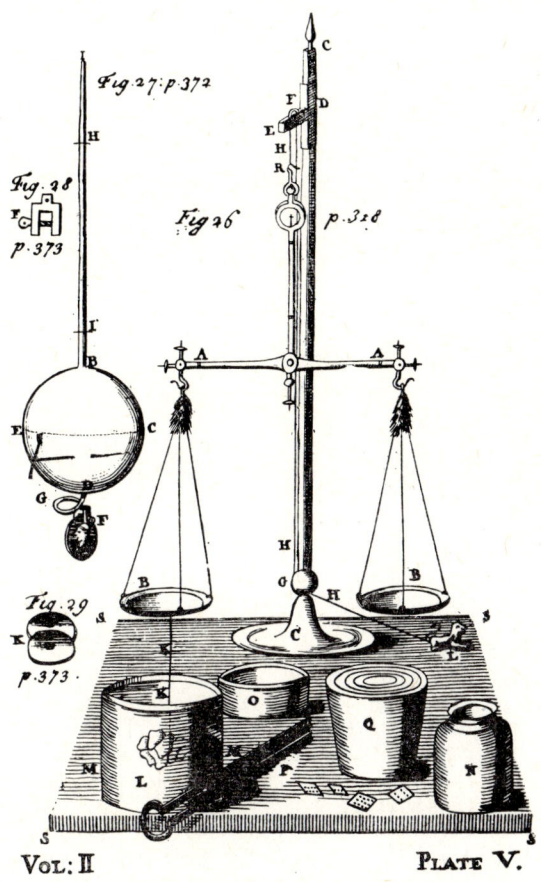

Fig. 12. Hydrostatic balance of Robert Boyle. (From his *Philosophical Works* [1725])

He also examined the sensitivity of the reaction between iron and the extract of gall nuts, and states that

> whence 'twas to conclude, that one grain of vitriolate substance ($FeSO_4$) would have impregnated six thousand time its weight of common water, so as to make it fit to produce with galls, a purple tincture [62].

Boyle, who knew nothing of the achievements of his predecessors in this field, reconstructed the areometer. He also designed a hydrostatic balance for the determination of specific weights (a drawing of this is shown in Fig. 12) [63]. It is probable that his ordinary weighing balances were constructed in a similar manner.

NOTES AND REFERENCES

1. BACON, F.: *Opera omnia.* Frankfurt (1665)
2. DANNEMANN, F.: *Naturwissenschaften in ihrer Entwicklung und in ihrem Zusammenhange.* Leipzig (1922) **1** 391 432
3. PALISSY, B.: *De l'art de terre. Oeuvres de Palissy.* Paris (1777) 5
4. The Accademia de Cimento in Florence was established in the year 1657, the Royal Society London 1660—1662. These were followed by the scientific Academies of Paris (1666), German-Roman Empire (1665), Berlin (1700) Petersburg (1725) and Stockholm (1739).
5. PARACELSUS, P. T. B.: *Bücher und Schriften usw.* Basel (1589) **1** 19
6. *ibid.* **7** 61
7. VAN HELMONT, J. B.: *Ortus medicinae id est inita physica inaudita etc.* Leyden (1656) 68
8. BOYLE, R.: *Philosophical Works.* (The sceptical chymist) London (1725) **3** 336—337
9. *ibid.* **3** 305
10. DANNEMANN, F.: *Naturwissenschaften in ihrer Entwicklung und in ihrem Zusammenhange.* Leipzig (1922) **2** 177
11. PALISSY, B.: *Des sels diverts et du sal commun. Oeuvres de Palissy.* Paris (1777) 203
12. ANGLI SALAE VICENTINI: *Opera medico-chymica etc.* Frankfurt (1647) 246
13. TACHENIUS, O.: *Hippocrates chymicus.* Venice (1666) 10
14. BOYLE, R.: *Philosophical Works.* (The imperfection of the chymical doctrine of qualities) (1725) **3** 425
15. BOYLE, R.: *Philosophical Works.* (Experiments and observations upon colours) (1725) **2** 71
16. LEMERY, N.: *Cours de chymie.* Paris (1675)
17. BOYLE, R.: *Philosophical Works.* (The chymical doctrine of qualities examined) (1725) **3** 436
18. KOPP, H.: *Geschichte der Chemie.* Braunschweig (1843) **2** 344
19. VAN HELMONT, J. B.: *Ortus med.* 89
20. GLAUBER, J. R.: *Furni novi philosophici.* Amsterdam (1648) **2** 14
21. GLAUBER, J. R.: *Opera chimica etc.* Frankfurt (1658) 50
22. GLAUBER, J. R.: *Furni novi philosophici.* Amsterdam (1648) **1** 147
23. VAN HELMONT, J. B.: *Ortus med.* 66
24. AGRICOLA, G.: *De re metallica.* Basle (1557) **7**. Introduction

25. AGRICOLA, G.: *ibid.* 190
26. BASILIUS, VALENTINUS: *Letztes Testament.* **2** 53 (Des Bas. Val. Chymische Schriften. Hamburg 1717)
27. *ibid.* Conclusiones
28. *ibid.* 467
29. BASILIUS, VALENTINUS: *Haliographia etc.* Eisleben (1603)
    Citation: HOEFER, F.: *Histoire de la chimie.* Paris (1866) **1** 481
30. GUDENUS, J. M.: *Historia Erfurtensis.* **2** (1675) 129
31. BUGGE, G.: *Das Buch der grossen Chemiker.* Berlin (1926) **1** 125
32. BUGGE, G.: *Der Alchemist.* Berlin (1943) 117
33. LIBAVIUS, A.: *De judicio aquarum mineralium.* (1597)
    KOPP, H.: *Geschichte der Chemie.* (1845) **2** 56
34. BOYLE, R.: *Philosophical Works.* (Heads for the natural history of mineral waters) (1725) **3** 495
35. *ibid.* 506
36. *ibid.* 507
37. *ibid.* 508; SZABADVÁRY, F.: *Talanta* (1958) **2** 156
38. *ibid.* 513
39. HOFFMANN, FR.: *Dissertatio physico-medico selecta.* Leyden (1708) **2** 187—198
40. *ibid.* 196
41. *ibid.* 200
42. HOFFMANN, FR.: *Observatio physico-chemic. select.* **2** 107 177
43. VAN HELMONT, J. B.: *Ortus med.* 56
44. TACHENIUS, O.: *Hyppocrates chymicus.* 115—117
45. *ibid.* 103
46. *ibid.* 34, 150
47. *ibid.* 28
48. *ibid.* 135
49. *ibid.* 47
50. *ibid.* 17
51. TACHENIUS, O.: *Antiquissimae medicinae Hippocratis clavis etc.* 6. Ed. Paris (1671) 137 141
52. GOCKEL, E.: *Miscell. Acad. Nat. Cur.* (1697)
53. GLAUBER, J. R.: *Opera chimica.* (1658) **1** 29
54. *ibid.* 187
55. BOYLE, R.: *Philosophical Works.* (Preliminary discourse) (1725) **1** 18 28
56. *ibid.* (The usefulness of philosophy) **1** 58
57. *ibid.* (The mechanical causes of precipitation) **1** 515
58. The word occult is not used in its present-day meaning. Scholastics attributed to substances two kinds of properties, elementary properties *(qualitates elementales)* and "hidden" properties *(qualitates occultae).* The former depend on the nature of the substance, for example density, colour, taste, weight, etc. To the latter were attributed all the properties the nature of which was not understood, for example, the medicinal properties, magnetism, the affinity or repulsion of substances, etc. These "hidden" properties were supposed to be incapable of explanation, or of being influenced in any way and were simply the arbitrary decision of the Creator. In other words external influences had no effect on these occult properties. The earliest experiments in natural science disproved the idea of occult properties and Boyle was fully aware of these developments.
59. BOYLE, R.: *Philosophical Works.* (1725) **1** 517
60. *ibid.* (Heads for the natural history of mineral waters) **3** 509—510
61. *ibid.* (Experiments and observations upon the saltness of the sea) **3** 231
62. *ibid.* (Memoirs for a natural history of mineral waters) **3** 516
63. *ibid.* (The hydrostatical balance) **2** 318

CHAPTER IV

THE DEVELOPMENT OF ANALYTICAL CHEMISTRY DURING THE PHLOGISTON PERIOD

1. THE PHLOGISTON THEORY

The so-called phlogiston period of the history of chemistry started at the end of the 16th century and lasted for about 100 years.

This time was a golden age for the monarchies who had absolute power throughout Europe. The aristocracy who flocked to the Royal courts enjoyed a glittering existence. But their luxurious life was completely divorced from the rest of the population. The monarchies had destroyed the effective political power of the nobility and instead of the feudalistic divided states of Europe, large areas governed by a central power came into being. This state of affairs assisted the development of commerce which was no longer hindered by thousands of despotic liege lords. The many voyages of discovery were soon followed by colonization. The aim of this colonization, unlike that of the Conquistadores was not the pursuit of gold but the production of raw materials, this search being promoted by merchant societies. The prosperity of the merchants during this time increased greatly. This was because the need for raw materials was always increasing with increasing industrialization and the growth of population. New economic methods and industrial processes were required to cope with this expansion, and banks and money exchanges flourished.

The changes that occurred in the methods of industrial production were to have a great influence on the development of the natural sciences, and to increase the rate of their development. This relation between industry and science can be seen most clearly in the history of analytical chemistry. The rapid development of industry resulted in many problems, and, in many cases, analytical methods were required to deal with them. This resulted in rapid progress being made in analytical chemistry in, for example, the metal industry and in metallurgical analysis.

In the 18th century the need for metals, especially iron, increased very greatly, and this resulted in the small scale methods of production being succeeded by industrial manufacturing processes which were often mechanized. The invention of spinning-frames and power-looms revolutionized the textile industry. These first looms were made of wood, but this was soon replaced by iron. The demand for iron increased so much that the industry could not cope with it. This was partly because iron production still used charcoal and the supply of wood became less. This shortage was most acute in England which had led the development of the iron industry. Attempts were made to substitute coal, which was plentiful, for charcoal and finally after many futile experiments Abraham Darby at last solved the problem in 1735.

It was found that iron made by this procedure was inferior to that made by the earlier process; it was fragile and brittle. In order to trace the causes of these faults it was necessary to develop accurate methods of analysis. A great many chemists applied themselves to this problem. It was not until the second half of the century, however, that Bergman was able to establish with sufficient accuracy the carbon content of various irons, and was able to understand the role that this plays, and also that phosphorus in iron is responsible for it being fragile.

The increase in the number of iron furnaces and in their capacity resulted in an increased demand for the raw materials, iron ore and coal, so that coal mining expanded rapidly. One of the major problems in mining was that of flooding, and to overcome this, water pumps driven by steam power were introduced. Steam engines at that time were used for a variety of tasks. The earlier and rather inefficient steam engines were eventually replaced when James Watt constructed his machine in 1769. This led to the development of the reciprocal steam engine which, in addition to replacing human and animal power, was later to be used to move vehicles. As more and more machines were built, the increased need for metal resulted in many more mines being opened and also in many new minerals being examined as a possible source of metals. This was a problem which required the help of chemical analysis for its solution, so it is quite understandable that during these times analytical chemistry enjoyed considerable prosperity. The most important contribution of this period was the development of methods for the analysis of minerals, and the composition of most of the known minerals was soon learned. In the course of these studies many new elements were discovered. Many new methods of analysis were developed and in addition to qualitative examinations, quantitative analysis by an aqueous method was attempted for the first time. It is obvious that the people who carried out this work were metallurgists or even miners, and as Sweden possessed the richest ore mines it is not unnatural that the science flourished most successfully there.

Sweden was to remain the centre of metallurgical analysis until the 3rd decade of the 19th century.

The introduction of volumetric analysis also took place about this time, and was at first mainly used in the textile industry, hence the origin of this branch of analysis is to be found in the countries having the most advanced textile industry, which at that time were England and France. This is discussed in more detail in Chapter VIII.

The new knowledge gained in the development of analytical chemistry had a profound effect on the structure of theoretical chemistry. A consequence of this was that research and technology were to be integrated during the phlogiston period, and this was to lead to the rapid improvement of industrial methods.

The development of all branches of science was accelerated by the newly formed scientific societies and academies of science. Many of these institutions enjoyed royal patronage, others were sponsored by the nobility and by wealthy people so that they were able to provide some financial support enabling research to be carried out. Of greater importance, however, was the fact that the regular meetings which consisted of discussions and polemics allowed the results of research to be rationally examined, and for scientists to defend their views before an intelligent and critical audience. Membership of many of these societies was conferred only as an acknowledgement of a distinguished career so that it was an honour to which many scientists aspired.

Before the formation of scientific societies the only method of disseminating scientific information and data was in the form of a book. The writing of a book, however, is a lengthy process and it was usual for a scientist to publish only one or two during his lifetime. Consequently, personal communication or correspondence was the only effective means for a scientist to make his work known and to get information about the results of other scientists.

With the introduction of periodicals, publication of society proceedings, etc. the situation improved and it became possible to publish single discoveries and the results of research in separate parts. This meant that scientists had access to the results of their contemporaries and could consider them in the light of their own work. Periodicals, therefore, played an invaluable part in the progress of science. The earliest periodicals were published by the academies, and contained the lectures delivered there. The *Journal des Savants* published by the Academy of Paris and the *Philosophical Transactions* published by the Royal Society originated in this way. Both are issued up to the present day.

The science of chemistry in the 17th century was to expand and break away from the patronage of the medical sciences. For people like Boyle and Glauber, chemistry was already a distinct branch of science, and it was soon recognized as an independent subject in the universities. The first Institutes of Chemistry were founded at the Universities of Marburg and at the Jardin des Plantes in Paris [1], and from then on the universities became the main centres for chemical research.

However, universities were still mainly concerned with teaching the humanities, so that the instruction tended to be of a theoretical nature. Instruction in laboratory practice did not exist at the universities before the 19th century, but at other institutions training in practical chemistry had started much earlier. The first of these was the Mining Academy at Selmecbánya. This academy, established in 1735, was the cradle of chemical research in Hungary, but its reputation spread beyond the country's border and in the 18th century it was one of the main scientific centres of Europe. Its importance is discussed in the work of J. Proszt[2].

The Selmecbánya Academy gave instruction in metallurgical analysis, in both the theoretical and practical aspects, beginning not only theoretically but also practically. Records show that a M. Weidacker who directed the practical classes received 50 forints for each student, and that there was a separate laboratory for teaching purposes. This laboratory training at Selmecbánya was the first organised laboratory course in Europe [3], and by the 18th century the Selmecbánya Academy had attracted students from all over Europe; many of them had already studied at other institutions, but specifically wanted to study analytical chemistry. Famous students of the Selmecbánya Academy were, for example, Manuel del Rio (1769–1841) the discoveror of vanadium, the d'Elhuyar brothers, the discoverers of tungsten, and also Ferenc Müller (1740–1825) who later became director of the Transylvanian mines, and was the discoverer of tellurium. The professors of chemistry in the academy were scientists famous throughout Europe. The first of them, such as Jacquin [4] came from abroad and he was later to become Professor at the University of Vienna. He was succeeded by Joannes Scopoli (1723–1788), who subsequently became Professor at Pavia. The next was Antal Ruprecht who studied with Bergman and whose name was well known in contemporary chemical literature through his articles.

The method of teaching at Selmecbánya was copied by the École Polytechnique of Paris, founded in 1794. This venture was welcomed by the French chemist Fourcroy, who commented

> Physics and chemistry have formerly only been taught from a theoretical basis. The Selmecbánya mining school has illustrated the usefulness of teaching the practical operations which are, after all, the basis of our science. By providing the students with apparatus and chemicals, they are able to reproduce for themselves the phenomena of chemical combination. The Committee for Public Welfare is of the opinion that these methods should be introduced at the École polytechnique [5].

We come now to a discussion of the concept which was to dominate this period in the development of chemistry and indeed was to give it its name. It is generally known that the phlogiston theory expounded that all substances are composed of a combustible and an incombustible component. The combustible part of all substances is identical, and is known as phlogiston, and when a substance is burnt the phlogiston escapes. The greater the amount of phlogiston a substance contains, the more readily is it combustible. Phlogiston is also lost when metals are combusted, and a metal-lime, or oxide or calx remains. The equation for this reaction being

$$\text{Metal} - \text{Phlogiston} = \text{Metal-lime}$$

The production of metals by reduction with coal is the following:

$$\text{Metal-lime} + \text{Phlogiston} = \text{Metal}$$

The combustion of sulphur can also be explained on the basis of the phlogiston theory. The combustion proceeds in two steps, in the first only part of the phlogiston leaves and the residue becomes phlogiston-poor sulphur (sulphur dioxide), then the remainder of the phlogiston escapes and phlogiston-free sulphur (sulphuric acid) is formed.

The origins of the phlogiston theory can be traced to Joachim Becher, in his work *Physica subterranea* published in 1681. According to Becher the earth-like (inorganic) substances are composed of three kinds of earth which are, respectively, the glass-like, the mercury-like and the combustible earths. When a substance is burnt this combustible or oily earth escapes.

Stahl later elaborated this theory in many of his works [6] published at the beginning of the 18th century.

Johann Joachim Becher (1635—1682) was born in Speyer, at the time of the Thirty Years War and was soon to become an orphan. In spite of the difficulties of his childhood he managed to educate himself. In 1660 the Archbishop of Mainz nominated him for the Professorship of Medicine at the University of Mainz, and also appointed him as his personal physician. Becher had a very restless disposition so that in spite of his ability he was not able to stay for very long in one place. From Mainz he went to Munich where he became the Court Physician to the Prince of Bavaria; from here he went to Vienna, to the post of Court Chancellor. However, a few years later he was in Holland and from there he went to England where he died.

Georg Ernst Stahl (1659—1734) was born in Anspach. He studied medicine in Jena, and after finishing his studies he remained at the university. In 1693 the newly established University of Halle appointed him to the Professorship of Medicine, a post which he held for twenty-two years. In 1716 he became the personal physician to Frederick I of Prussia and held this appointment until his death. Stahl also contributed to the development of medicine, in particular with his animistic theory.

The phlogiston theory was the first conscious attempt to explain different chemical phenomena from a unified point of view, the theoretical basis of alchemy having been rejected. The simplicity and clarity of the theory made it immediately acceptable and by 1740 it was the dominant theory of chemistry.

It has been a source of amazement to the scientists of the last century that the phlogiston theory survived for such a long time when it would seem obvious that it had so many weak links. It had been noted earlier that when a metal is combusted its weight increases, whereas according to the phlogiston theory its weight should have been less. The German historian Kopp in his vast History of Chemistry (*Geschichte der Chemie*) published in 1843—1847, is very critical of the phlogistonists. The fault of this extremely comprehensive book is that the author is far too critical of the earlier achievements in chemistry, and considered that his own generation was responsible for the sum total of scientific knowledge. Kopp's explanation of this great contradiction of the phlogiston theory is that the chemists of this period were reluctant to use the balance, and that they preferred to work qualitatively rather than quantitatively. This opinion was adopted later by other chemical historians and even today it is still widely held. This is, in fact, untrue; the balance had been in existence from a much earlier period (this has been mentioned previously) and was often used by chemists. The fact that metals increase their weight on combustion had also been known long before the phlogiston theory came into existence. It was referred to by Biringuccio in the 15th century, in his work the *Pirotechnia*, and also by many others; presumably it had already been noticed before this. But why should the phlogistonists take for an example the combustion of metals, when in everyday life the burning of wood or coal is a much more common occurrence? In the latter two cases it could be plainly observed that the burning substance becomes less, whilst something visibly leaves it. Consequently it was much more reasonable to associate combustion with the loss of a substance rather than to arrive at the much more difficult conclusion that combustion is supported by something from the surrounding air, the composition of which was unknown at that time. There already existed an explanation of the increase in weight which occurred when metals were combusted, this explanation, already a thousand years old, was confirmed experimentally by Boyle.

Boyle carefully weighed some metal (tin) and then combusted it in a closed space. He then reweighed it and noted that there was an increase in weight. As nothing could have entered the combustion flask Boyle attributed the increase in weight to the minute fire particles. Presumably he had opened the flask before carrying out the second weighing and allowed the air to rush in. This would account for the increase of weight [7].

FIG. 13. Georg Ernest Stahl (1659–1734). (From his *Opusculum Chymico-Physico Medium* [1725])

Finally it should be mentioned that in one respect the phlogiston theory is similar to the present-day electrochemical theory,

Reduction = taking up electrons = taking up phlogiston,

Oxidation = giving off electrons = giving off phlogiston

Accordingly redox phenomena could be explained on the basis of the phlogiston theory.

However, the phlogiston theory hindered the development of the ideas on the nature of elements and compounds, which were quite well advanced at that time; accordingly metals reverted to compounds and oxides became elements again.

But the progress of science, not least in the field of analytical chemistry, brought many new discoveries, and a large body of facts was observed which directly opposed the phlogiston theory. The most important of these was the complete failure to detect phlogiston analytically. Gradually the theory became more and more untenable, the original conception being modified to fit in with new facts, so that it eventually became a confused and complicated theory, which was not even interpreted in the same way in different places. For example Venel, and later Guyton de Morveau, found that phlogiston had a negative weight; thus when phlogiston escapes from a metal it becomes heavier. Guyton proved this by the following experiment:

> Let us place a balance under water, and put on both pans a lead ball of equal weight. Then a piece of cork of a suitable size is attached to one of the pans. It is observed that this pan, in spite of adding weight to it, rises! Cork is lighter than water, but phlogiston is also lighter than air, the taking up of phlogiston into the air is analogous with "rising of cork" in the water[8].

Later, after the discovery of hydrogen, this gas was thought to be phlogiston, especially since it is less dense than air and can also be seen to leave metals when they are dissolved in acids.

It is interesting to note that in the same century a book was published in which a process for the determination of phlogiston is described, actually it is for the determination of phlogiston dissolved in water. The title of the book is *Analyses aquarum Budensium* and it was published in Buda, in Hungary, and was the doctoral thesis of Josef Österreicher. The first part of the book is a description of the experimental procedure, whilst the second part gives the results of the analysis of certain medicinal springs. The method, however, was not original and had previously been devised by Jakob Winterl, who was Professor of Chemistry at the University of Buda. Österreicher makes due reference to this fact in his book.

Jakob Winterl was born in Steyrland in 1732, studied at Vienna University and died in 1809. Between 1760 and 1802 he was Professor of Chemistry and Botany at the University of Buda. Winterl although a very diligent researcher was possessed of a too vivid imagination, and this was to cause him to postulate many wild theories. One of the more mistaken of these became known all over Europe and was very detrimental to his prestige.

Winterl disputed the fact that elements were the simplest form of matter and maintained that the atoms of which the elements were composed were the simplest. These atoms consisted of either acidic or alkaline principles, varying amounts of each type of atom in an element being responsible for the difference in the properties of the elements. The positive and negative electrical charges are identical with the acidic and alkaline principles respectively. Water, an element, forms hydrogen with negative electricity, and oxygen with positive electricity. This is an explanation of the decomposition of water. So far the theory is sound, and is in agreement with the many other theories which resulted from the chemical experiments with the voltaic-pile. This theory of Winterl was accepted by a good many chemists. The difficulties only arose when Winterl attempted to prove his theory experimentally, and the results of his experiments are an object lesson for scientists. The problem of drawing conclusions from experimental results is made doubly difficult if one has any preconceived theories or ideas, as one always tends to read into them more information than they actually contain.

Winterl claimed to have discovered substances which were, according to him, more simple than the chemical elements. These substances he called *andronia* and *thelyke*. Andronia was a white powder which with water and oxygen gave carbonic acid and nitric acid, with hydrogen gave milk and with lime gave siliceous marl and with lead gave barium [9]. Guyton de Morveau repeated Winterl's experiments but without success; he found that the final product of the reaction was not the wonderful andronia, but was simply silicious marl. Winterl was not convinced and sent a sample of andronia to Paris, where it was examined by four eminent chemists, Vauquelin, Berthollet, Fourcroy and Guyton de Morveau. After examination of the material they concluded that andronia was composed simply of silicious marl together with traces of clay and iron. This conclusion disposed of Winterl's theory.

Many of Winterl's other discoveries were of importance; he discovered that iron gives a red colour with thiocyanate, although he did not know at the time that the colour was due to iron, and only remarked that cyanides gives this colour with iron under certain conditions (thiocyanic acid was not discovered until 1808 by Porret) and attributed this phenomenon to the formation of a certain "blood acid".

Winterl, long before Berzelius, elaborated a dualistic chemical concept which may have had some effect on Berzelius, although dualistic conceptions can also be found at a much earlier date. It is obvious that Winterl was a chemist with very great inventive power who, it would seem, did not take sufficient care in his experimental work, and drew conclusions from his experimental results which were too extravagant and often unwarranted. But the portrait of him left to posterity by Kopp, namely that he was a scientific swindler, is very far from the truth [10].

Josef Österreicher was born in Buda in 1756 and in 1782 qualified as a physician. He then worked at a hospital in Buda; afterwards he continued his medical practice in the country and finally in Vienna where he lived until 1831, when he died.

The basis of the determination of phlogiston is the supposition that it reacts with nitric acid and in doing so evaporates. Apart from this erroneous assumption, the method is extremely logical, and also the stoichiometry of the reaction is remarkable at a time when stoichiometry was absolutely unknown. It is a good example of Winterl's careful work and vivid imagination. The detail of the method is very carefully described, even a blank value for the phlogiston in distilled water was determined. This is the earliest reference to the use of a blank determination that the author has found. The method is described by Winterl as follows:

> To 400 cubic inches of distilled water was added as much nitric acid as is required to dissolve 1 oz of fluorite. The mixture was then evaporated to one third of its original volume. The nitric acid remaining could now only dissolve 14 gr of fluorite less than the original weight. This loss in weight is proportional to the phlogiston content of the water. The next stage in the procedure was the determination of equivalent weight of phlogiston and fluorite. To do this 4 gr of soot, which is practically pure phlogiston, is ignited in a closed crucible, mixed with 400 cubic inches distilled water, and added to it as much nitric acid as was required to dissolve 1 oz of fluorite, and again the mixture evaporated to one third of its original volume. After this the soot is filtered, and dried and found to weigh only 3 gr. The nitric acid remaining is able to dissolve only 2 drachms and 26 gr of fluorite, consequently the decrease is 5 drachms and 34 gr. Subtracting from this the 14 gr fluorite, i.e., that is "equivalent" to the "blank-value" of the distilled water then 320 gr of fluorite remain. This corresponds to the missing nitric acid which was oxidized by the 1 gr of phlogiston. Consequently 320 gr of fluorite is equivalent to 1 gr of phlogiston.
>
> It is then quite simple to determine the phlogiston-content of any water, nitric acid must be added which is equivalent to a known quantity of fluorite, and after the subsequent evaporation it must be examined to see how much less fluorite can be dissolved by the nitric acid remaining. The decrease is converted "stoichiometrically" into phlogiston, care being taken when alkali or alkaline earth metal salts are present in the water. When these are present a correction must be applied, as for example 15 gr potassium carbonate uses up nitric acid equivalent to 16 gr of fluorite, and 15 gr magnesia consumes an amount of acid "equivalent" to 30 gr of fluorite [11].

The phlogiston theory has been discussed in greater detail than is necessary, because during its period the theory was universal and all the great scientists explained their results on the basis of the phlogiston theory. The whole chemical literature of this time is written in "phlogiston language" which almost requires "translation" into the scientific language of today. Phenomena, chemical processes and chemical substances themselves were all explained on the basis of the theory; for example the solution process and the formation of salts were explained by the taking up or by the release of phlogiston. Even the newly discovered substances were named with some reference to phlogiston. Oxygen was called "dephlogisticated air", nitrogen "phlogisticated air", sulphurous acid was called "phlogisticated sulphuric acid", and even hexacyanoferrate was called "phlogisticated alkali".

2. THE BLOW-PIPE

The chemist of today is often unaware of the importance of that most simple instrument, the blow-pipe, which was so essential for chemical analysis in earlier times. With the help of the blow-pipe the qualitative composition of most minerals was ascertained and many new elements were discovered in the 18th century.

The blow-pipe, essentially a narrow tube with which air can be blown into the flame had been in use by goldsmiths since antiquity, and by the 17th century was widely used in the glass-industry. The journal of the Academia del Cimento in Florence for the year 1660 describes a small pipe with the help of which glass-blowers, using their mouths as bellows, blow into the flame, and by the use of which they can do extraordinary fine glass-work [12]. Kunckel describes this instrument in his book *Ars vitraria experimentalis oder vollkommene Glasmacherkunst* published in 1679. He mentions it in connection with the wonderful glass objects which had been made recently.

> The object of the blow-pipe is to admit air into the flame, and a very fine and concentrated flame is blown on to the surface of the glass which is being worked. The temperature of the flame is so high that it melts even the most resistant glass.

The lines with which Kunckel concludes his description of the blow-pipe are interesting:

> Also many possibilities are concealed in this art; in a chemist's workshop, for example, it might prove to be very useful. For example one use would be to melt a metal-lime in order to see what metal it contains; there are many other possible uses. It is necessary only to place the lime to be analysed into a hollow in a small piece of charcoal and to blow the flame on to it for a very short time [13].

Johann Kunckel was born in Holstein in 1630; his father was the court-alchemist to the Duke, and his son also chose the same profession. He eventually became the court-alchemist and apothecary to the Duke of Lauenburg, and later he served the Duke of Saxony in the same capacity. After he had to leave this post Kunckel taught at the University of Wittenberg for a short time, and then went to the court of Frederick William of Brandenburg. On the death of his master he bought an estate and lived there until he was offered a post as Minister of Mines by Charles XI of Sweden, and was later raised to the nobility as Baron von Loewenstern. He died in Stockholm in 1703.

The suggestion of Kunckel that the blow-pipe should prove a useful instrument for a chemist was accepted by Stahl, or at least he reports the use of it first; the effect on antimony and lead-lime (oxide) by treating it with a *tubulo caementario aurifabrorum* on a small piece of coal, *Lothröhrichen* he adds by way of an explanation [14].

A reference to the use of the blow-pipe is also to be found in a work of Johann Cramer (1710–1777). Cramer recommends the use of the blow-pipe to melt small quantities of substances. He used charcoal to ignite on but also investigated borax smelt. His blow-pipe was made of copper, and incorporated a small ball-like widening to collect the condensation from the breath during the blowing [15].

In Marggraf's account of his investigation of phosphorus and phosphoric acid (1740) there are several references to the use of a blow-pipe. Marggraf records that on the addition of phosphoric acid to gold the precipitate formed easily became an opaque glass when heated with a blow-pipe. Also with reference to silver phosphate: "it yielded a dark glass with the help of the blow-pipe",

Fig. 14. Glass blowing works in the 17th century. The person on the left works with a blow-pipe. (From Kunckel: *Ars vitraria experimentalis oder vollkommene Glasmacherkunst* [1679])

and on adding soda to it "the crystallized salts with a blow-pipe do not burst, but behave like borax" [16]. Marggraf does not explain the operation of the blow-pipe, assuming it to be sufficiently well-known.

The work of Pott is also of importance to this subject.

Johann Heinrich Pott (1692–1777) studied theology at the University of Halle, but was influenced by Stahl and became a physician. Primarily, however, he was a chemist, and he also taught chemistry at the *Collegium medico-chirurgicum* in Berlin. He was a diligent researcher and was outstanding in the field of practical examinations; he was a passionate arguer, who was always on bad terms with his fellow scientists, and eventually because of this he was forced to leave the academy.

Pott was commissioned by the King of Prussia to establish the composition of Meissen porcelain, so that the king could establish a factory similar to the one which was proving so profitable to his Saxon neighbour. Pott attempted to solve this problem by carrying out a systematic examination of the effect of fire on a large number of minerals, both singly and also in mixtures with other minerals and varying quantities of added salts. Although he carried out more than thirty thousand experiments he did not succeed in discovering the secret of the porcelain, nevertheless his work was not in vain. He conscientiously noted observable phenomena such as melting point, colour, etc., and published in 1746 an account of his investigations [17]. This volume, together with two subsequent accounts, contain much interesting information concerning the behaviour of different minerals when heated. This work contained considerable information on the value of the blow-pipe to analysts.

The Swedish metallurgists and mining analysts carried on the development of the use of the blow-pipe. According to Bergman [18], Swab [19] used it first in 1738 for the examination of antimony, but the report of this was not published until ten years later in 1748 [20]. However, Rinman [21] had previously reported the use of a blow-pipe in the examination of tin ores containing iron [22]. He also used it in the well known Rinman-green zinc test, and in addition examined cobalt salts with a blow-pipe test [23].

Most of the earlier work on the analysis of minerals using a blow-pipe is due to Cronstedt [24], who used sodium carbonate, borax and sodium ammonium phosphate as fusion mixtures. In his work *Försok till mineralogiens eller Mineral-*

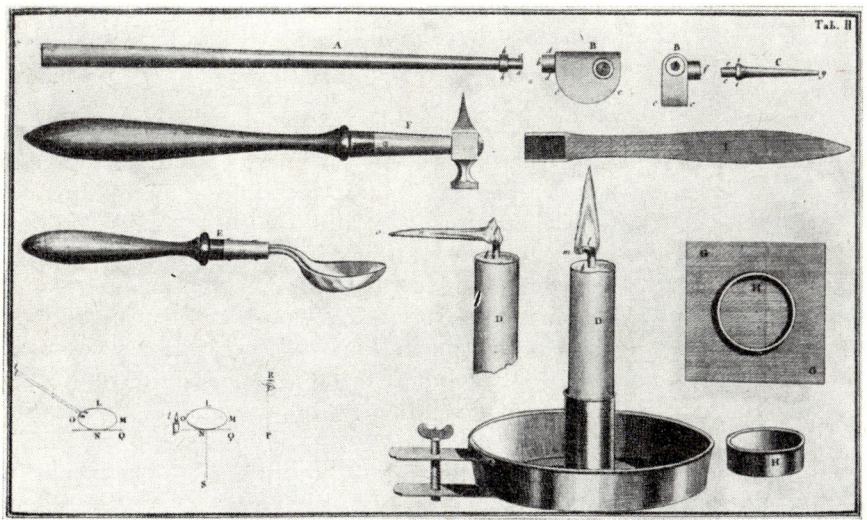

FIG. 15. Blow-pipe and its accessories in the second half of the 18th century.
(From Bergman: *De tubo feruminatorio* [1779])

Rikets upsällning published in 1758 he makes frequent reference to results obtained with a blow-pipe, but does not describe the technique of using it. When Engeström [25] translated this book into English, he added a chapter dealing with practical instructions on the use of the blow-pipe, and this was later published independently in Swedish.

Oxygen was first used in blow-pipe procedures by Achard in 1779 (according to Lampadius) and Lavoisier and Meusnier [26] also used an oxygen flame in 1782.

Bergman, who was the greatest analyst of the 18th century, made many references to the use of the blow-pipe, but in 1779 he published a book which contained all the existing knowledge on the use of this instrument. The title of this book was *De tubo feruminatorio*. He enumerates the many advantages of the method, namely that it does not require a furnace or any specialized apparatus; very little sample is required, and the operation is extremely rapid. Bergman draws attention to the fact that a quantitative estimation of composition is not possible

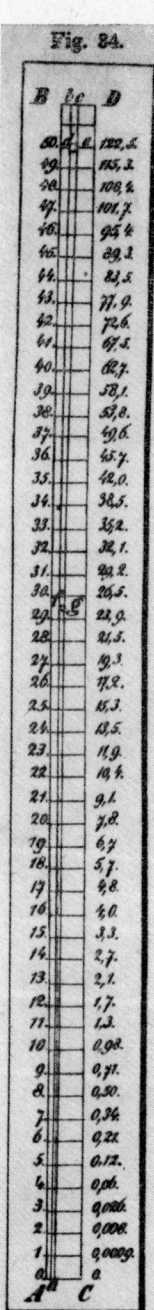

with a blow-pipe: *Ceterum prius queritur quid materia exploranda contineat, quam quantum* [27]. Here we find the first reference to the difference between quantitative and qualitative analysis. Bergman demonstrated the use of the blow-pipe by a series of illustrations (Fig. 15). He described the method of obtaining a suitable flame, i.e. a candle or a burner fed with vegetable oil, and also the substances suitable for fusing agents, such as sodium ammonium phosphate, sodium carbonate and borax. Finally he systematically recorded the reactions of the various groups of minerals with the blow-pipe. As well as charcoal Bergman used small silver or gold plates on which to fuse his samples. The technique of blow-pipe analysis was further refined by Bergman's assistant Gahn [28], who discovered the cobalt-pearl-test [29] and introduced the use of a platinum wire. The construction of the blow-pipe used by Gahn was to change very little through the years.

Berzelius [30] records that Gahn became exceptionally skilful in the use of the blow-pipe, and that no one could improve on his mineral analysis even with the use of wet methods. He mentions that Ekeberg, who discovered tantalum, asked Gahn to examine a sample of this metal for him. Gahn was able to detect the presence of tin, although the quantity of this metal did not exceed 1 per cent. Even more remarkable is the fact that he was able to detect copper in the residue of a quarter of a sheet of filter-paper, although it had not previously been thought that copper was a constituent of vegetable substances. Unfortunately, Gahn did not record any of his methods, but towards the end of his life he worked with Berzelius, who was able to publish some of his methods.

Berzelius's book on blow-pipe examinations published in 1820, under the title *Afhandling om Blasrörets anwändande i kemien och mineralogien,* was soon translated into most European languages. This book was widely read although it contained very little which was new, apart from the use of hydrogen fluoride as a reagent, and a description of some reactions of the newer elements.

It is probably an exaggeration to say that the blow-pipe, after its discovery in Sweden, was known only in other parts of Europe after the publication of Berzelius's

FIG. 16. Device for quantitative blow-pipe testing according to Plattner. (From his book: *Die Probierkunst mit dem Löthrohr* [1837])

work. Berzelius himself records, however, that whilst he was in France he was asked to write a book about the blow-pipe by several French chemists, in order for them to understand the use of it. However a book on this subject had previously been written, in 1794, by Saussure [31], a French Swiss.

Harkort [32] and Plattner [33] attempted to apply a quantitative method based on the use of a blow-pipe for the analysis of gold and silver. The principle of their method is that the weights of metal spheres are proportional to the cube of their diameters. They constructed an apparatus of ivory (Fig. 16) where $b\ a\ c$ form an isosceles triangle, the sides of which were divided into 50 equal parts. Besides the diameters belonging to the single schemes the weights were indicated (in "lots"), which were corresponding with them. With the help of the blow-pipe sufficiently ball-like metal pellets of gold or silver were to be made. This was to be pushed into the $b\ a\ c$ triangle, then it could be read with a lupe, which scheme corresponds with its diameter, beside of which the weight was to be read [34]. However this method was ousted soon by the granule-balance.

Since the last half of the 19th century, the blow-pipe has gradually fallen into disuse, partly due to the improvement of wet methods of analysis, but mainly due to the emergence of emission spectroscopy, which is far more accurate and depends less on the skill of the operator.

3. FURTHER LIGHT ON REACTIONS IN SOLUTION

Although the reactions between certain metals and acids such as $HCl + Ag \rightarrow AgCl$ and $Pb + H_2SO_4 \rightarrow PbSO_4$ had been used from a very early period, it was generally supposed that all metals gave a precipitate when treated with alkali. This was the basis of the separation of metals from the so-called earths. The term alkali covered a wide range of substances; in one text it referred to carbonate and in another to hydroxide. In many instances it is uncertain what the term alkali did refer to. It was not until the beginning of the phlogiston period that a closer study of the phenomena occurring in alkaline solutions was carried out. Stahl had observed that iron hydroxide dissolved in a large excess of alkali [35] (the blue colour of copper in ammoniacal medium had been known long before this), and this was followed by the work of Marggraf who systematically examined the behaviour of many different metals when treated with various alkalis [36].

Sigismund Andreas Marggraf was one of the most famous chemists of the phlogiston period. He was born in Berlin in 1709, the son of an apothecary, and studied pharmacology in Berlin and later at the Universities of Halle, Frankfurt and Strasbourg. He then went to the famous Mining Academy at Freiberg, and gave up pharmacy and concentrated on his chemical researches. In 1738 Marggraf went to the laboratory of the mathematical physics department of the Berlin Academy and in 1767 he became its director. He worked there until his death in 1782. He was a solitary quiet man who lived only for his work. Marggraf wrote many papers which were published in the proceedings of the Berlin Academy, the majority of them in French, as this was the official language of the academy. Analytical chemistry owes a great deal to Marggraf, as does the field of chemical technology. Distillation in a closed system which is used for the extraction of zinc is the discovery of Marggraf, who also discovered the process for extracting sugar from sugar beet.

Marggraf in his study of the metal-alkali systems established that the precipitate which was formed when gold or silver solution was treated with alkali hydroxide would not dissolve in excess alkali, but was soluble in ammonia. Zinc and bismuth hydroxides also dissolve in ammonia, the former being soluble in excess of alkali

FIG. 17. Sigismund Andreas Marggraf (1709—1782). (From Bugge: *Das Buch der grossen Chemiker*)

hydroxide. (The solution of zinc in ammonia had previously been mentioned by Boyle [37].) Lead and tin oxides are insoluble in any of the alkalis. Marggraf also examined the reactions of "alkali ignited with cattle-blood", which is in fact potassium hexacyanoferrate or potassium ferrocyanide [38].

This compound first appeared in the form of Prussian blue, referred to in an advertisement, published in the periodical review of the Berlin Academy in 1709, and was recommended as a paint. It was claimed to be superior and cheaper than ultramarine, and could be obtained from the librarian of the Academy. It is not certain who discovered it; possibly a paint-maker named Diesbach, who obtained from a merchant some potash, which gave a blue paint instead of the red paint intended. On investigation it appeared that the potash had been ignited with cattle-blood [39]. The industrial potentialities of this substance were immediately realized, the method of its manufacture was kept a closely guarded secret and the compound was sold under the name Prussian blue. It was not until 1728 that Woodward in England recorded the method of its manufacture [40]. Marggraf's method of preparation was to heat to incandescence a mixture of one part potassium carbonate, and two parts of dried blood. After cooling, the mixture was dissolved in a little water and filtered and washed. He also noted that gold and silver give a precipitate with this reagent which then redissolved in the excess, the precipitate of gold dissolving more rapidly than that of silver. Mercury behaves similarly [41].

The Prussian blue reaction was also adapted by Marggraf for analytical purposes, namely to test for the presence of iron. This is referred to in a work of his concerning the examination of water [42]. After having separated lime salts by fractional crystallization, he examined them carefully to ascertain that they were really lime salts. He found that they dissolved in hydrochloric acid with effervescence. Vegetable alkali (potassium carbonate) was ignited with lime salt. Marggraf describes this as follows:

> That in effect, this precipitate acquires all the properties and characteristics of the lime earth (calcium oxide). However it appeared that the precipitate also contains some iron particles. I shall now describe how the presence of iron particles was revealed. It is well known that Prussian blue owes its colour to iron. No doubt it is possible to detect the presence of iron particles in lime earths extracted from water by treating them with an alkaline solution of potash made incandescent with cattle blood, the preparation of which I have already described [43].

Marggraf heated the precipitate with sulphuric acid for one hour, by which time the acid had extracted the iron, and after allowing the vessel to stand for one hour he filtered the solution. To the filtrate he added a considerable quantity of potassium ferrocyanide solution drop by drop; the effect of this was to separate the iron as a blue precipitate. Greatly excited by his success he proceeded to test many other substances for iron by this method. His selection of samples for analysis was somewhat haphazard, and he was able to detect iron in limestone fluorite, human bladder-stone, sheepbone, human skull, coral, on the bile-stone of an ox *(pierre tirée du fiel de boeuf)*, and in a sample of spring water from Karlsbad. He could not detect iron in stalactite, whale-tooth, wild-boar tusk, oyster-shell, pearl, ivory, deer-antlers, egg-shell or in the carapace of lobsters [44]. To prove that iron is responsible for the formation of the Prussian-blue precipitate, he mixed some of the precipitate with fat and ignited it. The black powder remaining, after cooling, was attracted by a magnet [45]. He also examined different waters by an earlier method which he had developed, namely he separated the different groups of salts by fractional crystallization, and identified them by the characteristic forms of the crystals and also by various chemical reactions. Marggraf recorded all the results of his experiments.

Marggraf was also the first person to produce phosphoric acid. He found that 1 oz of phosphorus increases its weight by $3\frac{1}{2}$ drachms when it is burnt, the product so formed easily absorbs the moisture from the air, and also with water forms a sulphuric acid-like oil [46]. Marggraf confined his energies to practical experiments and never entered into any theoretical explanations. He worked on the basis of the phlogiston theory, but with some reservation as in the case of the increase in weight of phosphorus on burning. This fact was to be one of the most important points of the theory of Lavoisier. However, he did examine the effect of phosphoric acid on different metals, and also the reaction of phosphoric acid with various metal–salt solutions. His observations here are not very useful as his solutions were too acid for the metal phosphates to precipitate. A white precipitate was formed with mercury which mostly redissolved and in the case of lead a white precipitate was formed which would not dissolve [47]. With silver he did not obtain a precipitate although he had noticed in earlier experiments that silver in solutions of potassium ferrocyanide and ammonium chloride gave a red or yellow precipitate with phosphoric acid [48].

A further important discovery made by Marggraf is the observation that there is a difference between vegetable and mineral alkalis. During his investigation of rock salt to establish whether the alkali part is derived from alkaline earths [49], Marggraf made some interesting observations. First he removed the chloride with nitric acid, and examined the resulting sodium nitrate. He found that although it was similar to saltpetre (KNO_3) in some respects it was not the same. First, the shape of the crystals formed by evaporation was different, and second, when it was placed on glowing coal, although it decrepitated like saltpetre the flame became yellow whereas the flame of the latter was white [50]. In a later communication Marggraf records that "the flame of the former ($NaNO_3$) is yellow, while that of the latter (KNO_3) is blueish" [51]. This is the earliest reported use of a flame test applied to analysis. He confirmed this experiment by converting the nitrates

FIG. 18. Title page of complete works of Marggraf. Edited in Paris (1762)

to sulphate, chloride, and carbonate and in each case a noticeable difference in the two salts was observed, in the crystal form or in solubility, or else from the fact that potassium carbonate is deliquescent whereas sodium carbonate is not. Marggraf called the former *alkali fix minerale* (mineral alkali salt), and the latter *alkali fix vegetale* (vegetable alkali salt). Consequently Marggraf was able to differentiate between the two alkali metals, sodium and potassium. But Stahl in 1702 had hinted at the difference in the properties of the alkali constituent of sodium and potassium salts [52]. Duhamel de Monceau [53] also carried out experiments of a similar nature to those of Marggraf, and from the results of these he concluded that the basic part of rock salt is alkaline in character, but it is different from the alkali content of potash, and also that the alkali part of Egyptian natrum (soda) and of borax is the same as that of rock salt [54].

The chemical separation of potassium and sodium was first carried out by Winterl using tartaric acid, which forms the insoluble bitartrate with potassium [55]. Winterl also separated magnesium from iron by keeping the iron in solution with Seignette salt and precipitating magnesium as the hydroxide [56]. He was also able to separate calcium from magnesium with tartaric acid. That iron is not precipitated by alkalis in the presence of many organic acids, oxalic acid, tartaric acid, and amber acid was mentioned by Wenzel in 1777 before Winterl's method was published [57]. All these organic acids were first prepared by Scheele, who used oxalic acid for the separation of calcium.

In his examination of the two alkali salts Marggraf also noted that both the mineral and the vegetable alkali salts were able to dissolve sand or quartz when fused together. Two or three parts of salt when melted together with one part of sand, gave a mass which dissolved in water, and was called by Marggraf *massa pro liquore silicium* [58]. This is the earliest reference to the use of soda for indicating the presence of silicates, but reference is also made to the fact that this test had been carried out with potassium carbonate at a much earlier period. This is obvious, for glass had been manufactured since the very earliest times. It is a pity that Marggraf did not record in what kind of vessel he made the fusion.

The application of the microscope to analysis for the identification of substances is also due to Marggraf. He established with the microscope that beet-sugar and cane sugar are the same [59], and he also used the microscope for his tests with platinum [60]. This, in those days, was still a little-known metal and Marggraf was the first to examine its ores thoroughly, and in so doing he observed the reactions of platinum with various reagents.

The work of Marggraf was invaluable to the progress of analytical chemistry; it provided a wealth of new methods and experimental facts, and justified his reputation as one of the greatest analysts of all time.

During the phlogiston period many new elements and compounds were discovered and many of their properties were studied. Brandt [61] described many of the properties of cobalt, and separated cobalt from bismuth by hydrolysis of the latter [62]. Brandt also recorded the reactions of arsenic trioxide [63], and summarized the behaviour of various metals when treated with ammonia [64].

Borax was examined by Pott, and one of the interesting facts he uncovered was that it imparts a green colour to an alcohol flame [65]. For this experiment he used the so-called sedative salt, namely a solution of borax in sulphuric acid.

FIG. 19. Joseph Black (1728—1799). Oil painting.
(From Bugge: *Das Buch der grossen Chemiker*)

Numerous other observations and discoveries were made which were later to be applied to analytical chemistry; the most notable of these was due to Black who discovered the difference between strong and weak alkalis, i.e. between the alkali carbonates and hydroxides. The work of Black is of interest in that it is essentially quantitative, and it also leads into the field of gas analysis, which was to play a very important part in providing a new chemical theory.

Joseph Black was born in Bordeaux in 1728, the eighth child of a rich Scottish wine merchant. At the age of 12, he went to school in Belfast and afterwards studied medicine at the University of Glasgow and continued his studies in Edinburgh. It was here that he submitted his famous doctoral thesis. At this time the curing of gallstones and bladder stones was attempted by administering caustic alkalis to the patient, and in connection with this Black decided to examine the caustifying process. The influence of the ideas he put forward in his thesis was so great that barely a year after leaving the University he was awarded the Professorship of Chemistry at the University of Glasgow. This was in 1756 and ten years later he took a similar post at Edinburgh University where he worked until 1797. He died in 1799.

Although during his lifetime Black did not publish a great deal, his few publications were of great importance.

Black based his examination of alkalis on the tests of Hoffmann, and first showed that magnesium carbonate and limestone were two different substances. Although both substances effervesce when treated with acids the shape of the crystals obtained on evaporating the solutions are quite different. Magnesium carbonate does not form common lime when heated strongly, and on cooling, the residue is insoluble in water. This product of ignition (oxide) however forms the same salts with acids as does the original salt (carbonate) with the difference that no effervescence occurs. Black also observed that during ignition "air" (carbon dioxide) is lost, and supposed that to be responsible for the loss in weight as well as for the effervescence with acids. In order to confirm this he dissolved the magnesium oxide in sulphuric acid and then precipitated the magnesium with sodium carbonate. He found that the composition of the precipitate was identical with that of the magnesium oxide before ignition (carbonate) and also that the change in weight was only very small. He therefore concluded that alkali carbonates were not elemental substances as had originally been thought, because they give "air" to the magnesium oxide, and this same "air" is responsible for the effervescence with acids. Black then turned his attention to the examination of lime and limestone and applied similar experiments. He established that the air which leaves the carbonates is not identical with atmospheric air and is only a component of it. This air was called by Black "fixed air" and is that part of the air which is absorbed by lime and the alkali hydroxides. Therefore the relationship between the alkalis and fixed air is similar to that between alkalis and acids, in that alkalis are "in some measure neutralized" by the fixed air. However, the relation between acids and alkalis is stronger as the acid drives out the fixed air. Accordingly, the weak alkalis (carbonates) contain strong alkalis (hydroxides) and their corrosive properties result from their original form, which is lost after taking up the fixed air. The affinity of lime for fixed air is greater than its affinity for water, as limestone is precipitated by the action of fixed air on an aqueous lime solution. He also observed that the decrease in weight on heating limestone to drive out the fixed air was equivalent to the weight of acid which was neutralized by the corrosive alkali formed [66].

With this classical series of experiments Black revealed the difference between alkali carbonates and hydroxides as well as the methods of their interconversion. As well as solving many controversial problems Black's discoveries drew attention to the study of gases.

Black's views naturally aroused great controversy and he was fiercely attacked especially as his theory disproved the existence of the so-called "fire-substance" being present in lime. Jacquin, who was a teacher at the Academy of Selmecbánya, repeated Black's experiments with even greater thoroughness and completely confirmed Black's findings. He published the results of his examination [67] and Lavoisier remarked about this that "the theory of Black only became complete after Jacquin had examined it and published his book" [68]. Black also expressed appreciation for the work of Jacquin [69].

This interesting scientific debate, however, is not strictly relevant to a study of analytical chemistry so we cannot go into it in greater detail, but the experiments that Black carried out introduced the art of gas analysis, and in this field some of the most important discoveries of the phlogiston period were made.

4. THE BEGINNINGS OF GAS ANALYSIS

Many other gases apart from air are to be found in nature. Eruptions of natural gas were observed from very early times and the dangers of fire-damp in mines was soon realized. Pliny refers to air which cannot be breathed, and which is inflammable [70]. Alchemists often observed the evolution of gases during their experiments, and referred to them as *spiritus*, a term which also included acids. This has resulted in the meaning of many early texts not being quite clear. They supposed that the evolved gases were merely air, so they paid very little attention to them, and did not carry out any experiments to determine their nature.

Paracelsus mentioned that when iron is dissolved in sulphuric acid "air rises and breaks out like wind" (Luft erhebt sich und bricht herfür gleich wie ein Wind) [71]. It has been mentioned earlier that van Helmont was the first to realize that gaseous substances other than air exist. He observed that carbon dioxide, "*gas sylvestre*", is formed during the burning of oak-coal, and gave other instances of the formation of CO_2 from carbonate when treated with an acid, *acetum stillatitium dum lapides cancrorum solvit, eructatur spiritus sylvestris*, also during fermentation, and, further he remarks that this gas occurs in caves, cellars, and in mineral waters [72]. However, he was not able to identify this gas; the gas formed from the combustion of sulphur, and also the nitrous fumes which are formed when silver is dissolved in nitric acid, were called by Van Helmont *gas sylvestre* [73]. The word "gas" also originated from Van Helmont, who differentiated between gases and vapours, the latter being formed from water (liquid) by the effect of heat, and which condense to give water again on cooling, whereas the former are dry, air-like substances which cannot be changed into liquids. However, Van Helmont did not examine gases any further, because, as he writes

> The gases cannot be contained in a vessel, as they break out through all impediments and unite with the surrounding air *(gas, vasis incoercibile, foras in aerem prorumpit)* [74].

Toricelli [75] in 1643 determined the weight of air, and Boyle (also independently Mariotte in France) examined the relation between the volume of air and its pressure, and also established that combustion requires air. (This had already been noted by several earlier workers.) Boyle also observed the formation of other gases; he describes the formation of hydrogen for example and devised a way of containing the gases. He placed some iron nails in a narrow necked flask filled with sulphuric acid and then covered the neck of the flask with a vessel filled with water, and collected the bubbles which were evolved [76]. In spite of these experiments Boyle did not believe that the various kinds of gases were essentially different from air.

In 1655 Wren [77] isolated the gases which were evolved during the fermentation process by using a bladder equipped with a tap attached to the neck of the reaction vessel [78]. Mayow [79] constructed an apparatus for collecting gases similar to that of Boyle. A large vessel was filled with dilute sulphuric acid and a smaller one submerged in the liquid and then inverted; the granulated metal was then added and the bubbles allowed to rise into the smaller vessel [80]. Bernoulli [81] obtained the gas evolved when chalk is treated with acid in a similar type of apparatus, and concluded from this experiment that solid bodies may contain gas particles [82].

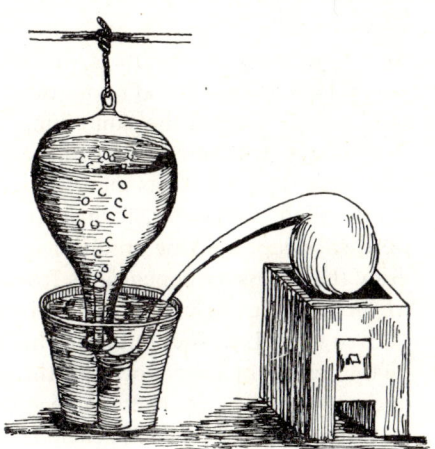

FIG. 20. Hales' gasometer. (From his book: *Vegetable staticks* [1727])

The characterization of the different gases was greatly assisted by the work of Hales [83], who constructed the earliest form of the gas-collection apparatus in use today. Previously, gases had been delivered and collected in the same vessel, but in Hales' apparatus two separate vessels were used, connected by a tube (Fig. 20) [84]. The advantage of this apparatus was that gases evolved by the action of heat on a substance could be isolated. Hales carried out many experiments with this apparatus and collected the gases evolved from the heating of wood, tobacco, oils, sulphides, limestone, fat, etc. He also measured the amount of gas obtained from a certain weight of substance, although the quantitative nature of his experiments was rather spoiled as he used water as the confining liquid. Although Hales succeeded in isolating many gases he did not identify any of them, believing them all to be air, and was satisfied to establish whether the air evolved from a certain substance was combustible or incombustible. The most important contribution of Hales was his gas-collecting apparatus which was to prove invaluable to subsequent workers.

Thus, it can be seen that in spite of the work of Van Helmont the general view in the 18th century was that all gases were composed of air. Black's discovery that although carbon dioxide is present in air it is a quite different gaseous substance, provided the break-through for the discovery of the gaseous elements.

The fact that carbon dioxide can be dissolved in lime-water was also discovered by Black, and provided a convenient method for separating the gas. The effect of carbon dioxide on various metallic salt solutions was examined by Bergman in 1774. He called carbonic acid *acidum aereum*, air acid, after testing the solution with litmus to establish its acidic character. In addition, Bergman determined the density of the gas and also measured its solubility in water and in alcohol. He also noted that an aqueous solution of carbon dioxide dissolves the alkaline earth metal carbonates and also zinc, iron, and manganese carbonates [85].

The greatest contributors in the field of gas-analysis, Cavendish, Priestley and Scheele all worked at about the same time, around 1770, and it often happened that they discovered the same things independently of one another. For example, oxygen was discovered by Priestley and Scheele at the same time. The discoveries that these three scientists made about oxygen, nitrogen, hydrogen and the composition of air and water, formed the basis of the new chemical theory, but they were so steeped in the traditions of the phlogiston theory that they could not draw the necessary conclusions from their discoveries and it was left to Lavoisier to incorporate them into his anti-phlogiston theory.

Even when Lavoisier published his new anti-phlogiston theory, which was based on the results of their work, all three opposed him and did not alter their views during their lifetimes. These three great scientists resemble one another in that none of them was educated as a chemist at a University, and that none of them was engaged in chemistry as a profession, only as a pastime. Scheele was a pharmacist, a self-taught scientist who carried out his experiments after his working day was finished. Priestley was a preacher, who had a large family and was constantly struggling against financial difficulties, and tried many other professions during his difficult life. It is amazing how Priestley found the time to carry out his experiments. Cavendish was an extremely rich member of the aristocracy who studied many branches of science as a hobby. Instead of giving a detailed account of the discoveries of these three men, it is proposed to examine their experimental methods as these are their main contributions to the progress of analytical chemistry. Unfortunately, none of them recorded their experimental methods in great detail, but even so sufficient information is available to assess the merits of their work.

Henry Cavendish (1731—1810) was born into one of the most ancient aristocratic families, and was even related to the British royal family. Although he studied at Cambridge University he did not pass any examinations. He dealt with scientific problems, drawn from almost all branches of natural science, from astronomy to chemistry. Cavendish lived in London, but he was a lonely and solitary man who avoided meeting people whenever possible. He seldom mixed in society, for his lack of grace and disregard for social conventions set him apart from the other members of his class. Cavendish had a very large library which he opened to the public. As the library was situated at some distance from his house, every time he required a book he would have to go there on foot, and he insisted on having a borrowing ticket, the same as anyone else. He was also very reluctant to publish the results of his work. As a result of his parsimonious economy he was able to leave one million pounds to his delighted heirs. According to one of his biographers Cavendish was the most scholarly of the rich, and the richest scholar.

Cavendish was an outstanding investigator in the study of gases. His powers of observation were very keen, and his experiments carefully designed and skilfully performed. His most important work was the *Experiments on Air* published in 1784. He also published many other articles. He is generally regarded as the discoverer of hydrogen as he was the first to establish that it is a separate gas quite different from air. Cavendish was also the first to measure the density of gases, and to use this property to differentiate between them. Assuming the density of air to be unity, Cavendish found the density of carbon dioxide to be 1·57, and of hydrogen to be 0·09 [86]. This experiment was carried out by filling bladders with the gas and weighing them.

Cavendish discovered that if air is passed several times over heated charcoal, and then passed through a solution of potassium hydroxide, an air-like substance remains which is lighter than air and

FIG. 21. Henry Cavendish (1731—1810). His own charicature drawing. (From Bugge: *Buch der grossen Chemiker*)

will not support combustion. Cavendish called this gas "phlogisticated air". Cavendish only communicated his discovery in a letter to Priestley, so that Daniel Rutherford [87] (who carried out the same experiment) was able to announce the discovery first. Soon after this Cavendish discovered nitrous oxide, and also its property of reacting with oxygen to form nitrogen oxide which when dissolved in water gives nitric acid. This reaction was used by Cavendish as the basis for establishing the composition of the air, and for this purpose he constructed a eudiometer [88]. After carrying out four hundred experiments at different localities and in varying weather conditions, he concluded that the atmosphere contains 20·84 per cent dephlogisticated air (oxygen).

The greatest of Cavendish's discoveries, however, was his discovery that water is a compound of hydrogen and oxygen, and not a primary element as had been thought for the past two thousand years. By passing an electric spark through a mixture of air and combustible air (hydrogen) Cavendish was able to observe the formation of water, and he later repeated the experiment using pure oxygen and hydrogen in the correct proportions [89]. With a similar experiment Cavendish

Fig. 22. Cavendish's gas explosion pipette. (Photograph in the German Museum in Münich)

found that nitric oxide, nitrous acid, and nitric acid are formed when a spark is passed through air. The great skill and accuracy which are characteristic of Cavendish's experiments is illustrated by this account of his experiment to determine whether or not all the nitrogen in the air could be converted into nitric air (nitric oxide). He passed a spark discharge through a mixture of air and oxygen until no further decrease in volume was apparent. The excess oxygen was then removed with "*hepar sulfuris*" and the residual volume of gas, which was only a very small bubble, occupied only about one hundred and twentieth part of the original volume of the nitrogen. Cavendish concluded that phlogisticated air (nitrogen) was not a homogeneous substance.

This fact was overlooked by chemists for more than one hundred years, and it was not until Rayleigh, in 1894, observed that the vapour density of nitrogen obtained from the air is greater than the vapour density of nitrogen obtained from ammonia, that the significance of Cavendish's finding was apparent. Rayleigh repeated Cavendish's experiment with the same result [90] and this led to the discovery of the inert gases.

Joseph Priestley (1733—1804), the son of a poor weaver, was orphaned at a very early age, and lived with a succession of relatives. He was self-taught and wanted to enter the Church, for this was the only profession which gave any opportunity for him to study. While a student at theological college it became apparent that he had a wonderful gift for languages, being able to speak nine, including German, French, Italian, Latin, Greek, Arabic, Syrian, and Hebrew. If it had not been for his passionate nature and outspoken manner, Priestley could certainly have had a successful career. However, he criticized the Church of England, and also the educational system to such an extent that he was forced to give up his position. He then took up a position as pastor at a very poor village dissenters' chapel, but very soon his outspokenness made it necessary for him to leave. His next appointment was teaching languages at a private school; he also taught natural science and lectured on electrical phenomena. His book *The History of Electricity* was a great success, as a result of which he was elected to membership of the Royal Society (1767). During this time he had also written an essay advocating liberal education which caused a great deal of controversy, and also resulted in his losing his employment. He married and took a post as a vicar in Leeds. In the neighbourhood was a beer factory. This turned his scientific attention to the process of fermentation, in particular to a study of the gases evolved, and he made many experiments to determine the nature of these gases. At the same time he also wrote essays about social conditions, and religious polemical treatises, with the result that he again lost his position.

FIG. 23. Joseph Priestley (1733—1804). Painting by Gilbert Stewart.

In 1773 he was offered a post as secretary to a rich aristocrat, which he accepted, and accompanied his employer on his journeys to the continent. In the course of one of these visits he met Lavoisier, and confided in him his discovery of "dephlogisticated air" (oxygen), and Lavoisier shortly published this fact as his own discovery, which of course produced a violent protest from Priestley. It was during his period of employment by Lord Shelburne that Priestley published his most important chemical work, but after he had published an essay criticizing the upper classes, in particular the aristocracy, he was again asked to change his employment.

After a period in which he experienced great poverty he at last obtained a post as vicar of a poor Free Church in Birmingham. In this city there existed a flourishing Scientific Society, which regularly held meetings. Several of its members were to play an important part in the development of science and technology: Wedgwood who was the owner of a large porcelain factory, and Watt who designed the steam engine were members of this society. Priestley soon became one of the leading members of this circle.* This pleasant interlude was not to last for very long, for as soon as the French Revolution broke out, Priestley openly sympathized with the cause of the Revolutionaries and also condemned the English form of government. When he and some of his friends attempted to celebrate the anniversary of the fall of the Bastille, his house was attacked and burnt by the Birmingham mob, and he had to flee to London, where his friends found him a new position as a pastor. Even in London he was unpopular because of his extreme views. The new French Republic elected him an honorary freeman, which was to cause him to become even more unpopular in England, even many of his closest friends forsaking him. In protest, Priestley resigned from the Royal Society and at the age of sixty-one emigrated to America. The University of Philadelphia offered him a professorship which he refused, preferring to spend the rest of his days on a small farm beside the river Susquehannah, occupying his time with writing essays on theological and social subjects.

Priestley's work in chemistry occupied only a small part of his life, mainly between the years 1770 to 1780. The results of his research were published, and extended to six volumes with the title *Experiments and Observations on Different Kinds of Air*. In the course of his work Priestley prepared oxygen by heating mercury oxide or saltpetre. The heating was accomplished by focussing the sun's rays with a burning lens. He also prepared hydrochloric acid gas, ammonia, nitric oxide, silicon tetrafluoride, sulphur dioxide and carbon monoxide. The reason that he was able to discover these gases was due to his ingenious idea of using mercury instead of water as the confining liquid in the Hales gas-collection apparatus.

Priestley also examined the properties of these gases, but although he was a brilliant practical chemist the theoretical possibilities opened by his discoveries often eluded him; moreover, Priestley himself realized that the true significance of his work had yet to emerge. The concluding sentence of his description of the preparation of oxygen *(dephlogisticated air)* illustrates this very clearly: "This series of facts, relating to air extracted, seems very extraordinary and important, and, in able hands, may lead to considerable discoveries" [91].

Carl Wilhelm Scheele (1742—1786) was born in the Prussian town of Stralsund which at that time was ruled by the Swedes. At the age of 15 he became apprenticed to an apothecary in Gothenburg. In the evenings he read all the chemistry books he could obtain, and carried out experiments with the limited number of chemicals available to him. He kept very detailed records of his experiments, which he took with him when he left Stralsund to become an apothecary's assistant in Malmö, and then later in Stockholm, and finally at Uppsala. It so happened that Bergman, who was Professor of Chemistry at the University of Uppsala, obtained his chemical supplies from the apothecary where Scheele was employed.

The manner in which Bergman first came to notice Scheele's great talent is of interest. The apothecary Loock, who was Scheele's employer, had noted the reddish-brown gas which was evolved when saltpetre was treated with acetic acid. Not knowing the reason for this he enquired from Gahn, who was then a young chemist working under Bergman. As Gahn could not supply the solution he turned to Bergman, who was also unable to give a reason for this phenomenon. When Scheele suggested that nitrous acid, related to nitric acid was formed, Bergman was very impressed, and a great friendship and collaboration developed

---

*The famous Lunar Society, so called because the meetings were held on the night of the full moon so that members who lived far away could see to drive home.

Fig. 24. Carl Wilhelm Scheele (1742—1786)

between them. Scheele began to publish the results of his research, and at the age of thirty-three, still an apothecary's assistant, he was elected a member of the Swedish Academy of Sciences. When Pohler an apothecary of Köping died in 1775, Scheele took over the shop, between them. Scheele began to publish the results of his research, and at the age of thirty-and in 1777 Pohler's widow sold it to him, and it was here that he was to show the world how great his work was in spite of his insignificant position.

Scheele resisted many offers of important and well-paid posts saying "I cannot eat more than enough, and I earn enough for me to eat here in Köping". Scheele's health finally deteriorated, but on the 19th May 1786, whilst on his sick bed, he fulfilled an earlier promise and married the widow of the apothecary. Two days later he died, only 44 years of age.

Scheele's greatest contributions to science were made in the field of organic chemistry, which in fact he originated. He prepared numerous organic acids, amongst them oxalic acid, which he employed as a reagent for calcium [92]. He also discovered molybdenum [93], tungsten [94], manganese [95] and barium [96]. Scheele discovered oxygen, hydrogen and nitrogen, independently of his English colleagues, and also analysed air. Into a vessel filled with water, a small dish containing a moistened mixture of iron granules and sulphur was placed and supported above the surface of the liquid by a small pedestal. A graduated cylinder was placed over this dish extending into the water. The absorption

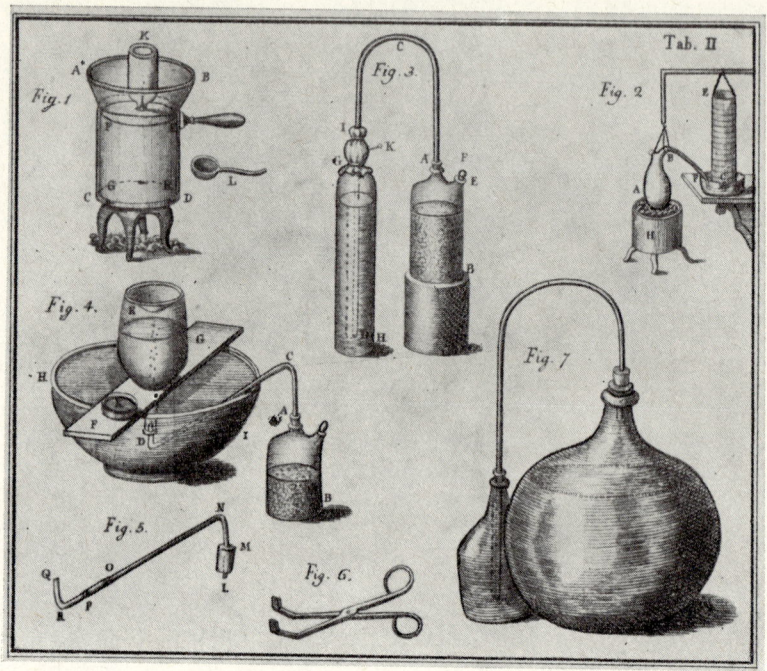

FIG. 25. A set for gas analysis according to Bergman. (From his book: *Opuscula physica et chimica* [1779])

of oxygen could then be measured by the rise of the water level inside the cylinder.

Scheele also discovered chlorine, which he prepared from manganese dioxide and hydrochloric acid, and investigated its properties [97]. Scheele isolated silicofluoric acid *(volatile fluorated earth)* from fluorspar and proved that it was composed of silica, the acid of fluorspar, and an alkali [98]. Scheele used a bladder for the collection of gases.

In *Analyses aquarum Budensium,* by Winterl and Österreicher (1781), descriptions of methods for the determination of oxygen, hydrogen sulphide, sulphur dioxide, carbon dioxide and nitrogen, dissolved in water are given. All are chemical, not gasometric methods. For example, carbon dioxide is determined with lime water, hydrogen sulphide by passing the gas through a solution of a silver salt, and oxygen by treatment with ferrous sulphate solution.

A typical example of gas analysis is Bergman's method for the determination of the carbon dioxide content of carbonates. He weighed the carbonate into a vessel, then after dissolving the sample in water, the vessel was reweighed (A). Into another vessel some acid was added, and the vessel weighed (B). Then acid was poured from the second vessel into the first until effervescence had ceased. Both vessels were reweighed. If the weight of the first vessel was a, and of the second, b, then the increase in weight of the first vessel should be equal to $B-b$. Therefore, the difference between $B-b$ and $a-A$ corresponded to the weight of carbon dioxide evolved [99].

This experiment supports the theory of the indestructability of matter. The methods of analysis of gases did not improve very much until the time of Bunsen.

5. TORBERN BERGMAN

Chemical analysis as it appeared in the course of the progress of chemistry, was not an independent science; there were no general methods of analysis. Analytical procedures were only published when they formed an integral part of some research or investigation, and usually the researcher had to devise the method of analysis for himself. Characteristic reactions of various substances, as well as chemical processes such as filtration, washing, evaporation, etc., were never described as a whole, only as they occurred in more specific problems. Only the methods for the examination of noble metals such as the fire-process, were described as a single subject, as were the methods of water examination. As the number of chemical elements and compounds increased, however, so did the amount of available information, so that some systematic method of classification became necessary.

In the field of analytical chemistry there was a great need for all the existing methods to be classified on the basis of their applicability to certain groups of substances, so that some source of reference would be available to help solve any problems of an analytical nature which arose. In other words a treatise on analytical chemistry was required.

The first person to attempt to overcome this problem was Torbern Bergman. Although the works of Bergman can be regarded as the earliest form of analytical text book, in that they give a methodical summary of the processes of analytical chemistry grouped according to the nature of the substance analysed, they are primarily a record of his own research.

No branch of science can be said to have originated from one discovery, or from the work of one man. Chemical analysis had been practised for two thousand years before Bergman, but it was he who gave it the status of a separate branch of science — Analytical Chemistry.

Torbern Bergman was born in 1735, in the territory of Westgotland in Sweden. His father who was a tax-collector would have liked him to become a priest or a lawyer. Bergman carried out his father's wishes and studied law at the University of Uppsala. Soon after this his interests turned to the natural sciences, but he also attended lectures at the faculty of medicine, and those given by Linne who was the Professor of Botany. For his parents' sake he still continued his law studies, but as a result of the extreme mental effort required he became ill and had to interrupt his studies for a while. After he had recovered he gave up his law studies and concentrated on medicine. He also studied mathematics and in 1761 he became assistant professor of mathematics. In 1767 the chair of chemistry became vacant, and such was the reputation of Bergman that he was given the Professorship without ever having published a single chemical paper. He then turned his whole interest towards chemistry and during

Fig. 26. Torbern Bergman (1735—1784). A picture in the University Library of Uppsala

the next sixteen years the record of his research filled five large volumes, the major part being concerned with analytical chemistry.

The results of his research were published regularly in the reviews of the Academies of Stockholm and Uppsala, but his collective work was published under the title *Opuscula physica et chemica*, written in Latin, and appeared from 1779 onwards. Bergman only lived to see the first three volumes published, the final two volumes edited by Hebenstreit were published in 1787 and in 1790 in Leipzig. The works were rapidly translated into French and German with the result that his reputation spread rapidly, and he soon became the leading authority on chemistry throughout Europe. Students from many countries were attracted to his research school, and many famous chemists were trained there. Gahn has already been mentioned, others were Gadolin [100] and the Spanish d'Elhuyar brothers [101]. The reputation of Bergman was so great that the few mistakes he made were allowed to go unchallenged for a very long time. This was apparent in some of his quantitative work, where many of his lesser known contempories obtained more accurate results, but it was not until long after Bergman's death that these were made known. Bergman was invited to work at the Academy of King Frederick the Great of Prussia but he refused the offer. In the last years of his life Bergman's health, which had never been good, became much worse, and he was unable to continue working in his laboratory. He died in 1784 at the age of 49 at Medev, the spa where he had gone for convalescence.

The works of Bergman which are of most importance to a study of the history of analytical chemistry are those which contain some form of classification of methods. His first book, *De analysi aquarum* [102], which gives a list of analytical reagents, considers the various salts which are to be found in waters. He states that water can be examined by two different types of procedures; it can either be tested in solution with a reagent, or it can be examined by the earlier method of fractional crystallization after evaporation. Bergman also defines reagents as: "Substances are called reagents, which — after being added to the solution — show the presence of foreign substances by changing the colour or the purity of it immediately, or after a very short time" [103]. After this he describes the physical properties of water which must be observed, such as purity, colour, smell, taste, specific weight, and temperature. His description of the available reagents probably represents all that were known at that time. For several of the reagents Bergman describes the preparation, as well as its application, and the sensitivity of the test. Several of these reagents are described in more detail:

Litmus: Sensitivity: 1 gr concentrated sulphuric acid turns 172·300 gr blue litmus solution red [104].

Fernambuc tincture, curcuma, violet-extract: All of these are for indicating the presence of acids and alkalis.

Gallus-extract: Gall nuts extracted with spirit until a saturated solution is obtained. The reagent is good if one drop of it makes one cantharus of water, in which 3 gr vitriol (iron sulphate) is dissolved, become noticeably red [105].

Alkali phlogisticatum (potassium hexacyanoferrate (II)): Dissolve 2 gr iron sulphate in 1 cantharus of water; one drop of the reagent should give the Prussian-blue colour *(coeruleum berolinense)*. The reagent also gives a red precipitate with copper and a white precipitate with manganese.

Oleum vitrioli (sulphuric acid) diluted: With barium *(terra ponderosa)* it gives a white precipitate. With waters containing carbonates it begins to effervesce.

Acidum sacchari (oxalic acid): The most sensitive reagent *(reagens maxime sensibile)* for the detection of lime. 1 gr lime in one cantharus water gives a slight cloudy precipitate with one drop of the reagent.

Sal microcosmicus (sodium ammonium phosphate): This is also a reagent for lime, but the precipitate formed is slower to separate than with oxalic acid.

Alkali vegetabile aeratum (potassium carbonate) precipitates the earths and the metals. If the water contains carbonic acid, certain carbonate precipitates dissolve, in these cases alkali must be used to separate the metals. However, one must be careful, because with the latter the heavy earth (barium) does not separate.

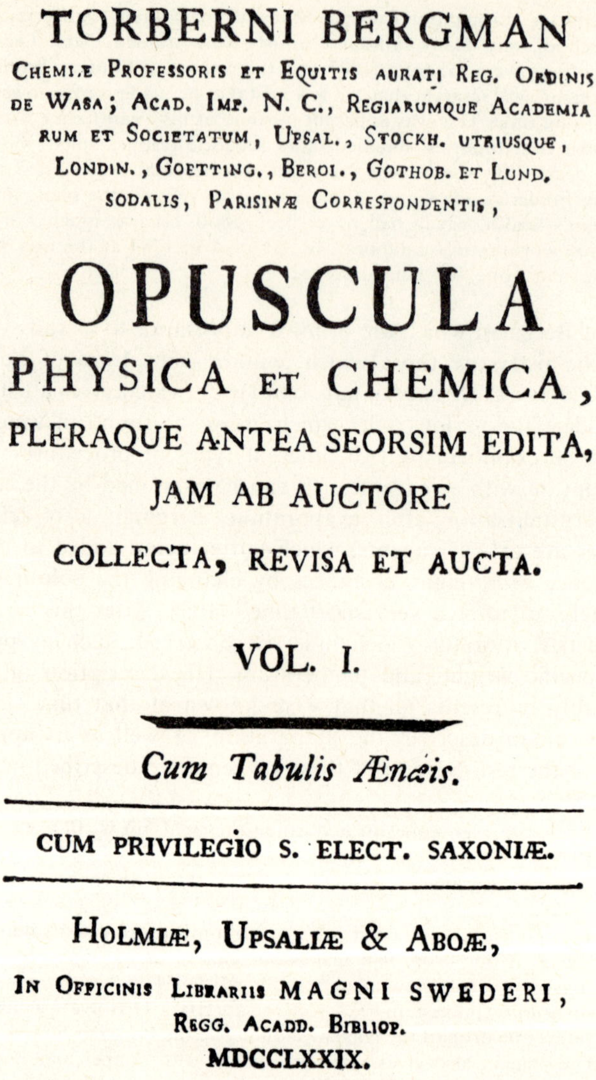

TORBERNI BERGMAN

Chemiæ Professoris et Equitis aurati Reg. Ordinis de Wasa; Acad. Imp. N. C., Regiarumque Academiarum et Societatum, Upsal., Stockh. utriusque, Londin., Goetting., Beroi., Gothob. et Lund. sodalis, Parisinæ Correspondentis,

OPUSCULA

PHYSICA et CHEMICA,

PLERAQUE ANTEA SEORSIM EDITA,

JAM AB AUCTORE

COLLECTA, REVISA ET AUCTA.

VOL. I.

Cum Tabulis Æneis.

CUM PRIVILEGIO S. ELECT. SAXONIÆ.

Holmiæ, Upsaliæ & Aboæ,

In Officinis Librariis MAGNI SWEDERI,
Regg. Acadd. Bibliop.
MDCCLXXIX.

Fig. 27. Title page of Bergman's *Complete Works*

Alkali volatile aeratum (ammonium carbonate): Precipitates the metals and the earths.

Alkali volatile (ammonium hydroxide): Metals are precipitated with this reagent. Of the earths, only magnesium, barium and calcium do not precipitate (strontium was still unknown). Also copper is not precipitated, but gives a blue colour.

Lime-water: the reagent of the *acidum aereum* (carbonic acid).

Terra ponderosa salita (barium chloride): This is the reagent for sulphuric acid and Glauber's salt. To a solution which contains 12 gr Glauber's salt in one cantharus of water, one drop of the reagent should give a colour, and three drops should give a precipitate. The reagent is more sensitive than litmus as a test for sulphuric acid.

Argentum nitrum (silver nitrate): This is the reagent for rock-salt. It cannot be used in sulphureous waters.

Hydrargirum nitrum (mercury nitrate) has various reactions; with *alkali vegetabile caustico* (alkali hydroxides) it gives a yellow precipitate, and with *alkali vegetabile aerato* (alkali carbonates) it gives a white precipitate. With sodium a yellow precipitate is formed which rapidly becomes white. With salt (chloride) a white, amorphous precipitate is formed. With sulphur a black precipitate is formed so that is cannot be used if sulphur is present.

Plumbum acetatum (lead acetate): Gives a precipitate with sulphuric acid in hydrochloric acid-free solution.

Vitriol (iron sulphate): With alkalis a greenish precipitate is first formed which remains in a closed vessel if the water contains carbonic acid, but if the vessel is open the precipitate rapidly becomes the colour of iron.

White arsenic (arsenic trioxide(?)): Gives a yellow precipitate with sulphur.

Soap: Does not dissolve in all waters. If free acids or metal and earth bases (*basi terrestri aut metallica* [106]) are present, then it will not dissolve. (Here is the first use of the word base.) If free acid is present then it combines with the alkali of the soap, and the fatty oil separates out. If either metal or earth bases are present in the solution, then when soap is added a precipitate separates. Waters of this type are called hard waters *(aquas duras vocari solent)*, these are good neither for washing or for cooking vegetables in. If soap is used as a reagent it is better to use a solution in alcohol rather than in water. One drop of a solution which contains 8 gr alum or magnesia or lime salt per one cantharus will give a turbidity with an alcoholic soap solution.

Hepar: This reagent is used to indicate acidic waters. A gas is formed similar to carbon dioxide except that it has a sulphurous smell.

Alcohol: If it is added in sufficient quantity will precipitate the salts which are dissolved in the water, such as sulphates, nitrates and chlorides.

By combining these reagents even more information can be obtained. For example litmus when treated with acid turns red. The acid can then be tested with silver nitrate, when hydrochloric acid gives a precipitate, and also with barium chloride solution when sulphuric acid gives a precipitate. Another example is when lime is removed with oxalic acid, then alkali carbonate solution precipitates magnesia and clay-earth (aluminium). If the precipitate is separated and tested with acid, and if it dissolves with effervescence it is magnesia, whereas if it dissolves without effervescence it is aluminium.

After the listing of the various reagents Bergman describes the course of the analysis. The first step is to isolate the volatile components [107]. The dissolved gases are boiled out into a mercury-filled gas-collecting apparatus. If the separated gas is shaken with lime water, then the carbon dioxide is absorbed and only air remains. If any hydrogen sulphide is present it can be recognized by its smell, and by the fact that it colours litmus solution red.

The next stage is the separation of the heterogeneous components [108]. One or more cantharus of water must be evaporated until its density has increased considerably. First of all the dissolved carbonic acid lime separates out from its abundant carbonic acid content. (Thus Bergman noted the difference between bicarbonates and carbonates.) Then silicic acid and calcium sulphate precipitate out, and, more slowly, the water-soluble salts in the reverse order of their solubilities *(ordine solubilitate congruo)*. The residue is then weighed and its components separated with different solvents. First one finger's breadth of alcohol must be poured on the salts, and this must be digested for a few hours, and then filtered. The insoluble portion is then treated with an eightfold volume of water and is again filtered. The residue from this operation is then boiled with a fourfold volume of water, and after filtration the residue is treated with cold acetic acid to dissolve the calcium and magnesium carbonates. The residue is then extracted with hot acetic acid to dissolve lime and magnesium. Calcium can be separated from the acetic acid solution by treatment of the hot solution with sulphuric acid. Magnesium can then be separated with alkali carbonate. Any insoluble portion remaining after these treatments is iron oxide, clay earth or siliceous earth. The first can be recognized from its colour, the second can be dissolved in hydrochloric acid, while silicic acid can be identified by a blow-pipe examination.

The next stage is the examination of the soluble portions.

The alcoholic solution can contain calcium and magnesium chlorides. This must be evaporated and the two salts can be separated by the process described previously. Iron can also be present, and can be revealed with the Prussian blue reaction.

The examination of the aqueous solutions is based on fractional crystallization, the fractions being examined both by their crystal shapes and also by their reaction with certain reagents. The metallic salt and the alkali content can easily be identified, but the situation is more difficult when any neutral salts are present.

By treatment with sulphuric acid, the salts of nitric acid can be driven out and can be identified by their smell and their brown colour. Sulphuric acid also removes the salts of nitric acid, which also can be identified by their smell and colour. Salts of hydrochloric acid can also be identified by treating with silver nitrate solution, salts of sulphuric acid by treating with lead or barium salts, and those of nitric acid by their explosiveness when heated. The salts of the two kinds of alkalis cannot be distinguished except by their crystal shapes.

At the end of this dissertation Bergman records the characteristic crystal shapes, solubility, and the percentage composition of a great many crystalline compounds, as a guide to their identification and quantitative estimation in the separated crystal-fractions. This is referred to in more detail later.

In one of his smaller works dealing with the examination of waters containing hydrogen sulphide, Bergman describes the properties of hydrogen sulphide [109]. He established that it turns litmus red, but does not have any effect on fernambuk-paper, and is not affected by acids. When it is passed through water containing chlorine, sulphur is precipitated. It also gives dark coloured precipitates with solutions of silver, mercury, lead, copper, and iron; a yellow precipitate being formed with arsenic solutions and a white precipitate with a solution of a zinc salt. It is strange that Bergman did not think about the analytical potentialities of these reactions and never actually used them.

The other important general work of Bergman published in 1780, has the title *De minerarum docimasia humida* which could be roughly translated as "The analysis of minerals by an aqueous process" [110]. It deals with the methods of analysis of silver, lead, zinc and iron ores by aqueous or "wet" methods and describes both qualitative and quantitative procedures. Bergman's intention in writing this work was to describe techniques for testing minerals based on entirely aqueous procedures. In the introduction he states

In our times the spagyric science has begun to analyse minerals by extraction with different solvents. However, it must be acknowledged that analysis by an aqueous process is used only in part, for the analysis of most minerals is carried out by mixed methods, partly by wet and partly by dry processes. The metals are separated by a wet process, but they are then reduced by heating. In the following I have tried to disclose methods which make the analysis of minerals possible by wet methods alone, without having to resort to fusion. This does not mean that dry procedures are in any way inferior, for in practice the most accurate and convenient methods must always be chosen ... But generally wet methods, although they take longer to perform, give more accurate and reliable results [111].

There is also a chapter dealing with practical instructions for carrying out the analysis. The mineral under examination must first be ground to a fine powder. The solvent used must be purified as much as possible; dilute acids should be used. Bergman gives the specific gravity of dilute sulphuric acid as 1·3; of dilute nitric acid, 1·2; and of dilute hydrochloric acid as 1·1. All precipitations must be carried out carefully in glass vessels. After the precipitate has been allowed to settle the clear part of the solution must be decanted off, then pure water is added and the solution shaken, and then allowed to stand until all the precipitate has settled out, when the clear solution is decanted. This washing must be repeated until the decanted liquid does not give a reaction when tested with a solution of the substance being determined. The precipitate is then filtered through a previously weighed filter paper containing a small amount of clay, and it is then dried at the temperature of boiling water. After cooling, the precipitate must be weighed, together with the paper, and the weight of the paper subtracted from the result [112]

In another section of this work Bergman deals with specific ores, and describes the method of analysis of gold, silver, platinum, mercury, lead, copper, iron, tin, bismuth, nickel, arsenic, cobalt, zinc, antimony and manganese ores.

A few extracts from this section are given below. In that part dealing with iron ores [113] Bergman states that "Iron ores can either be sulphuric, pyrite-like or calcinated (oxide) ores".

The simpler iron ores in a finely divided form, when heated with hydrochloric acid, "give their metal into the solution", in other words they dissolve. If the ore contains pyrites then a little nitric acid must be added. Iron can be separated from this solution with *alkali phlogisticatum* (potassium ferrocyanide). The precipitate must be dried and weighed, the quantity of the metal iron being 1/6th part of the weight of the precipitate. If manganese is present it is precipitated with iron, and in this case the precipitate must be heated with dilute nitric acid, which dissolves the manganese, the iron remaining behind.

Evidently Bergman considered the testing of iron ores rather simple, and it is interesting to note that only the determination of iron was of importance to him.

In the section describing the methods recommended for the analysis of lead ores [114], it is stated that lead occurs mainly in sulphuric ores which are usually contaminated with silver, iron or antimony. The ore can be dissolved in nitric acid, when the sulphur separates out. This must be filtered off, dried, and weighed. It is possible to determine its purity by dissolving it in a base. Lead is precipitated from the filtrate with soda. If lead only is present, and if the weight of the precipitate is a then the quantity of metallic lead it $100/132a$. If silver is also present, it will be precipitated with the lead so that it must be extracted from the precipitate with ammonia. If after drying and reweighing the precipitate, the decrease in weight is equal to b then the weight of silver present is $100/129b$.

If the ore also contains antimony, then if it is dissolved in concentrated sulphuric acid, the acid takes up the phlogiston of the antimony and antimony-lime separates out. This is then filtered off, dried and weighed. If the weight of the precipitate is c, then the amount of antimony present is $100/138c$. If iron is present in the sample the analysis is much more difficult. It must first of all be dissolved in hydrochloric acid, when most of the lead is precipitated. This must be dissolved in a very large volume of water, when lead dissolves and antimony precipitates. After this has been filtered off, iron plates must be placed in the filtrate in order to displace lead and silver. When this has also been filtered off, the iron is precipitated from the filtrate, either with soda or in the form of Prussian blue. The amount of iron dissolved from the iron plates must also be considered in the result. This can easily be found by reweighing the plates after the precipitation.

392   De Præcipitatis Metallicis.

100 part. Plumbi	alk. min aërato deder.	132
- - - - -	- cauſtico -	116
- - - - -	- vitriolaro -	143
- - - - -	- phlogiſtic. -	
- Cupri -	- aërato -	194
- - - -	- cauſtico -	158
- - - -	- phlogiſtic. -	530
- Ferri -	- aërato -	225
- - - -	- cauſtico -	170
- - - -	- phlogiſtic. -	590
- Stanni -	- aërato -	131
- - - -	- cauſtico -	130
- - - -	- phlogiſtic. -	250
- Viſmuti -	- aërato -	130
- - - -	- cauſtico -	125
- - - -	- phlogiſtic. -	180
- - - - aqua	pura -	113
- Niccoli -	- aërato -	135
- - - -	- cauſtico -	128
- - - -	- phlogiſtic. -	250
- Arſenici -	- aërato -	
- - - -	- cauſtico	
- - - -	- phlogiſtic. -	180
- Cobalti -	- aërato -	160
- - - -	- cauſtico -	140
- - - -	- phlogiſtic. -	142
- Zinci -	- aërato -	193
- - - -	- cauſtico -	161
- - - -	- phlogiſtic. -.	495
- Antimonii -	- aërato -	140
- - - -	- cauſtico -	138
- - - -	- phlogiſtic. -	138

FIG. 28. Table of equivalent weights. (From Bergman: *De praecipitatis metallicis* [1779])

Phosphoric acid ores of lead are also known. These must also be dissolved in nitric acid, in which they dissolve with difficulty, a residue of iron particles usually remaining. Lead is precipitated from this solution with sulphuric acid; it is then filtered, dried and weighed. If the weight of the precipitate is a, then the amount of lead is given by $100/143a$. (The present day value is $100/146a$!) The amount of phosphoric acid remaining in the solution can be determined after evaporation of the solution.

In his *De praecipitatis metallicis*, which deals with metal precipitates, Bergman summarizes the various solvents for the metals, as well as their precipitants, and the properties of the precipitates formed. He also gives results for the efficiencies of the various precipitants, namely for one hundred parts of pure metal he determines how many parts of metal are precipitated with the different reagents [115]. A few extracts from this work are given below:

> Solvents: Nitric acid dissolves most metals, with the exception of gold and platinum. Sulphuric acid is not as effective as nitric acid for this purpose, silver and mercury for example only dissolve slowly. If, however, it is evaporated to dryness then gold and platinum can be dissolved *(eludit aurum ac platina)* (?)
> Hydrochloric acid is a much less effective solvent than nitric and sulphuric acids, but if it is dephlogisticated (chlorine is removed), it attacks all metals most vigorously.
> The other acids, hydrogen fluoride, arsenic acid, and the organic acids have much weaker dissolving powers. When a substance is dissolved in acids gases are evolved; with nitric acid the red/brown coloured *aer nitrosus* is given off. Sulphuric acid, when zinc and iron are dissolved, gives off *aer inflammabilis* (hydrogen), but with other metals the *aer acidus vitriolicus* (sulphur dioxide) is evolved. Hydrochloric acid gives in most cases hydrogen, but in a few cases *acidum muriaticum dephlogisticatum* (chlorine) is evolved. Some metal limes (oxides) do not cause any gas to be evolved when they are dissolved in acids, with others *acidum aerum* (carbon dioxide) is given off. The solution of certain metals in acids is accompanied by colour formation, for example with gold, copper, iron, tin, nickel and cobalt, colours are produced.

The following chapter in his book is devoted to a discussion of the "theory" of solution and precipitation of metals. Bergman works on the basis of the varying phlogiston content of the pure metals, and from this he tries to explain their different properties. No comment need be made on this chapter.

In another chapter he describes the precipitants [116]. He states that all the metals can be precipitated with alkali salts, as these have a greater affinity for the solvent than does the metal.

> The following are classed as alkalis: the *caustic alkalis*, namely the hydroxides, the *fixed alkalis*, i.e. the alkali carbonates, the *volatile alkalis*, i.e. ammonia, and the *phlogisticated alkali*, i.e. potassium hexacyanoferrate (II). Mercury, lead and silver can be precipitated from nitric acid solution with hydrochloric acid. Tin and antimony are precipitated by nitric acid because it takes away their phlogiston. Metals can also displace other metals from solutions of their salts, and in this a certain sequence is to be found. The order is zinc, iron, lead, tin, copper, silver, and mercury; zinc will displace iron and so on. Bergman also remarks that this sequence depends on the solvent [117].

The colour of various metal precipitates is also recorded. For example: in nitric acid solution silver when treated with soda, gives a white precipitate, when treated with a base it gives a black precipitate, and with potassium hexacyanoferrate, a yellowish precipitate; hydrochloric acid gives a white precipitate which gradually becomes black on exposure to sunlight; mercury solution gives a reddish precipitate when treated with hydrochloric acid or soda; when treated with a base a yellow precipitate is formed and with potassium hexacyanoferrate a white precipitate; nickel gives a greenish white precipitate with alkali carbonate and a yellowish precipitate with potassium hexacyanoferrate. The work is concluded with a table of results for quantitative analysis, and this will be discussed in detail later.

Bergman—apart from these three comprehensive and systematic works, which were to be used as text books by the succeeding generation—also wrote many smaller works, and many of these contain interesting analytical data. For example Bergman mentions that there are five minerals which are difficult to dissolve, the heavy earth ($BaSO_4$), lime, magnesia, and the clay and silicious earths. The latter will only dissolve in one acid, and this is the acid obtained by treating fluorite with hydrochloric acid (hydrofluoric acid) [118].

The method for the estimation of silicates is more accurate than that of Marggraf; he describes it as follows:

> The substance must be ground to a fine powder and, after weighing, it must be mixed with twice its weight of pure *"alkali minerale"* (soda). The mixture must then be placed in an iron crucible which has previously been polished smooth inside so that no particles will come off during the ignition and contaminate the sample. The crucible must then be covered with a lid and placed in the furnace, above the sill, and heated moderately for two to three hours. The heating must not be too strong, or the mass may boil over, also if bellows are used the mass will swell. A few tests will find out how to regulate the fire correctly. When the crucible is taken out of the furnace it is allowed to cool and then the fused mass is broken up and ground in an agate mortar. It is then leached with water and then hydrochloric acid [119].

The solution is then filtered and the residue again extracted with hydrochloric acid for several hours. It is then filtered, washed, dried and weighed. This residue is silicic acid [120].

In another article Bergman mentions that manganese can be precipitated with oxalic acid in neutral solution [121]. He also recognized the different coloured manganese oxides, and attributed this to the varying phlogiston contents [122]. Bergman also investigated the different types of iron and concluded that their differing properties were caused by differences in their composition.

Bergman gives the following data for silicon and carbon contents of different forms of iron:

	Silicic acid %	Plumbago (Graphite) %
Ferrum crudum (cast iron)	1·0 — 3·4	1·0 — 3·3
Ferrum cusum (wrought iron)	0·05 — 0·3	0·05 — 0·2
Chalybis (steel)	0·3 — 0·9	0·2 — 0·8

He also states that the manganese content can be as high as 30 per cent in all of them [123], but the separation of iron and manganese was not very reliable at that time. Bergman also found that iron is very fragile if it contains a considerable amount of *sideros*. He gave this name to what he considered to be a new element, but Klaproth later proved that this *sideros* was in fact phosphorus.

It is clear from the preceding extracts from the work of Torbern Bergman that he contributed as much as anyone to the development of analytical chemistry.

NOTES AND REFERENCES

1. KOPP, H.: *Geschichte der Chemie.* Braunschweig (1844) **2** 18
2. PROSZT, J.: *Die Schemnitzer Bergakademie als Geburtsstätte Chemie-Wissenschaftlicher Forschung in Ungarn.* Sopron (1938) 4
3. SMEATON, W. A.: *Annals of Sci.* (1954) **19** 224
4. JACQUIN, NICOLAS JOSEPH, was born in Leyden, in the Netherlands in 1727. He took his medical degree at Löwen, and afterwards he worked in Paris as a surgeon. He was also interested in botany and became one of the first followers of Linne's system. In 1752 he was given an appointment in Vienna by the Empress Maria Theresa, and was responsible for the Imperial Gardens. In order to obtain new plants he led an expedition to the West Indies. On his return he was commissioned to establish the botanical gardens at the University of Vienna. In 1763 he was appointed to the Professorship of Practical Mining and Chemistry at the Mining Academy in Selmecbánya. In 1769 he was given the post of Professor of Chemistry and Botany at the University of Vienna. He worked there till 1797, when he retired. He died in 1817.
5. *Journal de l'Ecole Polytechnique* (1795) 1; *Gazette nationale ou Moniteur universal.* No 8 Oktidi 8, Vendemiaire
6. STAHL, G. E.: *Specimen Becherianum,* etc. Leipzig (1703): *Zufällige Gedanken und nützliche Bedenken über den Streit von den sogennanten Sulphur,* etc. Halle (1717)
7. BOYLE, R.: *Philosophical Works.* (Experiments to make fire and flame stable and ponderable) London (1725) **2** 388
8. GUYTON DE MORVEAU, L. B.: *Défense de la volatilité du phlogistique.* Dijon (1773) Quotation: KOPP, H.: *Geschichte der Chemie.* (1844) **3** 149
9. WINTERL, J. J.: *Prolusiones ad chemiam saeculi XIX.* Buda (1800); *Gehlens Journal* (1804) **4**
10. KOPP, H.: *Geschichte der Chemie.* (1844) **2** 284
11. ÖSTERREICHER, J. J.: *Analyses aquarum Budensium praemissa methodo prof. Winterl.* Buda (1781): 42
    SZABADVÁRY, F.: *J. Chem. Education* (1962) **39** 266
12. *Saggi di naturali esperienze fatte nell'Academia de Cimento.* Florence 1666. cf. KOPP, H.: *Geschichte der Chemie* **2** 44
13. KUNCKEL, J.: *Ars vitraria experimentalis oder vollkommene Glasmacher Kunst.* Frankfurt-Leipzig (1689) 399—400
14. STAHL, G. E.: *Specimen Becherianum comp.* cf. KOPP, H.: *Geschichte der Chemie* **4** 413
15. CRAMER, J. A.: *Elementa artis docimasticae.* Leyden (1739)
16. MARGGRAF, S. A.: *Miscellanea Berolinensa.* (1740) 54—60, Citation: MARGGRAF: *Opuscules chymiques.* (Sur les rapports du phosphore solide avec les metaux etc.) Paris (1762) **1** 20 21 24
17. POTT, J. H.: *Chemische Untersuchungen welche fürnehmlich, von der Lithogeognosia oder Erkennung und Bearbeitung der gemeinen einfacheren Steine und Erden, ingleichen von Feuer und Luft handeln.* Potsdam (1746). Continuation in 1751 and 1754.
18. BERGMAN, T.: *Opuscula physica et chemica.* (De tubo feruminatorio) Uppsala—Stockholm—Åbo (1779—1788) **2** 455
19. SWAB, ANTON (1703—1768), was employed at the Swedish mines, and later he was appointed director of the Royal Mining Office. He was a member of the Swedish Academy of Sciences.

20. SWAB, A.: Berättelse om en nativ regulus antimonii eller spetsglaskung. *Acta acad. reg. suec.* (1748)
21. RINMAN, SVEN (1720—1792) was the director of a silver mine, and later of the smelting-works in Hellefors; he was a member of the Royal Mining Council and the Swedish Academy of Sciences.
22. RINMAN, S.: Anmärkning om en art jernhaltig tenmalm ifran Dannemore socken i Upland. *Acta acad. reg. suec.* (1746) **8** 181
23. RINMAN, S.: Om grön mareförg af cobolt. *ibid.* (1780—1781)
24. CRONSTEDT, BARON ALEXANDER FREDERIK (1702—1765) the discoverer of nickel; he was a member of the Swedish Academy of Sciences.
25. ENGESTRÖM, GUSZTAF (1738—1813) a member of the Royal Mining Council and the Swedish Academy of Sciences. The title of the work in question: Beskrifning pa ett mineralogiskt ficklaboratorium och i synnerhet nyttan of blasröret. 1773 (In English 1765)
26. LAMPADIUS, W.: *Handbuch zur chemischen Analyse der Mineralkörper.* Freyberg (1801) 193

  ACHARD, C: *Nouv. Mém. Berlin* (1779); *Sammlung physisch. und chemisch. Abhandlungen.* Berlin (1784) 134

  LAVOISIER, A. L.—MEUSNIER, J. B.: *Mém. Acad. Paris* (1782) 457
27. BERGMAN, T. *Opuscula* **2** 452—505
28. GAHN, JÖNS GOTTLIEB (1745—1818) studied at the University of Uppsala and later worked with Bergman. Afterwards he became director of the copper mines in Kopperberg. He was a member of the Mining Council.
29. BERZELIUS, J. J.: *Die Anwendung des Löthrohrs.* 4. Ed. Nürnberg (1844) 6
30. *ibid.* 55
31. SAUSSURE, HORACE BENEDICT (1740—1799) Professor of Natural Sciences at the University of Genf. His work referred to: *Sur l'usage du chalumeau. Journ. de phys.* 45 (1794); also a shorter account in 1785. His method was not as well developed as Gahn's.
32. HARKORT, EDUARD (1797—1835), Professor of the Mining Academy in Freiberg; he later went to America, where he died. The method referred to *Die Probierkunst mit dem Löthrohre* (Silberproben). Freiberg (1827) **1** 54
33. PLATTNER, KARL FRIEDRICH (1800—1858) metallurgist, he later became Professor of Mining and Analysis at the Mining Academy in Freiberg. His work *Die Probierkunst mit dem Löthrohre* was first published in 1835, and was reprinted several times until 1900.
34. PLATTNER, K. F.: *Probierkunst mit dem Löthrohre.* 4 Ed. Leipzig (1865) 37
35. MARGGRAF, S. A.: *Opuscules* **1** 73
36. MARGGRAF, S. A.: *Mém. Acad. Berlin* (1745) **8**; *Opuscules* (Demonstration expérimentale de la solution de divers métaux, comme l'or, l'argent, le mercure, le zink et le bismuth par le moyen d'un alcali fix) **1** 72
37. BOYLE, R.: *Philosophical Works.* **1** 515
38. MARGGRAF, S. A.: *Opuscules* **1** 79
39. KOPP, H.: *Geschichte der Chemie.* **4** 369
40. WOODWARD: *Philosoph. Transact.* (1725) **35** 15
41. MARGGRAF, S. A.: *Mém. Acad. Berlin* (1745) 48; *Opuscules* **1** 72 76 79
42. MARGGRAF, S. A.: *Mém. Acad. Berlin* (1751) 131—158; *Opuscules* **2** 1 (Examen chymique de l'eau, etc.)
43. MARGGRAF, S. A.: *Opuscules* **2** 57
44. *ibid.* 63
45. *ibid.* 68
46. MARGGRAF, S. A.: *Opuscules* **1** 13
47. *ibid.* 22 23
48. MARGGRAF, S. A.: *Opuscules* **1** 84 comp. note 36

49. MARGGRAF, S. A.: *Opuscules* (Sur la meilleure manière de séparer la substance alcaline du sel commun) **2** 331
50. *ibid.* 338
51. MARGGRAF, S. A.: *Opuscules* (Preuves qui démontrent que la partie alcaline séparée du sel de cuisine est un sel alcali veritable) **2** 374—386
52. STAHL, G. E.: *Specimen Becherianum.* Refer on the base of KOPP, H.: *Geschichte der Chemie.* **4** 413
53. DUHAMEL DE MONCEAU, HENRY LOUIS (1700—1781) studied at the University of Paris, and was then employed by the Naval Office. He was a member of the French Academy of Sciences from 1740. He dealt mainly with meteorology and physiology.
54. DUHAMEL DE MONCEAU, H. L.: *Mém. Acad. Paris* (1736) 216
55. OESTERREICHER, J. M.: *Analyses aquarum Budensium* 147
56. *ibid.* 83
57. WENZEL, K.: *Lehre von der Verwandschaft.* Dresden (1777) 303 318 331
58. MARGGRAF, S. A.: *Opuscules* (Preuves qui démontrent que la partie alcaline séparee du sel de cuisine est un sel alcali véritable) **2** 397
59. MARGGRAF, S. A.: *Mém. Acad. Berlin* (1747) 79; *Opuscules* (Experiences chymiques faites dans le dessein de tirer un véritable sucre de diverses plantes qui croissent dans nos contrées) **1** 216
60. MARGGRAF, S. A.: *Mém. Acad. Berlin* (1756) 20; *Opuscules* (Essais concernant la nouvelle de corps mineral connu sous le nom de Platina del Pinto) **2** 238
61. BRANDT, GEORG (1694—1768) Swedish mining councillor; he was appointed director of the Royal Chemical Laboratory in Stockholm, and later Professor at the University of Uppsala.
62. BRANDT, G.: *Acta Soc. reg. Uppsal.* (1742)
63. BRANDT, G.: *Acta Soc. reg. Uppsal.* (1733) 39
64. BRANDT, G.: *Om det fiygtiga Alcaliska Saltet. Acta acad. reg. Sues.* (1747)
65. POTT, J. H.: *Observations ... praecipue Zincum, boracum, etc.* Coll. Berlin (1741) **2**
66. BLACK, J.: *Edinburgh Physical and Literary essays.* (1755) **11** 157; *Lectures on the elements of Chemistry.* Delivered in the University of Edinburgh by J. Black, Edinburgh (1803) **2** 52
67. JACQUIN, N. J.: *Examen chemicum doctrinea meyerianea de acido pingui et blackianae de aero fixo, respectu calcis.* (1769) Wien
68. LAVOISIER, A. L.: *Oeuvres.* Paris (1854) **1** 471
69. More about the life and role of Jacquin is to be found in the work of János Proszt mentioned under [5].
70. PLINIUS, C. S.; *Naturalis historiae libri.* **2** 110
71. PARACELSUS, T. B.: *Bücher und Schriften.* Basle (1589) 6 12
72. VAN HELMONT, J. B.: *Ortus medicinae.* etc. Leyden (1656) 68 261
73. *ibid.* 424
74. *ibid.* 68
75. TORRICELLI, EVANGELISTA (1608—1647) a disciple of Galileo, who later succeeded him as court-mathematician of the Duke of Toscana.
76. BOYLE, R.; *Philosophical Works* (Physico-mechan. experiments) **2** 432
77. WREN, SIR CHRISTOPHER (1632—1723) Professor of Astronomy at the University of London, and later of mathematics at Oxford. He was an architect, and the builder of Saint Paul's Cathedral. From 1665 he was Superintendent of the Royal Buildings. He was President of the Royal Society.
78. WREN, C.: *Philos. Transact.* (1665) **1** 122
79. MAYOW, JOHN, was a physician (1645—1679) in Bath, and a member of the Royal Society. About his life very little is known. In his book *Tractatus quinque*, he suggests some very interesting ideas on the nature of combustion, namely, that the increase in the weight of the metals by burning is caused by a saltpetre like component of the air.
80. MAYOW, J.: *Tractatus quinque medico-physici,* etc. Oxford (1674)

81. BERNOULLI, JOHANN (I) (1667—1748) one of the first members of the family which provided many famous mathematicians; he was Professor of Mathematics at the University of Basle.
82. BERNOULLI, J.: *Dissertatio de effervescentia et fermentatione*. Basle (1690)
83. HALES, STEPHEN (1677—1761) was a priest; he dealt with almost all branches of natural sciences. He was a member of the Royal Society, and tutor to the Princess of Wales.
84. HALES, ST.: *Vegetable staticks*, etc. London (1727)
85. BERGMAN, T.: *Opuscula* (De acido aereo) **1** 8
86. CAVENDISH, H.: Experiments on factitious air. *Philos. Transact.* (1766)
87. RUTHERFORD, DANIEL (1749—1819) a follower of Black, who later became Professor of Botany at the University of Edinburgh. The discovery of nitrogen is described in his work, *Dissertatio de aere fixo dictu aut mephitico*. Edinburgh 1772.
88. CAVENDISH, H.: On a new eudiometer. *Philos. Transact.* (1783)
89. CAVENDISH, H.: Experiments on air. *Philos. Transact.* (1784) 119
90. BUGGE, G.: *Das Buch der grossen Chemiker*. Berlin (1929) 262
91. PRIESTLEY, J.: *Observations of different kinds of air*. (1772)
92. SCHEELE, C. W.: *Opuscula chemica et physica*. Leipzig (1788/89) **2** 101
93. *ibid.* **2** 200. Scheele produced only molybdenum oxide. Hjelm was the first to reduce it to metallic molybdenum.
94. *ibid.* **2** 119. Scheele only produced tungstic acid. It was also produced by the d'Elhuyar brothers at about the same time, and these workers reduced it to the metal.
95. *ibid.* **1** 263
96. *ibid.* **2** 262
97. *ibid.* **1** 227
98. *ibid.* **2** 1 92
99. BERGMAN, T.: *Opuscula* (De acido aereo) **1** 19
100. GADOLIN, J. (1760—1852) Finnish chemist; he was Professor of Chemistry at the University of Abo and a member of the Swedish Academy of Sciences. He was also the discoverer of yttrium in the ore gadolinite, which was named after him.
101. D'ELHUYAR, FAUSTO and JUAN, Spanish chemists. The former (1775—1832) was the General Manager of the Mexican mines and later became a Spanish Minister. Together they discovered tungsten.
102. BERGMAN, T.: *Opuscula* (De analysi aquarum) **1** 68
103. *ibid.* **1** 89
104. *ibid.* **1** 93
105. *ibid.* **1** 97
106. *ibid.* **1** 106
107. *ibid.* **1** 110
108. *ibid.* **1** 117
109. *ibid.* (De aquis medicatis calidis) **3** 233
110. *ibid.* (De minerarum docimasia humida) **2** 399
111. *ibid.* **2** 403
112. *ibid.* **2** 406
113. *ibid.* **2** 342
114. *ibid.* **2** 424
115. *ibid.* (De praecipitatis metallicis) **2** 349
116. *ibid.* **2** 355
117. *ibid.* **2** 384
118. *ibid.* (De terra gemmarum) **2** 87
119. *ibid.* **2** 91
120. *ibid.* (De terra silicea) **2** 32
121. *ibid.* (De mineris ferri albis) **2** 224
122. *ibid.* (De minerarum docimasia humida) **2** 452
123. *ibid.* (De analysi ferro) **3** 1

CHAPTER V

THE ESTABLISHMENT OF THE FUNDAMENTAL LAWS OF CHEMISTRY

1. QUANTITATIVE ANALYSIS BEFORE LAVOISIER

That period of chemical history since the time of Lavoisier has been called the age of quantitative chemistry. It is quite common to read that Lavoisier introduced the use of the chemical balance. But already in this book there have been many references to the use of the balance, in fact the progress of chemistry might well have been impossible without it. It is indisputable that the principle of conservation of mass, the atomic theory of Dalton, and the new rules governing stoichiometry heralded a new era in chemical history, but these discoveries did not arise out of nothing, as some people would have us believe, but follow quite naturally from the work of the preceding generation of scientists.

In addition to the fire assay, which had been known since antiquity, the 17th century saw the first attempts at quantitative analysis by wet processes. The French chemist Lémery (1645–1715) was the first to provide quantitative results in his work.

He describes, for example, that when 1 oz of silver is dissolved in nitric acid and then precipitated with rock salt, the weight of filtered and dried precipitate is 1 oz and 3 drachms. This determination is quite accurate, because according to the Ag — AgCl stoichiometric equation, 1 oz and 2·6 drachms of silver chloride is formed from 1 oz of silver. According to Lémery the atoms of an acid are cone shaped while the atoms of a base are cavernous. The cause of the increase in weight in this reaction he attributed to the breaking off of the peaks of the HCl cones in the pores of the metal atoms. He also established that 12 $^1/_2$ oz of mercury can be obtained from 1 lb of cinnabar. By heating 16 oz of potassium nitrate with carbon he obtained 12 oz of potassium carbonate (11 oz is the theoretical yield for this experiment) [1]. It should be noted 1 oz = 8 drachms = 480 gr = = 25−35 g (weight of one g varied in different countries and often in different towns: see VI. 3).

Kunckel, who lived at about the same time as Lémery, also carried out many quantitative experiments. He found that 60 gr of silver is separated from a silver solution by 20 gr of copper. (Theoretically it should be 65 gr.) He precipitated silver in the form of silver chloride, and obtained 16 parts of *white silver lime* from 12 parts of silver treated with rock salt. (The theoretical value is 15·95 parts of silver chloride.) It can be seen that Kunckel's work is astonishingly accurate [2].

Wilson, in 1746, published many quantitative data in his text-book [3]. He determined silver as silver chloride, and from 1 oz of silver he obtained 1 oz and 3 drachms of silver chloride. He mentions that this salt is "to be composed of silver and the acid of sea salt" [4].

Marggraf was the first to propose that the determination of silver as silver chloride was more convenient than the very involved fire test:

> The silver must be precipitated from nitric acid—silver solution with rock salt solution; this must be added until the solution is no longer turbid. The solution must be allowed to stand overnight, and the next day the clear liquid must be removed. After this the precipitate must be washed and dried. From 2 oz of silver, 2 oz 5 drachms and 4 gr of precipitate is obtained. The increase in weight originates from the acid of the rock salt, consequently one oz of this precipitate contains 6 drachms and a few gr of pure silver. If the silver is not quite pure, the precipitate will have a smaller weight, because with this method only the silver separates, namely the copper remains in solution [5].

Black's tests with carbonates—as referred to in Chapter IV. 3—also rely on quantitative data. Upon ignition, 2 drachms of calcium carbonate decreased in weight by 52 gr, i.e. 1 drachm 8 gr (68 gr) of lime remains (theoretically 67 gr). When this lime was added to a solution of potassium carbonate, and the precipitate formed was dried and weighed its weight was found to be 118 gr (compared with the original 120). In order to neutralize 120 gr of calcium carbonate, 421 gr of dilute hydrochloric acid was required; 67 gr of lime prepared from 120 gr of calcium carbonate required almost the same amount of hydrochloric acid for neutralization, 414 gr.

He obtained a similar set of results from experiments with magnesium carbonate [6].

Reference has already been made to the quantitative analysis carried out by Bergman, Chapter IV. 4. For example, he gives the quantitative composition of numerous crystalline salts obtained by aqueous procedures, but without giving any detail of his methods or any record of his weighings. Several examples are given below (actual values given in brackets) [7].

Alkali vegetabile vitriolatum
(Potassium Sulphate)

52 parts alkali	(54·05 K_2O)
40 parts sulphuric acid	(45·95 SO_4)
8 parts crystal water	(0·0)

At a temperature of 15°, 1 part of salt dissolves in 16 parts of water. Its taste is rather bitter. It melts with difficulty and does not disintegrate on heating, but breaks up on vigorous heating.

Alkali minerale salitum
(Rock-salt)

42 parts alkali	(39·34 Na)
52 parts hydrochloric acid	(60·66 Cl)
6 parts water	(0·0)

At moderate temperatures 1 part dissolves in $2^{14}/_{17}$ parts of water, in hot water 1 part needs $2^{13}/_{17}$ parts. When heated in fire it breaks up and then melts; crystals are cube shaped.

Zincum vitriolatum, vulgo vitriolum album
(Zinc Sulphate)

20 parts zinc lime	(28·30 ZnO)
40 parts sulphuric acid	(27·85 SO$_4$)
40 parts mineral water	(43·85 H$_2$O)

It can be seen that the analytical results of Bergman are rather inaccurate.

Bergman also gives some instruction on the calculations of reaction stoichiometry, but they are rather involved. For example:

If m gr of magnesia is precipitated, it requires $\frac{33}{19} : \frac{47}{1000}$ m gr of sulphuric acid to neutralize it, $\frac{45}{19}$ m gr of magnesium sulphate being formed. For the precipitation $\frac{33}{27} \times \frac{45}{19}$ m gr of sodium sulphate, or $\frac{15}{27} \times \frac{33}{19} \times \frac{45}{100}$ m gr of alkali base, or $\frac{15}{27} : \frac{33}{19}$ m gr of sodium carbonate is required [8].

The method of calculation is rather obscure, even allowing for the complicated units of measure, and it is difficult to understand Bergman's use of such extraordinary fractions.

In the discussion on Bergman's *Docimasia humida* it was mentioned that the conversion factors for calculating the weight of metal from precipitate had been given for many metal precipitates. Bergman realized that it was not necessary to convert metals to their elementary state in order to determine them quantitatively; it was sufficient to separate them in the form of a precipitate, the composition of which was accurately known. Thus, in his work *De praecipitatis metallicis*, Bergman compiled a table in which he gives the number of parts of dried precipitate that is equivalent to 100 parts of pure metal for a variety of precipitants [9]. Table 2 and fig. 28.

It can be seen that a few of Bergman's results are completely wrong, but such was his prestige that these values were accepted for quite a long time, in spite

TABLE 2

THE AMOUNT OF PRECIPITATE OBTAINED
FROM 100 PARTS OF PURE METAL

Precipitate	The weight in parts of the precipitate	
	According to Bergman	Theoretical
Silver carbonate	129	127·8
Silver oxide	112	107·4
Silver chloride	133	132·8
Lead carbonate	132	128·9
Lead oxide	116	107·7
Lead sulphate	143	146·3
Copper ferrocyanide	530	266·0
Iron oxide	170	142·9
Prussian blue	590	384·6
Basic bismuth oxide BiO(OH)	125	115·7

of the fact that several of his contemporaries obtained much better results. Wenzel in particular obtained very accurate results for the composition of various salts.

Carl Friedrich Wenzel (1740—1793), was the son of a bookbinder at Dresden. Although he learned this trade his interest soon turned to another direction, and he left home to go to Holland where he became an army doctor. He also worked as a doctor in the navy and travelled throughout most of the world. In 1766 he returned to his native country, to Leipzig, where he studied chemistry and metallurgy. He was later employed at the mines in Freiberg and eventually became manager. In addition to his work at the mines he carried out research in chemistry and wrote several books on the subject, only one of which is worthy of note.

Wenzel's book was called *Lehre von der Verwandschaft*, and was published in 1777; in this work Wenzel records the composition of about two hundred different kinds of salts. This book represented a great deal of accurate research work and the results recorded are astonishingly accurate, much more so than those of Bergman. As Wenzel was quite unknown in the field of chemistry, whereas Bergman was the undisputed authority, little attention was paid to his work; consequently where his results differed from those of Bergman they were simply ignored. It was not until after Wenzel's death that Berzelius examined the results given in his book and was amazed at their accuracy. Table 3 will illustrate this more clearly.

Apart from this work Wenzel was far more concerned with synthesis rather than with analysis. He reacted either the metal or the metal oxide with acid, and

TABLE 3

A COMPARISON OF THE ANALYSES OF WENZEL AND BERGMAN

Compound		Composition (per cent)		
		According to		Theoretical
		Wenzel	Bergman	
Na_2SO_4 cryst	base	19·5	15	19·2
	acid	24·3	27	25·0
	water	55·2	58	55·8
$MgSO_4$ cryst	base	16·9	19	16·3
	acid	30·6	33	32·6
	water	52·5	48	51·1
K_2SO_4	base	54·8	52	54·05
	acid	45·2	40	45·95
	water	—	8	—

isolated the salt formed which he then weighed. The increase in weight he attributed to the acid.

Here is an example of his calculation

To half an ounce (240 gr) of sulphuric acid solution (which contained $75^3/_4$ gr "most acidic" namely sulphuric acid) Wenzel added 120 gr of magnesium carbonate. 20 gr of this were insoluble, therefore 100 gr must have reacted. He evaporated the solution in a vessel of known weight, and dried and weighed the residue, which weighed 247 gr.

Wenzel ignited 240 gr of magnesium carbonate and found that 140 gr of fixed air (CO_2) was given off. Therefore 100 gr of magnesium carbonate are equivalent to $43\ ^2/_3$ gr of earth. Consequently in the 247 gr salt there is $42\ ^2/_2$ gr of earth (base) and $75\ ^3/_4$ gr of acid, the remainder being water [10]. It can be seen from Table 3 that the result is very close to the exact composition of magnesium sulphate.

In the field of gas-analysis many scientists, notably Black and Cavendish, carried out some very accurate quantitative work. They established the composition of the air, and also the density of gases and the quantity of carbon dioxide evolved from carbonates (see Bergman, etc. Chapter IV. 4).

Thus, it can be seen that from about the middle of the 18th century quantitive procedures were in widespread use, and this will be made abundantly clear later (Chapter VIII. 1), when dealing with the origin of titrimetric methods.

It is obvious, therefore, that the techniques of quantitative analysis were well known before the time of Lavoisier, and by the end of the phlogiston period were already in general practice. However, the period in chemical history which was dominated by Lavoisier — a period in which the basic rules of chemistry which are still valid today were formulated, and which ended with the establishment of atomic weights — saw a great improvement in the scope and nature of quantitative examinations. This is a very good example of the mutual effect of related subjects; these rules were discovered on the basis of the results of chemical analysis, and their effect was to create improvement in analytical methods themselves.

2. THE PRINCIPLE OF THE INDESTRUCTIBILITY OF MATTER

The principle of the indestructibility of matter was to become one of the basic rules in chemistry after the work of Lavoisier. It is implicit, however, in earlier work, as obviously quantitative determinations of any nature must be based upon this assumption. The philosophy that "of nothing, nothing becomes, and what exists does not become nothing", which, in fact, expressed the indestructibility of matter goes back to ancient times. In this form it was proposed by Democritus, the originator of the atomist philosophy (circa 420 B.C.), but even before this Anaxagoras reached essentially the same conclusion (circa 500 B.C.), "Nothing comes into being and nothing disappears, everything is only the arrangement of such things, which existed also before." Empedocles (circa 450 B.C.) wrote the following:

> Only fools can believe, that something may come into being which did not exist, or something may disappear which exists.

According to Aristotle (384—222 B.C.) the substances that exist can neither be decreased, nor increased, and can only be changed. Thus it can be seen that on the question of the indestructibility of matter the various Greek Schools of philosophy were in agreement. The Romans also held similar views. Marcus Aurelius, the philosopher emperor (120—180 A.D.) drafted a law quite similar in content to Democritus: *De nihilo nihil* [11]. Lucianus of Samosta, a satirist (125—180 A.D.), expounds this thesis even more clearly, though not in an exact form, in one of his anecdotes. This tells of how a rather sarcastic questioner asked Demonax the philospher

"If I burn a hundred pounds of wood, tell me please, how many pounds of smoke leaves?"

To this the philosopher replied,

"Weigh the ash; what is missing, was the smoke" [12].

During the Middle Ages this idea was abandoned to be reborn in the age of the Renaissance. Telesio (1508—1588), an Italian philosopher, who was one of the first opponents of scholasticism proclaimed that "The substance can neither be increased, nor decreased" [13]. Giordano Bruno, who was burned at the stake in 1600 in Rome also stated the theory in the following way: "Nothing is as constant as matter." Francis Bacon in 1620 wrote,

> There is nothing more true in nature than the twin propositions: "nothing is produced of nothing" and "nothing is reduced to nothing" but the absolute quantum or sum total of matter remains unchanged without increase or diminution [14].

Mersenne [15] the French physicist reiterated this idea in 1634, "La nature ... ne perd rien d'un coté quelle ne le gagne de l'autre." (Nature loses on one side, only what it gains on the other [16]). Another French physicist, Mariotte [17], expressed the view that, "Nature does not create anything of nothing, and that matter does not get lost" [18]. René Descartes, the great French philosopher and mathematician (1596—1650), had already proclaimed that the kinetic quantum is constant, this being a declaration or more correctly an approach to the principle of the conservation of energy, which was limited only by the mechanical knowledge of that time. No attempt was made to prove this theory experimentally, not even by Descartes himself, but nevertheless it spread rapidly and was soon incorporated by the propagators of cartesian physics.

Hooke [19] also gave his views on the indestructibility of matter and the kinetic quantum, in 1765 [20], and stated that these are constant in their entirety, and that their quantity may neither increase, nor decrease. After this time the principle of the indestructibility of matter and of the conservation of energy are frequently mentioned in philosophical and physical books. Lomonosov wrote the following to Euler (1748):

> In nature every change occurs so that if something is added to a system then it must have been lost from another system in the same degree. Consequently a body must give up that substance by the same amount. This law is a general one in nature, and can also be applied to the rules of motion. If a body is made to move by another body pushing it, then the second body loses as much of its motion as it gave to the first [21].

Mihail Vasilievich Lomonosov (1711—1765) was fifty-four years of age when he died, but during this relatively short period he accomplished much. He was a poet, historian, sociologist, and a factory manager, and most of all he was a great natural scientist. Even though he was only an amateur in most of these fields, he made important contributions to all of them. This is illustrated by an understandable mistake made by Hoefer, the French chemical historian, in his book on the History of Chemistry (1866), when he refers to Lomonosov the chemist and adds that he should not be confused with the poet of the same name. Lomonosov was born into a fairly prosperous peasant family, who lived in Northern Russia near the shores of the Arctic Ocean. It was a considerable achievement for him to learn to read and write under these circumstances. When he was nineteen he left to go to school in Moscow, but it was rather embarrassing as he was considerably older than the other pupils. His years of study must have been very difficult, for later in life he referred to it,

> During my years of study there were many circumstances which made it difficult for me. On the one hand there was my father, for I was the only child, who reproached me for leaving him to look after the estate alone, and because the estate to which he had devoted his life's work would pass into the hands of strangers after his death. On the other hand there was my extreme poverty; I had only one altin per day, so all that I could afford to spend on food was half a kopek for bread and half a kopek for kvas. I lived in this manner for five years, while studying the sciences [22].

Afterwards he went to the academy in Petersburg, where he was awarded a scholarship to study natural sciences. A year later he was sent on a study tour in Western Europe, and during this time he studied under C. Wolff [23] in Marburg, and afterwards at the mining academy in Freiberg. He also travelled in Germany and Holland. After five years abroad he returned to his native country, and took up a post as lecturer in physics at the Academy in Moscow. In 1745 he was appointed Professor of Chemistry. In addition to his scientific investigations he indulged in the arts, and also managed the affairs of the academy. He also campaigned for the founding of the first Russian university, which is now named after him, in Moscow. He was also a popular member of the Imperial Court; the Czarina gave him

Fig. 29. Mihail Vasilievich Lomonosov (1711—1765). Oil painting in the Historical Museum, Moscow

a village with serfs, and at his request a factory for the manufacture of coloured glass was built, the management of which was entrusted to him. His extremely strenuous life, however, seriously undermined his health.

Besides the principle of the indestructibility of matter, Lomonosov also investigated the phenomenon of combustion, preceding the work of Lavoisier. He repeated the famous erroneous experiment of Boyle, according to which if lead is burnt in a closed vessel then the weight of the system increases. Lomonosov established that this was not true.

From my experiments I have concluded that the view of the famous Mr. Boyle is wrong, because if air from outside is not allowed to enter then the weight of the substance after combustion remains unchanged [24] (1756).

Unfortunately the experimental material of these tests did not survive. Twenty years later Lavoisier repeated the experiment with a similar result, and a radical change was brought about in chemical thought by the introduction of the theory of combustion which was based on this experiment. Lomonosov was also a pioneer of physical chemistry, and decisively marked out the framework of this branch of science (Chapter XII. 1).

However, even the energy of the greatest genius is not inexhaustible. The greater part of Lomonosov's chemical writings are only schemes or programmes containing splendid ideas which were never accomplished. However, Lomonosov dealt with so many different subjects that he did not have much time to carry out his experiments, or even to consider the significance of the results. Had he been able to devote his time entirely to chemistry, it is possible that he might have seen the significance of the principles of combustion and the indestructability of matter and provided a new chemical theory long before Lavoisier. How-

FIG. 30. Antoine Laurent Lavoisier (1743—1794). Copper engraving, 1790.

ever, that was not to be, and it was left to one of the greatest scientists in history, Antoine Laurent Lavoisier, to realize the profound significance of these facts.

It would need a separate volume to describe the life and work of Lavoisier in detail, so only a brief account will be given here. Of the many books written about him, some consider him as the greatest chemist the world has known, while others are bitterly opposed to him and belittle him. It is possible that both are right for Lavoisier was a man of many contrasts. He was without doubt the greatest brain and most talented scientist of his age and yet he was not without many human frailties.

His vanity was such that it often made him appear ridiculous. For example, in 1789 shortly after the fall of the Bastille, Lavoisier devised a mock-trial of the phlogiston theory. He invited a large and distinguished party, and enacted this trial in front of them. Lavoisier and a few others presided at the judicial bench, and the charge was read out by a handsome young man who presented himself under the name of "Oxygen". Then the defendant, a very old and haggard man who was masked to look like Stahl, read out his plea. The court then gave its judgement and sentenced the phlogiston theory to death by burning, whereupon Lavoisier's wife, dressed in the white robe of a priestess, ceremonially threw Stahl's book on a bonfire.

Lavoisier was also very proud and very desirous of glory, and even though his own work and ideas made him indisputably the leading scientist of his age, he was not ashamed to appropriate the discoveries of other people and announce them as his own. He was well favoured by fortune; his own inherited wealth was supplemented by the fortune of his beautiful and intelligent wife. At the age of twenty-five he was made a member of the Academy. In addition he was a director of tobacco and cordite manufacturing companies. His house was the centre for scientific society, and his dinners were famous throughout the whole country.

His position in formal society was as brilliant as the position he occupied in science. Lavoisier was the leader of the new chemical school of thought, the anti-phlogiston school. The essence of this theory, the correct explanation of the phenomenon of combustion, was attacked for a long time by many people, even in Lavoisier's own country. However, as time passed more and more people accepted the irrefutable logic and proofs which the new theory provided. Other countries were slower to accept these ideas, and this was not helped by the personal dislike that many scientists had for Lavoisier.

Lavoisier's brilliant career came to a dramatic and sudden end, with the outbreak of the French Revolution. A large part of his income came from his position as a chief tax-tenant. The chief tax-tenants hired out the indirect taxes and various customs for an amount paid in advance. The corporation of the chief tax-tenants, the so called *Ferme*, enforced the collection of taxes ruthlessly. This establishment was hated throughout the country, and when the Jacobins came to power the chief tax-tenants were called to account. Lavoisier was sentenced to death, and was executed in 1794, at the age of 51. According to a contemporary account Lavoisier boldly bowed his head under the guillotine. A witness of his death, who was not one of his friends, afterwards declared,

> I do not know whether I saw the last and carefully played role of an actor, or whether my judgement of him before was wrong, and a really great man has died!

These words illustrate the contrasting facets of Lavoisier's personality, but whatever the disputes over his personal character his reputation as one of the greatest scientific brains the world has known is indisputable.

Lavoisier did not make such great experimental discoveries as the earlier phlogiston-chemists, although he repeated many of the experiments, and in some cases he claimed the discovery as his own; in others he only made some passing sarcastic reference to the original discoverer. He decomposed water in a manner similar to that of Cavendish, prepared oxygen after Priestley and Bayen [25], and also carried out the experiment of Boyle previously referred to on the combus-

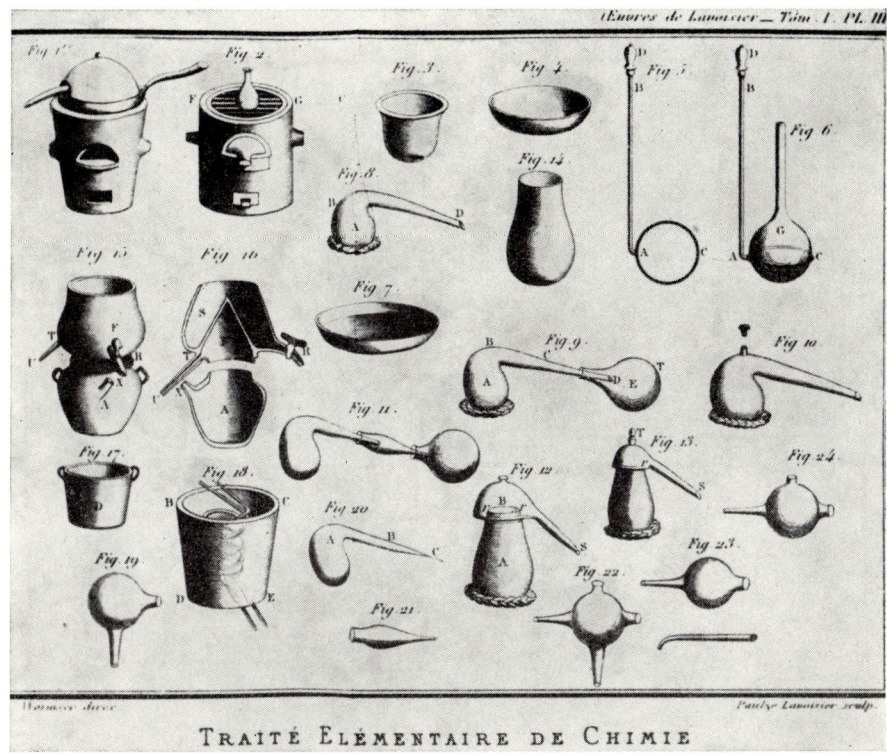

FIG. 31. Lavoisier's laboratory equipment. (From his book: *Traité élémentaire de chimie* [1789])

tion of a metal in a closed system. Presumably he did not know of Lomonosov's work on this subject, but he certainly knew of the work of the Italian chemist Beccaria [26], who had carried out similar experiments of his own fifteen years earlier, as in another work he refers to the periodical review in which the results of Beccaria's work was published. These slight aberrations do not detract, however, from the greatness of Lavoisier, as it is the conclusions that he draws from these experiments, not the experiments themselves, that make him famous. He first of all clarified the combustion process; he established that oxygen is needed for burning, and that as a result of burning non-metallic elements an acidic product is formed, and with metallic elements a basic product is formed. He also made the first clear distinction between elements and compounds, and explained the process of dissolving metals in acids. There are many other important aspects of Lavoisier's work, but these are concerned with more general chemistry so they cannot be dealt with here.

All Lavoisier's works are based on the principle of the indestructibility of matter, which he first postulated in 1789, this being considered the greatest of his

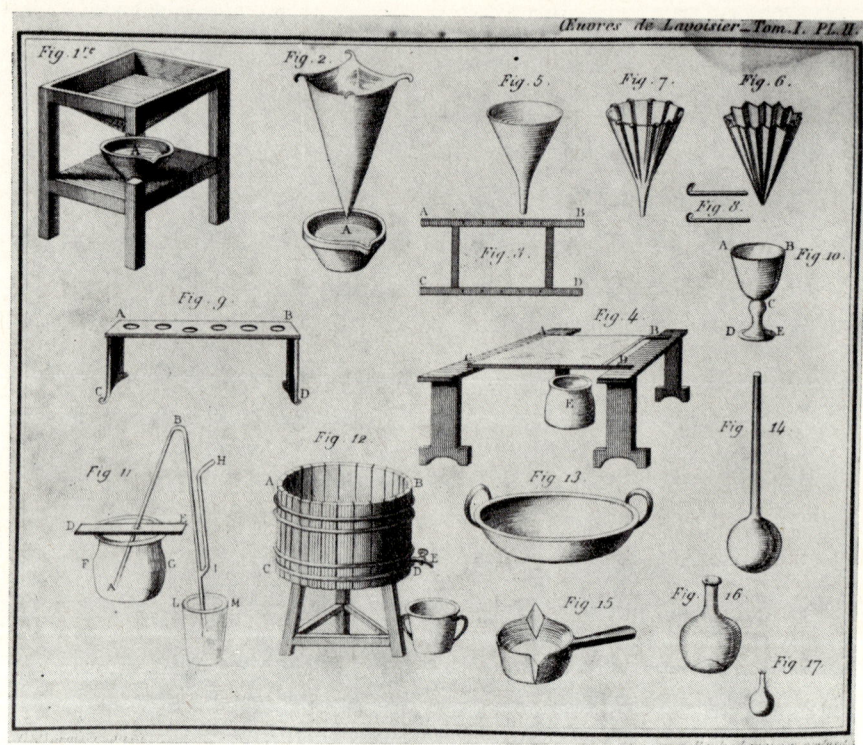

FIG. 32. Lavoisier's laboratory equipment. (From his book: *Traité élémentaire de chimie* [1789])

achievements. Although this theory came to be generally known through the work of Lavoisier, he himself did not publicize it or emphasize it in any way. As Lavoisier's tendency to place a great deal of emphasis on his own work is well known, it would appear that Lavoisier himself did not think that this theory was revolutionary, but rather a natural and well-known established fact. For example he writes about the fermentation of must, that must = alcohol + carbon dioxide

> because nothing is created, neither in natural, nor in artificial processes, and we can announce it as a principle that the quantity of the substance is the same before and after all operations, only changes and transformations are going on [27].

Surely this sounds like the scientific statement of a well known natural law? Lavoisier also approached another basic rule of chemistry, the question of stoichiometry, the nature of the numerical proportion which was apparent in chemical processes:

> The reacting and formed substances can be expressed in algebraic equations, and if one member of these is missing it can be calculated [28].

However he did not proceed any further with this problem, and much more work was necessary to establish the laws of stoichiometry.

3. STOICHIOMETRY

Stoichiometry, namely that part of chemistry dealing with the quantitative relationship between the composition and formation of compounds, could only develop when the concept of compounds was made clear. First of all the composition of salts became known. During the period of iatrochemistry it was discovered that from the reaction of acid and alkali a salt is formed; a salt being defined as a substance which has neither acidic not basic properties, and which consists of an acidic and a basic part. The process of neutralization was called "saturation" *(saturatio)*. Helmont recorded that silicic acid when dissolved in a base is precipitated by saturating the base with nitric acid *(quae tum saturando alcali sufficit)* [29]. And Boerhave [30] had previously referred to the saturation-point; by adding acid to a base a point is eventually reached where the basic character just ceases: *atque tum hoc punctum saturationis vocatur*. It is quite understandable that chemists soon became interested in the regularity and quantitative relations involved in salt-formation. The subsequent examination of this problem was to be responsible for the development later of stoichiometry and of the atomic theory.

Homberg [31] in 1699 added different acids to the same quantity of potassium carbonate until effervescence ceased. He recorded the amount of acid required and measured the increase in weight of the crystallized salt [32]. His results — calculated on a unified base — are shown in Table 4.

It can be seen that Homberg's results are nowhere near the correct figure. As the weight of salt that was obtained was almost the same in each case, Homberg concluded that the acids differ from one another only in their water-content but all are the same in their effective substance.

After Homberg there is a long break in the investigation of this subject, until 1767.

In 1767 Cavendish stated that a given quantity of alkali carbonate was equivalent to a certain quantity of lime because both neutralize the same quantity of acid [33].

In the previous chapter it was pointed out that quantitative analysis was carried out quite frequently during this period. Therefore the assumption must have been made that the composition of a compound is constant. This must have been regarded as self-evident. Wenzel, who was the most accurate analyst of this period, expressed this idea:

That all compounds must have definite and unchangeable composition which can neither be smaller nor larger, otherwise that nothing certain could be established from their comparison is self-evident [34].

TABLE 4

THE AMOUNT OF SALT (IN GR) FORMED FROM 480 GR OF POTASSIUM CARBONATE

The salt formed	The salt (gr)	
	According to Homberg	Theoretical
Potassium acetate	696	673
Potassium chloride	682	517
Potassium nitrate	696	701
Potassium sulphate	665	604

Consequently, Wenzel was postulating the law of constant proportions, but this was the limit of his speculation. This point must be emphasized because the literature of chemical history generally attributes much more to Wenzel than is his due. For example, Walden in one of his books [35], records several tables of data based on the results of Wenzel, and calculates the values of equivalent weight and atomic weight for various metals from them. He shows that the values obtained are astonishingly accurate. However, Walden is wrong in concluding that Wenzel was the original discoveror of the significance of atomic and equivalent weights. All these values can be calculated from Wenzel's data only because we can calculate stoichiometrically and because Wenzel's data are correct. Wenzel, however, did not have this knowledge, and did not attempt to draw any general conclusion from his analytical results. Also the so-called neutrality law which is based on the observation that the products of the double decomposition reaction of neutral salts are themselves neutral, is often erroneously attributed to Wenzel [36]. I found that this error originates from Berzelius who attributed this law to Wenzel. As Berzelius knew the work of both Wenzel and Richter, he evidently must have mistaken the two names. I have read Wenzel's book, and there is nothing in it to suggest that he was aware of the significance of the law of neutrality. The extent of Wenzel's understanding of the principles involved is clearly shown in the last chapter of his book which is headed *Anwendung der Lehre von der Verwandschaft der Körper auf besondere Fälle*. He raises the question of the amount of cinnabar that must be mixed with silver chloride in order "to separate the hydrochloric acid completely from the silver". His reasoning, based on his analytical results is as follows (the correct data given in brackets):

240 gr AgCl contains 180·916 (180·05) gr of silver. This amount of silver reacts with 26·75 (26·77) gr of sulphur to form Ag_2S.

In cinnabar, 65 gr of sulphur combine with 240 (407.4) gr of mercury, to form 305 gr of HgS. Consequently the 26·75 gr of sulphur required to react with 240 gr of AgCl is contained in 125·5 gr of cinnabar. However, Wenzel questions whether this 125·5 gr of cinnabar is enough to react with the chloride in the silver chloride. 240 gr of AgCl contain $53^7/_{16}$ gr of hydrochloric acid (chloride), and 240 gr of HgCl$_2$ contain $58\frac{1}{3}$ gr of chloride, together with 174 (173) gr of mercury. Consequently $53^7/_{16}$ of hydrochloric acid react with $159^2/_5$ gr of mercury, which is present in $202\frac{1}{2}$ gr of cinnabar.

Therefore, while $125\frac{1}{2}$ gr of cinnabar is required to react with the silver in 240 gr of AgCl, the chloride content is equivalent to $202\frac{1}{2}$ gr of cinnabar [37]. As to the fate of the excess sulphur Wenzel makes no comment.

Obviously there is an error in this reasoning, and it is easily traced to the composition of the cinnabar which is incorrect. Wenzel determined the composition of the cinnabar by heating two oz of the solid mixed with potassium carbonate in a glass retort, and collecting and weighing the mercury ($377^1/_2$ gr). The result he obtained was slightly less than half of the correct value [38]. It is unusual to find an error of this magnitude in Wenzel's work, and it is even more remarkable when it is considered that he analysed other mercury salts by this method and obtained accurate results. It is possible that Wenzel made a mistake and weighed only one oz of sample. However, the important point is that Wenzel was under the impression that by a double decomposition reaction to displace one part of a salt, a larger amount of salt is required to displace one component than is required to displace the other.

Thus Wenzel did not have any idea of the significance of the proportions of double decomposition.

Richter, who was the first to recognize the significance of the law of neutrality, and who established the basic rules of stoichiometry when faced with the conflicting results of Wenzel, would certainly have realized that something was wrong. He often queried the analytical results of his contemporaries when he found that the results did not satisfy the weight-proportions demanded by the neutrality law. Richter stated quite definitely that: "Stoichiometric formulae can be constructed quite independently, and from a priori principles" [39]. The word stoichiometry also originates from Richter.

Jeremias Benjamin Richter was born in Silesia in 1762, where his father was a merchant. He became an architect, and for seven years he served in the army in the corps of engineers. When he left, because of his objection to the discipline, he went to Königsberg, where he attended lectures on mathematics and philosophy at the university. The lectures on philosophy were given by Immanuel Kant who was the greatest philosopher of this period, and who wrote on the subject of science, "In the single branches of natural science, real science is only as much as the amount of mathematics that is in them." This thesis of Kant was to have a great influence on Richter.

His future interest is revealed by the title of his Ph. D. thesis: *De usu matheseos in Chemia* (The use of mathematics in chemistry). After obtaining his diploma he became an honorary lecturer at the University of Königsberg, but was unable to earn enough to support himself and had to obtain employment with a Prussian landlord on whose estate he carried out surface surveys. He built a small laboratory in which he carried out his experiments. His hopes of becoming a Professor at the University, or at least a teacher in a high school were never realized. When his appointment expired in 1795 he obtained, with much difficulty, despite the assistance of influential friends, a post as secretary at the Silesian Chief Mining Office. From one of his petitions we know that his salary was 300 thalers a year, and that

Fig. 33. Jeremias Benjamin Richter (1762—1807). (From Bugge: *Buch der grossen Chemiker*)

his lodgings was a small cubby-hole which was poorly heated, where he lived, together with a man-servant, and that all his clothes, books and drawings became mildewed, and his apparatus became rusty. He worked here in all his spare time, with that feverish activity peculiar to consumptives who feel that the rhythm of life is quicker than themselves. Later his circumstances improved and he obtained a post at the porcelain factory in Berlin, where his salary was 700 thalers a year, but by that time his health had been undermined. He died in 1807 at the age of 45, without experiencing the recognition and acclaim to which his chief work *Anfangsgründe der Stöchyometrie oder Messkunst chymischer Elemente*, published in 1792—

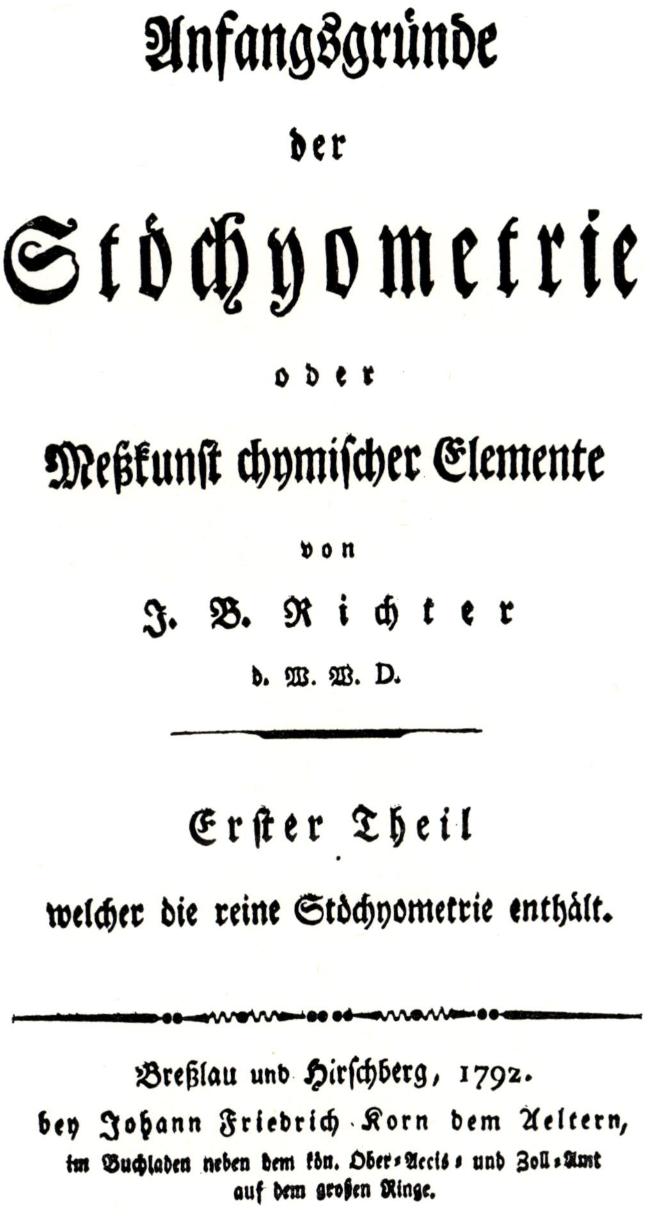

Anfangsgründe
der
Stöchyometrie
oder
Meßkunst chymischer Elemente
von
J. B. Richter
d. W. W. D.

Erster Theil
welcher die reine Stöchyometrie enthält.

Breßlau und Hirschberg, 1792.
bey Johann Friedrich Korn dem Aeltern,
im Buchladen neben dem kön. Ober-Accis- und Zoll-Amt
auf dem großen Ringe.

FIG. 34. Title page of Richter's book *Anfangsgründe der Stöchyometrie* (1792)

1793, entitled him. In his book Richter postulated the basic rules of stoichiometry which were to shape the future course of chemical progress.

The outline of Richter's work can be seen in his introduction:

> Mathematics regards as its own all sciences in which measurement occurs, so that every science belongs to the science of measurement according to the number of factors that must be measured. This fact often made me inquire in the course of my experiments as to the extent to which chemistry is a part of applied mathematics; this question was raised especially in the course of a very common observation: where two neutral salts displace one another then again neutral compounds come into being. Of this the direct conclusion can be drawn that between the components of the neutral salts certain definite volume proportions must exist. From then onwards I considered how to determine these proportions, partly by experiment and partly by the combustion of chemical and mathematical analysis [40].
>
> Because the mathematical part of chemistry is constituted mainly by such bodies, which are indivisible substances or elements, and because this science discusses the volume-proportions between them, I could not find a shorter and more appropriate name for it than the word "Stöchyometria", of the Greek στοιχειογ, which means something like indivisible, and of μετσειγ, which means the search for volume proportions ... [41]

The most important work of Richter is his *Anfangsgründe der Stöchyometrie*. In addition he published a booklet every year between 1791 to 1802, with the title *Über die neueren Gegenstände der Chemie*, in which he continued to develop the ideas proposed in his book.

The first part of the first volume of the *Anfangsgründe der Stöchyometrie* deals with so-called pure stoichiometry. The object of this introduction was to make known the fundamental concepts of the chemistry of that period, so that mathematicians could become aquainted with the new mathematical chemistry, and gain the necessary chemical knowledge. He then proceeds to impart mathematical knowledge to chemists for the same purpose. Either his opinion of the mathematical knowledge of the chemists of that period is very poor, or else they really were in need of some mathematical instruction for he begins this section:

> If a number is added to another number then the mark + (which is called plus) must be placed between them, whereas if we want to subtract them, then we apply the mark −, which is called minus. For example 19+424 means that we add 19 to 424, that is 443; and 424−19 means that we count back 19 from 424, that is 405. Just so 6+8+4 is 18, and +18−6−4 is 8 ... The mark + is not used in front of a number unless it is preceded by another [42].

Although he teaches very elementary mathematics at first, he soon arrives at the quadratic equations.

After this the section on "pure stoichiometry" follows. Here can be found the postulation of the idea of constant proportions,

> ... if we form a compound from two elements then, because the properties of the elements remain the same, then one will always require the same amount of the other, so for example if to dissolve 2 parts of lime requires 5 parts of hydrochloric acid then 6 parts of lime will require 15 parts of hydrochloric acid [43].

This is followed by the postulation of the neutrality law:

> If two neutral solutions are mixed with one another, and double decomposition occurs between them, then the products which are formed will also be neutral, almost without exception... The basis of this phenomenon is that the elements must have a definite volume-relation between one another [44].

From this he arrives at a conclusion, which is rather meaningless to a present-day reader, unless it is carefully examined when it will be found that it contains the fundamental principle of chemical calculation.

> If the weight of the mass of two compounds which displace one another is A and B, and the mass of one element in A is a, and in B, is b, then the mass of the elements in A are $A-a$ and a, in B, $B-b$ and b. The weight-proportions in the neutral compounds before the double decomposition are $A-a:a$ and $B-b:b$; but after the double decomposition the weight-proportions of the new products are $a+B-b$ and $b+A-a$, and the proportion of the elements in them $a:B-b$ and $b:A-a$. Consequently if the mass proportion is known in A and B compounds, it is known also in the products which are formed [44].

To carry out the calculation it was necessary to determine some relative values of a and b, which were called "mass numbers" by Richter. He gives an account of this in the second and third volumes of his book, titled *Applied Stoichiometry*, published in 1793. The first volume of this book, over a hundred pages in length, is best ignored, for it would have been better if Richter had not written it. It contains various artificial or supposed relations between density, mass-number, affinity, etc., which he intended to prove formally by algebraic reasoning. Although I have read extracts from this volume I was unable to decide what Richter hoped to prove by these demonstrations. Surely his contemporaries could not understand it, and probably abandoned the book without even looking at the second or third volumes. In the copy in the library of the Technical University, Budapest, which I used, the second and third volumes were not even cut open! This must surely be the reason why Richter's work had so little influence and why so little interest was aroused by it.

In the second and third volumes of this book Richter established the number of parts of alkali or alkaline earth required to saturate 1000 parts of sulphuric, hydrochloric, nitric and phosphoric acids. A few examples of his experiments and deductions are given:

from 2400 gr of $CaCO_3$ 1342 of CaO are formed, consequently $CaCO_3 : CaO = 1000 : 559$; to saturate 5760 gr of hydrochloric acid, 2393 gr of $CaCO_3$ were required, and after evaporation and ignition of the salt formed it was weighed and found to be 2544 gr. Therefore $HCl : CaCO_3 = 5760 : 2393$, and because $CaO = CaCO_3 \cdot \frac{559}{1000}$, we can write, that $HCl : CaO = 5760 : 2393 \frac{559}{1000} = 5760 : 1337$. Consequently this 1337 is the earth in the salt which was formed. Subtracting this from the salt obtained (2544 gr) then the acid is $2544 - 1337 = 1207$ part. Therefore the proportion of the earth and acid in the salt is [45]

$$1207 : 1337 = 1000 : 1107$$
$$HCl \quad CaO \cdot HCl \quad CaO$$

By similar procedures Richter established that 734 parts of aluminous earth, 858 parts of magnesia, 1107 parts of lime, 3099 parts of barium oxide, 955 parts of ammonia, 1338 parts of alkali (Na), etc., are equivalent, to use present-day terminology, to 1000 parts of hydrochloric acid. Richter's expression was "equal in mass number". Also with 1000 parts of sulphuric acid, 616 parts of magnesia, 2226 parts of barium oxide, 1053 parts of aluminous earth, 796 parts of lime are

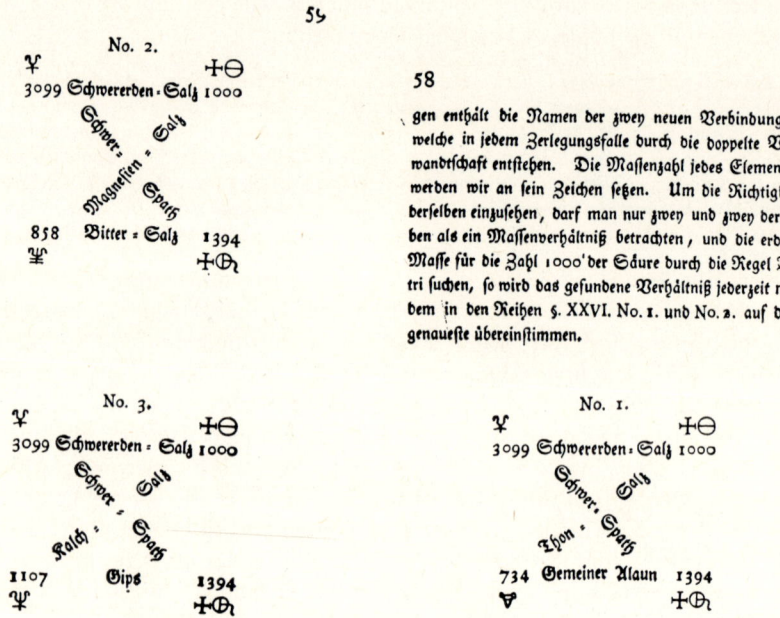

Fig. 35. Reaction equations of Richter. (From his book: *Anfangsgründe der Stöchyometrie* [1792])

"equal in mass number". He also established the relationships for other acids. All that was missing was to relate the values to a given quantity of a certain compound, in order to obtain a useful stoichiometric table of the "equivalent weights". Unfortunately he did not do this, although he had no doubts about its possibility, for in some of his equations he made this calculation. These are shown in Fig. 35. Let us take for example No 2. This is the reaction $BaCl_2 + MgSO_4 = MgCl_2 + BaSO_4$. The composition of the barium chloride: heavy earth (Ba) 3099, hydrochloric acid 1000 parts. This he writes above the horizontal line *(Schwererden-Salz)*. Underneath the line he writes the magnesium sulphate *(Bitter—Salz)*. 1000 parts of sulphuric acid corresponds to 616 parts of magnesia, but because 1000 parts of hydrochloric acid forms a salt with 858 parts of magnesia, then this corresponds to 1394 parts of sulphuric acid. These numbers are written at either end of the compound. Therefore all the values are calculated on the basis of hydrochloric acid = 1000. The reaction products are written across: magnesium chloride *(Magnesien-Salz)* the composition of which is automatically 1000 parts of acid and 858 parts of earth, and barium sulphate *(Schwer-Spath)*, in which 3099 parts earth contains 1394 parts of sulphuric acid [46]. These are the first entirely quantitative reaction equations encountered in the history of chemistry!

Richter did not calculate these on a unified basis, because he searched for far more mathematical regularity in chemistry than actually exists. He was convinced for example that the mass numbers of the different bases obtained with one acid would show some kind of regularity. He calculated for example that the alkaline earth mass numbers related to 1000 parts of hydrochloric acid form the following mathematical progression:

$$a, a + b, a + 3b \ldots a + 19b, \text{ where } a = 734, b = 124\,5.$$

In this way he obtained "mass-series" for different acids. He also calculated acid mass-numbers using different bases, and found that the acid mass numbers form a geometric progression. Richter regarded this as his greatest discovery! In the mass series of the alkaline earths there were many unfilled places between $3b$ and $19b$, but Richter explained this by attributing them to numbers of still undiscovered earths.

Naturally the contemporary reader soon became lost in the mass of complicated calculations and conflicting statements. The composition of many of the compounds established by Richter were very wrong, and his results were much worse than those of many of his contemporaries. This led to the not unjustifiable suspicion that Richter often corrected his results in order that they should fit better into his imagined progressions. This suspicion was supported by the supposed discovery of a new element, called *agust-earth*, in the mineral apatite, by a chemist called Trommsdorf. This element fitted into Richter's imagined proression perfectly, and he proudly proclaimed that this discovery proved his theory. Unfortunately, it was soon proved that the so-called *agust-earth* was in actual fact calcium phosphate. This little episode did not enhance the authenticity of Richter's work, and it was left to the keen brains of Dalton and Berzelius to realize its true significance and to develop its basic ideas further.

However, a chemist named Fischer [47] quickly realized that the many data of Richter must be brought to a unified basis, so he calculated the weight proportions corresponding to 1000 parts of sulphuric acid. He published this table (Table 5) in the German translation of Berthollet's book on affinity, which he had translated [48]. Berthollet included it in the notes of his later work, the *Essai de statique chymique* [49]. Although this book contradicted all the theories of stoichiometry proposed by Richter, it also assisted in spreading Richter's views, and was read by Dalton.

After this the problem of the composition of compounds and of weight-proportions gradually became more clear, Berthollet — the famous French chemist — unexpectedly entered into the discussion, and queried the more or less accepted

Table 5
The First Table of Equivalent Weights According to Fischer

Basis		Acids	
Magnesia	615 (411)*	Sulphuric acid	1000
Ammonia	672 (665)	Hydrochloric acid	712 (714)*
Lime	793 (572)	Carbonic acid	577 (632)
Sodium hydroxide	853 (816)	Phosphoric acid	979 (975)
Potassium hydroxide	1605 (1152)	Nitric acid	1405 (1285)
Barite	2222 (2380)	Oxalic acid	755 (897)
		Citric acid	1683 (1303)

*Present values calculated on $H_2SO_4 = 1000$ are given in parentheses.

view that compounds are formed from their components according to a certain definite weight proportion. Berthollet had been concerned with the nature of affinity and had considered the origin of the forces causing chemical combination. He considered that the mass of the combining substance was only one of several factors which decided the course of a reaction. (This idea is embodied in the well-known, but wrongly named, "Law of Mass Action" of Guldberg-Waage.)

Berthollet was one of the pioneers of physical chemistry, and he used dynamic principles in his reasoning. (Strangely enough neither he nor Ostwald, later, would accept the ideas of the atomic theory.) Berthollet's theory implied that two elements could form a compound, the composition could vary between maximum and minimum for all proportions. Although this theory, if correct, would have destroyed the whole theoretical basis of quantitative analysis, Berthollet's eloquent and inspired reasoning caused many chemists to support him.

Berthollet's theory was also supported by the fact that the analytical results of this period were unreliable, different chemists disagreeing over the composition of the same compound. The distinction between elements, compounds, and alloys was also not very clear, and this also appeared to support the validity of Berthollet's theory.

Claude Louis Berthollet was born in 1748 in Savoy, and after obtaining his medical diploma in Turin he went to Paris. Here he became the court-physician to the Duke of Orleans, who was a cousin of the king. During the revolution he occupied many important positions; he was the director of the mint, and he was afterwards sent to Italy with the delegation whose task it was to collect Italian art treasures and transport them to Paris. It was here that he made the aquaintance of a young general named Napoleon Bonaparte, and they later became close friends. When Napoleon came to power Berthollet occupied even more important posts, so that the revolutionary ended up as a Senator and a Count. After the restoration of King Louis the XVIII, Berthollet was appointed to membership of the Upper House, apparently because he was one of the first to oppose Napoleon in the Senate after the battle of Leipzig.

He died in 1822 as a respected and influential person, a man whose fame and importance increased throughout his life. This, however, is no reflection on his character. In his scientific views he was one of the first supporters of Lavoisier.

FIG. 36. Claude Louis Berthollet (1748—1822). Drawing by Wolf

Berthollet's views were opposed by Proust, who considered that a compound could only be formed according to strictly defined weihgt-proportions.

> On these unchangeable proportions, these constant characters, which characterize the natural and artificial compounds in the same way, these *pondus naturae* (natural weights) the chemist cannot exert an influence in any way [50].

The dispute between them was carried on for many years, although Proust who was an analyst had obtained experimental evidence from the formation of metal oxides and metal sulphides to support his views, whereas Berthollet argued more on theoretical grounds.

Joseph Louis Proust (1755—1826) was the son of a pharmacist, and later became one himself. His first experiments were carried out in the dispensary of a Paris hospital, and as a result of this work he was offered the chair of chemistry at the Artillery Academy in Segovia by the Spanish Government. He afterwards became Professor of Chemistry at the University of Salamanca, and then at the University of Madrid where his generous salary was provided by the king. During the Napoleonic wars, however, the French troops who were occupying Madrid destroyed his splendid laboratory and his valuable collections. Proust returned to France, and for many years he was a comparatively poor man, but by 1816 his circumstances had improved and he was elected a member of the French Academy.

Proust was the first to realize that if two elements form more than one compound with one another then the proportions of the two elements change in definite stages, so that the compounds have a definite composition and no product of an intermediate composition can exist.

Proust, who was a very good analyst, reported his results in terms of per cent of element sought, so that he recorded the analysis of the two tin oxides as follows:

	Tin (II) oxide	Tin (IV) oxide
Tin per cent	88·1	78·4
Oxygen per cent	11·9	21·6

If he had calculated his results to show how much of the one element combined with the same amount of the other element in both compounds, he would have obtained the following values:

	Tin (II) oxide	Tin (IV) oxide
Tin per cent	100	100
Oxygen	13·5	27·5

and in this case the law of multiple proportions, which Proust had come very near to establishing, would have been apparent.

The law of multiple proportions was eventually discovered by Dalton from the results of similar experiments with the oxides of carbon and nitrogen.

John Dalton was born in 1766 in the village of Eaglesfield in Cumberland. His father was a poor handloom weaver, who also farmed a small piece of land [51, 52]. At the age of twelve he was employed as a teacher in the village school, and when he was fifteen he moved to Kendal and became an assistant in a Quaker School there. After three or four years he became principal of this school, a post which he held for eight years, and during all this time he was constantly occupied in furthering his own scientific knowledge. In 1793 Dalton went to Manchester to teach mathematics and natural philosophy at New College. After six years in this post he resigned in order to devote more time to his research. He earned a little money as a private tutor, and as he was unmarried he was unhampered by domestic responsibilities. Dalton's earliest interest was in meteorology, and from the time he moved to Kendal until the day before his death he made daily observations of the weather, temperature, pressure and rainfall, etc. He was also interested in physics, and discovered the law of partial pressures of gases, and in the field of medicine he discovered the phenomenon of colour blindness which was named after him (daltonism). He was elected a member of the French Academy in 1816,

FIG. 37. John Dalton (1766—1844). Painting by Lonsdale

and reluctantly agreed to become a Fellow of the Royal Society in 1822. Many honours and distinctions were awarded to him, including a pension of Ł300 per year from the Government. He died in 1844.

The Law of Multiple Proportions was also proved in the year 1807 by Wollaston [53] who examined the salts formed from varying amounts of oxalic acid with the same amount of potassium, and established that these amounts are in the ratio 1 : 2 [54]. It can be seen that many chemists, even if they did not realize the law of multiple proportions, were unconsciously aware of some form of relationship of this type. Dalton, however, carried his investigations a stage further and searched for the reason for this phenomenon, and his conclusions were embodied in the atomic theory which is named after him. In a lecture given in Manchester in 1803 on the subject of the absorption of gases in water he refers to the influence of atomic weights. He stated that the degree of absorption is influenced by the weight of the smallest particle of the gases. Very little experimental detail was given in this lecture, but the results were given in a table. The lecture was also published in the form of a communication in 1805 [55]. In the table the value for hydrogen is 1, nitrogen 4·2, carbon 4·3, oxygen 5·5, phosphorus 7·2, and sulphur 14·4. One of his friends, Thomas Thomson [56], who was a chemist, afterwards outlined Dalton's atomic theory in his lectures, and also in his book *A System of Chemistry* published in 1807. This book aroused considerable interest for the then incompletely developed theory. Dalton finally published his detailed theory in 1808 in his book *A New System of Chemical Philosophy*.

The view that matter consists of minute, indivisible particles did not originate from Dalton; it was a very ancient concept. Many Greek philosophers, as well as physicists and chemists from Democritus to Boyle, Descartes and Richter, etc., held this view. Dalton himself wrote that his discoveries had led to the conclusion, which was later generally accepted, that all substances are composed of very large numbers of small particles. Even Dalton's idea that the smallest parts of the same element have the same weight and form is not original. But the essence of Dalton's theory is the idea that compounds are formed from the uniting of whole atoms, and that the atoms of different elements which are not of the same weight can be expressed by numbers, and on the basis of this the composition of the chemical compounds can be expressed quantitatively. He was of the opinion that an important task in all analyses was to establish the relative weights of the components of the substance. From these data the number and weight of the atoms in other compounds could then be determined. He described the object of his work as being to carry out these determinations [57].

Dalton's theory illuminated numerous weight ratios which had hitherto been inexplicable.

Dalton took hydrogen, the lightest element as the basis for his construction of the table of atomic weights. He then made the atomic weights of the other elements proportional to this. His table of atomic weights is shown in Table 6.

In this table all the alkali and earth metals refer to the metal oxide.

In 1809 Gay-Lussac published his law on the volume of reacting gases [58]. At first it appeared as if this law contradicted Dalton's theory, because the two ideas could exist side by side, if there is a proportionality between the atomic weight and the vapour density of the gases, namely if the density of the different gases was directly proportional to the atomic weight. However, this proportionality was not observed as the methods for determining both density and atomic weight at that time were very inexact. In fact the weakness of Dalton's atomic weights was their inaccuracy.

Table 6

Dalton's Table of Atomic Weights

Chemical element	Atomic weight	Chemical element	Atomic weight
Hydrogen	1	Stroncian	46
Nitrogen	5	Barita	68
Carbon	5	Iron	38
Oxygen	7	Zinc	56
Sulphur	13	Copper	56
Magnesium	20	Lead	95
Lime	23	Silver	100
Soda	28	Platina	100
Potash	42	Mercury	167

This was the regular pattern in the progress of analytical chemistry, and the reason why the understanding of the basic rules of stoichiometry took so long. The good analysts did not provide theories to explain their work, while the good theoretical chemists did not assist the spread of their views owing to their poor results. Eventually one man appeared who was outstanding in the fields of both theory and practice. This man was Berzelius.

NOTES AND REFERENCES

1. Lémery, N.: *Cours de chymie.* 5. ed. Paris (1675) 101 172 327
2. Kunckel, J.: *Laboratorium chymicum.* 2. ed. Hamburg (1722) 306 309
3. Walden, P.: *Mass, Zahl und Gewicht in der Chemie der Vergangenheit.* Stuttgart (1931) 39
4. Wilson, C.: *Course of practical chemistry.* London (1746) 29
5. Marggraf, A. S.: *Mém. Acad. Sci.* Berlin (1749) 16; cf. *Opuscules chymiques.* Paris (1762)
6. Black, J.: *Edinburgh Physical and Literary Essays* (1755) **2** 157
7. Bergman, T.: *Opuscula physica et chimica.* Upsala—Stockholm—Åbo (De analysi aquarum) **1** 133 134 137
8. *ibid.* **1** 138
9. *ibid.* (De preaecipitatis metallicis) **2** 391–393
10. Wenzel, C. F.: *Lehre von der Verwandschaft der Körper.* Dresden (1777) 68
11. Walden, P.: *Mass, Zahl und Gewicht in der Chemie der Vergangenheit.* Stuttgart (1931) 47
12. Lucianus: *Demonax. cap.* 39
13. Lasswitz, K.: *Geschichte der Atomistik.* Leipzig (1890) **1** 312
14. Bacon, F.: *Novum organon.* London (1620) **2** 40

15. MERSENNE, MARIN (1588—1648) a Minorite Friar. He lived in Paris, and taught philosophy and theology. He paid great attention to the natural sciences of his time and carried on correspondence with famous scientists. He himself carried out research in chemistry.
16. MERSENNE, M.: *Questions théologiques, physiques, morales et mathématiques.* Paris (1634) question 36.
17. MARIOTTE, EDME (1620(?)—1684) a French Benedictine Friar, physicist, and a member of the French Academy of Sciences. His name is perpetuated by the so-called Boyle—Mariotte law of gases. However Mariotte established it later than Boyle, but independently of him.
18. MARIOTTE, E.: *Oeuvres.* Paris (1740) **2** 656
19. HOOKE, ROBERT (1635—1703) educated at Oxford, and then became Boyle's assistant, and later a member and secretary of the Royal Society. He was a man of genius, who derived the law of gravitation, the interference phenomenon of light, etc., but generally did not expound his conceptions thoroughly.
20. HOOKE, R.: *Micrographia.* London (1665)
21. MENSUTKIN, B. N.: *Ann. der Naturphilosophie* (1905) **4** 223
22. KUDRAVSEV, P. S.: *A fizika története* (History of Physics) Budapest (1951) 326
23. WOLF, CHRISTIAN (1679—1754) Professor of Physics at the University of Halle. He was forced to leave for political reasons, and became the Professor of Philosophy at Marburg, but later went back to Halle as Professor of Mathematics. He was a supporter of Leibniz, the creator of the theologist philosophy.
24. MENSUTKIN, B. N.: *Ann. der Naturphilosphie.* (1905) **4** 223
25. BAYEN, PIERRE (1725—1798) French pharmacist, a member of the French Academy. He observed that when mercury oxide is heated a gas is formed.
    BAYEN, P.: *Opuscules chymiques.* Paris (1798) **1** 228
26. BECCARIA, GIACOMO BATTISTA (1716—1781) Italian Friar, the Professor of Physics at the University of Turin.
27. LAVOISIER, A. L.: *Traité élémentaire de chimie.* Paris (1789) 101
28. LAVOISIER, A. L.: *Oeuvres.* Paris (1854) **2** 339
29. VAN HELMONT, J. B.: *Ort. med.* (1656) 56
30. BOERHAVE, HERMANN (1668—1738) Professor at the Leyden University, and a very famous medical man as well as a chemist.
31. HOMBERG, GUILLAUME (1652—1715), was born in Batavia in Indonesia, his father was employed by the Dutch Society. Homberg read law and became a lawyer in Magdeburg. Later he left this profession and studied medicine. He travelled a great deal, and also worked in Boyle's laboratory for a time. He finally settled in Paris.
32. HOMBERG, G.: *Mém. Acad. Paris* (1699) 44
33. CAVENDISH, H.: *Philos. Transact.* (1767) 102
34. WENZEL, C. F.: *Lehre von der Verwandschaft.* 4
35. WALDEN, P.: *Mass, Zahl und Gewicht in der Chemie der Vergangenheit.* 80
36. KOPP, H.: *Geschichte der Chemie.* Braunschweig (1844) **2** 357; Daumas, M.: *Historie de la Science*, Paris (1957) 943
37. WENZEL, C. F.: *Lehre von der Verwandschaft.* 452
38. *ibid.* 397
39. RICHTER, J. B.: *Über die neueren Gegenstände der Chemie.* Breslau (1798) **10**
40. RICHTER, J. B.: *Anfangsgründer der Stöchyometrie oder Messkunst chymischer Elemente.* Breslau—Hirschberg (1792) **1** v.
41. *ibid.* **1** xxix
42. *ibid.* **1** lxxxviii
43. *ibid.* **1** cxxiii
44. *ibid.* **1** cxxiv
45. *ibid.* **2** iii
46. *ibid.* **2** lix
47. FISCHER, ERNST GOTTFRIED (1754—1831) Professor of Physics at the University of Halle.
48. BERTHOLLET, C. L.: *Untersuchungen über die Gesetze der Verwandschaft.* Leipzig (1801) 232

49. BERTHOLLET, C. L.: *Essai de statique chimique.* Paris (1803) **1** 136
50. PROUST, J. L.: *Ann. chim.* (1799) **32** 30
51. KOPP, H.: *Geschichte der Chemie.* **1** 363
52. BUGGE, G.: *Das Buch der grossen Chemiker.* (Ostwald W. Dalton) **1** 378
53. WOLLASTON, WILLIAM HYDE (1766—1828) physician. From 1800 he lived in London as a person of independent means, carrying out research in physics and chemistry. He discovered palladium and rhodium. He also discovered the ductility of platinum which made him quite rich, and also originated the use of platinum vessels for laboratory purposes.
54. WOLLASTON, W. H.: *Philos. Transact.* (1808) 96
55. DALTON, J.: *Memoirs of the Literary and Philosophical Society of Manchester, New Series* (1805) **1** 271
56. THOMSON, THOMAS (1773—1852) Professor at Glasgow University, was the most enthusiastic follower and propagator of Dalton's theory, and later he also helped to publicize Proust's theory.
57. DALTON, J.: *A New System of Philosophy.* (1808)
58. GAY-LUSSAC, I. L.: *Mém. Arcueil.* Paris (1809) **2** 207

CHAPTER VI

THE PERIOD OF BERZELIUS

1. THE RECOGNITION OF THE COMPOSITION OF MINERALS

Lavoisier's new theory of chemistry was gradually accepted, and by the end of the 18th century even England and Germany, where any ideas originating from revolutionary France were treated with grave distrust, had accepted the new theory. But at this time the laws governing stoichiometry, despite the valuable work of Richter, Proust and Dalton were not conclusive enough to be used for exact calculations. Dalton's atomic theory, and also the laws of stoichiometry, only became an integral part of quantitative analysis after accurate values of the atomic weights had been established. This work was carried out by Berzelius at the beginning of the 19th century, and his invaluable work marks the start of a new period in chemical history.

Berzelius's work was furthered in that he was able to rely on a large amount of analytical information, which had not been available to Wenzel or Richter. This was largely due to the rapid development of mineralogy between the years 1790—1810. During this time the quantitative composition of a large number of minerals and naturally occurring salts was established with a fair degree of accuracy. Mineral analysis was the primary concern of chemists throughout Europe at this time, but three names stand out above the others. Kirwan, Klaproth and Vauquelin, although only slightly more illustrious than most of their contemporaries, give a good impression of the manner in which analysis was carried out in Europe at that time.

Richard Kirwan was born in Ireland in 1735, where he first studied and practised law. However, being a rich man he soon retired and lived in London and Dublin, and at his country estate, where he engaged himself in the study of science as a hobby. His main interest was in analysis. He was a member of the Royal Society, and for a long time he supported the phlogiston theory in opposition to Lavoisier, but eventually realized that it was untenable. He died in Dublin in 1812.

Kirwan describes many types of salts, and the results of his analyses were extremely accurate, a fact which assisted both Richter and Berzelius in their investigations on stoichiometry. However, there was little in his work which was new, the main object being to make Bergman's methods of water analysis quicker and more simple. His work was published in London in 1799, under the title *An Essay on the Analysis of Mineral Waters*.

Kirwan's book is outstanding, mainly because of its comprehensive list of references, which summarize all the work carried out in the field of water analysis since the time of Bergman. The first part of the book describes the various compounds which can be found in water. Some of his observations in this section are remarkable. For example, he noted that:

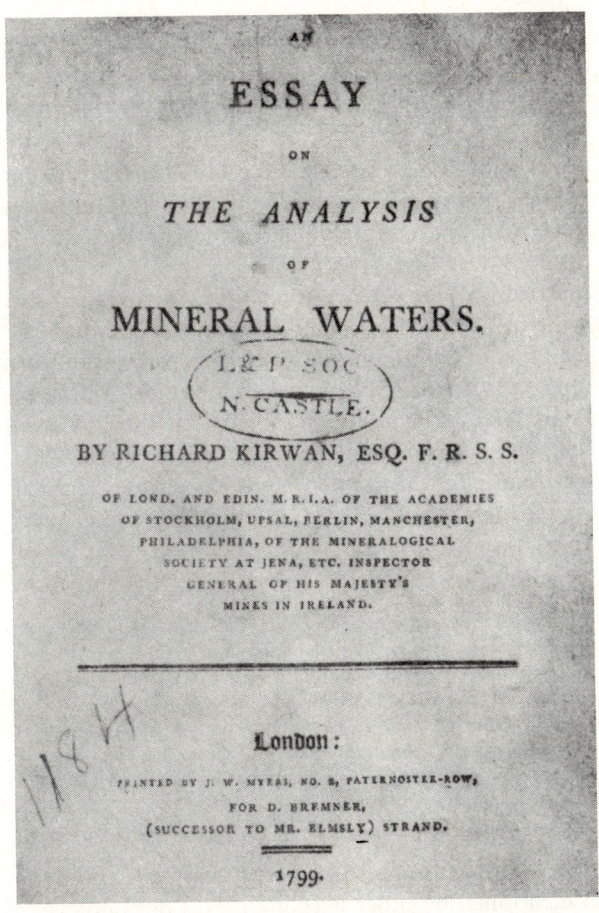

FIG. 38. Title page of Kirwan's book: *An Essay on the Analysis of Mineral Waters* (1799)

Aerated lime is itself soluble in an excess of fixed air (CO_2), as Mr. Cavendish first discovered, and here two questions arise important to subsequent calculations, namely what quantity of aerated lime can be held in solution by a given volume of water fully aerated... (p. 17).

He also compared the available information concerning this problem, but did not arrive at a definite conclusion.

In the second part of his book he describes the various tests that can be applied to identify the various compounds. There is little or no reference to any new reactions, the only difference being that the book is classified according to the different compounds, and alongside each is given the test for its identification. Carbon monoxide, which was discovered by Priestley, is also referred to in the section on gases;

> Heavy inflammable air or carbonated hydrogen. This air is distinguished by the fact that it burns without explosion when mixed with common air. It is not absorbed by lime-water, but water over which it is burned precipitates lime-water, as fixed air is produced (p. 42).

As can be seen, Kirwan had discontinued the use of the phlogiston nomenclature. One of the reagents that he used was boric acid, and he observed that it gave a precipitate with lead nitrate, but that carbonates and sulphates must first be removed (p.84).

Kirwan also carried out interference studies (p. 136) and records the salts which cannot be present for a given salt being determined, although he mentions the fact that in some cases small quantities of the interfering salt do not have much effect. As examples of salts which cannot be determined in the presence of each other, he gives the alkali carbonates and metal salts, the alkali sulphates and the alkaline earth salts, magnesium sulphate and calcium nitrate or chloride, and magnesium nitrate and calcium nitrate or sodium nitrate (?). The reasons for some of these salt pairs being classified as interfering is not very clear.

Kirwan suggests a simple procedure for the determination of the total salt content of water without the need for a quantitative examination:

> There is a method of calculating the quantity of salt in 1000 parts of a saline solution whose specific gravity is known, which, however inaccurate, is yet useful in many cases as the error does not exceed 1 or 2 per cent, and sometimes is less than 1 per cent. It consists simply in subtracting 1000 from the given specific gravity expressed in whole numbers, and multiplying the product by 1·4. It gives the weight of the salts in their most desiccated state and consequently freed from their water of crystallization. The weight of fixed air (CO_2) must also be included, thus for a solution of common salt having its specific gravity 1·079, I find the difference from 1000 is ... 79 and 79 × 1·4 = 110·6, then 100 gr of such solution contain 110·6 gr of common salt ... (p. 145).

He also mentions the error involved in the analysis carried out with "tests" which had been mentioned by other authors, that it does not give any indication which acid reacts with which base to form the salt. He concludes by saying that

> The usual method of applying tests is similar to that of sending out adventurers to an unknown country to see what discoveries they can make, hence that indications are vague and unconnected. Whereas if they were limited to some particular object, and so combined and arranged to prove or disprove its existence, their indications might be rendered certain and precise; and such information I have already shown them capable of conveying, when employed with single and definite views (p. 162).

He elaborates this view later in his book.

After this Kirwan deals with the practical aspects of analysis, and differences between qualitative and the quantitative examinations:

> The Analysis of a Mineral Water embraces, as I have already noticed, two objects: the discovery of the different species of ingredients contained in it, and the determination of the weight of each. How the species are discovered I have already set forth in the chapter of Tests. The quantities of solid ingredients I determine in most instances by estimation, as being the least laborious, in many cases equally and in many the most exact (p. 175).

Kirwan attempted to resolve the problem that he had previously stated: the problem of finding out which acid, and which base has reacted to form a salt. He attempts by using different reagents to determine the nature of the salt, as well as its concentration, but his method is rather unreliable and complicated. It is scarcely of practical importance, although Thomson discusses it a quarter of a century later in his work *System of Chemistry* [1]. Kirwan supposed that the single sulphates can be separated from one another with the help of calcium, barium and strontium salts, for example, magnesium sulphate precipitates with a strontium salt, while potassium sulphate does not.

Kirwan's analytical work was famous in his time, and his results were cited all over Europe. However, at the turn of the 18th century another great chemist, Klaproth played an even greater part than Kirwan. There are very few books published during this period which do not refer to some aspect of Klaproth's work.

> Martin Heinrich Klaproth (1743—1817) was the son of a provincial tailor; after leaving the grammar school for some unknown reason at the age of 15 he became, like so many great chemists before him, an apprentice pharmacist. He first worked at an apothecary's shop in Quedlinburg, but later became an assistant in Hanover, Berlin and Danzig. In 1771 he went to work at the pharmacy of Valentin Rose in Berlin. Rose at this time was a famous chemist, but died only four weeks after Klaproth had joined him. He entrusted the education of his children and the direction of his pharmacy to Klaproth, who faithfully carried out his last wish. He was in charge of the pharmacy for ten years, until Rose's son had grown up. In 1780 he married a niece of Marggraf, and used the not inconsiderable dowry to purchase his own laboratory. Previously he had only carried out his research in addition to his pharmacy, but he now gave up pharmacy and devoted himself entirely to research. Nearly all of his work was devoted to mineral analysis. In 1788 he was elected a member of the Berlin Academy of Sciences, and in 1800 he was appointed leader of the chemical laboratory of the Academy in succession to Achard. In addition he was a Professor at the Artillery Officer Academy. In 1809 King Friedrich Wilhelm founded the new University of Berlin, and Klaproth was called to be the first Professor of Chemistry there. According to the time-table of the university he lectured on "Practical chemistry" for four hours a week for the first term, in the second term, also for four hours a week, he lectured on the subject "Introduction to chemical analysis" [2]. In his later years he had many domestic troubles and was subject to fits of apoplexy. He was ill for a long time, and had to give up his lecturing. He died on New Year's Day 1817.

Klaproth lived in the same period as the great phlogiston chemists, and like many of them he was self-taught and did not attend a university. His scientific career did not really start, however, until he was about 40 years of age, and by this time Lavoisier had expounded the new *chimie française* so that Klaproth was not bound to the phlogiston theory. After confirming several of Lavoisier's experiments, he became an exponent of the antiphlogiston theory, and was the first important German chemist openly to accept the views of Lavoisier.

Klaproth was essentially a practical scientist; he had complete faith in his experiments, and mere suppositions did not interest him. He lived at a time when violent debates were going on in chemical society concerning the various chemical views. Klaproth never took part in these, and never gave any indication of his own opinions. Klaproth examined a wide variety of minerals with great thoroughness and published his results, but here his interest ended. The accuracy of his analysis, however, was an important factor in the establishment by Proust of his law of constant proportions.

Fig. 39. Martin Heinrich Klaproth (1743—1817). (From Bugge: *Buch der grossen Chemiker*)

Klaproth's point of view is best characterized by his debate with Ruprecht. Antal Ruprecht, who was Professor at the Selmecbánya Mining Academy, found that he could obtain metals from the earths (alkaline earth oxides) by reduction with carbon. This announcement aroused great interest in the world of chemistry and produced a bitter controversy because hitherto earths had been considered to be elemental substances which could not be decomposed further.

Klaproth carefully repeated Ruprecht's experiments, and proved that the metals formed did not originate from the earths but from the contamination of the chemicals, and concluded that the simple earths do not contain metals. Although Ruprecht's experiments were incorrect, the idea behind them proved to be right [3]. This was confirmed a few years later by Davy, when he succeeded in reducing the earths to the metals with the help of an electric current. Klaproth did not believe this. This was an often repeated and unfortunate occurrence in the history of the sciences, of the older scientist who is ingrained with the old concepts and refuses to accept the new developments. This was the fate of the phlogiston chemists, who had spent their lives creating a theory and were quite naturally reluctant to abandon it even in the face of irrefutable evidence.

A very clear appraisal of Klaproth's contribution to chemistry is the following:

> He was an incorruptibly accurate researcher, but was lacking only in ingenuity or intuitive thought; had he possessed this also then he would have taken his place among the greatest chemists [4].

He discovered three new elements, uranium, zirconium and cerium, and although titanium, strontium and tellurium had previously been observed by other workers, he examined these discoveries in more detail, and also gave a name to these elements. His contemporaries credit him with the discovery of tellurium although in fact this had already been discovered in 1782 by the Hungarian, Ferenc Müller [5], in the gold-ores of Transylvania [6], who had sent a sample of this ore for confirmation by Klaproth. The orator, who delivered the memorial speech for Klaproth, before the Berlin Academy in 1819, was wrong when he was carried away with the power of his rhetoric and said: "Klaproth increased the number of elements with which the Lord God created the earth, by four" [7].

Klaproth made many contributions to various periodicals, most of them giving an account of the analysis of a single mineral. He also published his collective results regularly in book form under the title: *Beiträge zur chemischen Kenntniss der Mineralkörper*, and this work extended to six volumes. The first of these was published in 1795, and the last in 1815, and in them Klaproth recorded the analysis of minerals originating from all parts of the world. In his books Klaproth departed from the earlier procedure of stating the experimental details very briefly, and often of publishing only the result without the inclusion of any relevant data such as sample weight, weight of precipitate, etc. He describes every step in the procedure in great detail, and in the introduction he explains his reason for doing so.

> For those who want to repeat my experiments, I have written the procedure in as great detail as possible without exposing myself to a charge of loquacity. Those who are acquainted with this topic will perhaps recognise my intention, that is to find simple and reliable procedures for the analysis of minerals [8].

In another book he states,

> the experiments must be carried out in such a manner that if they are repeated by other chemists, who work with the same accuracy, then the result will always be the same [9].

An example of the methods of Klaproth are now given, without the introduction. It concerns the analysis of zoisite:

> Chemical test: the specific weight of the ore is 3300. When heated on a carbon block with a blow pipe it shows a foam-like effervescence, and the colour of the ignited part becomes reddish and it is transformed into a fine powder. During the course of the ignition the loss is insignificant. It does not dissolve in acids.

BEITRÄGE

ZUR

CHEMISCHEN KENNTNISS

DER

MINERALKÖRPER

VON

MARTIN HEINRICH KLAPROTH,

Professor der Chemie bei der Königl. Preuss. Artillerie-Akademie;
Assessor Pharmaciae bei dem Königlichen Ober-Collegio medico;
Mitgliede der Königl. Preussischen Akademie der Wissenschaften,
wie auch der Akademie der Künste und mechanischen Wissenschaften zu Berlin, der Kurfürstlich Maynzischen Akademie der Wissenschaften zu Erfurt, der naturforschenden Gesellschaften zu Berlin
und zu Halle, imgleichen der Societät der Bergbaukunde;
und privilegirtem Apotheker zu Berlin.

Erster Band.

POSEN, BEI DECKER UND COMPAGNIE,
UND
BERLIN, BEI HEINRICH AUGUST ROTTMANN.
MDCCXCV.

FIG. 40. Title page of Klaproth's *Complete works* (1795—1815)

(*a*) 100 gr of finely powdered mineral I ignited moderately with 200 gr of potassium hydroxide in a silver vessel for two hours; I extracted the greenish white mass with water which was oversaturated with hydrochloric acid, and then I evaporated it to dryness. After extracting the residue with water I filtered the precipitated silicious acid, which after ignition weighed 44 gr.

(*b*) I precipitated the filtrate with corrosive ammonia (NH_4OH). I dissolved the precipitate which had previously been washed by decantation, in warm potassium hydroxide.

The iron oxide remained, contaminated with a little manganese oxide; after ignition the weight of this precipitate was $2\frac{1}{2}$ gr.

(c) I precipitated again the aluminious earth with sal ammoniac (ammonium chloride) from the solution made alkaline with potassium hydroxide. I filtered, washed and ignited it. Its weight was 32 gr.

(d) I decomposed the former filtrate by boiling with soda when $36\frac{1}{2}$ gr of carbonic acid lime separated, which is equal to 20 gr of lime-earth.

Accordingly the composition of the zoisite is:

Silicious earth	a	44
Aluminious earth	c	32
Lime earth	d	20
Iron oxide	b	2·5
Manganese oxide		trace
		98·5 [10]

If his results differed much from 100 per cent, then Klaproth searched for the cause of the divergence, and in this way he discovered new elements or found unsuspected components. In this way he found that potassium occurs not only in vegetables, but also in minerals. When examining the mineral leucite he found that there was a 21 per cent discrepancy in the total after the determination of aluminium oxide and the silicon dioxide. This was far greater than the possible experimental error, so he tested for the presence of borax, phosphoric acid and other less frequent components, but could not detect anything. Finally he evaporated and crystallized the residue remaining after the separation. The shape of the crystals formed was quadratic, and they did not affect litmus, nor were they affected by ammonia nor soda solution. Klaproth noted, however, that the taste of the crystals was similar to potassium chloride, so he evaporated with sulphuric acid and was able to obtain crystals of potassium sulphate, and potassium nitrate crystals with nitric acid. This established the presence of potassium in this mineral. He also discovered that this gives a precipitate with tartaric acid, which upon ignition forms potassium carbonate [11]. As a result of this discovery he proposed to change the now obsolete names alkali vegetabile (Pflanzenalkali) and alkali minerale (Mineralalkali) for *natron* and *kali* [12], which unfortunately were accepted neither in the French nor in the English languages.

Klaproth also originated the technique of alkali fusion, for which he used silver vessels. He also mentioned the use of platinum vessels in connection with the fusion with soda [13]. An indication of the exactness with which Klaproth carried out his analysis, can be judged from the fact that he carried out an analysis of the composition of the mortar he used to grind his samples, and thus found it was composed of almost pure silica. Also, after he had ground his sample for analysis he re-weighed it to ensure that it did not contain any of the material from the mortar, and if there was any increase in weight, then he subtracted an equivalent amount from the silicic acid content. Another important point that Klaproth was the first to note was that precipitates must be dried or ignited to constant weight, and in order to obtain the high temperatures sometimes required, Klaproth used the furnaces of the porcelain factory in Berlin [14].

It can be clearly seen how Klaproth developed his own ideas on stoichiometry. In the course of a silver ore test in 1795 he reduced silver chloride to silver metal, and he also reduced hydrated antimony oxide with potassium carbonate and carbon to pure antimony, and weighed them both in the metallic form [15]. Later he left out the reducing process, but still carried out an occasional reduction as a control test. For example, he dissolved 100 gr of antimony in 4 parts of hydrochloric acid and added some nitric acid to it, and after diluting with water found that he had obtained 130 gr of antimony lime (oxide). Using this proportion he calculated back his results to give metallic antimony [16]. After a while he must have realized that the control tests were superfluous as the results obtained were constant, so that in the last volume of his book he simply records that a certain weight of precipitate corresponds to a definite weight of metal or oxide. Moreover, he used these conversion factors for identification purposes.

An ingenious example of this is the proof that the mineral strontianite contains strontium and not barium. He placed 100 gr of strontianite in a small dish on the pan of a balance and saturated it with a weighed amount of hydrochloric acid. To the other pan of the balance he added an equal amount of hydrochloric acid. After the evolution of carbon dioxide was complete the weight of the sample was only 70 gr, therefore 30 gr of carbon dioxide had been evolved. If the substance had been barium carbonate then the weight of carbon dioxide evolved from 100 gr of sample would only have been 22 gr. The other proof he applied was that 100 gr of strontianite when treated with sulphuric acid formed 114 gr of strontium sulphate, of which $2^{1}/_{2}$ gr were soluble in 8 oz of hot water, whereas of the amount of barium sulphate formed from an equal amount of barium carbonate, none was soluble in hot water.

Klaproth also changed the established methods of water analysis, wherever possible entirely omitting the separations based on fractional crystallization

> ... because that does not give a sure result. So I have devised a more reliable method which first of all involves the saturation of the free mineral alkalis with acid, and then I decompose the neutral salts which are formed with a suitable reagent; at the same time I carry out a trial experiment in order to make clear the relationship, and on the basis of this I calculate the result [17].

An example of his trial experiment is as follows:

(a) He ignited 1000 gr of freshly crystallized sodium carbonate on a sand bath when 363 gr of dry powder were formed.

(b) 100 gr of this saturated 382 gr of a sulphuric acid solution, which contained 1 part of acid (density 1·85) and 3 parts of water.

(c) The weight of Glauber's salt formed, after evaporation and drying, was $132^{1}/_{2}$ gr.

(d) 1000 gr of crystallized Glauber's salt, after ignition, formed 420 gr of dry salt.

(e) 100 gr of this when precipitated with barium acetate gave 168 gr of barium sulphate. Therefore 1000 gr of barium sulphate are equivalent to $595\frac{1}{2}$ gr of water-free Glauber's salt.

(f) 100 gr of rock salt are equivalent to $233\frac{1}{2}$ gr of silver chloride, therefore 1000 gr of silver chloride are equivalent to $428\frac{1}{4}$ gr of sodium chloride.

After this he added sulphuric acid to the water, until it was just saturated (neutralized), and then precipitated the sulphate with barium acetate, and subtracted the amount corresponding to the added sulphuric acid from the weight of the precipitate. Then he precipitated the chloride with silver nitrate, and after filtration evaporated the filtrate and dissolved the residue in hydrochloric acid. Lime was precipitated from this solution with ammonium carbonate, and the iron with potassium ferrocyanide solution. The insoluble residue was silicic acid. On an analysis of 1 cubic inch (290 gr) of Karlsbad water, Klaproth obtained 39 gr of mineral alkali (Na_2CO_3), $34\frac{5}{8}$ gr of rocksalt, $70\frac{1}{2}$ of Glauber's salt, $2\frac{1}{2}$ gr of of silicic acid and $2\frac{1}{2}$ gr of iron.

For reasons of space, Klaproth's analytical methods cannot be described in their original detail. When examining silicate ores he separated the silicate after fusion by repeated evaporation with hydrochloric acid; he then separated the metals with ammonia and, finally, he precipitated magnesium in the form of the hydroxide. Ores containing sulphides he dissolved in nitric acid and filtered the separated sulphur, which he determined by ignition. He also questioned whether part of the sulphur does not form sulphuric acid during the solution process. He determined the sulphur content of a silver ore by this method, and then determined the sulphate-content separately, but was unsure whether any of the sulphate had been formed from sulphur. As the dissolution process was rather slow, and as only a very small amount of nitrous fumes was evolved, he considered

TABLE 7

KLAPROTH'S ANALYTICAL DATA

Precipitate weighed (gr)	Metal equivalent	
	According to Klaproth	Theoretical
46 AgCl	$34\frac{1}{2}$	34·6 Ag
93 $PbSO_4$	66	63·6 Pb
56 Sb_2O_4	46	44·3 Sb
12 Fe_2O_3	9	8·39 Fe
1000 AgCl	$428\frac{1}{2}$	407·7 NaCl
100 $SrCO_3$	114	125·2 $SrSO_4$
1000 $BaSO_4$	$592\frac{1}{2}$	608·4 Na_2SO_4

it unlikely. In order to solve the problem he determined the sulphate after dissolution of the silver in hydrochloric acid when no oxidation of the sulphur could take place, and found that this result corresponded satisfactorily with the former, so he concluded that the nitric acid does not change the sulphide into sulphate [18].

FIG. 41. Louis Nicolas Vauquelin (1763—1829). (From Bugge: *Buch der grossen Chemiker*)

Klaproth attempted a fusion of tin oxide, similar to the Freiberg method; he mixed the finely powdered cassiterite with an equal quantity of sulphur, and heated the mixture in a glass retort on a sand bath. However, the experiment was not very successful, so he repeated it using potassium hydroxide to fuse the sample [19]. He also observed the molybdenum-blue reaction, when he treated molybdenum ores in hydrochloric acid medium with tin [20]. He hydrolysed iron with sodium succinate in the presence of manganese and aluminium [21], and then precipitated the phosphate and arsenate as the lead salts, and the chromium as silver or mercury chromate. The mercury was removed from the latter by ignition, and the chromium weighed as the oxide [22]. There was no reference to the use of hydrogen sulphide in Klaproth's work, but he makes one reference to the use of ammonium sulphide for the identification of antimony [23]. Klaproth applied the corrosive action of hydrogen fluoride as a test for fluoride ion. He also utilized it for making glass micrometers for astronomical purposes [24]. Finally, we can see how the analytical data of Klaproth compare with present-day values (Table 7).

Louis Nicolas Vauquelin (1763—1829) was a peasant child, who also became an apprentice pharmacist at the age of 15. Later he went to Paris and worked as a laboratory assistant to Fourcroy, the famous chemist. He soon progressed from laboratory technician to become the co-worker of Fourcroy, and in 1791 he was elected a member of the Academy. He became a Professor of the Collége de France, and later at the University. He was the discoverer of beryllium and chromium, but in addition to his research he was engaged in the production of chemicals on an industrial scale. The novelist Balzac portrays him as a scientist and industrialist in one of his novels [25]. His chemicals were in great demand because of their purity and reliability, so that Vauquelin can be regarded as a pioneer in the production of analytically pure chemicals.

The work of Vauquelin is similar to that of Klaproth in that he analysed a large number of minerals by methods similar to those of Klaproth, except that there were fewer original methods. Vauquelin studied the reactions of chromium and chromate very thoroughly; he found that chromium becomes yellow when fused with alkali, and that this solution gives a red precipitate when treated with mercury, and a yellow precipitate with lead; and that the addition of a tin chloride (stannous chloride) solution causes the solution to become green again.

2. THE LIFE AND PERSONALITY OF BERZELIUS

Berzelius was born at the time when Lavoisier's book, the *Traité élémentaire de chimie* was published. By the time he had grown up and graduated from the university, the revolution that this book had caused with the bitter struggle between the supporters of the phlogiston theory and the new theory was at an end. Almost all practising chemists were engaged in this controversy, and just as after a great battle in war a period of relaxation follows, so it was in the field of chemistry. All the old heroes of the phlogiston period left the battle-field of this world, one after another, without relinquishing their convictions, so that only the young followers of the new theory remained in the lists of chemistry, but without a leader. The chemists who formed Lavoisier's entourage, the first followers of the new method, Fourcroy, Berthollet, Guyton de Morveau, etc., were not able to develop the new chemistry constructively, so that the turn of the 18th century was characterized by a period of stagnation in chemistry. However, towards the end of the first decade of the 19th century, a new source of inspiration was given to chemistry in the form of the phenomenon of galvanism, which was embodied in Davy's wonderful discoveries and in Dalton's atomic theory.

The development of many branches of science often needs someone with a systematic mind who is able to summarize and correlate the experiments and discoveries of others; someone who can observe the regular and irregular phenomena in that branch of science and can use these to construct a theory. Such a man was Lavoisier, as has already been mentioned; so also was Berzelius.

Berzelius became prominent among the leading chemists of the world while still quite young, impelled by his own ambition, and inspired by the new ideas and new problems which were continually arising. His most important work was carried out while he was between the ages of thirty and forty. During these ten years he proposed his dualistic theory, established a large number of atomic weights, and invented the chemical symbols which are still in use today.

In the later period of his life Berzelius became a great administrator and arbiter of chemistry. He summarized the scientific work carried out all over the world every year and reviewed it critically. Such was the nature of his comments that they were able either to encourage or to destroy the theories of others.

At last, in his old age he was a very sick man and unable to keep abreast of the more modern theories. He did not realize that the development of chemistry had progressed beyond his dualistic theory and stubbornly defended it against all critics. He chose rather to ignore the new facts which had arisen, and to oppose them not on the basis of experimental proof, but simply by the power of his authority. Naturally his struggle was in vain, and eventually he was left alone, embittered and offended.

> It is sad to see how the animating fire goes out by degrees. The water which formerly heightened the intensity of it, now extinguishes it, and a handful of ash is enough to cease its shimmering. Why does he not retire and relinquish the arena to those who can still gain from it?

These words written by Liebig in a letter to Wöhler describe the decline of Berzelius [26].

It is a very difficult task to assess the merits of Berzelius, or to describe his significance, because he occupied such a privileged, almost autocratic role in the sphere of chemistry; a role which had no parallel either before or after him. Autocracy has its disadvantages as well as advantages for its subjects, in this case chemistry. Berzelius's brilliance promoted, but his mistakes hindered, the progress of this science. In all branches of chemistry he made important contributions, but mainly in analysis. His methods and results arouse admiration even today, and more especially when one considers the nature of his equipment and working conditions. The amount of work he carried out is also remarkable for he worked alone, except for one or at the most two research students. Berzelius would never have more than two students in his laboratory, so that it was a great honour to work with him, and students from all over Europe competed for the privilege.

Jöns Jakob Berzelius was born on the 20th August 1779 in Wäfversunda in Sweden. His father, who was a teacher, died of tuberculosis in 1783. Two years later his mother married a widowed pastor named Ekmarck who already had five children, so that the family numbered ten. In 1787 his mother died, and the orphaned Berzelius had a meagre inheritance of 200 thalers. For a time he lived and was educated in the house of his stepfather, who was a kindly and cultured man. However, in 1790 Ekmarck married for the third time, and his wife who was a widow with children found that the family was a little too large for her and contrived to send the Berzelius children to live with their uncle, a Lieutenant Sjosteen.

This is how Berzelius wrote about his new home:

From a family where strict discipline and true fear of God were dominant, but where the children loved one another as real brothers and sisters, I went to a family which was also supposed to be God-fearing, but where the lady of the house became intoxicated by strong drinks quite early in the morning, as well as during the rest of the day, and often told her children that my inheritance did not cover the costs of my keep at all [27].

Although the head of the family did not make any difference between Berzelius and his own children and took adequate care of his teaching, his foster-mother and her children made him always feel that he was not welcome. As a result of this experience Berzelius sadly writes " ... this time left such an impression on me, that if ever anyone talked about the happiness of childhood I could never share their joyful recollections" [28]. In 1793 he went to a secondary school in Linköping. As he did not want to live with his foster-parents in the vacation, he found employment as a private tutor to the children of a landowner named Borre, and interrupted his studies for a year in order to continue at it. During this time he

FIG. 42. Berzelius in his laboratory. (From his *Selbstbiographischen Aufzeichnungen*)

also worked in the fields which was good for his feeble constitution although his room which was used for storing potatoes was bitterly cold at night. The room itself was heated, but Berzelius remarks:

> It would surely not have been heated had it not been used for the storing of sacks of potatoes which filled most of the room. However, it was only heated just as much as was needed to ensure that the potatoes should not become frozen [29].

During nights in this room he attempted to translate the Aeneid into Swedish, but without much success. He recommenced his studies and made satisfactory progress, but only the teacher of natural history succeeded in arousing his interest for his subject, and also for the works of Linné, the great botanist. Berzelius became an enthusiastic plant and bird-collector, and for this purpose he learned how to hunt. On one occasion he almost shot somebody, and only just avoided being expelled from the school as a result of it. Taking the advice of the natural history teacher he decided not to be a pastor, as his family had planned, but to become a physician, and in 1796 he finished his studies at the secondary school. His leaving certificate contained the following report " ... he possesses excellent talents but his morals are not good, so his future prospects are uncertain" [30]. He enrolled at the medical faculty at the ancient University of Upsala, but because of his lack of money he had to interrupt his studies after one term and take another appointment as a tutor. During this period he taught himself German, French and English, and later on in his life he had mastered these languages so perfectly that he was able to write his book on minerology directly in French, and also to rewrite his chemical book for schools in German. After this period of employment he managed to obtain a scholarship, and so was able to continue his studies with no financial problems.

Berzelius first became acquainted with chemistry in 1798, when he had to attend lectures on this subject and take an oral examination. When the examination was over the Professor told him that he would have failed had he not done very well in his other examinations, particularly physics [31]. The famous Chair of Chemistry, which had been occupied by Wallerius and Bergman was occupied by a Professor named Afzelius who did not make any notable contributions to the subject, and was a supporter of the phlogiston theory. His name would probably have been forgotten had it not been for the fact that he was Berzelius's chemistry professor.

In his next term Berzelius carried out practical exercises under Afzelius, his first experiment being the ignition of iron sulphate. Although this exercise was supposed to last for a week Berzelius finished it in a few hours, and when he asked for another exercise, the assistant professor reprimanded him for being too impatient [32]. It was in this laboratory that Berzelius first became interested in chemistry and in chemical analysis. In his autobiography he mentions that the professor did not take any part in the practical classes, and at the most he slipped through the laboratory a few times to make sarcastic remarks. Ekeberg [33] the assistant professor was in charge of the practical classes, and according to Berzelius he knew more chemistry than the professor, but unfortunately he was almost deaf so that the students were able to take advantage of him. It was Ekeberg who later discovered tantalum [33]. Berzelius soon realized that he could do whatever he wanted in the practical classes, so he took advantage of the opportunity and carried out either exercises for his own curiosity, or alternatively repeated experiments from the literature. The laboratory assistant allowed him to work in the laboratory whenever he pleased, in exchange for a small tip, and even supplied him with chemicals from the institute pharmacy [34].

The biographers of Berzelius often charge Afzelius with narrow-mindedness, jealousy, or malice against the "brilliant" student, but this is rather an exaggeration. In his time Berzelius was obviously no more than a diligent and talented student, who according to the words of one of his fellow-students "if he thought that he was right then he could be more obstinate than is proper for a young man" [35]. They also reproached Afzelius for postponing the final examination of Berzelius. Although this is true, it would appear that here Berzelius was the victim of a dispute between the Dean of the medical faculty and Afzelius [36]. On the other hand, Berzelius himself writes that Afzelius after discovering his intrigues for using the laboratory said that he did not approve of the hidden paths, but that he would be happy to see him if he came by the normal entrance which would never be closed to him [37]. Berzelius also mentions that "he treated me from that time with extraordinary good will and confidence" [37].

The first scientific works of Berzelius were made in the institute of Afzelius, before taking his degree, and the first of his published dissertations was the result of his work during a vacation. He was employed as an assistant by a physician at a health-resort, and analysed the mineral water of this watering-place on his return to Uppsala. He presented the results of this in a dissertation to the university (1800) [38], but he had already written three other dissertations before this, two dealing with nitrogen oxide and one with *"nitric acid naphtha"*. These he gave to Afzelius, who passed them on to the Academy of Science with the aim of publishing them. After this he did not hear anything about them for three years; when he enquired he was told that they had not been accepted for publication as they contained the new nomenclature proposed by Lavoisier, which had not at that time (1804!) been accepted by the Academy. These communications eventually appeared in 1807, in the Journal which had been established by Berzelius.

In the meantime Berzelius completed his studies, and in 1802 he received a diploma in medical sciences. He had to decide about his future, whether to be a medical practitioner or to enter the scientific profession, and he chose the latter. He obtained a post as unpaid assistant at the School of Surgery in Stockholm, where physicians (army doctors) were trained. Here he was free to carry out his research, but he could not afford to support himself. He invested his small capital in a scheme to produce artificial mineral water, collaborating with a man called Werner. Berzelius also wrote a paper on artificial mineral waters about this time. During this time he lived at the house of Werner.

Berzelius made friends with a foundry owner named Hisinger, who was interested in chemistry. At this period the Volta-column was the centre of interest, even though it had only been discovered a few years earlier and experiments were being carried out by chemists, physicists and physicians all over the world. Afzelius had an apparatus consisting of 60 discs, with which Berzelius carried out a few experiments during his university years. Berzelius examined the effect of the column for the treatment of different diseases, and wrote an account of this in one of his examinations. Together with Hisinger he examined the effect of the column on various salts, and in the course of their examinations they established that the combustible substances, the alkalis and the earths migrated to the negative pole, and oxygen, acids and oxidized products migrated to the positive pole. They concluded that salts decompose to metal oxide and acid anhydride. This communication was published in 1803 in German, under the name of Hisinger and Berzelius [39]. This article contained a great deal of the work which

Davy was to publish in his famous article three years later. (Davy's discoveries were by far the more important, however, because he established that the alkalis were compounds and was able to isolate the alkali metals from them.) This article shows the formation of the ideas which later led to the dualistic electrochemical theory.

At about this time he carried out a series of analyses of various parts of the human body and on body fluids, with the intention of writing a textbook for his students dealing with animal chemistry which is in fact physiological chemistry. He worked together with Hisinger on this and other problems, and in an investigation of the mineral felspar they discovered the oxide of a hitherto unknown element, cerium. Berzelius immediately sent a report of this discovery to Gehlen, in order for it to be published, but received the disappointing reply that he had only just received a communication from Klaproth describing the discovery of this oxide which he called ochroit earth. Klaproth's article was printed in the edition which had been published just prior to the receipt of Berzelius's communication [40].

As Berzelius still had no source of income he took a post as an official physician for the poor, which although it entailed the inconvenience of daily consultation brought in an income of 66 thalers a year [41]. In collaboration with Hisinger he published a journal [42], as a protest against the official publication of the Academy, and although this survived for six years it left him with a large debt. He also attempted to establish a factory for making vinegar, with the assistance of his partner Werner, and borrowed money to cover the cost. Berzelius recalls,

> I am sorry to say that neither he nor I had ever seen a vinegar factory before, and what was even worse was that I had no talent for applying science to industry, a disability which was to cost me a great deal of money during my lifetime. In our factory the acetic acid was only formed very slowly and was not strong enough [43].

Finally when they were heavily in debt Werner absconded to Russia, and left a deficit of 1,000 thalers for Berzelius to pay. For ten years almost all his income went to his creditors.

In 1807 the Professor of the School of Surgery [44] died and Berzelius was appointed as his successor. The salary of 166 thalers for this post supplemented his physician's salary. During this time he started to write his famous and most significant book: *Lärbok i Kemien* (the "Schoolbook of Chemistry"), the first part of which was published in 1808, but the last part not until 1818. In this book it is interesting to observe how Berzelius's conception of electro-chemical dualism and of the law of multiple proportions gradually crystallized with each succeeding volume. His financial position during this time improved slowly, and with the start of the war against Napoleon, the School took on a new character. More students were enrolled for shorter training times, and the professor, as any other army officer, received full pay, namely twice the peace-time salary. Later this institute became a Medical Institute and Berzelius, because of his better salary, was able to relinquish his physician's post.

In 1808 Berzelius was elected a member of the Swedish Academy of Science. During the war with France he had served on a committee which had been responsible for examining the possibilities of making saltpetre. (According to Berzelius the research was successful, but in 1818 the committee was dissolved, as

Fig. 43. Jöns Jakob Berzelius (1779—1848). Drawing by Krüger, 1842

the war had ended four years previously [45].) In 1812 Berzelius made a trip to England, where he made the aquaintance of many scientists, among them Davy, Watt, Wollaston, Herschel and Tennant [46]. All this time Berzelius worked very hard and produced a constant stream of publications, mostly dealing with the establishment of atomic weights.

Encouraged by Gahn, in 1808 he became a partner in a factory making sulphuric acid and vinegar, which they obtained for a very low price. Under the management of a talented young chemist the factory was profitable, but unfortunately in 1826 it was burned down and proved to be a great financial loss for Berzelius. Consequently, the only profit he obtained from this venture was the discovery of selenium in the mud around his factory. In 1818 Berzelius was raised to the nobility by the king, and in the same year he had a nervous breakdown, probably caused through overwork. For the sake of his health he was advised to take a trip abroad and visited Paris and England, and then travelled through France, Switzerland and Germany. The government gave him 2,000 thalers to finance his trip on the condition that he studied the techniques of gunpowder processing abroad. During his travels he met many famous scientists including Laplace, Gay-Lussac, Vauquelin, Berthollet, Dulong, Arago, Biot, Ampère, Chevreul, Thénard, Haüy, and Humboldt, and learned a great deal from them, particularly on the style of lecturing and lecture demonstrations [47]. He was invited to become director of the chemical institute in Berlin, in succession to Klaproth, but declined the offer. In 1820 he was elected secretary of the Swedish Academy of Science, a position which he held until his death. The Swedish Metallurgical Society awarded him a pension of 500 thalers a year in recognition of his work in analysis.

By this time Berzelius was in his forties, and having risen equally in rank, reputation, and financial position was now at the peak of his creative ability. His entire contribution to chemistry was recorded in his textbook which was translated into most European languages, and his dualistic concept wove the whole fabric of chemistry into an attractive and unified whole. No flaw in this pattern had yet been detected. At this time Berzelius was subject to severe attacks of migraine, and went to Karlsbad in search of a cure. There he met the poet Goethe, and in a dispute with him over the origin of volcanos proved that the poet's theory did not agree with the known facts [48]. In the following years he did not work quite so hard, although from 1821 onwards he published his famous *Jahresbericht* annually, in which he reported on the progress of the natural sciences. Between 1820 and 1840 Berzelius made many journeys in Europe. In 1832, after thirty years of teaching, he retired on a pension, and devoted himself to his scientific profession and to his work as secretary of the Academy. However, two years later in 1834 a violent cholera epidemic broke out in Sweden, and Berzelius was recalled to be President of the Public Health Committee.

Until now Berzelius's life had been dedicated to science. He did not marry, and lived alone. He writes that when he was much younger he had consulted a famous scientist (he does not give his name) who was much older than he. He asked him whether, from his experience, it was beneficial for a dedicated scholar to marry? The older scientist was very reluctant to advise him on this matter, but did say that for his own part, although having had a very happy marriage, if he was confronted with this problem again, knowing as he did the everyday problems which occur in family life, he would give the matter very careful consideration before making a decision. As a result of this conversation Berzelius says that "I then decided to give up any thought of marriage forever, and this decision has continued with me ever since" [49]. But as the years went by, and especially after seeing the ravages caused by cholera which was responsible for the death of so many, his opinion gradually changed and he wrote "I feel now for the first time that I am alone, and it would be better to get married just as soon as my financial position is secure" [49]. From his memoirs we can see that he was rather diffident in carrying out his plan, and awkwardly asked for advice from many of his friends. Eventually, however, he married the eldest daughter of his old friend Poppius who was a judge. He announced his forthcoming marriage to Liebig with some slight embarrassment and wrote

> the life of a bachelor especially in old age is a lonely and unhappy one, and this situation can only be avoided by getting married. My fiancee is much younger than I; she is in fact nearly 25 years of age, but she is considerably more intelligent than most older women, so I hope that this will compensate for at least one decade in our age difference [50].

Berzelius was married on the 19th December 1835, at the age of 57, and to mark the occasion the king honoured him with the title of Baron. Liebig wrote to him to congratulate him with a rather flattering but slightly ironic letter.

> You are now a married man. You are enviable because if you had married thirty years ago you would now have an old wife who could not make your life fresh with youth, and surround you with flowers. How well and happy you feel now when tender care fulfils all your needs. How miserable was your life before, being tired of work and unsatisfied by science! You had nobody who could make your evenings cheerful! How different is your life now! You can see now that there is no friend to compare with a loving wife ... I have been told that your wife is more beautiful, intelligent and cultured than all the other ladies of Stockholm. As a matter of fact I myself would have been happy to be your wife if nature had not predestined me for wearing trousers, because you have all the qualities required to make a lady happy and satisfied ... [51].

Berzelius's life became less strenuous, and although he still travelled it seemed as though his former vigour did not return. He makes a comment on this in a letter written in 1837:

> My health, thank God, is still as good as it used to be, and I pass the time happily, but that internal irrepressible force which drove me in my scientific study has now ceased. Sometimes stimulating ideas occur to me, but now instead of being fearful lest someone solves the problem before me, I prefer to allow someone else to carry out the experiments. Even though I wish to know the result passionately I have not the strength to carry it out myself. This I suppose is the sign of old age, which sooner or later comes to us all [52].

About this time he was awarded a pension of 2000 thalers by the Government, a considerable sum indeed in those days.

Meanwhile some flaws became apparent in the theories of Berzelius. Dumas established that hydrogen can be substituted by chlorine in organic compounds, whereas according to the electrochemical theory it would be impossible for the role of a negative element to be played by a positive one. Liebig presented his theory on hydrogen acids, and Daniell discovered that salts on electrolysis do not decompose to a mixture of metal oxide and acidic oxide, but to metal and acid residue. Berzelius attempted to defend his ideas in letters and in his *Jahres-*

bericht, first of all against Liebig, who was 25 years younger and who was later to become the leading figure in the field of chemistry. In the course of these arguments Liebig often became vehement and was also extremely tactless.

At this time Liebig was the editor of a journal called *Annalen der Chemie und Pharmazie* (today known as *Liebig's Annalen*), to which Berzelius was a frequent contributor. As Liebig was the editor he simply omitted those parts of Berzelius's papers which he considered obsolete or with which he disagreed, or on occasions made comments on them. This treatment quite naturally angered Berzelius who was at the same time engaging in disputes with Dumas and Faraday. Liebig commented on these arguments in a letter to Wöhler.

> Berzelius has been asleep while we have been working and the reins have now passed out of his hands. Now he is awake, but when an old lion whose teeth are not sharp roars not even a mouse is afraid [53].

Berzelius's health declined rapidly; he suffered from gout and apoplexy and finally his legs became paralysed. In a letter to Liebig, Wöhler mentioned that Berzelius's friends in Stockholm had denied the rumour that Berzelius had written his last *Jahresbericht* while in a fit [54]. Berzelius died in a wheel-chair while reading, at the age of 69 on 7th August 1848.

The most characteristic features of Berzelius were his tremendous capacity for hard work, always working to a plan, and his unbelievable energy. Berzelius produced no shining discoveries like Priestley or Davy, nor did he enunciate efficient hypotheses like Stahl or Dalton. His hypotheses were developed slowly, and were the result of much hard work. The most important of his hypotheses was the dualistic electrochemical theory which outlived nearly all the other hypotheses of science, in that it reigned and was discarded and finally after many changes began a second life in Arrhenius's hypothesis. The essence of the dualistic electrochemical hypothesis can be best illustrated by Berzelius's own words:

> Atoms contain both types of electricity, these being placed at different poles in them, but one type is dominant. Affinity is due to the effect of the electrical polarities of the particles. Thus, all compounds are composed of two parts, these parts differing in the nature of their electricity and are bound together by attraction. All compounds can therefore be divided into two oppositely charged parts irrespective of the actual number of elements from which the compound is composed. For example, sodium sulphate is composed of sulphuric acid and alkali, both of which are capable of further subdivision into an electropositive and electronegative constituent [55].

Of greater importance than this hypothesis is the practical contribution of Berzelius's work to the determination of the atomic weights of the elements. The analytical implications of this work are dealt with later, but first we must consider the circumstances under which this work was carried out.

At the time this work was done very few chemicals were available commercially; amongst these were sulphuric acid, a few metals and metal oxides, soda, sulphur, phosphorus, borax, ammonium chloride, rock salt, extracts of various plants, and a few others. Everything else had to be prepared from these. Berzelius even had to prepare his own hydrochloric acid from sodium chloride and sulphuric acid as the commercially available reagent was insufficiently pure. In 1808, only one platinum crucible existed in Sweden and this was owned by Hisinger. Although he was kind enough to lend this to Berzelius, it was found to be too heavy for Berzelius's balance [56]. Even when he received three porcelain crucibles which were an innovation in those days, the situation was little improved, Berzelius wrote "I have just received 3 porcelain crucibles from Heinrich Rose. Unfortunately, one was broken when they arrived, one I have had to give to Wachtmeister, and so I now have only one for myself" [57]. He had no help either personal or (apart from his salary) financial, and his chemicals and apparatus were bought with his own money. Even in his most active period, referred to previously, his financial position was desperate. From 1837 however he employed an assistant whom he paid himself [58].

Wöhler, who worked under Berzelius during the years 1823 and 1824 gave a description of Berzelius's laboratory in his memoirs. When, after a long journey, he arrived at Stockholm, he presented himself before his future master who at that time was secretary of the Academy and a famous scientist. He writes:

> I could hardly wait until the following morning to visit Berzelius who lived in a house belonging to the Academy. I rang the bell with a trembling heart. A well built and finely dressed gentleman opened the door; it was Berzelius. He accepted me kindly and said that he had been expecting me for a long time. He spoke the most perfect German as well as French and English. When he showed me his laboratory I thought that I must be dreaming, I could hardly believe that I had actually arrived in the place of which I had dreamt so much ... The next day I began work. I was given a platinum crucible, a balance and weights, and a washbottle for my own use. First of all I was instructed to make a blow-pipe, the use of which Berzelius considered was very important. The spirit used for the burners and the oil used for the glass blowing table I had to buy myself; the more common chemicals and glassware were available, but ferricyanide could not be obtained anywhere in Stockholm and I had to order it from Lübeck. I was the only one working there apart from Berzelius himself, but previously Mitscherlich and Rose had been working there. After I left Magnus came to replace me. The laboratory consisted of two ordinary rooms and contained only the most simple furniture. There was neither a furnace, hood, water nor gas in it. In one of the rooms there were two ordinary pinewood tables; at one of these Berzelius worked, while I worked at the other. On the walls there were a few shelves with reagents on them, and in the middle of the room there was a glass blowing table and a mercury vessel. In one corner there was a washing place, consisting of a stone-vessel with a tap and a pot where Anne, the housekeeper, washed the dishes every morning. In the other room there was a balance and a large cupboard containing instruments. In the kitchen where Anne did the cooking there was a small furnace, seldom used, and a sand-bath which was kept continuously heated. There was also a small workshop with a lathe.
>
> Berzelius was generally kind and would talk a great deal while working; he was also fond of telling jokes and of listening to them. He suffered quite frequently from headaches, and on these occasions he would not come to the laboratory for one or two days. His only other absences were when he was travelling and when he wrote his yearbooks.
>
> To start with I was given a zeolite to analyse, but he carried out the analysis himself first to teach me the method and a little technique. My next sample was a liverite, and I was made to repeat the analysis until the results became constant ... [59].

Berzelius's most important work was carried out under these conditions during the years 1807–18. His papers were written in a dry objective style, and only described the operations and the results. They were published in many different periodicals and journals, but his complete works were never published in one volume [60].

Berzelius taught only a few students, and these were not chemists but physicians. It is not known whether they were instructed in laboratory practice. He only had one, or at the most two, pupils in his laboratory, usually Swedes but occasionally Germans. They were first taught analytical chemistry, and it can be seen not without success. Nearly all of them were later to become leading figures in the development of chemistry. This was partly due to the fact that their instruction was so thorough, and partly to the fact that to have studied under Berzelius was a good recommendation. Berzelius looked after his pupils very well, even after they had left he still wrote to them, and whenever possible helped them in their careers. This extract from one of his letters shows this quite clearly.

> In the Spring I am to give up my position at the Carolina Institute (this was the medical school), although I shall only receive a pension from 1834 instead of my income. But I think this is my duty because otherwise Mosander, my student, would grow as old as my senior lecturer [61].

Mosander [62] in fact became his successor. He also recommended Mitscherlich to the Prussian Minister of Culture for the post of Professor at the University of Berlin, and Wöhler for a similar post at Göttingen, and Magnus as Professor at Berlin University. Gmelin [63], Mitscherlich [64], Heinrich and Gustav Rose [65], Arfvedson [66], Sefström [67], Wöhler [68], Magnus [69], Lagerhjelm [70], and Berlin [71] were his most famous students.

In addition to scientific and financial success Berzelius received most of the social awards of that time. He became a member of the Upper House, and was also a member of most of the scientific academies of Europe. He was received by the Kings of France and England, while the King of Sweden and the Czar of Russia visited him in his laboratory. He received awards from many countries, but in spite of all this success he remained modest, disliking flattery intensely, and led a quiet life. Liebig once wrote to Berzelius asking to be allowed to dedicate one of his books to him. The draft of the recommendation was enclosed written in a very effusive and flattering style [72]. Berzelius replied, thanking him for the dedication but added

> I can allow the dedication provided that the text that accompanies it is not printed. It is not advisable to write to me in this manner, as I always have the impression that anyone who praises so extravagantly must themselves be rather vain [73].

He also refused a request from Rose, who wanted to name a selenium-containing mineral "berzelite".

> I am an enemy of the bad fashion coming from England that minerals are named after living persons. All mineralogists introduce their friends into mineralogy, and as a result of this the most complicated and stupid names are formed. At the same time the desire to do this service for as many friends as possible results in many known minerals being renamed [74].

He objected also to naming elements in honour of various people, and when there was a suggestion to call tungsten "Scheelite" after Scheele, and a newly discovered element "Gadolinium" (this name was later to be adopted), he wrote:

> It is always ridiculous if personal names are introduced into chemical compounds, especially if the memory of the person is still fresh. How ridiculous, for example, is a name like gadoline scheeleate!... [75]

Berzelius had many great literary achievements, his textbook being edited five times during his own lifetime. He also wrote books on the theory of electrochemistry [76], mineralogical systems [77], and the previously mentioned book on the use of the blow-pipe [78]. In addition to these his *Jahresbericht* appeared regularly for 27 years, and after it was submitted to the Academy it was translated from Swedish into English, German and French. Each of these issues contained about 700—1000 pages. The material in this book was divided into chapters, each chapter dealing with the results of his research work done during the preceding year in various fields, inorganic chemistry, mineralogy and plant and animal chemistry. Apart from the presentation of the results, these works contained comment, criticism and discussion of the problems encountered. These reviews enhanced the already considerable prestige that Berzelius enjoyed, and that their publication was awaited with eagerness is confirmed by the letters of Liebig and Wöhler. After the death of Berzelius his *Jahresberichte* were continued in Germany, under the leadership of Liebig and with the help of many of his co-workers, and they were to appear regularly under different editors until 1912 when it became impossible to summarize critically the mass of rapidly expanding chemical literature.

Berzelius was involved in correspondence with all the leading scientists of his day, exchanging letters with Davy, Berthollet, Proust, Dalton, Liebig, Wöhler, Mitscherlich, Magnus, Rose, etc. Most of these letters have been published [79]. His correspondence is of great importance to historians as it gives an interesting picture of both the social and scientific life of those times. Berzelius was an accomplished letter-writer with a sense of humour, and his letters make interesting reading even today. The Swedish Literary Academy elected Berzelius a member because of his contributions to the development of Swedish technical language. On reading some of his books this election would appear to be justified.

As an example of the ingenuity of the style of Berzelius the following example is quoted. Wöhler had examined a mineral and thought that he had prepared a new element from it. He sent a sample of this to Berzelius with a letter saying:

> This substance I have called X was obtained from pyrochlor, and is either tantalic acid, in which case tantalic acid has some hitherto unknown properties, or alternatively it is the oxide of a new metal.

Berzelius later returned it with the following reply:

> Here I return your substance X. I interrogated him to the best of my ability, but he gave me inconclusive replies. I first asked him 'are you titanium?' and he replied 'Wöhler has already told you that I am not titanium.' 'Then are you zirconium?' I asked him. 'No!' he replied, 'because with soda I form a white glass, whereas zirconium does not.' 'Perhaps then you are tin?' I asked him. 'There is a little of the character of tin in me' he replied. 'Surely then you are tantalum!' I then asked him, but your X replied, 'I am related to tantalum, but potassium bisulphide slowly dissolves me and I am precipitated from this solution to give a brownish yellow coloured precipitate.' 'Who are you then, you son of a devil?' I asked him finally. 'I am afraid that I do not have a name yet' he replied. I am not quite certain about this, however, because he said it from my right hand, and as you know I am rather deaf in my right ear. Therefore, I am sending you back this fellow. Your ears are better than mine, try to examine him once more [80].

This, however, was unnecessary as Rose had, in the meanwhile, discovered this element and named it niobium [81].

Berzelius was also a very good, if somewhat malicious, judge of character. When he was a young man he visited Davy in London and describes his visit as follows:

> The French butler showed me into the dining room, where the breakfast table was already prepared. There I was left alone for a time, apparently in order for me to admire the luxury of my surroundings. The first impression I gained was one of a man who has acquired great wealth, and who has not yet adjusted his style of living to suit his circumstances. The butler then opened the door, and Sir Humphry entered. This impression did not last for very long for very soon Sir Humphry relaxed and became a very interesting chemist, called Davy . . . [82].

At times, however, his views were expressed openly and often with vehemence; in a letter to Liebig, for example, he writes,

> My dear Liebig, I would like to ask you, without any ill-feeling, to cease to be a chemical torturer. Your reputation is sufficient to stand alone, and does not need the demolition of other peoples reputations to enhance it [83].

He corresponded with Liebig over a long period, but their differences of opinion gradually became greater until personal issues became involved, and finally they disputed quite openly and publicly. Although in the majority of their disputes over scientific matters the younger Liebig was correct, it should be stated that he was also the more aggressive, and his passionate nature often caused him to overstep the barriers of proper delicacy and to become offensive towards the older scientist. We have already seen from two of his letters that some of his comments were very bitter, and even his good friends Wöhler and Rose cautioned him about his violent nature.

Berzelius attempted to moderate the disputes for a long time and to avoid any personal issues becoming involved. He states,

I have given my views, and if they are valid, they will be known for a long time without my doing anything further for them, while if they are wrong, they are bound to fail, even if I try to defend them ... [84].

You write that you do not know whether you can regard me as a friend in the future or not? ... You complain that I have turned away from you. In this there is some truth, but it is not true that I am not still your friend... I cannot demand that my friends shall have the same views on scientific problems as I have. Similarly, I cannot allow reasons of personal friendship to prevent my speaking out when I am in disagreement with your views [85].

Finally Berzelius's patience came to an end, and in the final review he published (1848) he condemned Liebig, asserting that

One professor (i.e. Liebig) uses his students only to disprove the results of older chemists and to publish such types of papers, even though the results are not always conclusive. These papers are published in the name of his students, because should they prove to be incorrect, the responsibility of publishing untrue scientific results would then lie on a young man, and not on the professor himself [86].

In this chapter I have tried to show Berzelius as a scientist and also as a man, if indeed this is possible after a century and a half, and although I have been concerned with his personality to a large extent, I think that it is justified in view of the extraordinary role which he played in the progress of analytical chemistry.

A very fitting tribute to Berzelius, and indeed one which sums up his achievements most plainly, was made by his student Rose in a memorial lecture to the Berlin Academy.

He was a man who showed exceptional qualities in his research work, and enriched all branches of science with important data as well as with many practical and theoretical discoveries, and summarized the whole of his work with the spirit of a philosopher. He treated his subject systematically, and enlightened it by the power of his critical examination. He was an incomparable example of a teacher of theoretical and practical chemistry who taught and inspired a large number of students. This man fulfilled the highest requirements of science to such an extent that his personality shall light as a glittering example for many centuries ... [87].

As a final tribute to Berzelius the comment of the young Bunsen who, after meeting Berzelius, said:

When one gets to know him well it is difficult to know what to admire most about him, whether his ingenuity, or his shyness, or his good heart! [88].

3. THE ESTABLISHMENT OF ATOMIC WEIGHTS

In 1807 while Berzelius was writing a textbook for students of medicine he consulted, amongst others, the works of Richter. He writes:

Of the several not very well known books that I consulted, the work of Richter was remarkable for the light which it shed on the subject of the composition of salts and the precipitation of metals. And yet nobody has taken any advantage of it! On the basis of the analytical results of Richter, it is obvious that from the composition of some salts the composition of other salts can also be established [89].

He resolved to continue the work of Richter.

When Dalton published his atomic theory and his law of multiple proportions, Berzelius had by then obtained many results which supported the new theory. He realized, however, that the atomic numbers of Dalton were not accurate enough to enable his theory to be of any use in practice, and that in order to prove its validity a large number of analyses must first be carried out. Berzelius decided to carry out this work, but soon found that it involved many problems. He first of all tried to apply the existing analytical data, but realized that these were not accurate enough and he had to start the work afresh.

For the determination of atomic weights two separate quantities had to be established in each case. First, the number of atoms in the compound had to be determined, and then the relative weights of the single atoms had to be measured. Berzelius recommended the use of oxygen as a standard instead of hydrogen (recommended by Dalton), because the determination of atomic weights was based mainly on the analysis of oxides, and because oxides are much more common than hydrides. He chose 100 as the atomic weight of oxygen, so that by adding 100 or a multiple of 100 to the atomic weight of the metal, the "atomic" weight of the compound could be found [90]. In some cases he calculated his atomic weights using hydrogen as a standard because in many places, especially in England, this convention was used. But Berzelius chose as his unit $2H = 1$, because opinion in England regarding the law of reacting gases was that one atomic weight of hydrogen was present in two volumes. Thus the values he obtained were one half of the present day values. The most difficult problem facing Berzelius was the determination of the number of atoms present in a compound. In the case of gaseous substances this was a relatively easy task as it was only necessary to find the volume ratios of the reactants from which the gas was formed. Berzelius was of the opinion (and this was before Avogadro!) that equal volumes of permanent gases at the same temperature and pressure, contain an equal number of atoms [91]. There was a slight uncertainty about this statement, as some English scientists were of the opinion that the combustible gases (H_2, N_2, Cl_2) only contained half as many atoms per unit volume as are present in the other gases. (Thus, the atomic weights of these gases were uncertain for half a century.) Berzelius established that salts are so composed that the number of oxygen atoms derived from the acid is a multiple of the number derived from the base [92]. Experience had shown that oxides are only composed of a very few ratios (1 : 2, 1 : 1, 2 : 4, 2 : 3) so that the number of possible atomic ratios was limited. The relative weight, i.e. atomic weight, of gases could easily be determined by vapour density measurements, and in the case of solid substances the atomic weight was easily calculated from a knowledge of the number of atoms present and from the composition. The determination of the number of atoms in a compound, however, was in many cases incorrect, so that the measured atomic weight proved to be either a fraction or a multiple of the true value. It was due to this fact that Berzelius often altered his atomic weights in the light of new experimental evidence.

Two important discoveries made about this time greatly assisted Berzelius in his correct choice of atomic weights. The discovery of Mitscherlich that compounds which contain the same number of atoms and have similar structures, exhibit similar crystal forms (isomorphism) was one of these, while the other was the law of Dulong and Petit [93], which stated that the product of atomic weight and specific heat for a metal was constant. (Berzelius did not regard Dulong and Petit's law to be valid in all cases, and was not inclined to alter his results to fit it.)

Whilst Berzelius was occupied with this work the English physician Prout [94] proposed a new theory which asserted that all the atoms of the different elements are composed of different numbers of hydrogen atoms, and therefore their atomic weights must be multiples of the atomic weight of hydrogen [95]. This was a new form of the ancient idea of an ancestor element. Prout's hypothesis was one of many discoveries in the natural sciences which were not proved until comparatively recent times, when it has been shown that Prout was practically correct in his assumptions. This theory found considerable support immediately it was published, because many of the atomic weights which were known at that time were multiples of the atomic weight of hydrogen. One of the supporters of this theory was Thomson who published the results of some work which he had carried out and claimed that they confirmed the theory of Prout [96]. This work did not greatly assist the acceptance of Prout's theory as the experimental atomic weights were, in actual fact, whole numbers. With reference to this work Berzelius made the observation, "The impression obtained was that in his analytical investigations he was led by the desire to obtain previously established results". It was not difficult for Berzelius to disprove Thomson's results by his own careful work! How could the unfortunate Thomson know that the natural elements are mainly mixtures of isotopes!

A few examples of Berzelius's methods of determining atomic weights will illustrate the problems involved. Of the three examples given, the atomic weight of chlorine was accurate, that of fluorine rather uncertain, and for silver the value obtained was incorrect.

> *Oxygen.* Its atomic weight must be taken as 100. According to the measurements of Dulong which correspond to mine its specific gravity is 1·1026, whereas that of hydrogen is 0·0688. Therefore, if we regard the double hydrogen atom as a unit, the atomic weight of oxygen is 8·013 [97].
>
> *Chlorine.* I established its atomic weight by the following experiments: (1) From the dry distillation of 100 parts of anhydrous potassium chlorate, 38·15 parts of oxygen are given off and 60·85 parts of potassium chloride remain behind. (Good agreement between the results of four measurements.) (2) From 100 parts of potassium chloride 192·4 parts of silver chloride can be obtained. (3) From 100 parts of silver, 132·75 parts of silver chloride can be obtained. If we assume that chloric acid is composed of $2Cl + 5O$, then according to these data 1 atom of chlorine is 221·36. If we calculate from the density of chlorine determined by Lussac (5·4252), the chlorine atom is 220. If it is calculated on the basis of hydrogen then it is 17·735 [98].
>
> *Fluorine.* Fluorine I have considered as a salt-forming substance, and I have determined its atomic weight on the basis of this. 100 parts of pure calcium fluoride gave in three separate experiments 174·9, 175·0, and 175·1 parts of calcium sulphate respectively.

The mean is 175·0, and from this when the calcium has been subtracted, the residue is fluorine. Whether there are one or two fluorine atoms in the compound it cannot be ascertained until it is known whether one or two atoms of hydrogen are combined with one atom of fluorine in hydrogen fluoride. We can suppose that fluorine behaves similarly to chlorine of bromine, but we cannot be certain. Gypsum is $\overset{..}{Ca}\overset{..}{S}$ so if we calculate on the basis of this composition we find 259·019, i.e. one atomic weight of calcium is combined with 233·81 fluorine. This corresponds to either one or two atoms of fluorine. If we consider it as two, then the atomic weight of fluorine is 116·90, or calculated for the double hydrogen atom, 9·367 [99].

Berzelius suggested the use of the first letter of the element's name as a symbol. He represented oxygen however by placing points over the element, and sulphur in many cases by hyphens similarly placed. For example $\overset{..}{Ca}\overset{..}{S}$ represents $CaOSO_3$

Silver. 100 parts of silver, as was mentioned in the determination of chlorine, form 132·75 parts of silver chloride. I have reason to suppose that its composition is $AgCl_2$, and further that silver oxide consists of one atom of silver and one atom of oxygen, because this is usual with the strongly basic metal oxides. The ease with which silver superoxide gives up oxygen indicates that the superoxide contains more oxygen than Ag + O. In this case the atomic weight of silver is 1351·607, or 108·305 calculated on the double hydrogen basis.

It must be noted however that Dulong and Petit obtained half this value in their experiments on specific heat. They found that the specific heat of silver was 0·0577 which, when multiplied by $\dfrac{1351·607}{2}$ gives the required value of 0·3764. It could be said that the composition of silver oxide is analogous to those of copper and mercury. Rose found that it has a lower degree of sulphurization (Cu_2S) and thought that silver sulphide, similar to copper sulphide, consists of two atoms of the metal and one atom of sulphur. But when we consider that silver oxide is so basic that it even gives an alkaline reaction on reagent paper, which does not occur with those weak bases that contain two atoms of metal and one atom of oxygen and at the same time if we compare silver with lead, from a consideration of the specific weights of their chlorides it would seem rather curious if the atomic weight of silver were half that of lead, so that in silver chloride twice as many metal atoms would be present as in lead chloride ... [100].

With this argument Berzelius finally accepts his original data, although the correct value is only a half of this, as was shown by the experiments of Dulong and Petit.

The first atomic weight table of Berzelius was published in 1814 [101].

The following table is taken from a later German issue [102]. These values mostly agree with those of the 1818 Swedish issue, but contain many changes from the original table. Its significance is that it shows the atomic weights calculated on the basis of the double hydrogen atom. Those data are correct therefore, which show half the present day values of atomic weights. In the 1814 table the following elements were still presented with double the values given below: As, Cr, W, Sb, Au, Hg, Ag, Cu, Ni, Co, Pb, Sb, Fe, Mn, Al, Mg, Ca, Sr, Ba, Na, K.

Nomina		Formulae	Pondera atomorum.		Partes centesimales.		
			O = 100.	H = 1.	+ E.	− E	H vol H.
Chloretum Cericum		1/2	825,78	66,17			
-	Cerosum	Ce Cl	1017,35	81,52	56,49	43,51	
-	Chromicum	Cr Cl³	2031,59	162,79	34,63	65,37	
		1/3	677,20	54,26			
-	hyper Chromicum	Cr Cl³	1237,12	99,13	28,44	71,56	
		1/2	618,56	49,57			
-	Cobalticum	Co Cl	811,64	65,04	45,46	54,54	
-	Cupricum	Cu Cl	838,35	67,18	47,20	52,80	
-	Cupricum tri basicum	Cu Cl + 3 Cu	2325,43	186,34	Cu 17,02	Cl 19,03	Cu 63,95
-	- - - c. aqua	Cu Cl + 3 Cu H	2662,87	213,38	Cu 14,86	Cl 16,62	H 12,67
					Ċu = 55,85		
-	Cuprosum	Cu Cl	1234,04	98,88	64,13	35,87	
-	Ferricum	Fe Cl³	2006,36	160,77	33,81	66,19	
		1/3	668,79	53,59			
-	Ferrosum	Fe Cl	781,86	62,65	43,38	56,62	
-	Glucinicum	G Cl³	1990,48	159,50	33,29	66,71	
		1/3	663,49	53,17			
-	Hydrargyricum	Hg Cl	1708,47	136,90	74,09	25,91	
-	Hydrargyrosum	Hg Cl	2974,30	238,33	85,12	14,88	
-	Iridicum	Jr Cl²	2118,80	169,78	58,22	41,78	
		1/2	1059,40	84,89			
-	nyper Iridicum	Jr Cl³	2561,45	205,25	48,16	51,84	
		1/3	853,82	68,42			
-	Iridosum	Jr Cl	1676,15	134,31	73,59	26,41	
-	hyper Iridosum	Jr Cl³	3794,95	304,09	65,01	34,99	
		1/3	1264,98	101,36			
-	Kalicum	K Cl	932,57	74,73	52,53	47,47	
-	Lithicum	L Cl	522,98	41,91	15,36	84,64	
-	Magnesicum	Mg Cl	601,00	48,16	26,35	73,65	
-	Manganicum	Mn Cl²	2019,73	161,84	34,25	65,75	

FIG. 44. Pages from Berzelius's tables of atomic weights. (From his *Lehrbuch der Chemie*)

Although there are several incorrect quantities in this table, nevertheless the majority approach the values of today quite closely, and they reflect great credit on Berzelius for the precision of his work; in fact, the work he carried out during those ten years would be a creditable accomplishment even for a modern, well equipped, research laboratory. In the same book Berzelius records about 2000 formulae of various compounds, their atomic weights and composition, according to the dualistic electrochemical theory [103]. These data are given in the same volume as the second table of atomic weights, and this table contains one or two alterations. The atomic weight of bromine is given as 39·20, while chromium is given as 69·39, but no explanation is given for these changes [104].

After Berzelius had presented his table of atomic weights, several other workers, using different methods, obtained values for some elements which did not agree with those of Berzelius, and numerous modified and invariably conflicting tables of atomic weights appeared. During the first half of the last century there was great confusion over the values of atomic weights; on one side were the group who accepted Berzelius's values, whilst the opposing faction used values which were exactly half these values, so that one chemist called an atom what another

TABLE 8

ATOMIC WEIGHTS OF BERZELIUS

		Fe;	64·25	Cd:	55·83
O:	8·013	Mo:	47·96	Fe:	27·181
P:	15·717	Ti:	24·332	Zr:	33·67
F:	9·367	Pa:	53·359	Mg:	12·689
Se:	39·631	Cu:	31·767	Li:	6·44
V:	68·58	Pb:	103·729	S:	16·20
Ta:	92·448	Co:	29·568	I:	126·567
Ir:	98·841	Th:	59·646	Si:	22·221
Hg:	101·431	Al:	12·716	Sb:	64·662
Sn:	58·91	Ba:	68·663	Ag:	108·305
Ni:	29·622	N:	7·093	Zn:	32·311
Ce:	46·051	Br:	78·392	Y:	32·254
Be:	26·544	B:	21·828	Na:	23·31
Sr:	43·854	As:	37·665	Cr:	28·191
K:	39·257	W:	94·795	Os:	99·722
2H:	1	Au:	99·604	Bi:	71·07
Cl:	17·735	Rh:	52·196	Mn:	27·716
C:	6·12	U:	217·26	Ca:	20·515

called a molecule. Gerhardt [105] introduced the concept of equivalent weights' but this only made the situation even more involved, so that the formula of water was variously given as H_2O, HO, or H_2O_2. The development of atomic weights is strictly beyond the scope of this book; it is sufficient to note that Cannizzaro [106] unified the conflicting concepts and values at the international chemical congress held at Karlsruhe in 1860. His proposals were widely accepted, the only exceptions being the French chemists, and even they eventually came to accept Cannizzaro's views. Although the values of the atomic weights are subject to slight changes, even at the present day, this has little influence on the progress of science. It can be concluded therefore that the basis of the atomic weights, as well as chemical formulae, and the development of analytical calculations based on stoichiometry, can be attributed to Berzelius today.

4. THE STATE OF ANALYTICAL CHEMISTRY IN THE AGE OF BERZELIUS

The chemical analytical methods of Berzelius can easily be assessed from his books. In his textbooks he describes systematically the behaviour of the various substances which he examined. In the last volume which appeared in 1818 he describes alphabetically the most important laboratory equipment, operations and the methods of analysis. As he described many of these in detail one can obtain a good idea of the appearance of a laboratory of that period, and the way in which work was carried out in it. Extracts from Berzelius's book (Vol. 10) [107] will illustrate this.

Vessels and apparatus were mainly made of glass. We know that Berzelius himself was a good glass-blower, and constructed his own apparatus. He invented several pieces of apparatus which are indispensable to the modern laboratory, such as the test-tube (p. 487) and the separating funnel. Berzelius used sand- and water-baths for evaporation, but the oil-bath was also in common use at this time. Also the syrupy solution of zinc chloride was often used as a heating bath for temperatures up to 160°C. Above this temperature, however, hydrochloric acid might be given off (p. 225.). There was also a metal-bath, which consisted of a low melting-point alloy, but this was not suitable for heating platinum crucibles. The clay crucible was not then in use, and the porcelain crucible was still a rarity in those days, so that the platinum crucible was used mostly for the ignition of precipitates. According to Berzelius's own text, platinum crucibles cannot be used for the ignition of caustic alkalis, nitric acid salts together with alkaline earth oxides, alkali sulphides, and alkali sulphates with carbon. Metals should not be melted in it, as alloys are immediately formed; one drop of molten lead or bismuth makes a hole in a few seconds. Apart from these, phospates should not be ignited, silicic acid should not be fused with carbon or *aqua regia* used in it, and samples containing manganese should not be heated with hydrochloric acid. Care must be taken to ensure that sooty flames are not used for heating platinum crucibles. Berzelius found that crucibles ignited in this way decreased in weight, and he was able to detect platinum in the soot from the bottom of the crucible (p. 515). For cleaning the crucibles *arena marina* can be used most advantageously, or potassium bisulphate. The prices of platinum crucibles are known from other sources; for example in 1821 a platinum crucible made in London, weighing 3 oz and $5^2/_2$ quentchen (about 110 g), costs £3—18—0. French ones were somewhat cheaper and more suitable for use, as they were generally thicker with more robust walls so that they lasted longer [108]. Porcelain crucibles which were used with increasing frequency were mainly broken on heating, according to Berzelius. The best were the thin-walled crucibles made in Berlin or Meissen. Porcelain crucibles could not be heated effectively with a spirit lamp, so that it was advisable to ignite them inside a platinum crucible. Clay crucibles, although widely used in the past, were hardly used by the chemists of Berzelius's day (p. 518).

Ignition was made in various types of furnaces, in which charcoal was used as a fuel (Fig. 45, 7). By blowing air into some furnaces, such a high temperature was obtained that the platinum residues and the coal were fused together into one mass (p. 455). Flames could also be used for ignition, and vegetable oils or spirit was used for fuel. With the Mitscherlich-type blow flame (p. 279) higher temperatures would be reached. Here ether was used as a fuel, and oxygen was blown through a thin pipe into the flame (Fig. 45, 6). Drying, however, was a difficult problem in those days; it was mainly carried out on a water-bath (Fig. 45, 4), or at room temperature in a desiccator-like vessel, in which the precipitate to be dried was supported over sulphuric acid or calcium chloride (Fig. 45, 5). Before this Berzelius used to dry precipitates by placing the filter paper containing the precipitate on a firebrick (later he used blotting paper) which absorbed the moisture (p. 522).

For filter paper Berzelius used unsized paper. He ordered this specially from paper mills, with the special requirement that it should be made with long fibres, and should be manufactured in the winter and allowed to freeze while wet. This was because the freezing expanded the pores so that it filtered efficiently (p. 259). For quantitative work paper as thin as possible was used. The Lessebo paper was stated to be very good, for 1000 parts of it contained only 1·962 parts of ash (0·2 per cent). The composition of this ash was 60·39 per cent SiO_2, 12·55 per cent CaO, 9·8 per cent MgO, 2·39 per cent Al_2O_3, and 16·08 per cent Fe_2O_3.

This can be subtracted, therefore, practically as pure silicic acid from the result of the analysis (p. 261). The round filter paper Berzelius used to fold similarly to the way it is done today. The filter paper must be dried previously at 100°C, weighed in a platinum crucible, and if the precipitate is dried, its weight must be subtracted from the result. When the paper is ignited the weight of the ash must be subtracted. Funnels were made of glass. They filter most rapidly if their conical axis is 60°. The paper should not stand out from the funnel,

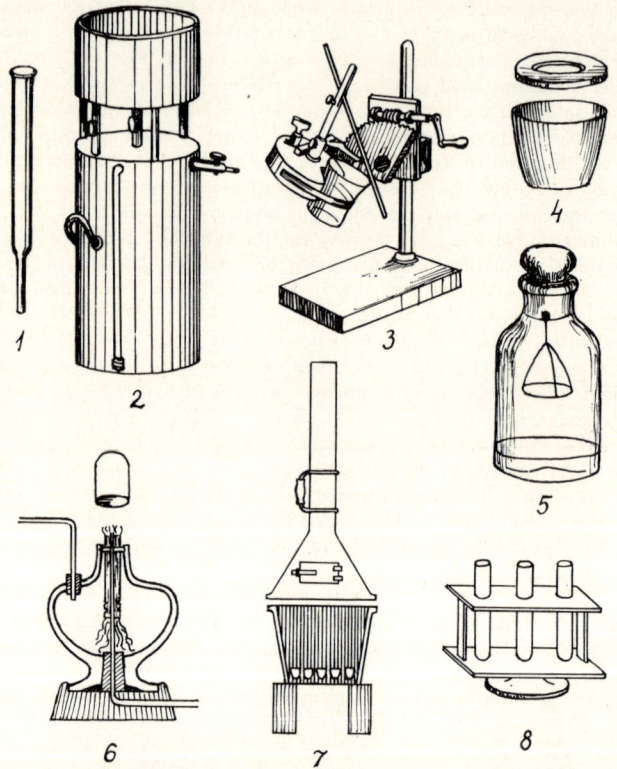

FIG. 45. Berzelius's laboratory equipment. 1. — Filter funnel; 2. — Gasometer; 3. — Filtering device; 4. — Water-bath; 5. — Desiccator; 6. — Blow tube; 7. — Igniting furnace; 8. — Test tubes. (From his *Lehrbuch der Chemie*)

because the solvent evaporates rapidly and therefore it is difficult to wash the precipitate. It is advisable to first make the filter paper damp with water, because if the turbid mixture is poured on to a dry filter paper it absorbs the mixture so rapidly that the pores of the filter become blocked (p. 267). The stem of the funnel must be placed against the side of the beaker to avoid splashing from the solution. Correct filtration is shown in Fig. 46, 4, and it can be seen that the same figure could be given even in a modern analytical textbook.

Berzelius filtered strongly acid solutions through broken glass, which he placed in a funnel, and filled the stem with slightly larger pieces. This is one of the earliest types of filter funnel, but an even earlier example is described in the book of Österreicher–Winterl mentioned previously (Chapter IV. 1). When the filter paper cannot be dried adequately then a ground glass filter pad is suggested, as is in use in all laboratories today (... *adhibeat vitrum in fundo sissuram habens, qualia in omni laboratorio semper haud pauca occurunt; haec aeque partem limpidam transmittunt, praecipitatum retinent*) (J. Oesterreicher: *Analyses aquarum Budensium*. Buda, 1781, p. 25).

Berzelius also used conical glass tubes filled with asbestos which had previously been purified with concentrated hydrochloric acid (pp. 187, 264; Fig. 45, 1). For the filtration

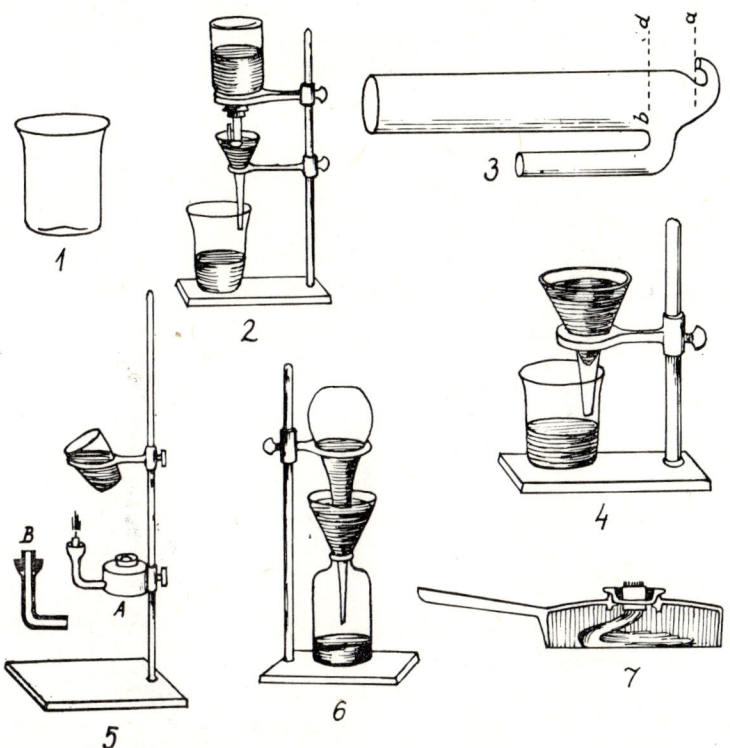

FIG. 46. Berzelius's laboratory equipment. 1. — Beaker; 2. — Device for automatic washing of precipitates; — 3. Capillary regulator of device 2; 4. — Filter holder with filter funnel; 5. — Ignition of precipitates; 6. — Automatic filtration apparatus 7. — Oil burner. (From his *Lehrbuch der Chemie*)

of solutions containing hydrogen fluoride he used platinum funnels. Beakers were similar to those in use today, but as can be seen from Fig. 46, 1, they had no pouring-spout. He placed a small amount of fat on the rim of the beaker where the solution was poured out, so that it could not run down the side of the beaker. During filtration the solution had to be directed down a glass rod, and the nearer the angle between the rod and beaker approached a right angle, then the more secure is the filtration. Nordenskjöld [109] designed a pouring device (p. 18) which was used by Berzelius and found to be very satisfactory (Fig. 45, 3). Its use, however, could not have proved to be a great advantage because it fell into disuse later.

The last traces of any precipitate adhering to the wall of the beaker was transferred with water and a small piece of filter paper, and the precipitate was then washed out of the filter.

Washing was carried out until a drop of the washing water evaporated without leaving any residue on either a glass plate or a platinum spoon (p. 220). Because washing often took a considerable time, an automatic device was constructed for this purpose (Fig. 46, 2). At the end of the flask, immersed into the filter, there was a device (Fig. 46, 3), which regulated the flow of the washing solution by capillary pressure. Its operating principle was given in detail by Berzelius. Haüy [110] had already constructed a similar automatic device for fil-

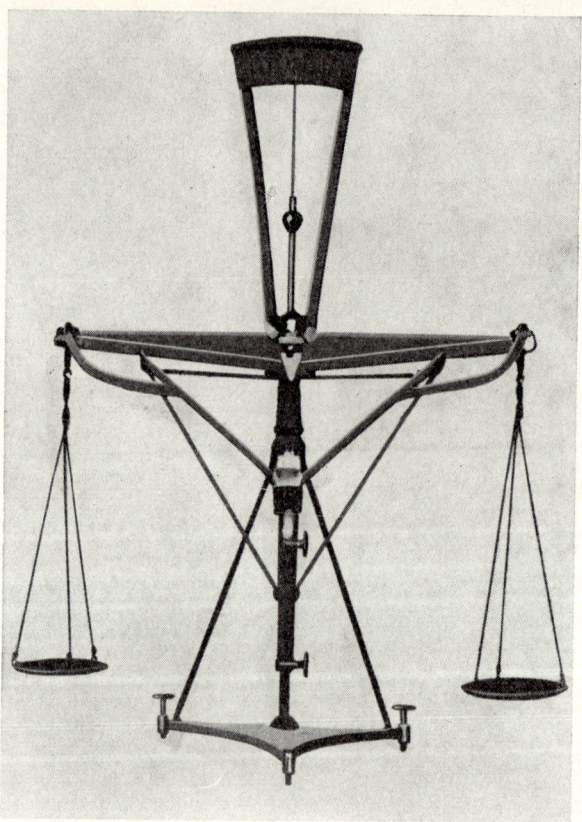

FIG. 47. Analytical balance of Berzelius

tration (Fig. 46, 6), and Eimbke [111] used a vacuum pump in combination with a similar device [112]. These automatic devices, however, did not become very popular, obviously because they complicated an already simple process. Pfaff mentions that the device of Eimbke makes filtration very difficult.

Fig. 45, 2 shows the gasometer of Pepys [113] based on the diagram from Berzelius's book. It can be seen that its form is quite similar today, and this is true of other devices shown in the figure.

Berzelius dealt in great detail with balances, mainly with regard to constructional problems, which are too involved to be dealt with here. The smallest weight he used was 5 mg, and this was a rider on the arm of a balance. (Pfaff's balance — as he mentions in his book — was sensitive to 1 mg if the loading was 10 g [114].) Berzelius was the first chemist systematically to use the French system of measures, i.e. the weights of the metric system. In his book however he refers also to the older weight systems, and gives conversion factors for the two systems. Earlier chemists had generally used the grain as the unit of weight, but it is very difficult to establish from most of their work exactly how many grams was represented by the unit, because the weight of the grain varied from country to country, and in some cases from town to town. Table 9 shows how difficult was the task of correct interpretation of analytical data in those days, because of the variation in the measures.

TABLE 9

EQUIVALENT VALUE OF 1 GR IN MG

Bern	61·9197	Holland	64·0630	Spain	59·9008
England	64·7804	Naples	55·6959	Sweden	61·8620
France	53·1148	Piedmont	53·3859	Venice	52·4349
Geneva	55·0533	Portugal	59·7588	Prussia	62·0944
Hanover	63·3538	Rome	58·8906	Austria—Hungary	72·9182

Berzelius had realized the advantages afforded by an international weight system, and advocated it in his books. In 1841 he criticized the English chemists for their use of out-of-date measures.

The first reference to the use of rubber tubing dates from about 1840. First of all the tubes were made by the chemist himself from rubber sheets, but at that time they were already produced by factories. According to Berzelius, the use of these tubes involved many complications (p. 483). Also at about that time the grinding of glass, and also the construction of glass stoppered bottles, are first mentioned.

Berzelius discussed the accuracy of analytical determinations and commented that slight errors are unavoidable; on the choice of a method he stated that the best method is that which depends least on the skill of the analyst. The margin of error for a complete analysis is 1 — 1·5 per cent, if a well-trained analyst carries it out, but for an analysis of two or three components then the error is only about 0·5 per cent (p. 154).

Berzelius carefully examined the analysis of the platinum metals, and devised a method for their determination. He also attempted to produce a scheme of qualitative analysis, at least for a limited number of metals. Thus, for a mixture of cerium, iron, manganese, aluminium, beryllium, yttrium and calcium, he proceeded as follows:

He first precipitated the metals with ammonium hydroxide. He tested the filtrate for calcium with oxalate. (For quantitative determination, however, he did not use the calcium oxalate precipitate, but dissolved this in sulphuric acid, and weighed the calcium sulphate after ignition.) From the precipitate he extracted beryllium and aluminium with potassium hydroxide, and on boiling this extract the beryllium precipitated out. The residue containing cerium, iron, manganese and yttrium, he dissolved in acid. He then immersed potassium sulphate crystals into this solution, when cerium potassium sulphate was precipitated on the surface of the crystals. Iron was then hydrolysed with ammonium succinate from the solution, and the manganese precipitated with oxalate and finally the yttrium with carbonate. He developed similar schemes for other systems, but did not work out any general scheme of qualitative analysis.

In his description of quantitative analysis Berzelius describes the analysis of single minerals or salts separately, but for reasons of space no detail of these can be given. They do not contain any reference to novel pieces of apparatus but several small improvements, such as manual simplifications, which made the analyses easier are mentioned. It is interesting to note that while Klaproth used a 5–10 g sample for his analysis, Berzelius carried out his operations with a 1 g sample (p. 53).

5. THE FIRST ANALYTICAL TEXTBOOKS

Many chemical guide-books were published during the 15th century, but these books were compiled for practical chemists rather than scientists. They contained practical instructions for carrying out various processes, as well as chemical prescriptions such as the recipe for gunpowder, or for certain medicines or ointments. Other books were mainly concerned with the distillation of spirit. Some of these books contain analytical observations. For example, one book written in the 15th century describes a test for the purity of sulphur which was to be used for making gunpowder. The test was quite simple, a handful of sulphur was made into a ball in the hand and then squeezed; if it gave out a creaking sound then the sulphur was pure. There were also more serious analytical prescriptions, mainly for docimastic gold analyses. During succeeding centuries the pattern of these books remained the same.

The first examples of truly scientific textbooks in chemistry were published in the 17th century. The oldest known chemical textbooks are Lémery's *Cours de chymie* (1675), Boerhave's *Elementa chimiae* (1732), from which Goethe obtained his chemical knowledge, and Jacquin's *Lehrbuch der allgemeinen Chemie* (Wien, 1783). These books, however, contained only a few analytical data. At the end of the 18th century the first books devoted entirely to analytical chemistry appeared. The first of these, so far as I know, is Göttling's book entitled: *Vollständiges chemisches Probekabinett* [115] (Jena, 1790). As with other books of this period Göttling's book was primarily concerned with the analysis of minerals and metals. Unfortunately, I was not able to obtain a copy of this book. Vauquelin's book, *Manuel de l'esseyeur*, which was published originally as separate parts in a journal (1799), deals mainly with noble metal analysis. In 1801 Lampadius [116], who was Professor at Freiburg, published a book with the title: *Handbuch zur chemischen Analyse der Mineralkörper*. This book, however, is not a general analytical textbook, and deals mainly with the analysis of minerals. His methods are generally no better than those of Klaproth, but there are parts in this book which have the style of textbooks of a much later period. For example, it gives a detailed list of equipment and apparatus required for analysis, as well as instructions for their use. It also describes some analytical methods and gives a list of reagents. Hydrogen sulphide, however, is not mentioned. The introductory part which deals with the preparation and purification of the various reagents is very long, and is interesting because it describes how some of the reagents were tested for purity. These descriptions are the earliest record of standard methods used for testing the purity of analytical grade reagents and in many cases are very similar to present day methods. For example:

> Hydrochloric acid. 1. Heavy earth solution (Ba) should not cause a turbidity, i.e. sulphuric acid earths should not be precipitated. 2. From potassium cyanide turbidity should not be caused. 3. On neutralization with potassium hydroxide it should remain clear. 4. After evaporation on a glass plate no residue should remain.

Even the purity of distilled water is checked:

> 1. Neither lead acetate nor heavy earth acetate should cause a turbidity. 2. It must remain clear after the addition of silver solution. 3. It should show a similar effect with potassium carbonate, and 4. similarly with potassium cyanide, and 5. similarly with lime water.

Before the description of the analysis of the individual minerals he gives a brief account of the properties of the individual acids, bases and metals. These accounts are very brief indeed, for example "Boric acid dissolves easily in alcohol, and this solution burns with a nice green flame".

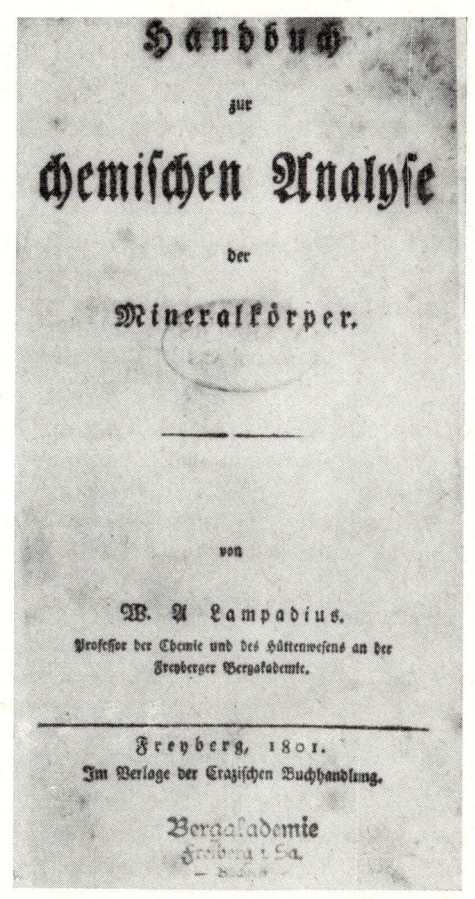

FIG. 48. Title page of Lampadius's *Handbuch zur chemischen Analyse der Mineralkörper*, 1801

He then describes the tests for the more important salts, but like Kirwan he attempts to identify the salt itself, and not the cation or anion. Thus:

> Copper sulphate dissolves easily in water, while heavy earth acetate gives an immediate turbidity to this solution. The clear part of the solution after the addition of ammonium hydroxide is deep blue. A piece of iron placed in copper sulphate solution becomes coated with copper.

This part of the book is similar in content to later analytical texbooks.

Lampadius did not realize, however, that it was sufficient to examine the reactions of the metals and the acids, and that there was no necessity to examine the individual salts; this only became apparent after the introduction of Berzelius's dualistic theory. This idea was later put forward by Pfaff in his book, which will be referred to again later. Several books on this subject were published before this which are of little interest as they do not contain any new material. Only their titles are listed here: J. Heinrich Kopp [117]: *Grundriss der chemischen Analyse mineralischer Körper*, Frankfurt, 1805; John [118]: (with a preface by Klaproth) *Geöffnetes Laboratorium*, Berlin, 1808–1814; Fabricius: *Anleitung zur chemischen Analyse unorganischer Körper*, Kul, 1810; Hermbstaedt [119]: *Anleitung zur Zergliederung der Vegetabilien*, Berlin, 1807; Thénard: *Traité de chimie élémentaire, théorique et pratique* (1813–1816).

All these books had a very limited content. The first, really general and comprehensive textbook was written by Pfaff [120] and entitled: *Handbuch der analytischen Chemie für Chemiker, Staatsärzte, Apotheker, Oekenomen und Bergwerks Kundige* (J. J. Hammrich, Altona, 1821). Pfaff, in this book, attempted to give a thorough summary of existing analytical knowledge, as he considered that there was no comprehensive analytical textbook available at that time, and states that "many famous people have written 'Introductions' (Anleitungen) to analytical chemistry" [121]. He aimed to provide a book which would be of value to the trained chemist as well as to the beginner. Pfaff took great care in selecting his analytical methods, most of which he tried out for himself beforehand. (This is the usual sentence in an introduction to an analytical textbook, although it is doubtful whether all authors are completely truthful.) His literature references are very accurate and he comments, "Berzelius's book would be much better if he gave a reference to the source of his information" [122]. In his book both the atomic theory and the theory of chemical proportions are used, although he wrote "many people doubted whether these theories were sufficiently proved to be incorporated in a textbook". Pfaff answered this criticism by saying that in his opinion these theories were valid because with their help many of the problems of chemistry became much clearer. However, he did not adopt the new nomenclature,

> because it makes the style very clumsy, and for those chemists who finished their studies before the discovery of alkali metals, it is very difficult having to read potassium oxide hydrate instead of potash, and calcium oxide instead of lime, or magnesium oxide instead of magnesia [123].

Pfaff begins his book with a discussion on the use of reagents, which takes up 264 pages. One of the reasons for this introduction being so extensive is that in most cases the preparation of the reagent is also described. As the analytical chemist of those days could not buy chemicals he had to prepare them himself. Only sulphuric acid, hydrochloric acid, acetic acid, some metals, and a few of their salts could be obtained commercially; the remainder had to be prepared from them. Some of the commercial chemicals had to be purified by complicated procedures; for example sulphuric acid had to be redistilled. Sometimes there are *circulos vitiosos* to be found in Pfaff's book, for example in the preparation

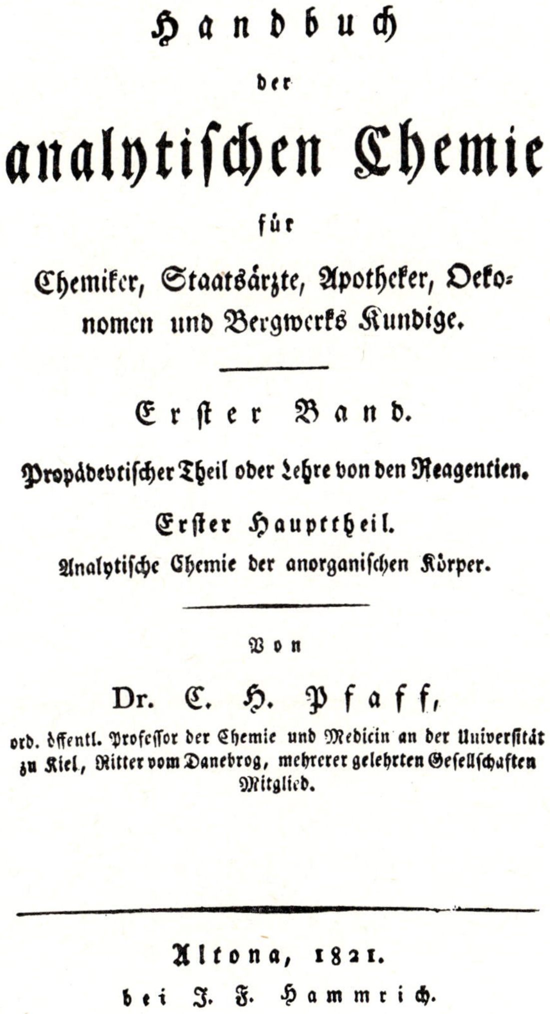

Handbuch

der

analytischen Chemie

für

Chemiker, Staatsärzte, Apotheker, Oekonomen und Bergwerks Kundige.

———

Erster Band.

Propädeutischer Theil oder Lehre von den Reagentien.

Erster Haupttheil.

Analytische Chemie der anorganischen Körper.

———

Von

Dr. C. H. Pfaff,

ord. öffentl. Professor der Chemie und Medicin an der Universität zu Kiel, Ritter vom Danebrog, mehrerer gelehrten Gesellschaften Mitglied.

———

Altona, 1821.

bei J. F. Hammerich.

FIG. 49. Title page of Pfaff's *Handbuch der analytischen Chemie* from 1821

of acetic acid he recommends lead acetate, while lead acetate is to be prepared from acetic acid [124].

After the description of the preparation Pfaff gave the composition of the reagent and its atomic weight according to Berzelius, which he marked M.G.

(Mischgewicht), and finally the uses of the reagent, and also in some cases their sensitivity. The definition of sensitivity was given as the amount of test substance required to give a perceptible change with the reagent. Sensitivity can also be measured for relative amounts as well for the absolute amount of the substance. Thus, he first expresses in how many parts of water one part of the test substance can be detected [125].

For example:

> Sulphuric acid sodium, natrium sulphuricum, sulphas natricus. $Na\dot{\ddot{S}}^2$ (Sodium sulphate).
> We prepare it from the substance which remains after the preparation of hydrochloric acid. The excess of acid which is occasionally present must first be neutralized with sodium carbonate, then filtered, evaporated and finally crystallized ... It forms large white prism-shaped crystals, which first give a cooling sensation when tasted, but also have a bitter taste. It crumbles in contact with air, and dissolves in its crystal water, but melts only after the evaporation of the latter, and heating to white heat. Solubility at 0°: 12·17 per cent; at 11·67°: 26·38 per cent; at 50·40°: 262·35 per cent. Above this temperature its solubility decreases. It is insoluble in alcohol.
> Its composition in the crystalline state is: 19·39 per cent sodium, 24·85 per cent sulphuric acid, 55·76 per cent water; in a dry state: 48·32 per cent sodium and 56·58 per cent acid. M.G.: 1784·16.
> (1) First of all it can be used for the precipitation of lead in the presence of nitric acid. Pure sulphuric acid, however, is better for this purpose.
> (2) It is very suitable for the precipitation of baryta. In one million-fold dilution of barium chloride 1 per cent sodium sulphate solution causes just a perceptible turbidity ... [126]

Pfaff's work records many new reagents such as hydrogen sulphide, ammonium sulphide, iodine, chlorine water, tin chloride, as well as platinum chloride for precipitation of potassium and mercury (I) nitrate for testing for ammonia [127]. He also used the latter reagent for the determination of chromate. He precipitated this in the form of mercury (I) chromate, then after evaporation and volatilization of the mercury, weighed the residual chromium trioxide.

The descriptive part on apparatus and practical details is essentially similar to that given by Berzelius. In gravimetric determinations Pfaff removed the bulk of the precipitate from the filter paper and ignited this separately, and after igniting the filter paper combined the two residues for weighing. This reduced the errors from reduction by the carbon of the filter paper.

The second part of the book deals with the analysis of minerals, salts, metals, mineral waters and gases.

In the section dealing with water-analysis he mentions that the evaporation–extraction method was inferior to the new method recommended by Murray [128], which consists in determining the main constituents in the water by the addition of various reagents without previous evaporation. First of all barium chloride must be added to the water, which precipitates sulphate and carbonate. The precipitate must be filtered, dried, and weighed, and then the barium carbonate must be dissolved out with hydrochloric acid leaving a residue of barium sulphate. This residue must again be filtered, dried, and weighed, when the sulphate and, hence by difference, the carbonate can be determined. Then chloride must be precipitated with silver nitrate, calcium with oxalate and finally magnesium with phosphate.

Pfaff's methods for the examination of the metals would be valid even today. His examination of cadmium shows how thorough his investigations were:

Cadmium. 1. *Physical properties and chemical behaviour.* (Here he gives a description of the metal, its specific weight, melting point, oxide formation and behaviour under the blow-pipe.)

2. *Reactions of the cadmium solution:*

(*a*) Zinc precipitates the metal.
(*b*) Alkalis cause the formation of a white precipitate which, according to some scientists, dissolves in an excess of the alkali.
(*c*) With ammonia it forms a white precipitate which dissolves in an excess of ammonia.
(*d*) With alkali carbonates a white precipitate is formed.
(*e*) With sodium phosphate a white, powder-like precipitate is formed which can be dissolved in ammonia.
(*f*) With hydrogen sulphide in acid medium, and also with alkali sulphides, a yellow precipitate of cadmium sulphide is formed.
(*g*) With blue acid iron potash (potassium hexacyanoferrate) a yellowish-white precipitate is formed.
(*h*) With the liquor of galls a dirty-yellow, voluminous precipitate is formed.

3. *Quantitative determination.*

(*a*) From a solution containing nitric acid it can be precipitated with ammonium carbonate. After filtration it must be ignited. 100 parts of the precipitate corresponds to 87.45 parts of the metal.
(*b*) It can be precipitated with zinc in the form of the metal, preferably from hydrochloric acid solution.
(*c*) Under certain circumstances it can also be precipitated with hydrogen sulphide. The precipitate must be dried. 100 parts of the dry precipitate is equivalent to 77·59 parts of the metal [129].

He then describes the separation of cadmium, and remarks that it can be separated from zinc with hydrogen sulphide. Finally he describes the analysis of cadmium ores.

Pfaff's book concludes with several chapters dealing with the analysis of gases and organic substances.

After Pfaff's book had been published it became customary from time to time for textbooks to be written which collected, summarized, and critically reviewed the existing methods of analytical chemistry. Up to the time of the first World War it was undoubtedly German scientists who were leading in this field. They were the great collectors, and the famous analytical textbooks of Rose, Fresenius, Treadwell, Mohr and Beckurts, which were published in many editions and in several languages, were the working textbooks of all analytical chemists. These books will be considered in more detail in later chapters.

NOTES AND REFERENCES

1. KIRWAN, R.: *An essay on the analysis of mineral waters.* London (1799) 145
   THOMSON, T.: *A system of chemistry.* 5th ed. London (1817) 3 227—235
2. DANN, G. E.: *Martin Heinrich Klaproth.* Berlin (1958) 48
3. PROSZT, J.: *Die Schemnitzer Bergakademie als Geburtsstätte chemiewissenschaftlicher Forschung in Ungarn.* Sopron (1938) 31
4. DANN, G. E.: *Martin Heinrich Klaproth.* Berlin (1958) 80
5. MÜLLER, FERENC (1740—1825) was the manager of a mine in Transylvania, and later became director of the Mining Authority in Vienna.
6. MÜLLER, F.: *Phys. Arb. d. einträcht. Freunde.* Wien (1783) **1** 57 59, (1784) **2** 49
7. BUGGE, G.: *Das Buch der grossen Chemiker.* Berlin (1929) **1** 338
8. KLAPROTH, M. H.: *Beiträge zur chemischen Kenntniss der Mineralkörper.* Posen—Berlin (1795) **1** 8
9. KLAPROTH, M. H.—JOHN J. F.: *Chemisches Laboratorium.* Berlin (1808) Preface
10. KLAPROTH, M. H.: *Beiträge zur chemischen Kenntniss der Mineralkörper* (1810) **5** 41
11. *ibid.* **2** 48
12. *ibid.* **2** 61
13. *ibid.* **3** 163
14. DANN, G. E.: *Martin Heinrich Klaproth.* (1958) 73
15. KLAPROTH, M. H.: *Beiträge zur chemischen Kenntniss der Mineralkörper.* (1795) **1** 146
16. *ibid.* **1** 173
17. *ibid.* **1** 332
18. *ibid.* **1** 150
19. *ibid.* **2** 253
20. *ibid.* **2** 268
21. *ibid.* **3** 63 140
22. *ibid.* **4** 132
23. *ibid.* **4** 86
24. KLAPROTH, M. H.: *Monatshefte der Akad. Künste u. mech. Wiss.* (1788) **1** 86
25. BALZAC, H.: *Histoire de la Grandeur et de la Décandence de César Birotteau.*
26. CARRIÈRE, J.: *Berzelius und Liebigs Briefwechshel.* München—Leipzig 185; Liebig's letter to Wöhler from 29. IV 1839
27. BERZELIUS, J. J.: *Selbstbiographische Aufzeichnungen.* Leipzig (1903) 4
28. SÖDERBAUM, H. G.: *Berzelius' Werden und Wachsen.* Leipzig (1899) 4
29. BERZELIUS, J. J.: *Selbstbiographische Aufzeichnungen.* 6
30. SÖDERBAUM, H. G.: *Berzelius' Werden und Wachsen.* 8
31. BERZELIUS, J. J.: *Selbstbiographische Aufzeichnungen.* 15 16
32. *ibid.* 17
33. EKEBERG, ANDERS GUSTAF (1767—1813) lecturer at the university of Uppsala, the discoverer of tantalum.
34. BERZELIUS, J. J.: *Selbstbiographische Aufzeichnungen.* 18
35. SÖDERBAUM, H. G.: *Berzelius' Werden und Wachsen.* 12
36. BERZELIUS, J. J.: *Selbstbiographische Aufzeichnungen.* 24 25
37. *ibid.* 18. The often cited autobiography of Berzelius appeared only in 1898, fifty years after his death. Up to this time it was preserved by the Swedish Academy of Sciences. His most important biographies were written before this date, without a knowledge of this work.
38. *Nova analysis aquarum Medeviensium.* Uppsala 1800
39. *Gehlens Journal für Chemie.* (1803) 115
40. KLAPROTH, M. H.: Chemische Untersuchung des Ochroits. *Gehlens Journal für Chemie.* (1804) **3** 303
    BERZELIUS—HISINGER: Cerium ein neues Metall aus einer schwedischen Steinart. *ibid.* (1804) **3** 397

41. BERZELIUS, J. J.: *Selbstbiographishe Aufzeichnungen.* 36
42. *Afhandlingar i Fysik, Kemi och Mineralogi.* (1807)
43. BERZELIUS, J. J.: *Selbstbiographische Aufzeichnungen.* 40
44. This is the Karolinska Institute, which still exists. This institute allocates the Nobel medical prizes.
45. BERZELIUS, J. J.: *Selbstbiographische Aufzeichnungen.* 53
46. *ibid.* 54
47. *ibid.* 72
48. *ibid.* 80
49. *ibid.* 98
50. CARRIÈRE, J.: *Berzelius und Liebigs Briefwechsel.* 107: Berzelius's letter to Liebig. Aug. 14, 1835.
51. CARRIÈRE, J.: *Berzelius und Liebigs Briefwechsel.* 111: Liebig's letter to Berzelius. Feb. 23, 1836.
52. CARRIÈRE, J.: *Berzelius und Liebigs Briefwechsel.* 117: Berzelius's letter to Liebig. Jan. 3, 1837.
53. CARRIÈRE, J.: *Berzelius und Liebigs Briefwechsel.* 165: Liebig's letter to Wöhler. May 18, 1838.
54. CARRIÈRE, J.: *Berzelius und Liebigs Briefwechsel.* 262: Wöhler's letter to Liebig. March 4, 1848.
55. BUGGE, G.: *Das Buch der grossen Chemiker.* (Söderbaum, Berzelius) **1** 444
56. BERZELIUS, J. J.: *Selbstbiographische Aufzeichnungen.* 46
57. SÖDERBAUM, H. G.: *Berzelius, Werden und Wachsen.* 189; Berzelius's letter to Mitscherlich. July 4, 1823.
58. *ibid.* 189; Berzelius's letter to Mitscherlich. Feb. 2, 1837.
59. WÖHLER, F.: *Ber.* (1875) **8** 838
60. Berzelius's papers were published in the following periodicals: *Afhandlingar i Fysik Kemi och Mineralogi* (1810—1818) **3**—**6**; *Gilberts Annalen der Physik* (1811, 1812, 1814, 1816) **37 38**; **40 42**; **46**; **53**; *Schweiggers Journal für Chemie u. Physik* (1811—1818) **6 7 11 15 16 21 23**; *Trommdorffs Journal der Pharmazie* (1812) **20 21**; *Annales de Chimie* (1811—1812) **78**—**83**; (1815) **94**—**95**; *Annales de Chimie et physique* (1816—1818) **5 6 9**; *Philosophical Magazine* (1813) **41**—**43**; *Thomsons Annals of Philosophy* (1813—1820) **2**—**20**
61. HJELT: *Aus Berzelius u. Magnus Briefwechsel.* (1900) 60; Berzelius's letter to Magnus. Feb. 21, 1832.

MOSANDER, KARL GUSTAF (1799—1858). The successor to Berzelius at the Karolinska Institute was a member of the Swedish Academy of Sciences. He initiated the studies of the rare earths. He discovered lanthanum, erbium and dydimium in cerite (from the latter Auer established later that it consists of two elements: praeseodymium and neodymium).
63. GMELIN, CHRISTIAN GOTTLIEB (1792—1860) was a well known chemist of his time and was Professor of Chemistry at the University of Tübingen from 1817.
64. MITSCHERLICH, EILHARD (1794—1863) the successor of Klaproth to the Chair of Chemistry at the University of Berlin (from 1824 he was a member of the Prussian Academy of Sciences). He made many important contributions to chemistry, especially in the field of crystallography where his law of isomorphism became very important in the determination of atomic weights. In the field of pure chemistry the discovery of permanganate was his most important contribution.
65. ROSE, GUSTAV (1798—1873) was a famous Professor of Mineralogy at the University of Berlin, and was a brother of Heinrich Rose.
66. ARFVEDSON, JÖNS AUGUST (1792—1841) was a Swedish mine and smelting (open hearth) factory owner who was also a member of the Swedish Academy of Sciences and was the discoverer of lithium.

67. SEFSTRÖM, NILS GABRIEL (1787—1845) Professor of Chemistry at the Mining College at Falun, and a member of the Swedish Academy of Sciences. He was the discoverer of vanadium.
68. WÖHLER, FRIEDRICH (1800—1882), first Professor at the industrial school of Berlin (1825—31), then at the industrial college at Kassel, while from 1836 he was Professor of Chemistry at the University of Göttingen. He became a member of the Prussian Academy of Sciences. A very diligent chemist who published a great many papers. He was the first to prepare metallic aluminium (1827), and later beryllium by reduction with potassium. He prepared urea synthetically in 1828, and this was the discovery that started the great development of organic chemistry.
69. MAGNUS, HEINRICH GUSTAV (1802—1870) Professor at the University of Berlin and a member of the Prussian Academy of Sciences.
70. LAGERHJELM, PER (1787—1856), inspector of the Swedish mines and a member of the Swedish Academy of Sciences. He assisted Berzelius in his calculations of the "atomic weights" of some 2000 compounds.
71. BERLIN, NILS JOHANNES (1812—1900) Professor of Chemistry at the University of Lund, and a member of the Swedish Academy of Sciences.
72. CARRIÈRE, J.: *Berzelius und Liebigs Briefwechel.* 234. Liebig's letter to Berzelius, March 28, 1842.
73. *ibid.* 237 Berzelius's letter to Liebig April 12, 1842
74. Berzelius's letter to H. Rose, Feb. 2, 1825.
75. BERZELIUS, J. J.: *K. Vetensk. Akad. Handl.* (1812) 66
76. *Försok att genom anwändenda of den Elektro-kemiska theorien,* etc. Stockholm 1814
77. *Nouveau systeme de minéralogie.* Paris 1819
78. *Afhandling om blasrörets anwändande i kemien och minerologien.* Stockholm (1820). (Published also in German, English, French, Russian and Italian.)
79. More than 3600 letters from Berzelius and more than 7000 letters written to Berzelius are in different Swedish archives. The most interesting were published by Söderbaum and Holmberg in 6 volumes: Jac. Berzelius Brev. V. 1—6+Suppl. 1—2, Uppsala 1912—41. A complete bibliography of works of Berzelius was given by Holmberg; published by the Swedish Academy of Science in Stockholm 1936—1953.
80. *Briefwechsel zwischen Berzelius und Wöhler.* Leipzig (1901) 2 120
81. ROSE, H.: *Pogg. Ann.* (1844) **63** 317
82. BERZELIUS, J. J.: *Selbstbiographische Aufzeichnungen.* 56
83. CARRIÈRE, J.: *Berzelius und Liebigs Briefwechsel.* 146: Berzelius's letter to Liebig, Feb. 20, 1838.
84. CARRIÈRE, J.: *Berzelius und Liebigs Briefwechsel.* 169: Berzelius's letter to Liebig, June 19, 1838.
85. CARRIÈRE, J.: *Berzelius und Liebigs Briefwechsel.* 249—251: Berzelius's letter to Liebig, Nov. 14, 1843.
86. BERZELIUS, J. J.: *Jahresbericht* (1848) 27 592
87. ROSE, H.: *Gedachtnisrede auf Berzelius.* Berlin (1851) 59
88. CARRIÈRE, J.: *Berzelius und Liebigs Briefwechsel*, 223: Bunsen's letter to Liebig, Oct. 29, 1841.
89. BERZELIUS, J. J.: *Lehrbuch der Chemie.* Dresden—Leipzig (1836) **5** 24
90. *ibid.* 98 99
91. *ibid.* 43
92. *ibid.* 83
93. DULONG, PIERRE LOUIS (1785—1838) Professor of Physics at the University of Paris, and a member of the French Academy; PETIT, ALEXIS THÉRÉSE (1791—1820) preceded Dulong to the same chair.
   DULONG, P. L.—PETIT, A. T.: *Ann. chim. phys.* (1819): **10** 395
94. PROUT, WILLIAM (1786—1850) was a practising physician in London, and a member of the Royal Society.

95. PROUT, W.: *Annals of Phil.* (1815) **6** 321;
    (anonym) (1816) **7** 111
96. THOMSON, T.: *An Attempt to establish the First Principles of Chemistry by Experiment.* London (1825)
97. BERZELIUS, J. J.: *Lehrbuch der Chemie.* **5** 104
98. *ibid.* 106
99. *ibid.* 107
100. *ibid.* 120
101. BERZELIUS, J. J.: *Annals of Phil.* (1814) **3** 362
102. BERZELIUS, J. J.: *Lehrbuch der Chemie.* Dresden—Leipzig (1836) **5** 104
103. *ibid.* 132—433
104. *ibid.* 170 204
105. GERHARDT, CHARLES (1816—1856) French chemist and a pupil of Liebig. He was Professor of Chemistry at the University of Montpellier; later he opposed the educational authorities, and had to resign. He then established a private teaching laboratory in Paris, and later, with some difficulty, obtained a post at the University of Strasbourg. He was one of the first workers to treat organic chemistry in a systematic manner.
106. CANNIZZARO, STANISLAO (1826—1910) Italian chemist. Because of his collaboration in the Italian War of Liberty he was condemned to death in his absence. During this time he studied in Paris under Chevreul, and was only able to return to Italy in 1856. He became Professor of Chemistry first at Palermo, and later in Rome.
107. BERZELIUS, J. J.: *Lehrbuch der Chemie.* (Chemische Operationen und Geräthschaften) Dresden—Leipzig (1841) **10**
108. PFAFF, C. H.: *Handbuch der analytischen Chemie.* Altona (1821) 1 271
109. NORDENSKJÖLD, NILS GUSTAF (1792—1864), intendant of Finnish mines.
110. HAÜY, RENÉ JUST (1743—1822) was an abbot, who was also Professor at various colleges, and finally became Professor of Mineralogy at the University of Paris. He was a member of the French Academy, and one of the foremost authorities on mineralogy and crystallography in his day.
111. EIMBKE, GEORG (1771—1843) was a pharmacist in Hamburg.
112. EIMBKE, G.: *Schweiggers Journal. Neue Reihe* **1** 90
113. PEPYS, WILLIAM (1775—1843) director of an English gas company. He was a member of the Royal Society. The gasometer referred to was described in *Phil. Mag.* **13** (1802).
114. PFAFF, C. H.: *Handbuch der analytischen Chemie.* **1** 284
115. GÖTTLING, JOHANN FRIEDRICH (1755—1809) was a pharmacist who later became Professor of Chemistry at the University of Jena, and was the author of several chemical textbooks.
116. LAMPADIUS, WILHELM AUGUST (1772—1842) was a pharmacist who later became Professor of Chemistry at the Mineral Academy at Freiberg. He travelled a great deal including a visit to Russia.
117. KOPP, JOHANN HEINRICH (1777—1858) was a physician. He was the father of Hermann Kopp, the famous chemical historian.
118. JOHN, JOHANN FRIEDRICH (1782—1847) was a Professor in Moscow between 1802—1804, and later at the University of Frankfurt (Oder), and finally settled in Berlin.
119. HERMBSTAEDT, SIGISMUND FRIEDRICH (1760—1833) was a court pharmacist in Berlin who became Professor at the Medical School in Berlin. He later became Professor of Technology at the University of Berlin, and was elected a member of the Prussian Academy. He was one of the first pioneers of agricultural chemistry.
120. PFAFF, CHRISTIAN HEINRICH (1773—1852), was a physician. He first worked as court physician to a Count, but later became a practising physician in Heidelberg. In 1797 he was appointed Professor of Chemistry and Pharmacy at the University of Kiel. He published many papers and more chemical textbooks.
121. PFAFF, C. H.: *Handbuch der analytischen Chemie* **1** 5
122. *ibid.* **1** 7

123. *ibid.* **1** 10 315
124. *ibid.* **1** 88 214
125. *ibid.* **1** 33
126. *ibid.* **1** 150—151
127. SAUSSURE: *Gilberts Neue Ann.* **19** 129
128. MURRAY: *Neues Journal der Physik* **2** 387; THOMAS, T.: *A system of chemistry.* 5th ed. London (1819) 231
129. PFAFF, C. H.: *Handbuch der analytischen Chemie.* **2** 354

CHAPTER VII

FURTHER DEVELOPMENTS IN QUALITATIVE AND GRAVIMETRIC ANALYSIS

1. INTRODUCTION OF SYSTEMATIC TESTS FOR THE IONS

Most of the reactions of qualitative analysis, with the exception of those involving synthetic reagents, were discovered before the beginning of the last century. Wet processes, as we have seen, were mainly used for water analyses, and most of the reactions were discovered during the course of this work. The most important field of chemical analysis at that time was the analysis of minerals. It was usual for qualitative tests to be carried out first mainly using only the blow-pipe, and after this preliminary experiment a quantitative analysis was carried out. Mineralogy was, at this time, one of the most important auxiliary fields of chemistry. Often from an examination of the crystal form, and other mineralogical features, it was possible to decide which elements were present. In the case of silicates, it was usual to determine silicic acid, iron, aluminium, calcium and magnesium, and only if the results of these analyses did not approximate to 100 per cent were any other elements tested for. It was from the anomalous behaviour of certain minerals in the course of analysis that elements such as chromium, beryllium and tantalum were discovered. Only the main components of the mineral were analysed (> 1 per cent), as no suitable methods were available for testing smaller amounts than this. Even in this case large errors were sometimes made even by the most celebrated chemists. Thus, Klaproth did not recognize the presence of phosphate in the mineral wavelite, because he incorrectly identified the precipitated aluminium phosphate as aluminium hydroxide. This error was later corrected by Fuchs [1, 2].

Most of the fusion methods were also known. As we have seen, Marggraf carried out fusions with alkali cabonate, while Klaproth used alkali hydroxide, and Berzelius used hydrogen fluoride. Fusion with potassium hydrogen sulphate was first carried out by H. Rose.

The application of most of the chemical reagents has been described in the previous chapter. But one very important reagent, hydrogen sulphide, was only used to a very limited extent. In Pfaff's book it is referred to as hydrothionic acid. Boyle also used hydrogen sulphide for the qualitative test for tin and lead in water (Chapter III. 3). His discovery did not find any support so that hydrogen sulphide fell into disuse, and had to be rediscovered later. F. Hoffmann gives an account in 1772 of the gas having a bad smell which is evolved from alkalic sulphur solutions on treatment with acids [3]. Several other accounts are given regarding this gas; for example Rouelle [4] mentions that this gas is evolved when acids attack iron or zinc [5]. Obviously these metals contained a considerable quantity of sulphur.

We have referred to Bergman's observation that hydrogen sulphide gives precipitates with many metals, but surprisingly, these reactions were used only for the detection of hydrogen sulphide. Winterl pointed out that lead can be completely precipitated with hydrogen sulphide, and that after filtration of the precipitate lead cannot be detected in the filtrate with sodium sulphate [6]. This was mentioned only in a note, however, and he did not use the reaction for his analyses. For the detection of lead in wines, Fourcroy and Hahnemann [7] almost at the same time recommended the use of hydrogen sulphide [8, 9]. The latter used hydrogen sulphide in water as the reagent, prepared in the following way:

calcium sulphide was shaken with a solution of tartaric acid in a stoppered flask, and then allowed to settle. The supernatant liquor was then used as the reagent.

This rather difficult method of preparation is somewhat curious because Scheele had mentioned earlier that by the ignition of iron filings with sulphur "sulphurated iron" is formed, "which yields with acids bad-smelling sulphur-air" [10].

The main problem was the production of a uniform flow of the gas so that originally a solution was the easiest way of dispensing the gas. The earliest form of the hydrogen sulphide generator, which is still in use today, was devised by the English chemist Griffin [11]. This was later modified by the Dutch P. J. Kipp (1808−1864) who was the owner of a firm producing scientific apparatus, and the celebrated Kipp's apparatus first appeared in 1864 [12]. Fourcroy was the first to realize that this compound contained hydrogen and sulphur. The analytical uses of hydrogen sulphide, however, were only developed very slowly, firstly because of the inconvenience of using a gas, and secondly because its reactions, despite the investigation of Berthollet [13], were very uncertain, especially in the case of manganese, cobalt, nickel and iron, etc. It was left for Gay-Lussac to place hydrogen sulphide in the important position which it now occupies in analytical chemistry [14].

Gay-Lussac first prepared iron sulphide by igniting a mixture of iron filings and sulphur, and by treating this with acid he produced the gas. Gay-Lussac established that the effect of this reagent is dependent on the acid strength of the medium, some metals being precipitated from strongly acid solutions, while others were only precipitated from weakly acid solution. He also discovered that hydrogen sulphide can reduce certain metals in their highest oxidation state, sulphur being precipitated in the process. He made another interesting point, namely, that metals which cannot be precipitated from strongly acid media, are precipitated if potassium acetate is added to the solution, because "by the effect of this the metal is transformed into an acetic acid compound". This is the earliest application of the buffer-principle. In ammoniacal medium, as Gay-Lussac pointed out, all metals give precipitates with hydrogen sulphide (arsenic was not regarded as a metal).

In his book, Pfaff writes that in acid medium gold, platinum, silver, mercury, lead, tin, copper, bismuth, uranium, molybdenum, tellurium, arsenic, antimony and zinc form precipitates with hydrogen sulphide, while manganese, titanium, cobalt, nickel and chromium do not [15]. He also records the sensitivities of these reactions. Ammonium sulphide precipitates all metals with the exception of arsenic and chromium(?) [16]. (It must be mentioned that the alkaline earths, although they could be prepared pure, were not really regarded as metals, but were always referred to as earths and alkalis.)

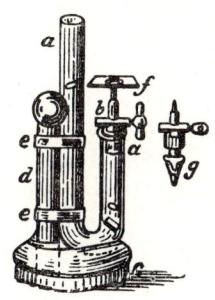

Fig. 50. Marsh's device for arsenic testing, 1836

If a new element was discovered, its reactions were immediately examined, but they were not generally used as reagents. There are only few exceptions to this rule, for example, the halogens, bromide and iodine, which, soon after their discovery, found important applications mainly in volumetric analysis.

Iodine was discovered by Bernard Courtois or, to be more accurate, by his cat. Courtois (1777—1838) first worked as an apprentice pharmacist, and then became an assistant to Fourcroy and Thenard. Eventually he left Fourcroy and opened a small saltpetre factory. But he could not compete with the Chile-saltpetre and in 1815 he became bankrupt. From then on he made a very poor living by producing other chemicals. It is said that his cat once pushed over a vessel containing sea-weed. The liquid mixed with something that had been spilt on the floor, and violet vapours appeared.

The characteristics, compounds and reactions of iodine were described by Gay-Lussac, and his famous paper dealing with this is often quoted as an example of chemical drafting [17]. The blue colour of the iodine-starch reaction, this very important phenomenon of analytical chemistry, was discovered by Stromeyer [18] in 1815 [19]. This reaction was at first used only for the testing of iodine or starch, and did not acquire its important role as an indicator in iodometric titration until twenty-five years later. Even today there is no comparable indicator.

Bromine was discovered by Balard [20] in 1825. He wanted to test for iodine in the ashes of sea plants, and in order to do this he added chlorine water to it. He observed that over the blue layer of iodine-starch there was another layer with an intense yellow colour. After distillation he obtained a red liquid. He examined the reactions of the new element and established that it could be extracted with ether [21].

The most sensitive test for arsenic (used as a poison since the oldest times) was the Marsh test. Its discoverer, Marsh [22] published his method in 1836, and a picture of his first apparatus can be seen here (Fig. 50) [23]. The method and the apparatus have since been modified many times, but the principle remains the same. Arsenic is reduced to hydrogen arsenide with sulphuric acid and zinc. Thompson soon discovered that the test is not specific, and that antimony gives the same reaction [24]. Marsh himself tried to distinguish between the two elements. He placed a drop of water on the piece of porcelain on which the spot was developed and held it some distance away from a flame, when the arsenic was oxidized to arsenious acid. On treating the water drop with silver nitrate it gave a yellow turbidity whereas antimony gave no reaction. In the next few years a large number of papers were published dealing with distinguishing between antimony and arsenic. The widely used sodium hypochlorite method was first recommended by Wackenroder [25] in 1852 [26].

Another test for arsenic, is the Gutzeit test. This was published by Gutzeit in 1879, and is based on the fact that hydrogen arsenide gives a yellow colour to a paper impregnated with silver nitrate [27]. The reaction itself, however, was known before this. The test for arsenic with tin chloride, the so-called Bettendorf [28] test, has been known since 1870 [29].

Of the classical inorganic reagents ammonium molybdate was the last to be discovered. It was recommended by Svanberg [30], in 1848 as a very sensitive test for phosphorus [31].

Thus, the reagents for qualitative analysis were available, and all that was required was to arrange the reactions in some systematic order so that a scheme of analysis could be developed. It was impossible for a practical analyst to sort out the large number of available reactions as there were no guiding principles. Berzelius only described separate analyses, and Pfaff, although he had noted the reactions of all the different reagents, comments on the qualitative examination of the elements as follows:

> The chemical behaviour of the substance must be determined first of all by a preliminary (qualitative) test. This can be done in two ways, either by wet or dry methods; generally the latter is sufficient.

For the wet tests he takes up two pages in his book, which can be summarized as follows:

> The substance to be tested must be dissolved in a strong solvent. Important conclusions can be drawn from the behaviour of the substance during the dissolution, namely whether it dissolves or not, if so, whether with ease or with difficulty. It must also be noted whether any gas is evolved during the dissolution, of, if the solvent is hydrochloric acid, whether any oxidized hydrochloric acid (chlorine) is formed, or if nitric acid is used whether any nitrous fumes are evolved. He states that the colour of the solution is also important from this point of view, but does not give any details. The solution must then be examined with reagents. The reagents used must be such that especially characteristic reactions with the groups of substances occur, for example, hydrogen sulphide solution, ammonium sulphide potassium hexacyanoferrate (II). These reagents are to be used for the testing of metals. Other reagents are ammonium carbonate, ammonia, potassium hydroxide and potassium oxalate, and their reactions are described in the introductory part of the book [32].

This is all the information that Pfaff gives and apparently it did not mean a great deal to the practising chemists of that time. However, we can now see that these few lines summarize the group separation of the cations, as this is based on the use of these reagents.

After Pfaff's book the next important textbook of analytical chemistry was written by Heinrich Rose. His book, *Handbuch der analytischen Chemie* was published in Berlin in 1829. It had reached four editions by 1838, while in 1851, with a new title and a more detailed text, it was published as *Ausführliches Handbuch der analytischen Chemie*. It was also published in French and English.

Heinrich Rose (1795—1864) was born into a family which already had many connections with chemistry. His grandfather, Valentin Rose, was assistant to Marggraf and later had Klaproth as his apprentice. His father, bearing the same name, who was a pupil and student of Klaproth, was also a pharmacist and published several chemical papers. Thus, it was not surprising that both grandsons, Heinrich and Gustav, also studied chemistry. They were both educated in Danzig and Berlin, and after completing their studies they worked for two years in Stockholm under Berzelius. They indeed had a very favourable start to their careers as scientists. In 1823 Heinrich was appointed (at the age of 25) Professor of Chemistry at the University of Berlin, while his brother Gustav became Professor of Mineralogy in 1826 at the same university. They worked there until their deaths. Heinrich Rose discovered niobium and his name is preserved by the crucible named in honour of him and used for ignition in a gas atmosphere.

Heinrich Rose used a new concept in his book; he did not treat the different reagents separately and give their reactions with the various elements, but used the present day system of presenting the different elements in separate chapters, and giving a summary of their reactions, starting with potassium. His treatment of the subject is very dull and the printing of the book makes it very tedious to read. This was improved in later editions by spacing out the text. In the fifth edition of this book there is a report of a new element, called pelop, which is black and shows similar properties to tantalum. On combustion it forms pelopic acid, which is insoluble in *aqua regia*, but is soluble in a mixture of nitric and hydrofluoric acids. The aqueous solution of sodium pelopate does not react with hydrogen sulphide, but gives white precipitates both with barium chloride and silver nitrate; the latter turns brown when ammonium hydroxide is added, etc. [33]. Pelop was considered to be a new element, but further work showed that this assumption was incorrect and in the 1871 edition of Rose's book there was no mention of pelop. Such erroneous discoveries were common in the last century.

Rose presents a general scheme of analysis. First of all he tests with hydrochloric acid, and identifies silver, mercury and lead. Then the next group of elements is precipitated and separated with hydrogen sulphide, and a further group with ammonium sulphide and then finally with potassium hydroxide. The elements in the different groups were examined, mainly by the same methods that are in use today. Rose also notes the solubility of the sulphides of arsenic, antimony and tin in ammonium sulphide.

The order in which Rose examines the elements corresponds to their chemical reactions with the individual reagents. (The order starts with the present day 5th group and ends with group 1.) After listing the series of reactions he describes a scheme of analysis, which in general is similar to that used today: first of all hydrochloric acid was added to the solution, when silver, mercury and the majority of the lead present were precipitated. Hydrogen sulphide was then bubbled into the slightly acid solution, and the precipitate extracted with ammoniacal ammonium sulphide to redissolve the gold, antimony, tin and arsenic. Gold was detected with tin by forming Purple of Cassius, while tin was similarly tested with gold. If the solution became turbid on dilution then antimony was present. The residue contained cadmium, lead, bismuth, silver and mercury. By treating with nitric acid, all these elements with the exception of mercury sulphide were dissolved. To a portion of this solution ammonia was added and if a blue colour was formed then copper was present. By testing with hydrochloric acid, silver, and with sulphuric acid, lead could be detected. If a turbidity was caused when the solution was diluted, then bismuth was present. Cadmium could be detected with the blow-pipe. To the filtrate from the hydrogen sulphide precipitation, ammonium sulphide was added, when iron, nickel, cobalt, zinc, manganese and aluminium were precipitated. This precipitate was then dissolved in hydrochloric acid (in the case of cobalt and nickel the solution was digested with nitric acid). Iron and aluminium were precipitated from this solution with ammonia, and the precipitate redissolved and treated with alkali which precipitated the iron. If the filtrate from the precipitate with ammonia was blue, then nickel was indicated, while if it was pink, cobalt was present. These two elements were precipitated with alkali, and cobalt identified with the blow-pipe by the cobalt-blue reaction. Zinc and manganese were precipitated from the filtrate with ammonium sulphide. Barium, strontium and calcium were precipitated with ammonium carbonate from the first main filtrate, and after filtration of this precipitate the magnesium was precipitated with sodium phosphate. Potassium was detected with platinum chloride [34].

Rose's book, therefore, was the first to describe the reactions of the individual elements, and also the first to present a systematic course of analysis for the different elements. This scheme was probably the work of Rose himself.

With Rose's book, however, it was possible for a beginner to become confused by the mass of information, and the lack of order in the presentation. Remigius Fresenius in the preface of his book says,

> I could perceive that in the great wealth of material, presented in the classical work of Rose, beginners are losing their way, as this very fine book is not very clear to them [35].

Fresenius rectified this characteristic defect of qualitative analysis when he proposed his scheme of cation analysis. He chose the most important elements and separated them into groups on the basis of their characteristic reactions. When we read this book, we have the impression that it is a modern textbook of qualitative analysis. This feeling is quite natural as qualitative analysis even today is written, taught and carried out according to the system of Fresenius.

Fig. 51. Carl Remigius Fresenius (1818—1897). (From *Zeitschrift für analytische Chemie* [1897])

His book, *Anleitung zur qualitativen chemischen Analyse* was undoubtedly one of the most successful books ever written on analytical chemistry. The first edition of this book appeared in 1841, the second in 1842, the third in 1844, the fourth in 1846, and by 1852 the seventh edition had been published. It is extremely rare for a book to reach so many editions during 10 years, and this success was certainly due to his clear presentation of the subject matter. By the time of the author's death in 1897 the book had reached 16 editions. Each edition was revised and brought up to date, but no major revision of the text was ever required. Only three years after its first appearance, English, French, Italian, Dutch, Spanish, Hungarian, even Chinese and, later, Russian editions were published [36].

Fresenius originally wrote his book for his own use when he was a university student in Bonn. In those days there was no practical laboratory instruction (university laboratories did not exist) so he worked in the private laboratory of his professor, Marguart. As he had to work mostly alone, with no instruction in analytical chemistry, he attempted to teach himself by examining all the analytical processes. His professor found his notes interesting and persuaded Fresenius to publish them in book form. This he did and was probably surprised at the tremendous enthusiasm with which it was received.

Carl Remigius Fresenius was born in Frankfurt am Main in 1818. His father was a solicitor. After finishing at the secondary school he worked for several years as an apprentice pharmacist before going to the University of Bonn. In 1841 he became a lecturer at the University of Giessen and worked under Liebig. This university, due to Liebig's insistence, was one of the few where the students were given instruction in practical chemistry. In Liebig's preface to Fresenius's book he states that all the mineral analyses in this laboratory were carried out according to this text. He became a senior lecturer in 1843, and in 1845 he was invited to become Professor at the Agricultural College at Wiesbaden. There was no laboratory here, however, so Fresenius bought a house and built his own private laboratory where students who were interested in analytical chemistry could come and practise. This private laboratory was founded in 1847 and work was begun with 1 assistant (Erlenmeyer [37]) and 5 students. In 1852 he had thirty students and in 1855 he had sixty students, The institute soon began to flourish and its name and reputation became well known. Many industrial firms consulted the institute about their problems. The small principality of Nassau in which Wiesbaden is located, agreed that any time spent by a student at the institute be recognized as university studies. The work of the institute was interrupted in 1861 when Prussia occupied the small state, but this setback was only temporary and the institute soon resumed its important role in the development of analytical chemistry. The institute still exists today.

Fresenius directed his institute and published his journal up to his death, and there are few analytical chemists who attained such a reputation. As well as a scientist he was a respected public figure and a wealthy property owner. He died in 1897, in his 79th year.

In 1862 Fresenius founded the first journal entirely devoted to analytical chemistry, entitled *Zeitschrift für analytische Chemie*, which is still published, and he was the chief editor until his death. Most of the chemical journals were mainly filled with papers dealing with the analysis of minerals or ores. The rapid progress of organic chemistry resulted in descriptions of new compounds and their synthesis occupied more and more space in these journals, so that slowly organic chemistry became the dominant subject. For example, *Annalen für Chemie* (Liebig's *Annalen*) had a marked analytical and inorganic chemical character at the beginning of the 19th century, but by the middle of this century it was mainly an organic chemistry journal.

ZEITSCHRIFT
FÜR
ANALYTISCHE CHEMIE.

HERAUSGEGEBEN

VON

Dr. C. REMIGIUS FRESENIUS,
HERZ. NASS. GEH. HOFRATHE, DIRECTOR DES CHEMISCHEN LABORATORIUMS ZU WIESBADEN
UND PROFESSOR DER CHEMIE, PHYSIK UND TECHNOLOGIE AM LANDWIRTHSCHAFTLICHEN
INSTITUTE DASELBST.

ERSTER JAHRGANG.

MIT 43 HOLZSCHNITTEN, EINER LITHOGRAPH. UND EINER FARBENTAFEL

WIESBADEN.
C. W. KREIDEL'S VERLAG.
1862.

FIG. 52. Title page of the first volume of the first analytical journal: *Zeitschrift für analytische Chemie* (1862)

The foundation of Fresenius's journal marks the end of "universal" chemistry, as this was the first journal which dealt only with a small branch of chemistry. Other journals dealing with specialized fields of chemistry soon appeared. Fresenius wrote in his announcement of the journal:

The establishment of this journal characterizes the importance of analytical chemistry. It requires little knowledge to realize that all the major developments in the field of chemistry are connected in some way with the developments in the methods of analytical chemistry. The development of suitable methods of mineral analysis resulted in the establishment of stoichiometric laws; improvement in the methods of inorganic analysis made possible the accurate determination of atomic weights; after the foundation of organic analysis a rapid development of organic chemistry occurred, while spectrography immediately led to the discovery of new elements. The development of analytical chemistry, therefore, always preceded the development of general chemistry. Just as new routes lead to new discoveries so improved analytical methods always yield new chemical results. The development of analytical chemistry, however, is important to other sciences and professions. For example, the study of blow-pipe reactions became very important in the field of mineral analysis, development of organic analysis led to improved methods for testing drugs and therefore was advantageous to pharmacy. Alkalimetry, chlorimetry and other rapid volumetric methods of analysis made possible the production of chemicals of guaranteed composition. The development of a method of analysis for nitrogen led to a wealth of discovery in the physiological and agricultural sciences, and together with other tests on body fluids provided an important new means of medical diagnosis. Methods devised for the testing of poisons and blood became dangerous enemies of evil-doers. Together, therefore, analytical methods represent an important contribution and a worthy scientific treasure for mankind [36].

Other countries soon followed this example and *The Analyst* published in London in 1875 was the next to appear.

To refer back to Fresenius's book, this book was expressly intended as a textbook for students and beginners. He tried therefore to establish a clear, easily followed system, which has since become very well known.

Fresenius divided the metals, or more accurately the metal oxides, into six groups:

Group 1. Potassium, sodium, ammonium. Their sulphides and carbonates are soluble in water. Aqueous solutions of their oxides make litmus turn blue.

Group 2. Baryta, strontianite, lime and magnesia. Their oxides dissolve with difficulty, while their sulphides dissolve easily in water. They can be precipitated with alkali carbonates and phosphates.

Group 3. Alumina and chromium oxide. They are insoluble in water. They cannot be precipitated with hydrogen sulphide, but with ammonium sulphide their hydrated oxides can be precipitated.

Group 4. Oxides of zinc, manganese, nickel, cobalt and iron as well as iron oxydule. They cannot be precipitated by hydrogen sulphide in mineral acid medium, also in neutral solution they are either not precipitated, or only precipitated to a very small extent. In alkaline medium their sulphides are precipitated.

Group 5. Silver oxide and mercury oxide, mercury oxydule, lead oxide, bismuth oxide, copper oxide, and cadmium oxide. Their sulphides can be precipitated with hydrogen sulphide from acid, alkaline and neutral solution. The group can be sub-divided further, according to the reactions of the metals with hydrochloric acid (silver oxide, mercury oxydule and lead oxide are precipitated, while the others are not). Lead chloride is appreciably soluble so that it can also go into the second sub-group.

Group 6. Gold oxide, platinum oxide, antimony oxide, tin oxide, tin oxydule, arsenious acid, and arsenic acid. In acid solution they are precipitated with hydrogen sulphide, but the sulphides are soluble in alkali or ammonium sulphides. They can be further sub-divided into two parts: some of the sulphides are soluble in hydrochloric acid (gold and platinum), while the others are not.

It can therefore be seen that the grouping of the metals is essentially the same as is in use today, only the order has been changed so that Fresenius's first group is now the last. Fresenius only examined the more important metals and did not consider the less common metals (in later editions they are mentioned in the appendix). This selection of metals by Fresenius is still valid in some places, so that many chemists have no idea of which group metals not referred to by Fresenius, such as vanadium, tellurium, tantalum or osmium fall into.

Fresenius used fewer reagents than his predecessors, but the ones he used were selected with care, and most of them are still in use today. One or two have gone out of fashion; for example, hydrogen silicofluoride which was used for the identification of barium in the presence of the other alkaline earth metals.

For an example of his analysis we can consider the fifth group, second subgroup.

The sulphide precipitate which consists of mercury, lead, bismuth, cadmium and copper, must be dissolved in hot nitric acid. If the dissolution is complete, apart from any sulphur, then generally mercury is absent. The excess of nitric acid is removed from the filtrate by boiling. To one portion of this solution sulphuric acid is added and the solution is heated. If no precipitate is formed then lead is absent. To another portion an excess of ammonia is added and the solution heated. If no precipitate is formed then bismuth is absent, while if the solution becomes blue, then copper is indicated. Small amounts of copper, however, cannot be identified in this way, so that the solution must be evaporated to dryness, acetic acid and potassium hexacyanoferrate (II) must be added, and if copper is present a red precipitate is formed. The residual solution is tested for cadmium with hydrogen sulphide. If copper is present then the copper sulphide obscures the cadmium sulphide. In this case the solution must be treated with potassium cyanide to dissolve the copper sulphide when the cadmium sulphide remains.

Thus it can be seen that during the course of the last century no important alterations have been made to this process.

Acids (anions) were also examined by the methods derived by Fresenius; these also are still in use at the present day.

It is interesting to examine the equipment recommended by Fresenius for the use of his students:

A spirit burner as recommended by Berzelius, blow-pipe, one platinum crucible, one platinum sheet and 3—4 platinum wires, a test tube stand with 10—12 test tubes, several beakers and flasks, one porcelain dish and a pair of porcelain crucibles, several glass filter funnels in various sizes, a wash-bottle, several glass rods and watch-glasses, one agate mortar, several iron spoons, a pair of steel or brass pincers, a filtration stand made of wood and one iron tripod stand [38].

We can also find some pedagogical directions in Fresenius's book. After becoming familiar with the reactions of the single ions, the student must examine unknown samples. It is important that the student should later be told whether his result was correct or not, therefore he must always receive substances, the composition of which is known by the teacher. The latter should answer "yes" or "no", but never "perhaps" or "probably". He considered that 100 practical qualitative analyses must be carried out in order to obtain an adequate knowledge of this subject.

The first twenty samples should contain only one water-soluble salt, of which only the cation need be identified. In the next thirty samples one anion (acid) and one cation must be identified. Among these samples there should be in addition to soluble salts several samples which require fusion. The next fifteen samples should contain several cations in aqueous solution, followed by a further fifteen samples of solid salt mixtures. The last twenty analyses should be carried out on naturally occurring substances, water, earth, minerals, alloys, etc. [39].

The development of qualitative analysis was virtually complete, the only further radical change was in the size of the sample, with the introduction of microanalysis.

We now come to the introduction of the organic reagents. In modern analytical literature, organic reagents are always referred to as if they were comparatively new. The chemist thinks first of dimethylglyoxime, introduced by Chugaev [40] in 1905 [41], the beautiful red precipitate being so characteristic for nickel. Some textbooks refer to dimethylglyoxime as the first organic analytical reagent.

In fact, the first analytical reagent of which there is record, the gallnut liquid, recommended by Pliny for identification of iron, is an organic reagent. Several other organic reagents were in common use long before dimethylglyoxime was discovered, for example, oxalic acid, tartaric acid, succinic acid, and starch.

Nowadays, when organic reagents are referred to, it is usually the synthetic and not the naturally occurring ones which are meant. Very often the term is limited to a reagent which is specific for one metal (this criterion is only fulfilled by a very few reagents). Even in this limited sense there were predecessors to dimethylglyoxime.

The Griess—Ilosvay reagent, amixture of α-naphthylamine and sulphanilic acid can be regarded as the first specific organic reagent; with nitrite ions this reagent gives a red colour.

The reaction was recommended by Griess for the identification of nitrite in sulphuric acid medium in 1879 [42]. Griess was primarily an organic chemist, and in his paper he only briefly mentions the analytical application of this reagent. He referred to two earlier papers which described the use of diazobenzoic acid [43] and phenylenediamine [44] as reagents for nitrite ion. In actual fact reference [43] is a paper by Griess on diazobenzoic acid, but no analytical applications are mentioned, while phenylenediamine, as Griess himself mentions, is not a very good reagent for nitrite. Under the conditions given by Griess and his co-workers, the α-naphthylamine—sulphanilic acid reaction was not sufficiently sensitive. In 1889 Lajos Ilosvay made a study of the reaction and recommended the use of an acetic acid medium. He also used the reagent for the detection of nitrate after reduction with zinc [45].

Peter Griess (1829—1888) was born in Germany into a wealthy peasant family. He was educated at the technical school at Kassel, and later at the Universities of Jena, Marburg

Fig. 53. Lajos Ilosvay (1851–1936). Photograph

and Munich. His studies continued over a rather long period as he was absorbing more beer than knowledge. He completed his studies in 1856 and was employed in an aniline dyestuff factory. He quickly became an industrious worker, mainly with the object of clearing the considerable debts incurred during his student days. In 1858 he discovered the basis of diazotisation, and during this time he became acquainted with W. Hoffmann, the famous organic chemist who visited him during his travels. His impression of Griess was not very favourable:

> Instead of the young man I expected, I found a man who had passed his first youth, and I could not discern very much intelligence in his pale, inexpressive face... He gave me the impression of a man who spent more time in beer-cellars and fencing-planches than in cultural pursuits... His manner was rather discourteous; he treated me as if I was an unwanted visitor who had interrupted his work... He only became cheerful when I mentioned his recently published work... From then on it was impossible to talk about any other subject. He talked only about his reactions and his great hopes in connection with his investigations [46].

Shortly afterwards Hoffmann invited Griess to London. He appeared in an old top-hat, with an enormous red scarf around his neck, wearing a brown coat and sea-green trousers. In London he continued his research on diazotisation. It was about this time that a French university professor mentioned in one of his public lectures that the English breweries added strychnine to their beer to increase the bitter taste of their products. When the newspapers in England reported this, there was an immediate decrease in the amount of beer consumed. One of the leading breweries commissioned Griess to prove that the accusation was false. After this Griess continued to work for the brewery at Burton-on-Trent where he stayed until his death.

Lajos Ilosvay was born in Dés in 1851. He was educated in Kolozsvar, but finished his studies before the last year of the gymnasium course, and became an apprentice pharmacist. After being apprenticed for four years he went to the University of Budapest, and eventually obtained a doctor's degree both in pharmacy and philosophy. Then he became a lecturer at the same university and carried out research work. In 1880 he was awarded a fellowship and spent two years abroad, first of all in Heidelberg where he worked under Bunsen, then in Munich under Baeyer, and finally in Paris under Berthelot. In 1882 he became Professor of General Chemistry at the Technical University of Budapest. He remained director of this institute for more than fifty years, until 1934, and during his life he was prominent in the university, scientific and public life of his country. Ilosvay was also the Vice President of the Hungarian Academy of Sciences. He died in 1936.

Diphenylamine was an organic reagent known before α-naphthylamine and sulphanilic acid. It was recommended for the detection of nitirte; the blue colour formed in sulphuric acid medium was first discovered by Hoffmann [47, 48]; Kopp [49] later recommended it as a reagent for qualitative and colorimetric analysis (1872) [50]. This reaction, however, is not specific, a number of oxidizing agents giving the same colour.

Since then the number of synthetic organic reagents which have been recommended for qualitative tests has become enormous. The main objective of specificity, however, has only been achieved by a very few. Two good examples of specific and selective organic reagents, similar to the Griess—Ilosvay reagent, are dimethylglyoxime and starch.

2. GRAVIMETRIC ANALYSIS

Gravimetry is the oldest branch of quantitative analysis, and is also the most logical method because the determinations are made by direct weighing. In the earliest times the determination of metals was carried out by converting them to the elementary state. A big step forward was made when the metals were determined in the form of easily precipitable compounds, the amount of the metal present at first being found by experiment and later with the help of stoichiometric equations. Most of the classical gravimetric forms for the determination of the individual elements or radicals are taken from one of their characteristic qualitative reactions. Gravimetric methods were the first quantitative methods of analysis to be developed. In the earlier part of the last century only gravimetric methods of analysis were used, even though some volumetric methods had been in existence for fifty years. Both Klaproth and Berzelius used only gravimetric methods and even the analytical textbooks of Rose and Fresenius do not describe any analytical methods other than gravimetric ones. It was not until the 1860's that volumetric methods became accepted in analytical chemistry and appeared in the standard textbooks.

There has been no essential change in the basic technique of gravimetric analysis since the middle of the last century. It is interesting to observe how a gravimetric analysis was carried out at that time; contemporary textbooks such as those of Rose and Fresenius give suitable illustrations.

Rose's book was published in German in 1829, but the examples given below are taken from the French edition of the book (*Traité pratique d'analyse chimique*, Paris 1832). Rose gives a list of elements, together with the form in which they should be determined. He describes the procedure for the determination of the pure element, together with the modifications required if other elements are present. It is remarkable that there are no better methods even today; in many analytical textbooks the methods of separation are often confusing.

There are no experimental details in Rose's book, only a general outline of the various methods. The following extract describes the determination of magnesium and the separation of iron from manganese.

> Magnesium: If no other elements are present in the solution, then it must be evaporated with sulphuric acid. In the presence of alkalis it is precipitated in the form of the carbonate.
> Determination in the presence of calcium: Ammonium chloride is added to the solution in sufficient amount to prevent the precipitation of magnesium hydroxide. Ammonium hydroxide is then added and the calcium precipitated with oxalic acid. Although magnesium oxalate is not very soluble, in the presence of sufficient ammonium chloride it does not precipitate, but it can be precipitated from the filtrate by the addition of either potassium carbonate or sodium phosphate [51].
> Separation of iron from manganese: This is a difficult problem for an analyst. Ammonium chloride must be added to the solution (this is not necessary if hydrochloric acid is present in any quantity) and then ammonia added carefully until iron hydroxide just begins to appear. This must be redissolved, and then a neutralized solution of succinic acid (sodium succinate) is added when iron succinate is slowly precipitated. Care must be taken to prevent the succinic acid reducing the iron during the ignition. This can be avoided by moistening the precipitate with ammonia, when the hydroxide (hydrous oxide) is formed, this being indicated by the colour change and the decrease in volume [52].

At the end of the book there are almost 100 pages of analytical tables, and these are arranged under the headings of substance sought and substance weighed in the manner used at the present time. It would appear that the analysts of the last century attempted to avoid the use of calculations whenever possible, because these tables dispense with the need for making any calculations. The tables have the following form:

Substance weighed	Substance sought	Atomic weights									
		1	2	3	4	5	6	7	8	9
potassium sulphate $KOSO_3$	potassium oxide KO	0·54067	1·08134	1·62201	2·161268	2·70335	3·24402	3·78469	4·32536	4·86603

In order to calculate the metal oxide content of 2·6589 g of the potassium sulphate precipitate, the following calculation is made:

$$\begin{array}{r} 1\cdot08134 \\ 0\cdot32440 \\ 0\cdot02703 \\ +0\cdot00432 \\ \hline 1\cdot43709 \end{array}$$

the last figure gives the potassium oxide content of 2·658 g of potassium sulphate.

"Thus the chemist", Rose writes, "who is not accustomed to the use of logarithmic tables" need only make additions. Rose considered that by using this calculation the results obtained were more accurate than those obtained with the use of logarithmic tables. The use of a slide rule is far too inaccurate and must never be used. The originator of this type of table, Rose tells us, was Poggendorff [53]. From this example it can be seen that the weighing process was accurate to 1 milligram.

Fresenius's quantitative analytical textbook was published soon after his famous qualitative work, in 1846. This book, in contrast to Rose's textbook, gives very detailed experimental instructions. The extracts from this book, given below, are taken from the second edition, published in 1847 [54].

The preface to this book gives some very useful advice to the beginner in the study of chemistry, advice which is also of value to the trained analyst:

> Knowledge and ability must be combined with ambition as well as with a sense of honesty and a severe conscience. Every analyst occasionally has doubts about the accuracy of his results, and also there are times when he knows his results to be incorrect. Sometimes a few drops of the solution were spilt, or some other slight mistake made. In these cases it requires a strong conscience to repeat the analysis and not to make a rough estimate of the loss or apply a correction. Anyone not having sufficient will-power to do this is unsuited to analysis no matter how great his technical ability or knowledge. A chemist who would not take an oath guaranteeing the authenticity, as well as the accuracy of his work, should never publish his results, for if he were to do so, then the result would be detrimental not only to himself, but to the whole of science [55].

The book starts with a description of the weighing process and we can see that the sensitivity of the balance was about the same as it is today, i.e. 0·1 mg [56]. Fresenius also describes apparatus for the measuring of liquid volumes, but makes no reference to the volumetric flask or the pipette although both of these were known at this time. Fresenius considered that drying was more prone to error than was ignition. For drying he used steam-jacketed metal containers, as well as drying by means of a warm air current.

The first operation in an analysis consists of drying the sample which is then dissolved and finally precipitated. The more insoluble the precipitate, the more accurate is the procedure, although it must be realized that all precipitates are soluble to a certain extent, but that this solubility is so small in many cases as to be negligible. Solubility can also be decreased by evaporation of the solvent, or in some cases, for example, with lead chloride, calcium sulphate and potassium chloroplatinate, by the addition of alcohol.

The filtration and washing of the precipitates was carried out in the same way as it is today. The final determination was made either by drying or ignition, and in both cases the weight of the filter paper had to be taken into consideration. In the first case the weight of the filter paper itself had to be subtracted, while in the second case the weight of the ash had to be subtracted from the result. Even with the best filter paper the ash content was 0·3 per cent. The filter paper was cut from the arch of the Mohr-form to a suitable shape, and in order to determine the ash content accurately, ten of these pieces were ignited and the ash weighed.

The automatic devices for filtration, decantation, and washing, so often recommended by Berzelius, were abandoned as they did not improve the accuracy of the methods.

The next chapter deals with gravimetric methods for the determination of the individual elements. Fresenius records the percentage composition of the precipitates, as well as their "equivalence weights", i.e. their molecular weights as we know them. His calculations are based on an equivalent weight of 100 for oxygen [57]. The majority of the values correspond to those of Berzelius, but many of them were the more recent and correct values determined by Dumas, Erdmann and Marchand, Marignac, Rothoff and others. To give a few examples:

For the determination of potassium Fresenius recommends sulphate, chloride, nitrate and chloroplatinate [58], for barium he recommends sulphate, carbonate and silicofluoride [59] and for calcium, sulphate and carbonate. He states that calcium oxalate should be weighed after ignition as calcium carbonate. Calcium oxalate contains 1 molecule of crystalline water which it loses at 180—200°C, while at dull red heat it loses carbon monoxide to form the carbonate. The carbonate on vigorous heating loses carbon dioxide to form the oxide, but this process is not quantitative even at high temperatures [60].

Although this information could only be derived from thermal investigations, he does not give any detail of how he carried out these experiments. He also recommends that magnesium be determined as the pyrophosphate or the oxide. The formula of the former is given as Mg_2PO_6 [61]. For aluminium, chromium, nickel, cobalt, only the oxides are known as the weighing forms, while for zinc, manganese, copper, bismuth, the sulphides and carbonates are mentioned as weighing forms, but only after drying. Silver can be determined as the chloride, sulphide, cyanide or as metallic silver. With regard to silver chloride, Fresenius records that it is soluble in ammonia, turns violet on exposure to air, and melts at 260°C. Silver chloride must be ignited very carefully because the carbon monoxide formed from the combustion of the filter paper may partly reduce it to metallic silver. Therefore the bulk of the dry precipitate must be ignited separately and then the residue combined with the ash of the filter paper. The whole residue is then moistened with a few drops of nitric acid, and then a few drops of hydrochloric acid added, and the precipitate carefully reignited [62]. Silver cyanide can only be determined after drying.

Among the "acids" chromate can be determined as chromium oxide or lead chromate, arsenate as lead arsenate, phosphate as lead phosphate, magnesium pyrophosphate, basic iron phosphate or silver pyrophosphate, while bromide and iodide can be determined as their silver salts. It had already been known for some time that in the precipitation of aluminium with ammonia, any large excess of ammonia will cause the precipitate to be redissolved, and also that in the precipitation of magnesium with sodium phosphate it is necessary to have a considerable amount of ammonium chloride present in order to avoid the precipitation of magnesium hydroxide.

For the complete precipitation of lead as sulphate it is necessary either to add alcohol, or to evaporate the solution to a small volume. Alcohol is also used to wash the precipitate.

Table 10 gives the results obtained by Fresenius on the composition of the precipitates compared with the data of today. These values can also be compared with earlier results, which were shown previously (Chapters V. 1, VI. 1).

The formulae and symbols were originally used by Fresenius. It can be seen that, with very few exceptions, these values agree fairly well with the values of today.

The next chapter of the book deals with separations, and there is only space enough here to give a few examples.

In the separation of calcium and magnesium, calcium must first be precipitated with ammonium oxalate from ammoniacal medium. From the filtrate magnesium is precipitated with sodium phosphate in the presence of ammonium chloride. If phosphate is also present in the solution, calcium can be precipitated in acetic acid medium with oxalic acid, while magnesium can be precipitated as magnesium ammonium phosphate [63].

In the separation of iron (III) from zinc or manganese, barium carbonate is added to the acid solution. The precipitate of iron hydroxide and barium carbonate is filtered off, and dissolved in hydrochloric acid. The barium is then precipitated with sulphuric acid, and after filtration the iron can be precipitated from the filtrate with ammonia. In the filtrate from the original solution, the barium can be precipitated as the sulphate, and then zinc, manganese and nickel can be precipitated in a suitable form. The results for iron are generally high, owing to co-precipitation of other metals. Consequently the results obtained for the other elements are found to be low by a corresponding amount [64].

Fresenius also records the separation of cobalt and nickel, which was devised by Liebig:

The metal solution must be made strongly acid with hydrochloric acid, and then potassium cyanide must be added until the first precipitate redissolves. The solution must then be boiled in a flask, held at an oblique angle with occasional additions of acid until all the hydrogen cyanide has been evolved. An excess of potassium hydroxide is then added, when nickel hydroxide is precipitated, while cobalt remains in solution as potassium cobalticyanide. The solution must then be evaporated to dryness with nitric acid, and heated strongly. The

Table 10

Compositions of Analytical Precipitates according to Fresenius

Compound		According to Fresenius	Theoretical
Potassium sulphate	KO	54.08	54.06
	SO_3	45.92	45.94
Sodium chloride	Na	39.32	39.33
	Cl	60.68	60.67
Sodium carbonate	NaO	58.47	58.47
	CO_2	41.53	41.53
Barium sulphate	BaO	65.64	65.70
	SO_3	34.36	34.30
Calcium oxalate	CaO	38.36	38.37
	C_2O_3	49.32	49.31
	1aq	12.32	12.32
Aluminium oxide	2Al	53.19	52.92
	3O	46.81	47.08
Zinc oxide	Zn	80.26	80.33
	O	19.74	19.67
Silver chloride	Ag	75.28	75.26
	Cl	24.72	24.74
Lead sulphate	PbO	73.56	73.59
	SO_3	26.44	26.44
Arsenic sulphide	As	60.95	60.89
	3S	39.05	39.11
Lead chromate	PbO	68.94	69.05
	CrO_3	31.06	30.95
Silicic acid	Si	48.03	46.74
	2O	51.97	53.26

residue is washed with hot water and dissolved in hydrochloric acid; from this solution cobalt oxide is precipitated by the addition of alkali. This procedure was rather unusual and was soon dispensed with when organic reagents were introduced [65].

For the separation of copper and mercury, Fresenius used sodium formate to precipitate the mercury. It is interesting to note that this reagent was not used for the separation of lead and bismuth; here the lead was precipitated as sulphate. He also describes a more complicated method:

Lead and bismuth are precipitated as carbonates and then dissolved in acetic acid. The solution is placed in a stoppered bottle together with a lead rod, and allowed to react for

12 hours with occasional vigorous shaking. The precipitated bismuth is filtered, and the precipitate dissolved in nitric acid. From this solution the bismuth is precipitated as the oxide. The lead is determined in the filtrate and the weight of lead dissolved from the leadrod subtracted from this amount.

The separation of tin and antimony was achieved by reducing the antimony to the metal with zinc. There was also a description of the simultaneous determination of chloride, bromide and iodide. Iodide was precipitated as palladium iodide and from the filtrate the chloride and bromide were determined by an indirect method.

Thus, indirect methods of analysis were in use at this time but there is no evidence to suggest how this type of procedure originated. According to Fresenius an indirect method is used only if there is no suitable direct method available.

The two substances must be precipitated with a common precipitant. For example with a mixture of chloride and bromide the following method can be used:

First of all the halides are precipitated with silver nitrate. Let us assume that the weight of the precipitate is 20 g. The precipitate is then heated strongly in a stream of chlorine gas for about 20 min; this transforms the silver bromide to silver chloride. After cooling, the precipitate is again weighed. Let us now assume that the decrease in weight of the precipitate is 1 g. The "equivalent" weight of silver chloride is 1792·21, while that of silver bromide is 2346·64, the difference being 565·54. This difference is proportional to the equivalent weight of silver bromide, similarly the difference in weight of the precipitate is proportional to the bromide content of the salt mixtures. Thus

$$556\cdot 43 : 2348\cdot 61 = 1 : x$$

$$x = 4\cdot 221$$

Thus the mixture contained 4·221 g of silver bromide and 20 − 4·211, i.e. 15·779 g of silver chloride.

Fresenius gives a similar example for the determination of potassium and sodium in mixtures [66].

Finally Fresenius gives details of the method of calculation of mean values, and also the method for the determination of the formula of a compound from its composition. The results shown below were obtained in a determination of the composition of a salt:

sodium oxide	17·93 per cent
ammonium oxide	15·23 per cent
sulphuric acid (SO$_3$)	46·00 per cent
water	20·84 per cent

The "equivalent weight" of NaO is 391, and as oxygen is 100, then it is 4·53 per cent from 17·93.

The "equivalent weight" of NH$_4$O is 325, and as oxygen is 100, then 15·23 per cent contains 4·68 per cent oxygen.

The "equivalent weight" of SO$_3$ is 500, of which 300 is due to oxygen, therefore 46·00 contains 27·60 per cent oxygen.

The "equivalent weight" of HO is 112·5 of which 100 is due to oxygen, thus from 20·46 there is 18·25 per cent oxygen.

The results obtained for oxygen in the substance are: 4·58 : 4·68 : 27·60 : 18·52, which, ignoring any small variations due to experimental error, are in the ratios 1 : 1 : 6 : 4.

Therefore the formula of the salt is

$$NaOSO_3 \cdot NH_4OSO_3 \cdot 4aq \quad [67]$$

This type of problem was encountered much more frequently by chemists in those days than it is today.

Finally the book describes the analysis of some naturally occurring substances (water, soil, etc.) and concludes with tables which give the stoichiometric factors in the same manner as is used today (sought/weighed). These tables are based on the equivalent weights established by Gerhardt. In most cases these values correspond with modern atomic weights, but a few results are double the present day values. In these instances the formulae were modified so that HO was used instead of H_2O and NaO instead of Na_2O.

It can be seen from this brief summary that gravimetric analysis was already well developed by 1847. The methods and techniques were almost the same as used today, and many of the original precipitation forms and separations are still in use. The technique of filtration, however, has been improved since then. In the earliest gravimetric procedures the ash content of the filter paper had to be taken into account. Many attempts were made to reduce the ash content, and in 1878 Austen [68], by using a hydrochloric and hydrofluoric acid treatment, prepared a filter paper which gave practically no ash on ignition [69]. Initially, therefore, it was possible for chemists to purify the filter paper themselves, but in 1883 the firm of Schleicher and Schüll (Düren) produced for the first time ash-free filter papers. This greatly simplified the techniques of gravimetry and since 1898 filter papers have been available which have an ash content of less than 0·1 mg [70], which is negligible.

The earliest form of filter crucibles, devised by Österreicher [71], who used a glass tube packed with broken glass, has already been referred to (see page 146.). The first robust and durable filter crucible was the Gooch crucible, recommended by Gooch [72] in 1878 [73]. Sintered glass crucibles were only introduced in the 1920's by the glass firm of Schott in Jena.

The use of synthetic organic reagents considerably increased the scope of gravimetric analysis. The specificity of the first reagents gave rise to the hope that a specific reagent could be discovered for each element, and so dispense with the need for separation. During this century a large number of organic reagents have been recommended for use in analysis, the majority of which have ended their career immediately after their publication, as in practice they proved to be unsuitable. It is often said that the first catch is the best catch, and it would also seem that very few of the more recent organic reagents can compare with α-nitroso-β-naphthol, or dimethylglyoxime.

It is the author's opinion that the enormous potential of organic reagents in analysis has not yet been realized. When a new organic reagent is introduced it is seldom used immediately by chemists in practice, partly because there is, as yet, no scheme of analysis for silicates, or for metals, which is based on the use of organic reagents. Classical analysis had its Fresenius, and titrimetry had its Mohr, who critically examined and correlated the varied procedures. In the field of organic reagents there is yet to appear a systematizer who could correlate all the existing knowledge. Experience has shown that such a work would result in a rapid spread of interest in this subject. It is true to say that in recent times only a very few books have been written which have fulfilled this requirement, not only in the field of gravimetric analysis, but in the whole of analytical chemistry generally. Recent monographs have reviewed the published methods without subjecting them to any practical examination, although admittedly a critical review of this nature would have involved a tremendous amount of work.

The first synthetic organic reagent to be used in the field of gravimetric analysis was α-nitroso β-naphthol, recommended by Ilinski and Knorre [74] in 1885 for the determination of cobalt in the presence of nickel [75]. The precipitate, obtained in acid medium, was ignited in a Rose crucible in a stream of hydrogen, and the residue of cobalt weighed. The organic metal complex itself was first used as a weighing form by Brunck [76] in 1907 [77]. Soon after this Baudisch [78] introduced cupferron for the separation of iron and copper [79]. After the first World War the number of synthetic organic reagents increased tremendously.

Gravimetric analysis developed in a completely empirical manner although precipitation, which controls the accuracy of the method, is influenced by several factors. The accuracy of analytical methods has increased during the last century to such an extent that slight errors due to variation in these factors must be taken into account. The development of crystal chemistry and the study of colloids made possible the general examination of the conditions for precipitate formation, as well as giving some information on the phenomena of co-precipitation, occlusion and inclusion, etc., and this information was used to establish the optimum conditions for precipitation. A considerable amount of research is still being carried out on the mechanism of precipitation.

One of the most notable workers in this field was Lajos Winkler who attempted to develop methods in gravimetric analysis with the highest attainable accuracy. He critically examined the problems of precipitation, filtration, effect of heating, and tried to select experimental conditions which would give the minimum error. Winkler also estimated the relative size of errors, but did not try to compensate for them as he realized that these random errors were variable. However, he did work out corrections for improving the results of analyses. Winkler devised many methods for the determination of ions, all of which gave very good results [80].

Lajos Winkler was one of the greatest personalities in Hungarian analytical chemistry. He was born in Arad in 1863, where he later became an apprentice to a pharmacist. Afterwards he studied pharmacy in Budapest, and after graduating he remained at the university as an assistant to Károly Than. After Than's death his original institute was divided, and Winkler, as Than's successor, was put in charge of one part of it. With only a very modest array of instruments, he achieved many important analytical results, many of which wil be referred to later. The most important of his 200 or so original papers deal with the deter-

Fig. 54. Lajos Winkler (1863–1939). Photograph

mination of dissolved oxygen in water, iodine and bromine numbers, absorption of ammonia in boric acid solution, and the determination of chlorine and iodine in water. Winkler wrote several books, mostly in German, and mainly describing his own methods, and he also wrote the section on water analysis in the final (1921) edition of the famous Lunge—Berl technical analytical handbook [80]. Winkler was one of the greatest exponents of classical analysis, and probably the most accomplished scientist in this field during this century. He was very much averse to physico-chemical methods of analysis, possibly because he considered that analytical chemistry was an art rather than a science, and that arts cannot be subjected to automation. He said that in order to obtain results, a certain amount of basic equipment is necessary just as a tailor needs a sewing machine, but even the finest equipment cannot replace the skill and intelligence of an expert chemist [81]. Winkler was for a long time the Director of the Chemical Institute at the University of Budapest, but in 1933 he contracted pneumonia and sepsis, from which he never completely recovered. His health gradually became worse, and he died in 1939.

For the control of precipitation many other methods were used and examined. One idea was to carry out the precipitation from an extremely concentrated solution, which causes crystals of great internal tension to be formed. These can easily be purified by recrystallization after dilution of the mother liquor. This technique was used by Kolthoff, Njegovan [82], and Marjanovich [83]. The opposing idea is to carry out the precipitation from an extremely dilute solution. In this method there is only relatively slight supersaturation, so that a crystalline, easily filterable precipitate is formed (Hahn [84, 85]). Precipitation from a homogeneous solution is another technique and is based on the slow generation of the precipitating reagent in the solution itself. An example of this is given by the slow hydrolysis of urea in boiling solution which gradually increases the pH of the solution until iron is precipitated as the hydroxide. The main advantage of this method is that a dense crystalline precipitate of high purity is formed. This technique was first described by Chancel [86], who recommended sodium thiosulphate for the precipitation of aluminium as hydroxide [87], but of course it was not referred to as precipitation from homogeneous solution (P.F.H.S.) in those days. A systematic study of P.F.H.S. has only been carried out in recent years, pioneered by Moser [88, 89] and, later, Willard [90]. Gordon and his school [91] represent the direction of this field of research today.

These methods are only the beginning of the study of the theoretical background to gravimetric analysis, and in this respect gravimetry lags behind other branches of analytical chemistry, although modern research may soon rectify this.

The examination of the effect of heat on precipitates was initially studied in order to improve the accuracy of gravimetric methods. It was important therefore to establish the temperature over which a precipitate was stable. In order to carry out these investigations it was necessary to construct a balance which could measure the weight of the sample during ignition. The first experiment to construct a thermobalance was completed by Nernst and Riesenfeld in 1904 [92]. The earliest crude thermobalance for practice was designed by Honda [93], and consisted simply of suspending a porcelain crucible from one side of the balance beam into a furnace. From this early thermobalance the study of thermogravimetry began, and it is now a specialized branch of analytical chemistry. The design of the thermobalance has also improved greatly, both weight and temperature now being recorded automatically.

The fundamental principles of thermogravimetry were established by Duval [94] who, together with co-workers, examined nearly a thousand analytical precipitates and determined the temperature ranges over which they showed a constant weight [95]. Duval has used this technique not only to find the optimum temperature for the drying or ignition of gravimetric precipitates, but also to carry out differential analyses, based on the different decomposition temperatures of two precipitates. For example, if a mixture of calcium and magnesium are precipitated as the oxalate and the precipitate is heated on a thermobalance, then as the decomposition of the two oxalates takes place at different temperatures, an examination of the weight v temperature curve will indicate the original amounts of calcium and magnesium [95]. This method is called by Duval "automatic gravimetric analysis".

Derivative thermogravimetry (L. Erdey, F. Paulik [96], J. Paulik [96]) is a further development of thermogravimetry. The derivative of the thermogravimetric curve is obtained instrumentally by applying the principle of magnetic induction and shows the rate of change of weight of the precipitate as a function of temperature, and gives a more precise indication of the starting and finishing temperatures of the various transitions [97]. Most of the recent investigations of F. Paulik, J. Paulik and L. Erdey have been concerned with a completely automatic instrument called a derivatograph, which, in addition to recording the thermogravimetric and derivative thermogravimetric curves, also records the change in enthalpy during the process [98].

3. MICROANALYSIS

Microanalysis is concerned with the identification and determination of small amounts of substances (it is often referred to as microchemistry) which cannot be accomplished by conventional macromethods. As on the macro scale, microchemistry has several branches, but in some cases it consists simply of scaling down macro methods and apparatus in order to cope with a small sample. In volumetric microanalysis, micro-pipettes and micro-burettes are used. Qualitative microanalysis, however, which is the classical microchemistry, is quite different both in principles and in practice from common wet qualitative analysis. Here the identification of the compounds is not based on the formation of a characteristic precipitate but on the examination of crystal form under the microscope of the compound formed with different reagents. Therefore reagents are chosen which give characteristic crystal forms and a high molecular volume. In wet analysis for example, silver can be detected by the silver chloride reaction, because the latter is insoluble and therefore the reaction is sensitive for silver. The microanalyst prefers to use silver chromate, because although this is less sensitive, the volume is much larger and therefore is more suitable. Separation, evaporation, and filtration, etc., on the micro scale, cannot be carried out by simply scaling down the techniques of macroanalysis, but needs an entirely different technique. In qualitative microanalysis the microscope, discovered by Leuwenhoek, is the most important piece of apparatus.

Antoine van Leuwenhoek (1632—1723) was an accountant in a draper's shop, who studied optics as a hobby. He discovered the microscope and carried out many microscope examinations; he discovered bacteria, spermatozoa, and blood corpuscles. Later he gave up his accountancy and concentrated on his scientific researches, despite difficult financial circumstances. He was a very modest man who was worried by his inability to speak Latin. His investigations were described in a book of seven volumes, written in Dutch [99].

In the discussion of Boyle's method of water analysis, reference was made to the fact that he examined one drop of water under a microscope and counted the particles moving in it (Chapter III. 2). He did not give any opinion on the nature of these particles. Marggraf established with a microscope that beet sugar is identical with cane sugar, and he also examined other substances under the microscope (Chapter IV. 3).

The investigations of Tobias Lowitz were essentially of a microanalytical nature. Lowitz examined various chlorides and other salts under the microscope and recorded their crystal forms and other characteristics, as a basis for their identification with a microscope [100]. The characteristic features of each substance were used for its analytical identification. The following extract is taken from Lowitz's examination of a platinum salt.

> It is known that in this compound platinum is united with hydrochloric acid, and it has to be established whether it also contains mine rock salt (NaCl). To establish this accurately, I took a small amount on a silver spoon and tested it with a blow-pipe. The platinum was obtained as the metal by this process after the evolution of hydrochloric acid fumes. I then placed the residue on a glass plate and mixed it with 4—5 drops of water and allowed it to dry. A salt deposit remained which was visible to the eye, but under the microscope it could be seen that rock salt crystals were present. In order to be quite certain that this deposit was rock salt, I treated it with two drops of diluted sulphuric acid and a few drops of water and rubbed it with my fingers and then heated the glass plate over a candle. I could smell the hydrochloric acid given off. I then mixed it with a few drops of water and after heating it to dryness the characteristic shape of the crystals of Glauber's salt was observed. The whole experiment which only required a small piece of the salt, and completely confirmed the presence of rock salt, did not take more than half an hour to perform [101].

Tobias Lowitz came from a scientific family. His father was Professor of Mathematics at the University of Göttingen, and later went to Petersburg in 1767 at the invitation of the Czar, where he became a member and Professor of the Academy. Tobias Lowitz, born in Göttingen in 1757, was apprenticed to an apothecary at the Court in Petersburg, and later studied at universities in Germany. He then returned to Russia and took charge of the apothecary shop at the Court, and in 1793 he was elected a member of the Russian Imperial Academy. He published many chemical papers, some in the journal of the Academy of Petersburg, and some in Crell's *Annalen*. He died in 1804 at Petersburg. He did a great deal of pioneering work in the fields of chemistry and technology, and was the first person to prepare glacial acetic acid. He also discovered gas-adsorption phenomena on *carbo medicinalis*. In analytical chemistry he carried out the separation of the alkaline earth metal chlorides with alcohol, and also noted their characteristic flame colours.

The next reference to the use of the microscope in analysis is by Raspail in 1831. In his book entitled *Essai de chimie microscopique appliquée à la physiologie*, Raspail recommends the use of the microscope for the identification of substances, particularly in physiology. He points out that, "to work with small amounts of substances is not only economical, but involves a new technique".

François Vincent Raspail (1794—1878) was a catholic priest and a Professor of Theology who later left the ecclesiastical order and devoted himself to the study of the natural sciences and politics. He was a fervent republican and, although he had received an award from the French Academy after the publication of his textbook of organic chemistry, he was arrested several times for his political views. In 1840 he suddenly became rich, owing to his discovery of a camphor cigarette, which he had patented. In 1853 he was expelled from France and was only able to return in 1871 when the Empire had fallen.

Physicians and biologists began to use the microscope more and more in their work, and in the course of this the descriptions of the characteristics of a great many compounds were recorded, but without any analytical application. There are a great many of these isolated references to the early use of the microscope, most of them given in zoological or botanical publications.

In chemistry, the microscope begins a new life in the second half of the last century. This arose from its use in mineralogy, along with the blow-pipe. Microscopy was quite well established in mineralogy at that time, and a complete journal was devoted to it. Sorby [102] in 1869 used a microscope for the identification of the products formed with a blow-pipe [103]. As mineralogists have always been searching for methods which would displace chemical analysis as a method for the identification of minerals, they welcomed the introduction of the microscope. The idea of basing a new method for the chemical identification of minerals and ores with a microscope originates from Emanuel Boricky. In 1877 he published a book in Prague, *Elemente einer neuen chemisch-mikroskopischen Mineral- und Gesteinanalyse*. Samples of the mineral were treated with hydrogen silicofluoride (silicofluoric acid) on a microscopic slide, and the slide covered with Canada balsam. He examined the new crystals formed; the silicofluorides of the alkali and alkaline earth metals form characteristic, easily distinguishable crystals. In the case of other metals, however, this method did not prove to be too successful. Treatment with hydrochloric acid and hydrogen sulphide did not give improved results.

Fig. 55. Tobias Lowitz (1757—1804). Drawing

Boricky was born in Millin (Bohemia) in 1840. He studied at the University of Prague, and became an assistant there. For several years he was a teacher in a secondary school, but later returned to the university. In 1871 he became an assistant professor and in 1880 he was appointed Professor of Mineralogy. The publication of his book was only the first stage of his plans, but he could not fulfil them as he died in 1881 at the early age of 41.

His imaginative work, however, had created some interest and many scientists continued his work on the microscopic examination of crystals and the analytical application of this information. Reinisch [104] wrote in 1881:

> The use of the microscope in analytical tests becomes more and more important, and approaches the spectroscope in the field of analysis of small samples. It gives even more information about the substance, because the approximate amount of the latter can also be determined [105].

It is extremely probable that this opinion is rather optimistic even today, and especially so when it was published. Very little publication of microscopic examination was made in those days, and no account of quantitative microscopic analysis could be found. Reinisch prepared crystals on the microscope slide by evaporating a few drops of the solution on it. In the conclusion to his paper he notes that from the more important substances he had prepared specimen crystals, and offered to analyse twelve samples for 10 marks, which was undoubtedly a high price for the job in those days.

In 1885 Haushofer [106] published his book *Mikroscopische Reactionen*, in which he had attempted to fulfil the ambitions of Boricky. He describes characteristic microscopic reactions for almost all the elements, with the shape of the crystals illustrated. The problem had not been too difficult because all that he had to do was to choose a suitable macro reaction for microscopical purposes. (In a few cases this process was reversed, microchemical reactions being adapted for use in macro analysis. Thus uranyl acetate was first used for the determination of sodium on the micro scale [107].) Haushofer developed some new microanalytical techniques; he used a small test tube for the preparation of the crystal instead of a microscope slide and was able to obtain a better shaped crystal [108]. If the amount of sample available was very small, then he carried out the reaction on a microscope slide, allowing the reagents to mix slowly through a wool strip. As the whole operation was carried out on the slide, no losses through transference could occur. In this way the techniques of microchemistry were developed.

Haushofer also describes a method of separation; he placed the solution containing the precipitate on a watch glass, and placed one end of a strip of wet filter paper into it; the other end was placed on another watch glass at a slightly lower level. The clear liquid was drawn through the filter paper [109]. He also used this technique for washing precipitates. Haushofer designed a device for filtration (Fig. 56) where the test-tube receiver was closed, apart from a side-arm, and the filter funnel connected to the tube by a strip of filter paper. Filtration was assisted by a very slight suction applied to the side-arm with the mouth.

One year after the publication of Haushofer's book, another book appeared, this time published in France, which covered almost the same field [110].

One of the most important contributors to the field of microchemistry was Theodor Behrens who was responsible for the development of a great many microchemical techniques. Behrens selected reliable, accurate, and reproducible methods which he classified and published in his book, *Anleitung zur mikrochemi-*

schen Analyse. The volume dealing with inorganic reactions appeared in 1894, and the volumes dealing with organic compounds over the years 1895–1898. The book was subsequently revised and was published in several editions. The crystal properties of 59 elements are described as well as a number of new operations; thus the first reference to the use of a centrifuge for the separation of precipitates is to be found here.

Theodor Heinrich Behrens was born in Büsom in 1843. He studied natural sciences at the University of Kiel, and later specialized in meteorology. It is interesting to note that, as was the custom, his doctoral theses were published in Latin. He first of all taught physics at the Nautical Academy at Kiel, but distinguished himself by his publications in petrography and mineralogy. In 1874 the Technical High School of Delft (Netherlands) invited him to be Professor of Mineralogy and Geology, and it was here that he designed and built the first microchemical laboratory in 1897. He died in 1905.

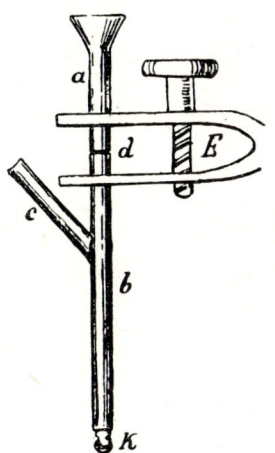

Fig. 56. Micro filtration device from Haushofer's book *Mikroskopische Reaktionen* (1885)

These workers were the pioneers of microscopic qualitative analysis, and many famous scientists used these methods and also made their own contributions to the subject. Nevertheless, much of this work is of unimportance to the general development of analytical chemistry as no great advantages or innovations were made to the existing methods. For reasons of space it is impossible to describe the techniques of microcrystallography in greater detail.

The development of quantitative microanalysis resulted in the need for increased sensitivity of the chemical balance. Nernst designed, for the first time, a microbalance, based on the torsion in a quartz fibre. This balance had a capacity of only a few centigrams, but was extremely sensitive [111]. Later Kuhlmann's microbalance became very popular, as it was capable of weighing up to twenty grams with an accuracy of ± 0.001 mg [112]. This is the extreme limit of accuracy of the microbalance even today.

Almost all analytical methods have been adapted to the microscale and where this adaptation is of special importance it will be referred to alongside the macromethod.

There are two important subdivisions of microanalysis which have no analogies on the macro scale. The first of these is catalytic analysis. The phenomenon of catalysis has been used by mankind since a very early period, an obvious example which springs to mind being the process of fermentation which has been known since the beginning of recorded history. The first conscious attempts to find a catalyst can be seen in the alchemists futile attempts to obtain the "philosopher's stone". The alchemists imagined that this perfect catalyst would be capable of transforming large quantities of base metals to gold. The true nature of catalysis was only established at the beginning of the last century, when the role of acids in the production of ether and in the conversion of starch to sugar was studied. Both the name and the concept of catalysis were originated by Berzelius, who defined catalysts as those substances which accelerate a chemical reaction without taking part in it (1836).

The analytical application of catalytic processes is based on the fact that small amounts of substances can be detected or determined by their catalytic effect on certain reactions. The first application of this principle was made by Guyard, who identified vanadium by means of its catalytic effect on the formation of aniline black. Guyard recommended aniline and potassium chlorate as reagents for vanadium (1876) [113].

Witz and Osmond in 1885 attempted to apply this reaction to the quantitative determination of vanadium, by assuming that the aniline black colour formed over a given time is proportional to the amount of vanadium. They developed the colour by adding oxycellulose threads, which had previously been dipped into a vanadium solution, to the aniline black. The colour obtained was compared with the colours of solutions containing known amounts of vanadium. The most dilute of these standard solutions contained 0·00005 mg of vanadium, but no mention is made of the degree of dilution. These workers also examined the effect of temperature and various added salts, but did not publish any of their experimental data [114]. After this publication there is no mention of analytical methods based on catalysis until half a century later Kolthoff and Sandell [115] developed the first practicable method based on this principle. These workers found that iodide ions catalyse the reaction between cerium (IV) ions and arsenite ions. They measured the time elapsing from the mixing of the solutions until the colour of the redox indicator changed. This reaction time is inversely proportional to the iodide concentration. The evaluation was based on measurements with solutions containing known amounts of iodide [116]. This method can only be applied when the reaction is well-defined, and the end-point can be detected by a suitable indicator and where the reaction velocity is infinitely slow in the absence of the catalyst.

The number of reactions which fulfil these requirements are very few, but most of the catalysed redox reactions are connected with a colour change. This type of reaction can be adapted to the method of Szebellédy and Ajtai [117] who measured the initial colour intensity of the system with a spectrophotometer, and then measured the time taken for the system to arrive at a given optical density [118]. Many catalytic reactions can be applied in this way, and Szebellédy and his school have carried out a great deal of work in this field.

The other specialized microchemical technique is the field of spot analysis. This is essentially a very old technique, for Boyle himself made spot tests when he detected the presence of acids or bases with litmus paper (Chapter III. 3). For the end-point detection in redox titrations, at the end of the last century similar spot tests were used with papers impregnated with suitable reagents. (Chapter VIII. 6).

The spot tests in use today are based on the fact that a drop of solution placed on a filter paper will be distributed so that the dissolved substance will be concentrated in a relatively small area; and at the same time capillarity enables a separation to be made of very small amounts of sample. The history of this method originates from Schiff [119] who in 1859 used a filter paper impregnated with silver carbonate to identify urea in urine [120].

Schönbein [121], in 1861, observed that if a drop of an aqueous solution is placed on a filter paper the water spreads faster than the dissolved substance, and that the relative suction heights of the latter vary according to its properties. Schönbein pointed out that this was a possible method for the separation of salts [122].

Bayley [123] placed drops of a solution on a filter paper and found that it is distributed so that various rings are formed, containing the various components and the solvent. He also established that this phenomenon is dependent on the nature and concentration of the dissolved substance. Bayley also describes a method for the identification of copper and cadmium by a spot test [124].

Another worker in this field, Lloyd, noticed that there is a limiting distance of travel on the paper for a dissolved substance [125].

Trey [126] published a method in 1898 for the separation of copper and cadmium by spot analysis. He also describes a simple device whereby very small spots from a capillary can be placed on the paper. The tube of a filter funnel was drawn into a capillary and bent to a shape, and by pouring the solution into the funnel, and touching the end of the capillary on the paper, a drop is formed which wets the paper evenly [127]. These experiments are the basis of paper chromatography, too.

In 1918 Feigl [128] began a systematic study of the methods of spot tests and as well as selecting the best available reactions and discovering new ones, he recorded the limits of sensitivity for each colour reaction. His results were published in a large number of papers, and finally in book form [129]. Among the pioneers of spot tests, mention must be made of Tananaiev [130] who, in the 1920's, devised many methods for the simultaneous detection of various metals. He published the first book on this subject in 1927.

A further development of spot colorimetry is the ring-oven technique, which was developed by Weisz [131] in 1954 [132]. The ring-oven evaporates the solvent in a ring around the spot of the paper, so that the dissolved substance is concentrated in a narrow band. It is often possible to carry out separations by fixing the component as an insoluble precipitate at the centre of the spot, and then extracting it.

In recent years ultramicroanalysis has also developed a great deal. This technique is capable of analysing samples weighing between 10^{-3} and 10^{-6}, and in many cases volumes less than 10^{-3} ml are used. The methods of handling such small quantities of substance were developed by Benedetti-Pichler [133] around 1930 [134], who introduced many new methods and original apparatus (micromanipulators, etc.) and made observations under a microscope. Ultramicroanalysis now has almost as many branches as analytical chemistry itself. Many of the more recent improvements in methods and apparatus for ultramicroanalysis have been described by El Badry [135] and Wilson [136, 137] (1954). Alimarin [138] and Petrikova combined this technique with chromatography and electrolysis, and were able to identify as little as 10^{-8} to 10^{-10} g of certain compounds.

NOTES AND REFERENCES

1. FUCHS, JOHANN NEPOMUK (1774—1856) physician, Professor of Chemistry and Mineralogy at the University of Landshut (later at Munich), he was the author of many analytical chemical publications. For example, he devised the acetate separation of iron and manganese.
2. FUCHS, Y. N.: *Schweig. Journ.* (1818): **34** 121
3. HOFFMANN, F.: *Observationes chymico-physicae.* Halle (1722) 537
4. ROUELLE, GUILLAUME FRANÇOIS (1703—1770) Professor of Chemistry at the Jardin des Plantes and a member of the French Academy, as well as a teacher of Lavoisier.
5. ROUELLE, G. F.: *Journal Medecine* (1773) **39** 454
6. ÖSTERREICHER, J.: *Analyses aquarum Budensium.* Buda (1781) 153
7. HAHNEMANN, CHRISTIAN SAMUEL (1755—1843) physician; he practised in several places, and finally settled in Paris. Famous on account of his medical recovering method, the so-called "homöopathie".
8. FOURCROY, A. F.: *Mém. Acad. Paris.* (1787) 250
9. HAHNEMANN, S.: *Crells Ann.* (1788) **1** 291
10. SCHELEE, C. W.: *Chemische Abhandlung vom Feuer und Luft.* Uppsala (1777) AYNSLEY, E. E., CAMPBELL, W. A.: *Journ. Chem.* (1958) **35** 347
11. GRIFFIN, JOHN JOSEPH (1802—1877) English chemist and manufacturer of laboratory equipment. The Griffin beaker is named after him. He translated the books of Rose into English and also published several papers.
12. AYNSLEY, E. E.—CAMPBELL, W. A.: *J. Chem. Ed.* (1958) **35** 347
13. BERTHOLLET, C. L.: *Ann. chim.* (1798) **25** 233
14. GAY-LUSSAC, J. L.: *Ann. chim.* (1811) **78** 86; **80** 205
15. PFAFF, C. H.: *Handbuch der analytischen Chemie.* Altona (1821) **1** 115
16. *ibid.* 185
17. GAY-LUSSAC, J. L.: *Ann. chim.* (1813) **91** 5
18. STROMEYER, FRIEDRICH (1776—1835) physician, Professor of Chemistry and Pharmacy at the University of Göttingen. He discovered cadmium and published many papers on the analysis of minerals.
19. STROMEYER, F.: *Gilberts Ann. Neue Folge.* (1815) **19** 146
20. BALARD, ANTOINE JEROME (1802—1876) Professor of Chemistry at the University of Montpellier, and later at the University of Paris.
21. BALARD, A. J.: *Ann. chym. phys.* (1825) **28** 178
22. MARSH, JAMES (1790?—1846) chemist at the British Royal Arsenal, who was also an assistant to Faraday at the Military Academy.
23. MARSH, J.: *New Edinburgh Phil. Journ.* (1836) 229
24. BERZELIUS, J. J.: *Lehrbuch der Chemie.* Dresden—Berlin (1841) **10** 202
25. WACKENRODER, HEINRICH WILHELM (1798—1854) pharmacist; from 1828 he was Professor of Chemistry and Pharmacy at the University of Jena.
26. WACKENRODER, H.: *Arch. Pharm.* (1852) **70** 14
27. GUTZEIT: *Pharm. Ztg.* (1879) **24** 263
28. BETTENDORF, ANTON JOSEPH (1839—1902) chemist; he was a man of independent means who financed his own research.

29. BETTENDORF, A. J.: *Z. anal. Chem.* (1870) **9** 105
30. SVANBERG, LARS (1805—1878) Professor of Chemistry at the University of Uppsala from 1853.
31. SVANBERG, L.: *Vetensk. Akad. Handl.* (1848)
32. PFAFF, C. H.: *Handbuch der analytischen Chemie* **1** 289—290
33. ROSE, H.: *Ausführliches Handbuch der analytischen Chemie.* Braunschweig (1851) **1** 298
34. ROSE, H.: *Traité pratique d'analyse chimique.* Paris (1832) **1** 446—452
35. FRESENIUS, R.: *Anleitung zur qualitativen chemischen Analyse.* Braunschweig 7. Ed. (1852) 8
36. FRESENIUS, H.: *Z. anal. Chem.* (1897) **36** 10
37. ERLENMEYER, EMIL (1826—1909) Professor of Chemistry at the University of Heidelberg, and later of Munich. He was primarily an organic chemist, in which field he made many important contributions; his name, however, is preserved by the flask which he designed and constructed.
38. FRESENIUS, R.: *Anleitung zur qualitativen chemischen Analyse.* (1852) 18
39. *ibid.* 287
40. CHUGAEV, L. A. (1873—1922) Professor of Chemistry at the Academy of Artillery, and then at the University of Petrograd. He worked mainly as an organic chemist, and studied, among other problems, the composition and optical properties of complexes. He was also interested in industry. In 1917 he recommended the foundation of the Platinum Institute, which he directed up to his death.
41. CHUGAEV, L.: *Ber.* (1905) **38** 2520
42. GRIESS, P.: *Ber.* (1879) **12** 427
43. GRIESS, P.: *Lieb. Ann.* (1861) **120** 333
44. GRIESS, P.: *Ber.* (1878) **11** 624
45. ILOSVAY, L.: *Bull. soc. chim.* (1889) **2** 347
46. BUGGE, G.: *Das Buch der grossen Chemiker.* Berlin (1929) **2** 222
47. HOFFMANN, AUGUST WILHELM (1818—1892) was one of the greatest personalities of organic chemistry. He studied under Liebig, and in 1847 he was invited to London where he became the head of the College of Chemistry. In 1865 he was appointed Professor of Chemistry at the University of Berlin, as a successor to Mitscherlich.
48. HOFFMANN, A. W.: *Lieb. Ann.* (1864) **132** 160
49. KOPP, EMILE (1817—1875) Professor of Chemistry at the Pharmaceutical College in Paris. Because of his socialist views he was dismissed during the second Empire. He emigrated to England where he worked in several factories. In 1868 he became Professor of Chemical Technology at the University of Turin, and in 1871 at the Technical High School at Zurich.
50. KOPP, E.: *Ber.* (1872) **5** 284
51. ROSE, H.: *Traité pratique d'analyse chimique.* Paris (1832) **2** 29
52. *ibid.* **2** 64
53. POGGENDORFF, JOHANN CHRISTIAN (1796—1877) pharmacist professor at the University of Berlin, editor of *Annalen der Physik*, shortly Poggendorff's *Annalen*.
54. FRESENIUS, R.: *Anleitung zur quantitativen Analyse.* Braunschweig (1847) 2. Ed.
55. *ibid.* 3
56. *ibid.* 17
57. *ibid.* 483
58. *ibid.* 82
59. *ibid.* 87
60. *ibid.* 90
61. *ibid.* 92
62. *ibid.* 100 159
63. *ibid.* 237
64. *ibid.* 246
65. *ibid.* 250

66. *ibid.* 229 286 359
67. *ibid.* 364
68. Austen Townsend Peter (1852—1907) American chemist, Professor at the Rutgers College, New Jersey.
69. Austen Townsend, P.: *Chem. News.* (1878) **37** 146
70. Gawalowski, A.: *Z. Anal. Chem.* (1898) **37** 377
71. Österreicher, J. M.: *Analyses aquarum Budensium.* Buda (1781) 42
72. Gooch, Frank Austin (1852—1929) American chemist, he worked for a Railway Company. From 1886 to 1918 he was Professor of Chemistry at Yale University.
73. Gooch, F. A.: *Chem. News.* (1878) **37** 181
74. Knorre, Georg (1859—1910) was born in Nikolaievsk, Russia. His father was a famous astronomer. He studied in Germany, and later (1898) became Professor of Electrochemistry at the Technical University of Berlin. M. A. Ilinski (1856—1941) was a professor in Russia.
75. Ilinski, M.—Knorre, G.: *Ber.* (1885) **18** 699
76. Brunck, Otto (1866—1946) gave up his studies at the Technical University of Munich and went to work with Fischer in Erlangen, where he worked on organic chemistry. He later became an assistant to Winkler at the Mining Academy at Freiberg, and carried out research into analytical chemistry. In 1896 he became an extraordinary, and in 1902 an ordinary Professor of Chemistry as successor to his father-in-law, Winkler. In 1935 he retired. He published several papers and books, mainly about metal analysis.
77. Brunck, O.: *Z. angew. Chem.* (1907) **20** 834
78. Baudisch, Oskar (1881—1950) was born in Bohemia and studied in Zurich. He later worked at the Radiation Institute in Hamburg, and eventually became director there. In 1920 he was appointed to a professorship at the University of Yale, and later worked in the Rockefeller Institute.
79. Baudisch, O.: *Chem. Ztg.* (1909) **33** 1298
80. Winkler, L.: *Z. anal. chem.* (1917) **30** 302
    Z. angew Chem. (1918) **31** 187 211 214
    cf. Winkler, L., *Ausgewählte Untersuchungsverfahren für das chemische Laboratorium.* Stuttgart (1931) 93
81. Rom, P. A.: *A Gyógyszerész.* (1954) **9** 101
82. Njegovan, Vladimir, was born in 1884. Since 1919 he has been professor at the Technical Department at the University of Zagreb.
83. Njegovan, V.—Marjanovich, V.: *Z. anal. Chem.* (1928) **73** 271
84. Hahn, F. L.—Keim, R.: *Z. anorg. Chem.* (1932) **206** 398
85. Hahn, F. L., was born in 1888 in Oels (Silesia) and studied in Berlin. He worked at the University of Frankfurt where he lectured in analytical chemistry and in 1922 he became Professor of Chemistry there. In 1933 he emigrated to France and then to Equador, Guatemala and finally to Mexico, where he worked both for the state and as a consultant chemist. Since 1958 he has been Professor of Analytical Chemistry at the University of Mexico.
86. Chancel, Gustave (1822—1890) Professor of Chemistry at the University of Montpellier.
87. Chancel, G.: *Compt. rend.* (1858) **46** 987
88. Moser, Ludwig (1879—1930) studied at the Technical University of Vienna, and after working for a short period in industry he returned there. In 1911 he became the head of the analytical institute. He was killed, together with his wife, in a car accident.
89. Moser, L.: *Monatshefte* (1929) **53** 39
90. Willard, Hobart, H., was born in 1881 in Erie, and studied in Michigan and at Harvard University. In 1909 he went to the University of Michigan where, in 1922, he became Professor of Analytical Chemistry. He retired in 1951.
91. Gordon, Louis, was born in 1914 in New York. In 1946 he became a professor of The Ohio State University and later at the Syracuse University (1948) and the Case Institute of Technology, Cleveland, Ohio (1957).

92. NERNST, W.—RIESENFELD, E. H.: *Ber.* (1904) **36** 2086
93. HONDA, KOTARO (1870—1954) Professor of Metallurgy at the University of Tokyo.
    HONDA, K.: *Sci. Reports. Tohoku. Imp. Univ.* (1915) **4** 97
94. DUVAL, CLEMENT, was born in 1902. He graduated as a secondary school teacher of physics. Between 1936—1943 he was a Professor at the École des Sciences, Rouen. Now he is the Professor in microanalysis at the Sorbonne, and is a Director at the National Research Centre.
95. DUVAL, C.: *Inorganic thermogravimetric analysis.* Amsterdam (1953)
96. PAULIK, FERENC, chemical engineer, was born in Budapest in 1922. He works as a research fellow at the Institute for General Chemistry at the Technical University of Budapest. His brother, JENŐ PAULIK, born in 1927, is a chemist, and works at the same Institute.
97. ERDEY, L.—PAULIK, F.—PAULIK, J.: *Nature.* (1954) **174** 885
98. PAULIK, F.—PAULIK, J.—ERDEY, L.: *Z. anal. Chem.* (1958) **160** 241
99. LEUWENHOEK, A. V. (1685—1718): *Sendbrieven outledingen en ontdekkingen,* etc. Delft
100. LOWITZ, T.: *Nova Acta Imp. Petropoli.* (1794—1795—1798) **8 9 11**
101. LOWITZ, T.: *Technologichesi Journal* (1804) **1** 27
102. SORBY, HENRY (1826—1908) a man of wealthy means, member of the Royal Society.
103. SORBY, H.: *Monthly Microsc. J.* (1869) 1
104. REINISCH, EDGAR HUGO (1809—1884) Professor of Chemistry at the Technical School of Zweibrücken, and later at the Agricultural High School at Erlangen.
105. REINISCH, E. H.: *Ber.* (1881) **14** 2325
106. HAUSHOFER, KARL (1839—1895) Professor at the Technical University of Munich.
107. STRENG, A.: *Jahresb. f. Mineral. Geol. Paleontol.* (1888) **2** 142
108. HAUSHOFER, K.: *Mikroskopische Reaktionen.* Braunschweig (1885) 3
109. *ibid.* 159
110. KLEMENT—RENARD.: *Réactions microchimiques à cristaux.* Bruxelles (1886)
111. NERNST, W.: *Z. f. Electrochem.* (1903) **9** 622
112. EMICH, F.: *Lehrbuch der Microchemie.* München (1926) 73
113. GUYARD A.: *Zentralblatt.* (1876) 120
114. WITZ, G.—OSMOND, F.: *Bull. soc. chim.* (1885) 45 309
115. SANDELL, ERNEST BIRGER (born in 1906) Professor of Analytical Chemistry at the University of Minnesota since 1946.
116. SANDELL, E. B.—KOLTHOFF, I. M.: *Microchim. Acta* (1937) **1** 9
117. AJTAI, MIKLÓS, was born in 1914. He gave up his studies at the University of Budapest, where he worked later as an assistant. From 1938 he worked in industry, and since 1945 for the State authorities. He is now the president of the National Planning Institution of Hungary.
118. SZEBELLÉDY, L.—AJTAI, M.: *Mikrochem.* (1939) **26** 87
119. SCHIFF, HUGO (1834—1915) was primarily an organic chemist. From 1868 he was Professor of Organic Chemistry at the University of Florence.
120. SCHIFF, H.: *Lieb. Ann.* (1859) **109** 67
121. SCHÖNBEIN, CHRISTIAN FRIEDRICH (1799—1868) was a well known chemist, and the discoverer of shooting wool and ozone. He was Professor of Chemistry at the University of Basle.
122. SCHÖNBEIN, C. F.: *Pogg. Ann.* (1861) **114** 275
123. BAYLEY, THOMAS (1854— ?) Professor at the Mining High School at Bristol.
124. BAYLEY, T.: *J. Chem. Soc.* (1878) **33** 304
125. LLOYD, J. M.: *Chem. News.* (1885) **51** 51
126. TREY, H. P. (1851—1916) was born in Dorpat and studied there, and later worked at the University of Riga, where he became Professor of Chemistry in 1903.
127. TREY, H. P.: *Z. anal. Chem.* (1898) **37** 743
128. FEIGL, FRITZ, was born in Vienna in 1891. He studied at the Technical University of Vienna, and also worked there, and in 1936 he became a Professor. Since 1940 he has lived in Rio de Janeiro, where he is the head of a microchemical laboratory at the Ministry of Agriculture. Talanta Prize winner for Analytical Chemistry in 1962.

129. FEIGL, F.: *Z. anal. Chem.* (1918, 1921) **57** 135 **60** etc.;
     Spot Tests. Amsterdam (1954)
130. TANANAIEV, NIKOLAI ALEXANDROVICH (1879—1959) Professor of Chemistry at the University of Kiev.
131. WEISZ, HERBERT, was born in 1922. He was first a private dozent at the Technische Hochschule of Vienna, but since 1960 he has been Professor of Analytical Chemistry at the University of Freiburg.
132. WEISZ, H.: *Microchim. Acta* (1954) 140 376 460 785
133. BENEDETTI-PICHLER, ANTON was born in Vienna in 1894, and studied in Graz and later worked at the Technical High School there. From 1924 he lived in the United States and worked at various high schools. He was advisor on microchemistry to the General Motors Company. He died in 1964.
134. BENEDETTI-PICHLER, A.: *Ind. Eng. Chem. Anal.* Ed. (1937) **9** 483
135. BADRY, HAMED EL, was born in 1918 in Dumiat, Egypt. He studied in Cairo, and has been, since 1961, Professor at the Mining and Petroleum Engineering University, Cairo.
136. WILSON, CECIL, was born in 1912 in Northern Ireland, and studied at the University of Belfast. He worked at the University of Glasgow, and at the John Cass College, London. Since 1958 he has been professor of Analytical Chemistry at the University of Belfast.
137. BADRY, H. E.—WILSON, C.: *Microchim. Acta* (1954) 121 218 230
138. ALIMARIN, IVAN PAVLOVICH, was born in 1903. He has been Professor of Analytical Chemistry at the Lomonosov University, Moscow, since 1954, and the head of the microchemical laboratories of the Geochemical Research Institute of the Soviet Academy of Sciences. Winner of Talanta Prize for Analytical Chemistry in 1965.

CHAPTER VIII

VOLUMETRIC ANALYSIS

1. ANCIENT HISTORY OF TITRIMETRY

The word "titre" in French means the purity of noble metals. Gay-Lussac, in 1835, devised a volumetric method for the determination of silver with a standard sodium chloride solution, in which the amount of sodium chloride solution consumed gave a direct measure of the purity of the silver for a one gram sample. This method was soon used all over the world, and it would be convenient to assume that the term "titration" originated from this method of Gay-Lussac. If, however, one reads earlier French chemical papers, it is obvious that the word "titre" was used even before the introduction of this method by Gay-Lussac. It was, in fact, applied to any substance under examination, and meant that its quality was adequate, so that to refer to a satisfactory titre of saltpetre meant that the saltpetre was pure.

At the beginning of the 19th century, many papers dealing with volumetric analysis use this term, and it was always in some way connected with the amount of standard solution; in other words as a measure of the purity of the substance. The word "titre", however, also means "title", so that *liqueur titré* may mean, also, entitled, or labelled liquor. This is a question to be settled by the French themselves. As we shall see later, in many English or German translations of contemporary French papers the translators could not translate this word. By the middle of the last century, all those analytical methods which were based on the measurement of a substance required for the completion of a reaction (and not by the weighing of a precipitated substance), were called titrimetric methods. The measurement of the consumption of the substance was made either by weighing, or by volume measurement. In recent times, volumetric analysis and titrimetry have come to mean the same, although according to the original definition, volumetric analysis was really only a branch of titrimetry. The other branch, based on weighing (weighing burettes, etc.) has fallen into disuse, although in the historical development it preceded volumetric methods.

Titrimetric analysis originated during the middle of the 18th century, accompanying the rapid development of industry, where it was used originally as a rapid and simple method of quality control.

Industrial chemistry also began in the 18th century, and the first industrial chemical process was the lead-chamber process for the manufacture of sulphuric acid. The discoverer of this process is not known. By the middle of the 18th century there were lead-chamber plants in operation in England. In one of Dossie's [1] books (1759) it is recorded that a patent had been obtained for the production of sulphuric acid by ignition of sulphur with saltpetre. Dossie only mentions the use of glass apparatus. According to another source [2], Roebuck [3] first used lead chambers in his sulphuric acid plant in Birmingham in the year 1746. Gunpowder manufacture, as well as the purification of saltpetre, was also carried out in quite large-scale plants in those days. For this purpose potash was used. This chemical was also in great demand by the textile industry where it had begun to be used for the bleaching of linen. Potash had to be imported from America because Europe had not enough forests to provide for the great demand. This proved rather expensive. Natural soda was also used for this purpose, but this also was not available in sufficient quantity. The artificial manufacture of soda became an important problem, and the French Academy announced a competition, promising an award of 12 000 francs for the solution. (It must be mentioned here that when the payment became due, the French revolution had broken out, the Academy was dissolved and, therefore, the award was never given although many people had worked hard on the problem.) Many attempts were made, but a suitable process was not to be found easily.

Malherbe transformed rock salt into Glauber's salt, and heated this with charcoal and iron, and in this way obtained soda. De la Métherie [4] transformed sodium sulphide, obtained from Glauber's salt, to the acetate and managed to obtain soda by the ignition of the latter. Finally Leblanc discovered his famous method after a long investigation and this method was still in use at the beginning of this century.

Nicolas Leblanc (1742—1806) founded a factory with the financial assistance of Phillip, Duke of Orleans. When the latter was executed in 1793, his estates—among them this factory—were confiscated. Leblanc's patent was seized and published, and at the age of 52 Leblanc became a ruined man. He lived from occasional employment and tried many ways of raising some money, but in vain. Misery remained his only friend, his wife and daughter died, and his strength left him. In 1806 he shot himself in the aisle of the Church of Saint Denis.

The Leblanc soda-process became an important industry, mainly in England. The textile industry needed large quantities of soda. The invention of spinning and weaving machines caused a revolution in the textile industry, and also textile finishing processes improved. Scheele discovered chlorine, and Berthollet and Tennant [5] used chlorine and hypochlorite for bleaching in the textile industry. For the production of chlorine, hydrochloric acid had to be prepared, so that this process, too, was also greatly improved.

Sulphuric acid, hydrochloric acid, soda, and chlorine water were the main products of the chemical industry of the 18th century. Mostly these products were bought by other factories, so that any inadequacy in the quality of these chemicals resulted in a financial loss. The strength of hypochlorite solution was very important to a textile factory, and similarly the strength of soda was important for a soap factory. The raw materials purchased from other companies were therefore examined, and in this way quality control was developed. Although the quality of leather or textiles could be judged by its appearance and texture, in the case of raw materials, such as chemicals, the situation was not quite so simple. In order to determine the purity of soda or acid a chemical method was required and also a laboratory. Thus a laboratory became an important part of the factory, and its importance rapidly increased. The importance of analytical chemistry similarly increased, and whereas it had formerly been the pastime of scientists at university, it was now an everyday practice of many chemists. This was very important with regard to its development, primarily in its practical applications.

After this brief excursion we must return to that period when industrial analysis was only concerned with the determination of acids, alkali carbonates and hypochlorites.

Lampadius, in his analytical textbook (1801) wrote as follows:

> For work in all natural sciences persistency is needed, but especially in analytical chemistry. At the beginning of my analytical work many analyses were wrong, because I could not wait until the reagent dissolved the mineral, or until the filtration, washing or similar operations were finished according to their order. Those who cannot wait weeks and months for the results, should never begin analytical work [6].

Industry, however, could not wait weeks and months for the results! Their need was for rapid and simple methods. Titrimetry itself developed from the need for rapid and simple methods for the determination of the three chemicals mentioned above. In their original forms, naturally, these methods did not yield absolutely accurate results. The principle of the methods however can be observed; to the substance to be determined a reagent which reacted with this substance was added, and some method of end point detection showed the completeness of the reaction. The amount of the reagent added was then measured by weight or volume. Originally this test was used only to establish whether the substance was suitable or not. If it consumed empirically established amounts of "standard solution" it was good, if it consumed less, then the substance was rejected.

In the ancient history of titrimetry there are to be found the names of people who never played an important role in the development of chemistry, indicating that titrimetry developed mainly in an ancillary industry. The further development of this technique, so that it became of importance to other fields of chemistry, occurred when scientists began to study this branch of analytical chemistry.

The history of titrimetric analysis is at the same time a test of the maxim that in science everything has its preliminaries, and that every new discovery is related to something which was very similar. Therefore, for the historian of analytical chemistry it is difficult to say with certainty that some observation was the earliest example of its kind. Consider for example the question of the first titration. In the second half of the last century, on the basis of Mohr's book, textbooks mentioned mostly Gay-Lussac as the founder of titrimetry. Koninck, who published a paper in 1901 on the history of titrimetry [7], referred to Descroizilles as the discoverer of this method, who carried out measurements which can be considered as titrations some decades before Gay-Lussac. Lüning [8] later found an even earlier date, namely the method of Francis Home in 1756, which also showed the characteristics of titrimetric analysis. More recently Rancke-Madsen has very thoroughly studied the chemical literature of the 18th century in order to find the origins of titrimetric analysis, and has published a book on the results of his work. This excellent book [9], dealing with the history of titrimetric analysis in the 18th century, provided much important information for the compilation of this chapter. According to Rancke-Madsen the year 1729 can be regarded as the birthdate of titrimetric analysis.

However, an even earlier method, which has some resemblance to the elements of titration in the method used for endpoint detection, can be found in one of Glauber's books in 1658. The description of the preparation of pure saltpetre from nitric acid and potash states that the latter

> must be added dropwise to the *spiritum nitri*, until on further addition bubbling ceases and both lose their inimical characteristics and kill each other [10].

Thus, the use of an indicator, in this case the evolution of bubbles, was known in those days but "standard solutions" were not yet in use.

In 1729 Claude Joseph Geoffroy presented a dissertation to the French Academy on the concentration of vinegar. In this dissertation there is a description of a method which contains several features characteristic of volumetric analysis. Here is the text:

> Among the vinegars we must then choose by chemical operations that which contains most acid. For this purpose several methods can be used ... For my investigations I take out from each type of vinegar about 2 drachms, and pour it into a glass and then weigh it accurately. Then I add finely powdered, dry potash until the bubbling ceases. The acids of the vinegar then disappear and the solution becomes salty, and if the bubbling has ceased I can determine the strength of the vinegar from the amount of potash used for the absorption of the acid. This is usually about four grains in the case of the weaker Parisian vinegars, while it can be as much as eight grains in the case of the stronger ones. The vinegar of Orleans may consume even eleven grains, moreover, according to my experience, a vinegar from Orleans, may reach twelve also [11].

Thus, according to our recent knowledge, Geoffroy was the first to use the process of neutralization for analytical purposes. His method is a characteristic titrimetric one. He used a "standard solution", i.e. solid potassium carbonate, added in this case to the substance to be determined, until the indicator (effervescence) showed the end of the reaction. The strength of the acetic acid he characterized by the amount of consumed "standard solution".

Claude Joseph Geoffroy was born in Paris in 1683, the son of a pharmacist. His father was also interested in science and at his house gathered several of his friends who were interested in the natural sciences. Claude Joseph Geoffroy, who also became a pharmacist, is often called Geoffroy junior, and should not be confused with his brother, Francois Etienne Geoffroy, who became rather more famous in the history of chemistry than his brother, by constructing the first table of affinities. The younger Geoffroy worked diligently, and became a member of the Academy, but was mainly interested in pharmaceutical problems. He died in 1752.

A considerable period elapses after the publication of Geoffroy before the next method which can be regarded as a titrimetric one is found. In 1747 Monnier [12] presented his results of the analyses of mineral waters. He evaporated 60 oz of each water sample, and then

> to know exactly the amount of alkaline earth in each residue I placed 10 gr from each into a small glass, and added to each 40 drops of sulphuric acid. The first ceased to effervesce after the addition of the 12th drop, the second after the 25th drop while the third only after the 39th ... [13].

The consumption of the standard solution therefore indicated the relative carbonate contents of the waters.

Venel [14] in 1750 gives an account at the analysis of the mineral water from Seltz. Here the titrimetric principle is even more marked. He used a standard solution, together (for the first time) with an indicator, the extract of violets. Sulphuric acid was used as the standard solution.

> A volume of sulphuric acid, equal in volume to 1 oz of water, weighed 1 oz and 6 gr ... This I diluted with six parts of distilled water, and then I added this to half a libra of Seltz water, slowly, and mixing the solution and heating it constantly; the violet syrup turned to red when 28 gr of this dilute acid were used, which corresponds to 6 gr of concentrated acid [15].

He concluded from this experiment that the water was not alkaline and the standard solution was only consumed by the indicator, because if the experiment was repeated with melted snow, the same amount of acid was required.

At about the middle of the 18th century Scheffer determined the strength of nitric acid with an aqueous solution of potassium carbonate, which he added drop by drop to the acid, until the effervescence ceased. Only a relative, comparative evaluation was possible from this experiment [16].

In 1756 Francis Home (1720–1813), who was Professor of Pharmacy at the University of Edinburgh, published a book entitled: *Experiments of Bleaching*. This book was later published in French (1762) and in German (1777). In this book Home describes the examination of potash by the following methods:

> In order to discover what effect acids would have on these ashes, and what quantity of the former the latter would destroy, from which I might be able to form some judgment of the quantity and strength of the salt they contained, I took a drachm of blue pearl ashes, and poured on it a mixture of one part spirit of nitre, and six parts water; which I shall always afterwards use and call the acid mixture. An effervescence arose, and, before it was finished, 12 teaspoonfuls of the mixture were required. This effervescence with each spoonful of the acid mixture was violent, but did not last long [17].

The novelty of this method is that it is really a volumetric one, because Home measured the consumption of standard solution by volume, even if it was only with a teaspoon. Those samples which consumed less that 12 teaspoons were considered to be of insufficient strength.

We can also find that the earliest examples of precipitation titrations originate with Home (1756). For example he determined the hardness of water in the following way:

> Let a certain quantity of alkaline salt be dissolved in a certain quantity of soft water. Into a certain quantity of hard water in a glass pour in the solution gradually, so long as the milky colour is on the increase. When that is at the height, let the water stand till it becomes pellucid. Try it again with a few drops of the solution; if no whiteness arises in the water, it is then soft; if there does, go on drop by drop until no more white clouds arise. By this means it is known what quantity of salts is necessary to soften that quantity of water; and, consequently, how much any given quantity of water will require [18].

In 1767 a monograph was published in London with the long title: *Experiments and Observations on American Potashes with an Easy Method of determining their respective Qualities*. This work by William Lewis describes the examination of eight samples of American potash.

At this time there were three chemists with this name in England. One of them was a member of the Royal Society, a physician, and author of several chemical works, in the list of which this work is not mentioned. This William Lewis died in 1781. The second chemist with the name William Lewis died in 1814, but the date of his birth is not known. There was a third William Lewis who was born in Jamaica, and died in 1823. From the year of publication of this book it is probable that the author was one of the first two, but it could equally well be a fourth William Lewis.

Lewis also added acid to the potassium carbonate in a similar manner to the methods previously mentioned. The method of end-point detection in his procedure was not the cessation of effervescence, but the colour change of an added indicator paper. A more important improvement was that his method gave an absolute result, and was not merely a relative measurement. In order to achieve this he established the amount of hydrochloric acid solution required to neutralize a known weight of pure potassium carbonate, i.e. he standardized his solution, so that in fact Lewis used a standard solution of hydrochloric acid. He measured the amount of standard solution consumed by weighing. For an example in his text:

> Having extracted the whole of the saline matter from the several Potashes, it was judged that the quantity of true alkali in the salts might be discovered by their power of saturating acids, compared with that of an alkali of known purity; and this method succeeded so well, that it is hereafter proposed for the assaying of Potashes, and the manner of procedure described at large. The true strength of Potashes, as it depends on their quantity of pure alcaline salt, must be estimated from characters which belong to alcaline salts only; among which, the neutralization of acids is one of the chief. Several persons have, on this principle, endeavoured to ascertain the comparative strength of different Potashes, but not their

absolute strength; nor do they seem to have always attended to certain circumstances necessary for the accuracy of the experiment...

If the whole substance of the Potash is used for mixture with the acid, a considerable error may arise from the earthy part; for the earthy matter in genuine potashes, or lime, or vegetable ashes fraudulently mixed, will saturate a portion of acid, as well as the true alcaline salt. The salt must therefore be separated from the earth, by solution in water and filtration...

The quantity of acid, necessary for the saturation of the lye, should be determined, not by drops or tea-spoonsful, but by weight; and the point of saturation, not by the ceasing of the effervescence, which it is extremely difficult, if not impracticable, to hit with tolerable exactness, but by some effect less ambiguous and more strongly marked, such as the change of colour produced in certain vegetable juices, or on paper stained with them.

The finer sort of purplish blue paper used for wrapping sugar in, answers sufficiently well for this purpose; its colour being changed red by slight acids, and afterwards blue or purple again by slight alcalies. What I have chiefly made use of, and found very convenient, is a thick writing paper stained blue on one side with an infusion of lacmus or blue archil, and red on the other by a mixture of the same infusion with so much dilute spirit of salt as is sufficient just to redden it. The paper is washed over with a brush dipt in the respective liquors, two or three times, being dried each time, till it has received a pretty full colour, and afterwards cut in slips a quarter of an inch or less in breadth; a bit of the end of one of the slips being dipt in the liquor to be tried, the red side turns blue while any of the alcali remains unsaturated, and the blue side turns red when the acid begins to prevail. If either the acid or alcali considerably prevails, the paper changes its colour immediately on touching the liquor; if they prevail but in a low degree, the change is less sudden. The part dipt is always to be cut off before a fresh trial...

Take a quantity of spirit of salt, and dilute it with 10 or 12 times its measure of water; fill with this mixture a vial that will hold somewhat more than 4 oz of water: the vial which I find most commodious is nearly of the shape of an egg, with a broad foot that it may stand sure, a funnel-shaped mouth for the convenience of pouring the liquor into it, and a kind of lip or channel at one side of the mouth, that the liquor may be poured or dropt out without danger of any drops running down on the outside. Hook the vial, by means of a piece of brass wire tied round its neck, to one of the scales of a balance; and counterpoise it, while filled with the acid liquor, by a weight in the opposite scale.

Pour gradually some of the acid from the vial into the solution of salt of tartar, so long as it continues to raise a strong effervescence; then pour or drop in the acid very cautiously, and after every small addition, stir the mixture well with a glass cane, and examine it with the stained papers. So long as it turns the red side of the paper blue, more acid is wanted: if it turns the blue side red, the acid has been overdosed. That there may be means of remedying any accident of this kind, without being obliged to repeat the whole preceding part of the experiment, it will be proper to reserve a little of the alcaline solution in another vial: this is always to be added towards the end, and washed out of the vial with a little water.

When the liquor appears completely saturated, making no change in the colour of the paper, hook the vial on the scale again, to see how much it wants of its first weight: this deficiency will be the quantity of the acid liquor consumed in saturating the two grains of alcaline salt [19].

Lewis standardized the acid with 1/8 of an oz of pure potassium carbonate, and diluted the acid solution so that it required 4 oz exactly to neutralize this weight of carbonate. From a knowledge of these data it was not difficult to calculate the alkali content of the samples.

By consulting the well known chemical encyclopedia, compiled by Macquer (1766), which was translated into several languages, we can conclude that the theory of neutralization titrations was well known at that time. He writes about water analysis, that

> if the water contains free acid or alkali, this can be identified by its taste, or by its effect on the colour of litmus or violet liquor, as well as by the addition of just as much acid or base to it, as is needed to reach the point of saturation [20].

From this sentence it is not quite clear whether the final reference is to a titration, or merely serves for a qualitative test, although it is probable that it refers to titration. This conclusion is supported with a statement by Lavoisier in 1778 as follows:

> All chemists know that if a mineral water contains small amounts of alkalis, which cannot be separated by crystallization, its amount can be determined by the dropwise addition of acid to it until saturation. If we suppose that we used up 25 gr of acid for the complete neutralization of the alkalic water. Now we must determine how much crystallized soda is needed for the saturation of 25 gr of the acid. The same amounts of alkali were present in the water. This method is used daily, and it is very accurate because it is independent from all inaccurate factors [21].

It would appear that the use of this primitive form of acid–base titration was quite common in water analyses at the end of the 18th century. We can find another reference to it in a book by Gioanetti, published in 1779. The residue, after the evaporation of water, containing mainly sodium sulphate, sodium chloride and sodium carbonate, he redissolved and evaporated once more and then added acetic acid solution to saturation. The acetic acid solution he standardized with sodium carbonate and thus from the amount of acetic acid required for saturation (which he probably measured by weighing), he calculated the amount of sodium carbonate present in the original residue which weighed 139 gr. Gioanetti was also able to establish whether the sodium carbonate used for the "standardization" was pure enough. In order to do this he neutralized a known amount of sodium carbonate with acetic acid, and then extracted the sodium acetate with alcohol, evaporated to dryness and ignited the residue to form sodium carbonate again. Comparison of the weight of this residue with the original weight of carbonate indicated the purity. Gioanetti also determined the sodium chloride content of this residue by mixing it with alum and distilling off the hydrochloric acid and absorbing this in water. He then added litmus to this solution and added sodium carbonate until the colour changed to blue. He then evaporated the solution to dryness and determined the weight of the sodium chloride residue. Since he used an indicator in this determination it can be assumed that the end-point was also indicated by litmus in the first determination [22].

> Vittorio Amadeo Gioanetti was born in Turin in 1729. He studied medicine, and soon started to conduct chemical experiments for himself. He was appointed by the King, as director of the porcelain factory at Vinovo. The factory was in a very bad financial state, and it was Gioanetti's job to reorganize it. The Turin Academy of Sciences elected him a member in 1783. During the wars with France he worked at the porcelain factories of Sevres, and after the reinstatement of the House of Savoy he continued his work in the factory at Vinovo. He died in 1815.

Wiegleb [23] determined the alkali content of salt mixtures with sulphuric acid, instead of by fractional crystallization, and he also determined the strength of the acid with soda [24].

Österreicher's book, referred to earlier, dealing with the analysis of the mineral waters of Buda, was published in German just two years after the original Latin version had appeared. In it he describes the determination of the alkali content of the water which is similar to the previous method:

> Die Menge eines jeden blossen Laugsalzes wird geschäzt; wann die eingedikte Auflösung mit einer genau genommenen Säure gesättiget wird, dem andern gleichen Theil der nehmlichen Säure aber, wird bis zur Sättigung ein solches Alkali in einer ebenfalls genau gewogenen Menge zugesetzt: Die Menge des zu bestimmenden Alkali wird mit derjenigen Menge die nehmliche seyn, welche zur Sättigung der Saure angewendet worden ist [25].

A similar method is also described by Bergman [26], and this was probably a reason for its widespread use.

This method of water analysis was known to Ferenc Nyulas, who described the problem of separation of sodium chloride and sodium carbonate. In his book, he describes the titration of sodium carbonate with sulphuric acid, the end-point being reached when effervescence had ceased and by the "dis-appearance of the alkali".taste of He also describes the standardization of the acid with soda [27].

In the works of Guyton de Morveau, also, there are to be found the basic principles of volumetric methods.

> Louis Bernard Guyton de Morveau was born in 1737 in Dijon, where his father taught law at the university. He himself studied law there, and later continued his studies in Paris. It would appear, however, that this subject did not occupy him completely, for he started to study literature, and began to write long satirical poems. In 1760 he returned to Dijon, where he was appointed a lawyer. He continued to write poetry, and soon, either because of his literary merits or because of his family's influence, he was elected a member of the Academy of Dijon. At one of the meetings of the Academy a lecture on chemistry was given, and afterwards Guyton made some critical comments on it. The lecturer replied that a lawyer and poet should not criticize subjects of which he had no understanding. This made Guyton rather angry and in order to justify his comments he began to study chemistry. He became so engrossed in this subject that he continued to study it for the rest of his life. He began to carry out experiments, and published a number of papers in the journal of the Academy, partly with chemical and partly with juridical contents, occasionally interspersed with a poem. In 1776 he gave lectures on chemistry, and in 1777 he published a book on the subject but he also turned his knowledge to practical use. In 1778 he became a partner in a saltpetre factory, which he expanded in 1783 by adding a synthetic soda plant. By this time his name had become quite famous, and he was asked to write the chemical section in the Great French Encyclopedia. He became so involved with his chemical and industrial occupations that in 1783 he gave up his position as a lawyer.
>
> Guyton de Morveau at this time was a supporter of the phlogiston theory although Lavoisier tried to convert him to his own views, and paid several visits to Dijon. Eventually Lavoisier's journeys met with success and Guyton accepted the new theory of combustion, and became an ardent supporter and propagator of the new ideas. In 1789 Lavoisier and his friends tried to elect him to membership of the French Academy, but without success. Guyton de Morveau finally came to Paris during the revolution, not as a member of the Academy but

as a member of the National Assembly. As a member of the convention, in 1793 he voted openly for the execution of King Louis XVI. He later took part in the Belgian military expedition, and was in charge of the balloon troops. In the Battle of Fleurs he went up in a balloon in order to observe the enemy troop movements. On returning to Paris he became a teacher at the École Polytechnique, and later was appointed director of this institute as well as director of the National Mint. The Emperor Napoleon (who used to reward his scientists with aristocratic titles and ranks; with a few exceptions, these were accepted and these titles were used even though the holders had previously been revolutionaries) gave Guyton de Morveau the title of baron in 1811. With the coming of the restoration, however, Guyton's vote was still remembered and he became discredited and was not allowed to hold public lectures. He died in 1816.

In order to make his saltpetre plant more efficient, Guyton investigated the methods for obtaining saltpetre from the mother liquor. During these investigations he used indicators for the first time in industrial technology for the neutralization of the nitric acid. For this determination he attempted to find a method "which can also be used in the hands of a less intelligent worker". Morveau devised a method for the examination of the mother liquor which contains several important features of volumetric analysis. He published this method, as well as later improved modifications of it, in several papers. The eventual procedure he described as follows:

1. First take out from the mother liquor two portions each of 3 cubic inches.
2. Dissolve in a given amount of water a known amount of alkali carbonate.
3. Dilute one of the samples of the mother liquor, place two paper strips into it, thfirst having been treated with *fernambuke*, and the second with *corcume* extract. To this add the alkali solution, prepared under 2., until the papers indicate that the point of saturation is reached.
4. Now weigh the residual alkali. From the difference the amount of acid present in the mother liquor can be calculated. This corresponds to the total amount of the two acids, nitric and hydrochloric, present in the mother liquor.
5. Dilute similarly the second mother liquor sample.
6. Prepare lead nitrate solution by dissolving a known amount of lead in nitric acid, and determine its weight.
7. Then add this lead nitrate solution in small portions to the mother liquor until no further turbidity occurs. This can be observed at the end by allowing the solution to clear before adding another drop.
8. The lead nitrate solution must again be weighed and from the difference the amount of hydrochloric acid can be calculated. (First of all this was carried out by weighing the lead chloride precipitate, but another, later paper records that it was calculated from the amount of lead nitrate consumed.) From the investigations of Wenzel we know that the amount of hydrochloric acid required to neutralize 11 parts of potassium carbonate combines with 16 parts of lead; and from this we can calculate the amount of alkali consumed in section 4. and its proportion to the amount of lead consumed in section 8., and hence to the amount of hydrochloric acid. By subtracting this from the total amount of acid, the amount of nitric acid can be calculated [28].

This method, in which Guyton de Morveau solves a quite complex analytical problem, is an example for the solution of difficult analytical problems. As well as the alkalimetric determination there is the introduction of a new precipitation titration of chloride, using lead nitrate solution. It is interesting that he mentions silver nitrate at the beginning of his paper and that it can also be used for the determination of hydrochloric acid, but he gives no information on the manner in which it should be applied. The amount of potassium carbonate, equivalent to lead was also determined by Morveau, and for this determination he introduced another precipitation titration. To the lead solution he added potassium carbonate solution in the presence of litmus and *curcuma* indicators, and the end-point was detected, not by the completion of precipitation, but by the colour change of the indicator.

Guyton de Morveau, this diligent chemist, also describes another interesting method (1784), this time for the determination of dissolved carbon dioxide in water. The principal of this method is based on the phenomenon, discovered by Cavendish [29], that water containing carbon dioxide first of all makes lime water turbid, but if a sufficient quantity of carbon dioxide is added the precipitate redissolves to form bicarbonate. The accuracy and applicability of this method are very doubtful, but it is of interest because a device is described which can be considered as the forerunner of the burette. Morveau calls this a *"gasometre"* or gas measuring device.

> This is a cylindrical glass tube, on the outside of which we glue a paper strip, on which a scale can be drawn. One part of this scale corresponds to the volume of the small glass used for volume measuring. ... Then two volumes of lime water must be poured into this device, and then three times this volume, i.e. six volumes of water at 10 °R saturated with carbon dioxide are added. At this temperature the water can dissolve nearly its own volume of carbon dioxide. The milk-like turbidity disappears only after the addition of the sixth volume of water ... It follows therefore that water, which contains only half this amount of carbon dioxide, will require 12 volumes, while that containing only a quarter of the amount will require 24, similarly water containing one eighth part of this will require 48 volumes to clear the turbidity from two volumes of lime water. We know how many cubic inches of carbon dioxide the six volumes of saturated water contains, thus it is only necessary in the case of an unknown water sample to find the volume required to clear the turbidity from a known volume of lime water and then on the basis of the inverse proportions the amount of dissolved carbon dioxide can be calculated [30].

This method therefore is really volumetric, because the principle of the calculation is based on the measurement of the volume of the solution.

In the 18th century we can also find potassium hexacyanoferrate(II) used as a standard solution. Kirwan in 1784 was the first to use this reagent in this way, for the determination of iron. The method is a precipitation-titration, the end point probably being detected by the cessation of precipitation. One of the interesting features of the method is that the solution was standardized against metallic iron, of which 1 gr was accurately weighed, and dissolved in sulphuric acid. The weight of standard solution required for the "titration" of the iron was recorded by Kirwan on the label of the hexacyanoferrate(II) solution [31]. Slightly later Gadolin, probably independently of Kirwan, used the same standard solution for the determination of iron, but the method he uses, however, is much more complicated than the former [32].

In addition to acid–base and precipitation titrations a third type of titration appeared at the end of the 18th century. This was the redox titration, which was originally developed by Henri Descroizilles. With his name a new trend began in the development of titrimetry.

2. FROM DESCROIZILLES TO GAY-LUSSAC

In the previous section we find that most of the contributors to the development of titrimetry have been French, and in this section the pattern is repeated. This is because for several decades this very convenient new method was used solely in France. This was no mere coincidence, but was due to the close connection between industry and science in that country after the revolution. Many of the best chemists were employed by the government and naturally had to work on any industrial production problems which were of importance to the state. Many such problems arose in France during the Napoleonic wars, and resulting blockades. Thus Berthollet was closely concerned with the textile industry, and Vauquelin established a factory for the production of chemicals, and Guyton de Morveau occupied himself with saltpetre plants. Descroizilles and Gay-Lussac, those two great personalities of early titrimetry, developed their pioneer methods for practical purposes. Descroizilles was also one of the forgotten names of chemistry, but his life and work have been described by Duval [33].

François Antoine Henri Descroizilles was born on 11th June, 1751 in Dieppe. His family had been pharmacists for four generations, his father having studied under Geoffroy in Paris, and then specialized in botany. Descroizilles had 15 brothers and sisters, of whom nine had died in early childhood, but one of his brothers later became a famous botanist. Henri Descroizilles studied at Dieppe, and then his father sent him to Paris to study, where he worked in the laboratory of Rouelle. In 1777 he finished his studies and obtained a certificate of *demonstrateur royal de chimie*, which means some kind of chemistry teacher, and also after passing a few other examinations, it was possible for the owner to open a pharmacist's practice. The parliament of Rouen consulted Descroizilles for expert advice on the prevention of the adulteration of apple wine. This adulteration was a very common occurrence in Normandy, and consisted of adding lead to the wine in order to improve its clarity. The seriousness of this practice can be judged from the fact that anyone convicted of it was fined 500 francs by the authorities, and even the church condemned it. Descroizilles developed a reliable method for the detection of this adulteration: he estimated the lead content of the apple wine by testing it with heparine. His method was published in the journal *Archives de Normandie* and was discussed before the French Academy by Lavoisier in 1787. Descroizilles gave public lectures on chemistry at this time, but presumably did not find them financially rewarding, because he studied pharmacy and eventually opened an apothecary's shop. Madame Roland, whose husband later became Minister for Internal Affairs, and who was executed during the Jacobin period, described Descroizilles in a letter as follows: I think that he makes court to the world for financial purposes. This is no doubt a very astute appraisal of Descroizilles' character and the fact he named his first and most famous volumetric methods *Berthollimetry*, indicates that he was a flatterer. Descroizilles applied his knowledge to many other fields, for example, it was customary at that time in Dieppe to place lighted tar torches along the harbour at night for the guidance of ships. It was quite easy for ships to mistake the lights of houses along the shore for the harbour, so Descroizilles had the idea of constructing a series of light-houses which could be illuminated in a definite rotation using a watch system which he had invented. He also designed a coffee-boiling machine which was, in effect, a primitive form of the espresso machines in use today.

Soon after Scheele discovered chlorine in 1774, Berthollet found that it could be used for the bleaching of textiles, either in the gaseous form or in alkaline solution. Before this bleaching had only been carried out by exposure to sunlight, a long process much dependent on the vagaries of the weather. Berthollet tried to disseminate his new method among the textile industrialists, and one of his pupils, Grandcourt, went to Rouen to give a lecture together with some practical demonstrations before the assembled industrial experts. It was unfortunate that during the practical demonstration only part of the material being bleached actually became white, the rest simply disintegrated. Descroizilles however became interested, and realizing the value of the process persuaded the Lord Mayor of Fontenay to finance the construction of a pilot plant. He soon discovered that the most important factor in the method is the concentration of the hypochlorite liquor, and devised a titrimetric method of analysis for its assay. This was the earliest example of a redox titration. Once this method was established the development of techniques of bleaching became very rapid and many bleaching plants were built around Rouen, which soon became the centre of the textile bleaching industry. In 1806 Descroizilles was awarded a gold prize for the introduction of the best bleaching process based on the Berthollet reaction. Apart from this it would not appear that he gained much financially from this discovery. When Roland became Minister of Internal Affairs, he appointed Descroizilles Inspector of markets in the Dieppe area, a rather soul-destroying occupation, as it involved inspection of all raw materials and finished products. He was later arrested for some unknown reason but was released after several months, and was appointed as an inspector of gunpowder and saltpetre factories in the area. He, himself, built a saltpetre plant in a disused church. This occupation lasted until 1795, and during this time he married and a son, who also became a chemist, was born in 1793. Between 1793 and 1806 there is no record of his occupation, although it is probable that he worked in the field of textile bleaching. In 1806 he was living in Paris, and published his second important paper on alkalimetry. (According to this paper, he described this method in a lecture to the Academy of Rouen in 1805 [34]). In Paris he worked both in the fields of science and industry, he designed an alcohol meter, and a portable alcohol distillation apparatus (weighing 3·35 kg) and also a fire-extinguisher. It was during this time that he invented his coffee-machine referred to previously. His official occupation during this time is unknown; according to one source he was the secretary of the Assembly of Manufacturers (Conseil général des Manufactures) [35], but according to his death certificate [36] he was a clerk in the Ministry of Home Affairs. He died in Paris in 1825.

Fig. 57. Henri Descroizilles (1751–1825). (From a miniature in the Museum of Rouen)

Descroizilles' method for the determination of hypochlorite was quite widely known by the time of its publication. Berthollet had described the method in 1789 in a paper [37] dealing with the general methods of chlorination. In this paper he writes,

> ... a skilful chemist from Rouen, M. Descroisille, who made attempts to start a plant in the town, published in the *Archives de Normandie*, that he had found a new method for the cheap preparation of chlorine ... To avoid all accidents, owing to the too great strength of the liquor, we need an adequate method by the use of which it can be measured. M. Descroisille used for this purpose indigo solution in sulphuric acid. He dissolved one part of finely powdered indigo in eight parts of concentrated sulphuric acid, heating it on a water bath for several hours until dissolution was complete, and then diluted it with a thousand parts of water. To examine the strength of the chlorine water he placed a given amount of the indigo solution into a tube mounted with marks, and added chlorine water until it decolourized. Now it must be determined that in the case of a chlorine water, which proved to be suitable for textile whitening, how much is needed for the decolourization of the given amount of indigo solution. This quantity will serve for the comparison of various chlorine water samples, and these can then be used for the comparison of others.

Berthollet probably heard about this method indirectly. This is indicated by the fact that he describes the method in reverse to the original description by Descroizilles, and he also gave the latter's name incorrectly. In actual fact Berthollet used this method in 1788, together with Lavoisier, for the relative comparison of the colour strength of samples of indigo. They added chlorine water to the powdered indigo and measured the amount of this solution required for decolourization [38]. It is not certain whether the amount (the word used was "parts") referred to volume or weight measure. It is extremely probable that they already knew of Descroizilles' method, for if they had not Berthollet would later have referred to his own method in preference to that of Descroizilles. Descroizilles' method, after the description by Berthollet, became widely known, so that after a couple of years it is mentioned in English and German journals. In 1791 Roe, in England, referred to this method, which he called the "Blue Test". He found it to be very good, but disagreed with the use of the calibrated glass tubes [39]. Kirwan also praised this method.

In 1792 Weinlig described a similar method, without giving any reference to Descroizilles, so it is possible that he also discovered the method. Weinlig who worked in Berlin describes the method rather differently from Descroizilles.

> Um nun die Stärke der dephlogiscirten Salzsaure zu prüfen, so giesst man in eine Glasrohre eine kleine Portion der Indigauflösung hinein, und vermischt sie mit ungefähr achtmal so viel Dephlogistici, man lasst es eine Zeit lang stehen; zerstört das Desphlogiston die Farbe des Indigs in einer Zeit von einer Viertelstunde, oder auch etwas langer, so ist das Wasser zum Bleichen gehörig geschickt gemacht, und hat seinen gehörigen Grad erreicht [41].

Therefore this method is based on time measurement, i.e. a chronometric one. In 1793 another German author, Tenner [42], describes this method. He poured a given volume of indigo solution into a cylindrical glass vessel, but mentions that if these are not available then a beer-glass is suitable. He marked the volume of the indigo solution and then added chlorine water to the solution until decolourization occurred and again marked the volume. Knowing the volume of a standard solution of chlorine water required for decolourization, it was then possible to determine the concentration of chlorine water samples, and either reject or dilute as necessary. Tenner mentions finally that litmus or other plant extracts can also be used in the determination in place of indigo [43].

Descroizilles finally published the method himself, in a rather obscure journal in 1795. The title of the paper was: *Description and use of the Berthollimeter, an instrument which can be used to test chlorine water, indigo and manganese dioxide; with notes on the marking of glasses with hydrogen fluoride* [44].

The "Berthollimeter" was the graduated glass tube in which the determination was carried out. We have no idea why Descroizilles called this instrument after Berthollet, whether it was because of his high regard for Berthollet or was simply flattery. He also referred to chlorine water as Berthollet-solution, which was later abbreviated by the workers of the bleaching plant to *berthollet*, and the bleaching plant itself was called the *berthollerie*, and the plant workers *bertholleur*, while the verb *bertholler* meant to "make white". Not everyone approved of this nomenclature, and one unknown translator of Descroizilles method referred to the

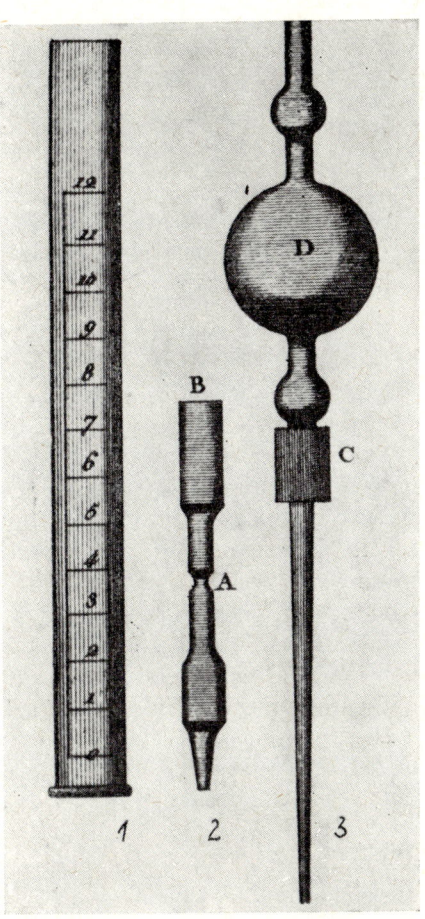

FIG. 58. The first titration assembly of Descroizilles, 1795. (From Rancke-Madsen's book: *The Development of Titrimetric Analysis till 1806.*)

berthollimeter as a "whitening water measure", while Klaproth, in his chemical dictionary, describes the instrument and commented that some people called this instrument by the very inadequate name of *berthollimeter* [45].

The pieces of apparatus designed by Descroizilles are shown in Fig. 58. The first picture shows the "burette", a vessel which resembles the present-day measuring cylinder. The second picture is of a pipette which he called a *measure*. This was immersed into the chlorine water, and then the upper end sealed by placing a finger over it. The solution level was adjusted to the mark A, and this volume of solution run into the cylinder. The cylinder was graduated by means of the measure (pipette), chlorine water reaching up to the zero mark. Then by sucking the pipette-like device shown in Fig. 3 the bulb D was filled with a 1 per cent solution of indigo in sulphuric acid, and he then

> cautiously let the blue solution pour into the berthollet-solution, which destroys the colour immediately and the solution turns redish. Interrupt some times the pouring of the solution from the mouth-pump and shake the solution. Take care that no foam or bubbles should be formed, as this obscures the meniscus. The reaching of the saturation point can be observed by the formation of a pale olive-green colour, which is the resultant of the red and the small amount of blue. The latter disappears towards the end more and more slowly... Water, saturated with chlorine at 10° above freezing point shows 8 degrees in the bertholli-meter, if the indigo is of a good sort and the test solution was prepared well [44].

Descroizilles mentions the reverse of this method, i.e. the examination of indigo with saturated chlorine water, and records that it was not quite so effective.

This method was in use by the end of the century in almost all bleaching plants in Western Europe. Descroizilles' method is entirely a volumetric method, and the two essential pieces of apparatus used in this technique, the pipette and the burette, appear in their earliest form. Rancke-Madsen, however, has found a picture of a pipette dating from 1785, this has almost the same shape as a modern pipette, and was used by Achard [46], but only for physical examinations [47].

To return to acid–base titrations, Tobias Lowitz used an aqueous solution of potassium carbonate for the titration of acetic acid, the end point being detected, not as in earlier methods by the cessation of bubbling or by an indicator, but by the appearance of a precipitate. "Instantly a cloudy precipitate occurs", he writes, so it is possible that calcium was present in the acetic acid sample [48].

It is worth mentioning the paper of Fordyce [49] (1792) which describes for the first time the use of alkali hydroxide for the determination of sulphuric acid, instead of alkali carbonate. The end point was detected by the colour change of violet liquor. He mentions that the colour change is more easily observed if the acid is titrated with alkali than vice versa. Fordyce only used this titration to follow the course of an experiment and he did not calculate any absolute values for the acid strength. He measured the amount of titrant consumed by weight difference [50].

One of Black's published papers is of importance to this subject. Black, who was an extremely accurate worker, found and eliminated two small errors which were not really considered by analytical chemists until a far later date. He determined the alkali content of water by titrating with dilute sulphuric acid in the presence of litmus indicator, and as the colour change in the direct titration was not very sharp, he added an excess of the acid, and back-titrated with potassium carbonate solution. He first of all boiled the solution to be titrated, in order to expel "fixed air" i.e. carbon dioxide, and he also established that the litmus indicator itself contains some slight alkaline impurity. He measured the amount of sulphuric acid which was required to change the colour of the amount of litmus solution used in the titration, and subtracted this amount from the total consumption of sulphuric acid. Black, therefore, was the first to observe the interference of carbon dioxide, as well as the indicator error, and introduced the use of a correction to eliminate the latter. He was also the first to use a back-titration, and measured the amount of titrant by weight [51].

In Lampadius's book, *Handbuch zur chemischen Analyse der Mineralkörper* (1801), referred to previously, acid–base titration methods are described. For example, he titrates sodium carbonate with sulphuric acid using curcumin paper as an indicator. He measured the consumption of sulphuric acid by weight difference, but his method gave absolute results, for he determined the amount of pure sodium carbonate equivalent to the consumed sulphuric acid [52].

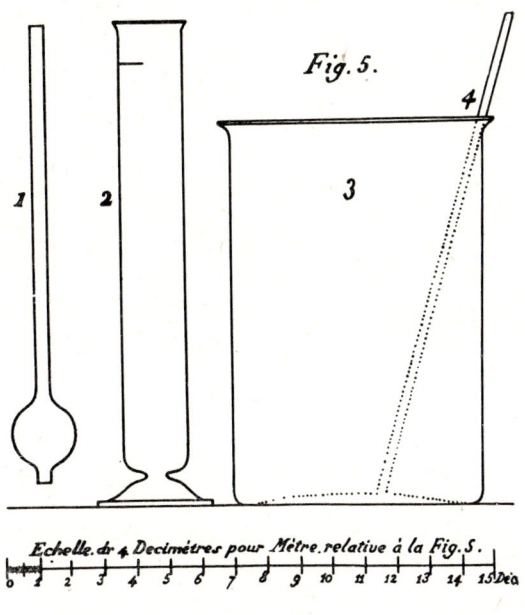

FIG. 59. Titrimetric assembly of Welter. (From Berthollet: *Elements de l'art de la teinture* [1804])

Berthollet, in one of his books, describes the method of Welter [53] for the determination of acids and bases. In this method several pieces of volumetric apparatus are described (Fig. 59). There is, for example, a measuring cylinder (similar to that of Descroizilles), a pipette measuring one-fiftieth part of the latter, and also a glass beaker. This apparatus was also used for the determination of sodium carbonate and chlorine water. He poured the soda solution from the cylinder into the beaker, then added dilute sulphuric acid from another cylinder and mixed the solution with a glass rod and then spotted one drop of the solution on to a litmus paper. If this did not turn red, he then added more acid with the pipette, testing with the litmus paper after each addition. The determination was completely empirical, the soda having a grade of 50, 51, 52, etc., according to the amount of acid used The original French expression was; "la potasse est au titre 56" [54], so we can see that the word "titre" was in use before Gay-Lussac. Welter determined the amount of acid consumed by volume measurement.

The principle of volumetric measurement in acid–base titrations became well known after the publication of Descroizilles' paper in 1806 entitled: *Notices sur les alcalis du commerce* (About commercial alkalies) [55]. He was commissioned to develop a method for the examination of potash because

> among the American potashes there are three qualities, known as first, second and third grades. The bleaching plants pay for these according to their alkali content.

Therefore he devised a simple method, which "the ordinary pharmacist can easily perform".

The most important contribution of this paper is the apparatus called an *alkalimeter*, which led directly to the development of the burette. The alkalimeter, shown in Fig. 60, was 20–22 cm long, and 14–16 mm in diameter. At the top there was a lip to assist the pouring of liquids, and a small hole (b) to allow the air to escape. The vessel, filled to the mark with a standard solution (1 : 10 sulphuric acid – 66 Baume grade) holds 38 g.

It can be seen that by alkalimetry, Descroizilles meant the determination of alkalis with acids, while acidimetry is the reverse process. The present day terminology is slightly different as the methods are named after the standard solution used, and not after the substance being determined. We therefore speak of permanganometry, argentometry, ascorbimetry, reductometry, etc., and this rule also applies to acidimetry and alkalimetry. In this discussion alkalimetry refers to a method utilizing a standard solution of an alkali.

Descroizilles calibrated his "burette" by weighing the standard solution, so that one division corresponded to half a gram of the solution.

To carry out the determination the following are required: a small balance, a half decilitre vessel, common table glasses, wooden rods or matches (those without sulphur) and a solution of extract of violets.

We weigh one decagram of the potash and place it into a glass, and dissolve it in about 40 ml of water. Dissolution is assisted. The glass which contains the aqueous solution must then be placed on a plate, and on this plate several drops of the violet extract must be spotted. The "burette" is filled with the "test solution" up to the zero mark, and held in the left hand with one finger placed over the hole at the top. The solution is poured in a slow steady flow, or dropwise into the glass by gradually allowing air to enter by moving the finger slightly. If the hole is completely closed by the finger air cannot enter the "burette" and the flow of solution ceases. The solution in the glass must be mixed with a match with the right hand, and when the test solution reaches the 40 mark, a drop is taken out with the match and spotted onto the violet extract and the colour noted. This must be repeated frequently as the titration continues, until the colour of the violet drop turns from green to red, when the saturation (neutralization) point is reached. The average value of potash is 55, which means that 55 parts of sulphuric acid are needed for saturation (i.e. 55 volume parts; 2 parts were almost equal to 1 g or 1 ml).

Thus, even in this method Descroizilles did not calculate the absolute value of the strength of the potash; he was satisfied with a comparative result which was quite sufficient for practical purposes in the factory. The only information required for control purposes was whether a 10 g sample of the potash required 55 volumes of sulphuric acid for neutralization; if it consumed less than this volume then it was not of sufficient quality. This shows that Descroizilles was essentially an industrial chemist, who had little interest in the theoretical side of his work.

He mentions, however, that on the basis of this method sodium carbonate, sodium hydroxide, tobacco ash, etc., can also be determined, and he published the average value for the consumption of acid by these substances,

which is the result of several thousand experiments, that I have made during the past twenty years,

which would seem to indicate that Descroizilles may have used this method from about 1786.

This "burette", in addition to its use as an "alkalimeter", could also be used as a "berthollimeter", i.e. for the determination of the strength of chlorine water. For this purpose there were two engraved scales, the one on the left having 18 graduations. The unification of these two methods was the first step towards the generalization of titrimetric methods!

Laurens published a work in which he determined the alkali content of sodium carbonate with hydrochloric acid. The end-point of the reaction was indicated by the cessation of effervescence. It is not quite clear whether the consumption of acid was measured by volume or by weight in this method, because the author only recorded that "we determine the amount of absorbed hydrochloric acid". It is probable, however, that he measured the difference in weight. He standardized his solution with pure sodium carbonate, and so was able to obtain absolute results. This paper is of interest because it shows that titrimetric methods, in those days, were only used for industrial purposes. The title itself indicates this: *Concerning the soda consumption of soap factories in Marseille.* Laurens mentions that by using this method, soap manufacturers, even though they are not chemists, can protect themselves from buying inferior quality raw material [56]. One year later (1809) Descroizilles describes a very important piece of new apparatus, the volumetric

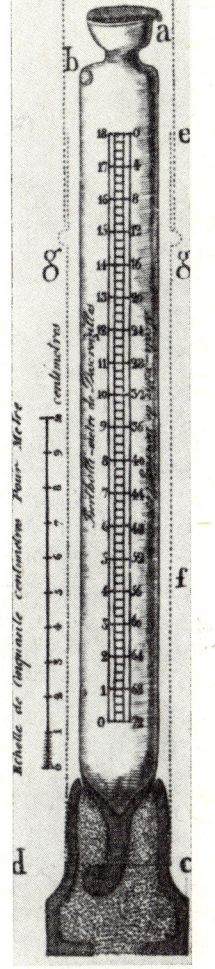

Fig. 60.
Descroizilles'
burette for
alkalimetry, 1806.
(From his original paper)

flask. He describes this in a paper dealing with the analysis of commercial potashes:

> a small jug, with a volume slightly greater than 2 dl, on the neck of which a mark is made with a diamond needle, indicating the place, up to which the vessel contains just 2 dl.

By this time principles of acid–base titrations were well established, and further development only occurred in the improvement of the existing apparatus and refinements in the technique, as well as a rather more unified system of standard solutions. No radical changes were introduced until the end of the 19th century when synthetic indicators began to be used, and the fact that these indicators had colour changes over a wider range of pH considerably increased the scope of titrimetric analysis.

In the field of precipitation titrations, the original methods of Home for the determination of the hardness of water, and of Guyton de Morveau for the determination of chloride with lead nitrate, was followed by Bartholdi [57].

Bartholdi published a method in 1792 for the determination of the sulphate content of pigment extracts. He precipitated the sulphate with lime water, which he had previously standardized against pure magnesium sulphate [58]. There is no reference to the method of end-point detection in this procedure so presumably it was simply the cessation of precipitation. In the description of a water analysis some time later, Bartholdi mentions that he dissolved the residue after evaporation in water, and after determining the carbonate content with acetic acid he determined the sodium sulphate with barium acetate:

> I added this until no more barium sulphate was formed; I noted how much of the barium acetate was needed for the decomposition, used this same amount to decompose recrystallized sodium sulphate, and the barium sulphate formed in this case weighed as much as before [59].

Here, therefore, the principle of precipitation titrations can be seen quite clearly, but the determination of the amount of reagent consumed was made by weight difference. Bartholdi then filtered off the barium sulphate, and added silver nitrate solution to the mixture, until no more silver chloride was formed. The result was calculated as before.

Bartholdi, therefore, was a pioneer of argentometry, but his method was rather inaccurate owing to the uncertainty of the end-point detection.

In the tenth volume of the *Annales de Chimie* a paper was published by the Powder and Saltpetre Directory *(Administrateurs-Generaux des Poudres et Salpetres)*. This volume was published in the year 1802, and not in 1799 as is recorded in Beckurts's book [60]. The alteration in the calendar after the French revolution was probably responsible for this mistake. In this paper [61] reference is made to a prescribed method for the examination of potash. The analyst who carried out this work is not named. Calcium nitrate was used as the titrant; this precipitates both the carbonate and the sulphate. It is mentioned that strontium or barium nitrate is better for this purpose than calcium nitrate, strontium being preferred to barium "because its use is harmless".

In the determination the standard solution was poured into a measuring cylinder, until the sample of potash (20 gr) was saturated. The cylinder was graduated with 100 divisions, and the determination of the quantity of titrant was volumetric. The carbonate content was measured in a separate sample by treating it with nitric acid. Using the same measuring cylinder 20 gr of potash required 88 volumes of nitric acid for neutralization. The solutions of strontium nitrate and nitric acid were standardized with pure potassium sulphate and potassium carbonate respectively, and hence the carbonate and sulphate content of the potash samples could be determined. In this paper the word "titre" is mentioned:

> On connaitrait par la différence qui pourrait se trouver entre le titre total indiqué par cette dernière liqueur et celui resultant de l'emploi de l'acide nitrique, ou par l'égalite de ces deux titres...

From the French text it would appear that the meaning of this word is the same as it is today. Roughly translated it reads as follows:

> From the difference that is found between the consumption (titre) of the latter solution (strontium nitrate) and that of the nitric acid, or from an equality of the two consumptions (titre).

The word consumption is used in place of the word titre. The word "titre" was rather uncommon and often difficult to understand, and this is clearly indicated in the English and German translations of the text. The English translator replaced the word "titre" with the word "quality" [62], while the German translator transcribed the whole of the text and left out the word titre altogether [63].

The basis of precipitation methods, therefore, were worked out during the 18th century, and the earlier inaccurate methods were replaced by the very accurate and precise procedures of Gay-Lussac, notably his method for the determination of silver which was very widely used.

The only redox titration used for a long time was the determination of chlorine water and hypochlorite. Dalton, whose name was best remembered in connection with the atomic theory, also worked in the practical field. During this period all the most notable chemists attempted to develop a method for the determination of hypochlorite or alternatively to improve on existing methods. Dalton introduced the use of a standard ferrous sulphate solution for this purpose. The description of his method [64] is rather difficult to follow; Fe(II) ions are transformed to "red iron oxide" [Fe(III)] when treated with the "oxidized hydrochloric acid solution" (hypochlorite). This rather uncertain method, however, was used for quite a long time, especially after Otto [65] had recommended the spot-test end-point detection method using potassium ferricyanide [66].

Welter attempted to improve the method of Descroizilles by standardizing the indigo solution with chlorine gas [67]. Quite naturally this was a very inaccurate method, because as Gay-Lussac points out, the amount of hypochlorite solution consumed also depends on its rate of addition to the indigo. Welter tried to standardize the conditions of the reaction by adding the total amount of hypochlorite solution all at once [68]. Although with this method he was able to get quite accurate results, it needed a considerable degree of skill. Gay-Lussac soon found a better method, titrating hypochlorite with arsenious acid in the presence of indigo as a redox indicator. This method started the development of redox indicators.

In all the branches of titrimetry we meet with the same name, Joseph Gay-Lussac. The work of Gay-Lussac provided the firm basis for the development of titrimetry, and it was largely due to the widespread use of his methods that this subject changed from an industrial method to a distinct branch of science.

Joseph Gay-Lussac's name is also famous for his discoveries and contributions to other fields of science, notably in physics and in general chemistry, where one of the most important laws is named after him. In all the branches of analytical chemistry we can find his contributions. The use of hydrogen sulphide in qualitative analysis, the systematization and improvement of titrimetric analysis and the development of the first practicable method of elementary organic analysis. All these are the work of Gay-Lussac, but they represent only the most important of his activities. It is difficult to rank the scientists of the old days, as the conditions under which they worked, and the circumstances in which they lived are so vastly different from our own. Yet if we consider that Berzelius was the most accomplished analyst the world has yet seen, then surely Gay-Lussac was not far behind him.

Joseph Louis Gay-Lussac was born in 1778 in the small town of Saint-Leonard of the ancient Limousin, near the Auvergne where his father was a judge and it was here that he was first taught by a priest. Later he went to a college in Paris to continue his studies. With the outbreak of the French revolution all the students with the exception of Gay-Lussac were sent home. Gay-Lussac by this time had shown a great talent for mathematics, and the head of the college asked him to stay. In 1796 he became a student at the newly founded École Polytechnique. His father had lost his occupation and all his property during the revolution, so he could not provide any financial support for his son's studies, but the young Gay-Lussac earned enough money to live on by giving private lessons. Gay-Lussac's exceptional ability had been noted, and he was offered the post of laboratory assistant to Berthollet, a great honour, for at that time Berthollet was at the height of his career. Berthollet asked Gay-Lussac to carry out a small experiment, and when Gay-Lussac informed him of the result he said,

Young friend, your destiny is to make discoveries, from now on you shall eat at my table; I want to be your father in scientific matters, and I know that I shall have reason to be proud of it some day.

In 1802 he became the assistant to Berthollet at the École Polytechnique, and Berthollet allowed him to give many of his lectures. Gay-Lussac became famous as the result of a very exciting and dangerous experiment which he undertook. At that time Dalton had given his opinion that the composition of the air in the upper atmosphere is different from that of the air in the lower atmosphere. There was also a diversity of opinion about the behaviour of a magnet at high altitudes. The French Academy charged Gay-Lussac and Biot (who later became a famous physicist) with the solution of these problems. These two young men decided to take suitable instruments up in a balloon, in order to examine the upper atmosphere. Laplace, the famous mathematician and physicist, obtained a small reconnaissance balloon from the army. This had already proved successful during the war in Egypt. On the first ascent the two scientists reached a height of 4000 m, while on the second ascent, which Gay-Lussac made alone, he reached a height of 7000 m. The results of their experiments were quite remarkable. They established that Dalton's

Fig. 61. Joseph Louis Gay-Lussac (1778—1850). A lithograph.
(From Bugge: *Buch der grossen Chemiker*)

supposition is untrue, the composition of the air, at least at this relatively low altitude is the same as it is near ground level. They discovered, however, that the temperature of the atmosphere decreases by 1°C for every increase in elevation of 174 m. In addition they carried out various magnetic measurements. As a result of this work Gay-Lussac achieved considerable fame.

In 1805—1806 Gay-Lussac, in the company of his close friend Humboldt made a trip to Italy and Germany, and they happened to observe one of the more violent eruptions of Mount Vesuvius, Gay-Lussac carried out a series of geological investigations in the vicinity of the crater. In 1807 the French Academy elected him to membership, and shortly after this he married. According to his biographer, Arago, Gay-Lussac saw a pretty young girl sitting in a lingerie shop, deeply engrossed in a book. When he asked her what she was reading he was told that it was a chemistry book, so quite naturally he married her. Whether this story is true or not Gay-Lussac and his wife were happily married for more than forty years.

Gay-Lussac's scientific achievements increased rapidly, and in 1808 he established the gas laws, which still bear his name. In this same year he became Professor of Physics at the Sorbonne, while in 1809—in addition to this post—he occupied the chair of chemistry at the École Polytechnique. Later he was appointed a director of the mint, and he also became a professor at the Jardin des Plantes. In spite of the burden of his numerous occupations and honorary posts, Gay-Lussac still found time to continue his research work, both in physics and in chemistry. The most important of his discoveries in the field of chemistry is his method for the determination of vapour densities, but his investigations on the nature of solubility, and capillarity, and his study of the properties of iodine are also of extreme importance.

After Davy's original preparation of sodium by electrolysis, Gay-Lussac was the first to produce this metal by an alternative method. He was able to produce quite large quantities of the metal by reduction with iron filings and carbon. It was during this experiment that an explosion occurred, in which Gay-Lussac was quite badly injured, and nearly blinded. When Berthollet died he willed his sword, the most important part of his uniform as *Pair de France*, to Gay-Lussac. This meant in effect that Gay-Lussac had succeeded to the title, but official recognition of this fact was only made several years later. The reason for his reluctance, according to Arago, was the fact that the great chemist worked every morning in his laboratory with his hands, and this was rather incompatible with the dignity of the title *Pair de France*.

Gay-Lussac was a rather shy and solitary man, but nevertheless he was very brave. He was never offended if anyone contested his theories, but he also stated his own opinions clearly and openly. After the fall of the 100-day rule of Napoleon there was political opposition to Arago [69] on the grounds that he had supported Napoleon and was therefore no longer worthy to teach at the École Polytechnique. Gay-Lussac opposed the authorities over this, saying that if his friend Arago was dismissed then he and several of his colleagues would resign. Towards the end of his life a long and painful disease prevented Gay-Lussac from working, but the nature of his affliction is not known. Even while he was very ill Gay-Lussac tried to keep pace with the rapid development of science, mainly with the study of electricity. It is reported that his last words were "It is a pity that I must depart now, when it begins to become so interesting". He died in 1850.

The first titrimetric work of Gay-Lussac [68] was a critical examination of the method of Welter, earlier referred to, recommending certain modifications. For example, the standardization of indigo solution was carried out with chlorine gas at 0°C and 1 atm pressure in the following way: He found that 1 l. of chlorine gas is evolved from 3·980 g of manganese dioxide under ordinary conditions. This gas was passed into lime water, and the chlorinated lime solution was used for standardization. The indigo solution was diluted so that 10 l. of the solution was just decolourized by 1 l. of chlorinated lime solution. This part of his paper is not very interesting, for it is obvious that the method was not very accurate. In the next section, however, Gay-Lussac describes a "chlorometre" which was used in this method, this being the new Gay-Lussac burette. This is shown in Fig. 62, together with the titration apparatus. Its use is described as follows:

> F is a small measure or pipette *(petite measure ou pipette)* with a volume of 2·5 cm^3...; this is used for the measurement of the chlorinated lime solution. To fill it, it must be immersed into the solution of chlorinated lime, and the solution allowed to rise above the mark n,

which indicates the volume. When it is filled the first finger, which should neither be too wet nor too dry, is placed over the upper end of the pipette, which is then taken out of the solution. The end of the pipette is then placed on the lower part of the cylinder, as shown in *G* on the figure, and by gently lessening the pressure on the finger the solution

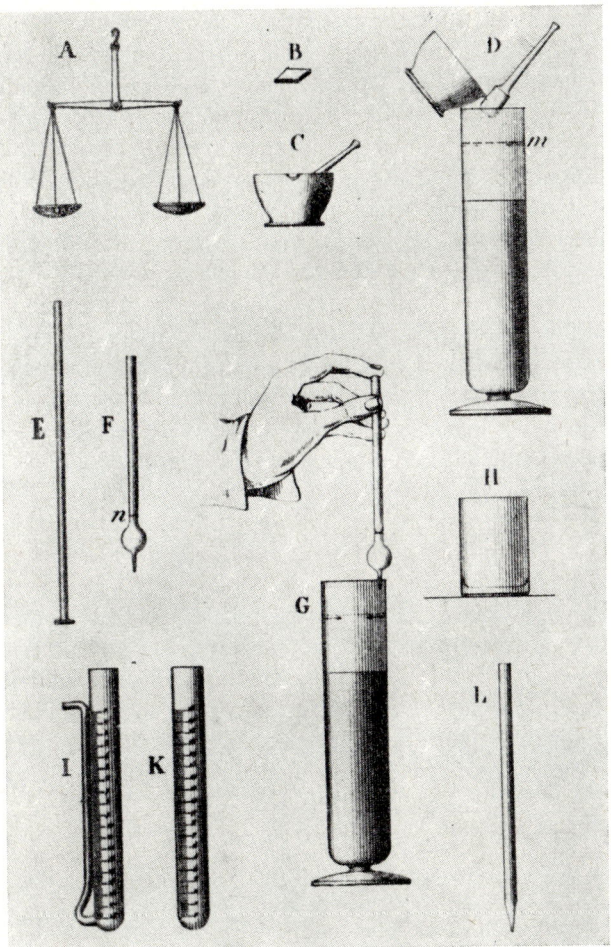

FIG. 62. Titrimetric device of Gay-Lussac, 1824.
(From his original paper)

flows into the cylinder. When the lowest part of the concave arc of the solution surface reaches the mark, then the pressure is reapplied with the finger when the flow of solution ceases. Then the contents of the pipette are run into the beaker *H*... *I* is a burette, which is used for the measurement of the test pigment *(burette destinée à mesurer la teinture d'épreuve)* the large marks or degrees correspond to the volume of the small pipette *F*, and these large divisions are further subdivided into five parts. This is adequate for practical purposes, but for the sake of calculation the large divisions were divided into tenths. The burette must be filled with the test liquid up to the more solution and then running off the excess. The end of the tube is carefully lubricated with wax, in order to control the flow of the drops [68].

In 1828 Gay-Lussac wrote a paper on the examination of commercial potashes [70], and in it he refers to Descroizilles who was the first to examine alkalis in this way. He mentions that the method of Descroizilles contains many difficulties. In this paper the word "titre" is again mentioned. The gravimetric titre of an alkali *(titre ponderal)* is, according to Gay-Lussac, the amount of alkali contained in 100 kg of the material, while its alkalimetric titre *(titre alcalimetrique)* is the amount of acid of given strength that is neutralized by 100 kg of the crude potash. For this procedure he uses the burette described above and also a pipette which is very similar in construction to the one in use today. He also mentions that a volumetric flask, of one litre capacity, is used to dissolve the sample, and from this solution, 50 ml are taken for the titration.

Gay-Lussac obtained an absolute determination by this method but the calculation was simplified for practical purposes. In order to do this the strength of the solutions was adjusted so that the consumption of the reagent gives a direct value for the component being sought. It was probably for this reason that Gay-Lussac did not make any general classification of titrimetry. As in the method of Descroizilles, Gay-Lussac prepared a solution of 100 g of sulphuric acid (sp. gr. 1·84) and diluted this to 1 l. (i.e. approx 2 N). He found that 50 ml of this solution are required to neutralize 4·807 g of potassium carbonate. This acid he referred to as normal acid *(acide normal)*, and on his burette 100 divisions correspond to this 50 ml. Therefore, by weighing 4·807 g of the potash sample the acid titre gave a direct reading for the alkali carbonate content. In order to avoid any slight errors due to the heterogeneity of the sample he weighed out ten times this weight (48·07 g), dissolved this in a half-litre volumetric flask, and took a 50 ml aliquot. Gay-Lussac preferred to use an immersion pipette rather than a suction pipette, and in this paper we can read a description of the titration with the Gay-Lussac burette. Here we find the first use of the verb 'to titrate'.

> Take the beaker and pour into it the alkaline solution with one pipette ... pour into it so much litmus solution that should be definitely blue, hold the beaker over a white paper, to perceive better the colour change of the litmus. Fill the burette to the mark 0 with the normal acid. Hold the burette in one hand and the beaker in the other, and pour the acid into the alkaline solution, and move the latter constantly in small circles in alternative directions. At the beginning the colour does not change, but if the potassium carbonate is examined after 11/20 parts of the saturation has taken place then the solution turns to wine red, owing to the liberated carbonic acid. Care must be taken to avoid passing the saturation point, if the acid when it is added to the solution only causes a slight bubbling, then continue the addition of the acid dropwise, testing after the addition of each drop by placing a drop on litmus paper ... As the saturation point is passed the wine colour of the solution turns to mushroom-red, and the colour of the paper turns to red, and remains at this colour ... [70].

This method of titration was not easy to carry out, although with practice it could be improved.

Gay-Lussac also describes the colour transition of the indicator in different titrations.

Firstly, in the case of caustic soda, i.e. alkali hydroxide, the colour of litmus changes only at the end of saturation, and the blue colour will turn immediately to mushroom-red.

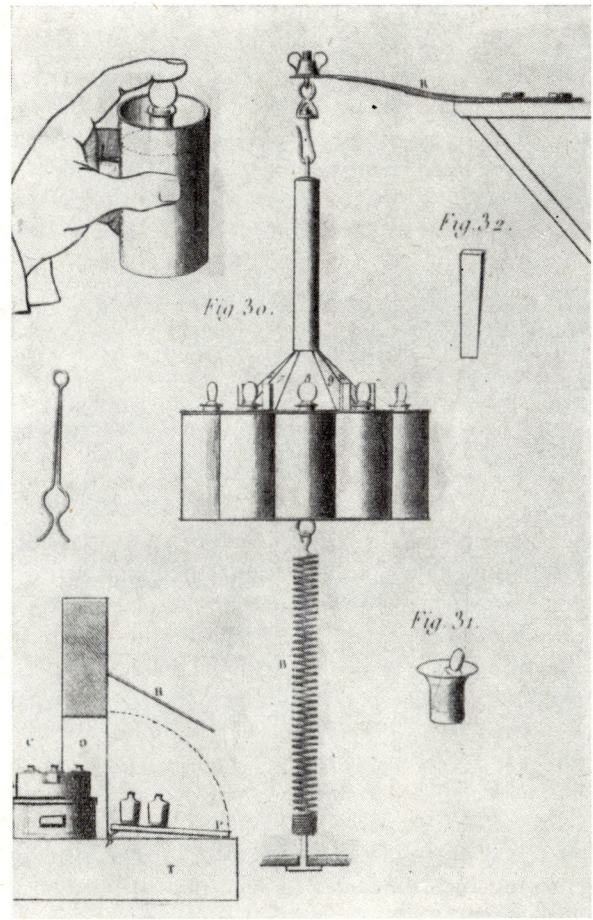

FIG. 63. Shaking device of Gay-Lussac for his silver determination, 1832. (From his work: *Vollständigem Unterricht über das Verfahren Silber auf nassem Weg zu probieren*)

Secondly, in the case of carbonate at 11/20 fold neutralization the bubbling becomes vigorous and the indicator turns to wine-red colour, and only changes to mushroom-red when neutralization is complete.

Thirdly in the case of bicarbonates the colour of the litmus turns to wine-red, and remains so until the neutralization is complete.

In many cases the potash also contained sulphate, and Gay-Lussac, in the same paper, describes a titrimetric method for the determination of sulphate. The method is a precipitation titration, using a standard barium chloride solution. He dissolved 248·435 g of barium chloride in 1 l. of water, and assumed that this precipitates 48·07 g of potassium sulphate. By dissolving 4·807 g of potash, then the amount of barium chloride solution consumed will give a direct percentage sulphate content. Theoretically, 61·6 g of barium chloride are required to precipitate 48·07 g of potassium sulphate, so even allowing slight variation for the impurity of the reagent, these results of Gay-Lussac are impossible to understand. The end-point detection was by the observance of the cessation of precipitation, this being carried out by constantly filtering the solution and titrating the filtrate.

Also in the same paper, he describes another interesting method which was not this time a titration [70]. According to this, potassium and sodium chloride can be determined by the following procedure.

If 50 g of sodium chloride are dissolved in 200 g of water at a given temperature, then the temperature of the solution decreases by 11·4° while if the same amount of potassium chloride is dissolved then the temperature is only 1·9°. If, therefore, 50 g of a mixture of the two salts are dissolved, then from the temperature decrease the composition can be calculated. He provided detailed tables increasing in tenths of a degree giving the correlation of temperature decrease with salt concentrations.

Although the method is very original, it did not prove very useful, for neither this nor similar methods can be found in the later literature.

In 1829 Gay-Lussac published a method for the determination of borax; he titrated the alkali content with sulphuric acid in the presence of litmus. He mentioned that the litmus does not change to a similar shade of red as in the titration of strong acids, and only shows a wine-red colour at the end-point. He also measured the amount of acid required to change the colour of the same amount of indicator as is used in the actual determination (indicator correction) [71].

Gay-Lussac published his most famous method, the precipitate-titration of silver, in 1832. This method, named after him, is still used today, and its precision is not inferior to any of the indicator methods [72]. This paper aroused a great interest, which was not surprising as it claimed that the method was much more accurate than the cupellation method which had been in use for hundreds of years. Gay-Lussac claimed that the cupellation method gives a low result, so that is effect the Government was losing money. To test his assertion the French Mint prepared accurately alloyed silver samples, and sent them to be examined at several places in Europe. The deviations in the results completely proved Gay-Lussac's supposition [73].

The method was worked out by Gay-Lussac both as a gravimetric and a titrimetric procedure. He was of the opinion that the method involving a weighing burette is rather more precise, but that the volumetric procedure is simpler. The principle of both methods is almost the same. He prepared a sodium chloride standard solution and the concentration of this was such that 100 ml precipitated slightly less than 1 g of silver. Another sodium chloride solution, called tenth-solution, of which 1 l. would precipitate 1 g of silver, was also prepared. 1 g of silver was accurately weighed, and after solution 100 ml of the concentrated

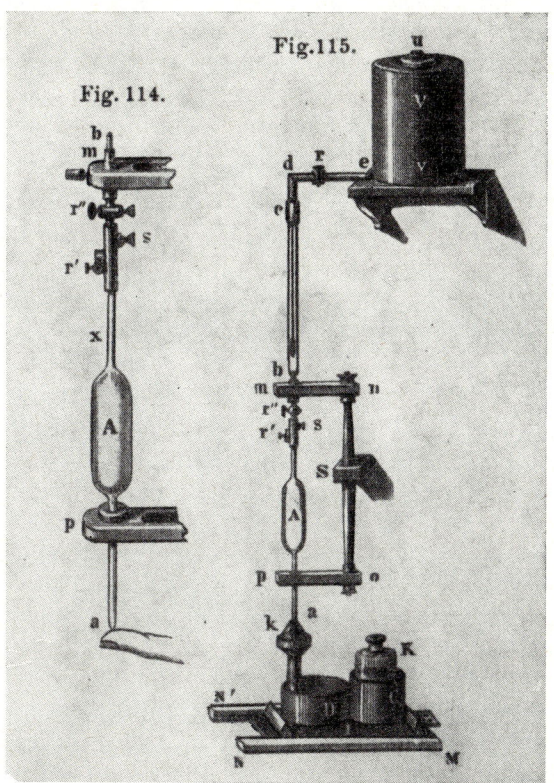

FIG. 64. Pipette filling device of Gay-Lussac. (From his work: *Vollständigem Unterricht über das Verfahren Silber auf nassem Weg zu probieren* [1833])

sodium chloride solution were added, and the precipitate allowed to settle. Then the dilute sodium chloride solution was added in 1 ml portions, and after each addition the sample was shaken, and then the precipitate allowed to settle. This was repeated until a further addition did not cause precipitation, then this last excess of chloride was back-titrated with a tenth silver nitrate solution.

Gay-Lussac also realized that the volume of the standard solution increases with temperature, so he calculated temperature corrections. The pipette was similar in construction to that used today, but suction with the mouth was not recommended by Gay-Lussac. He constructed a device for filling pipettes (Fig. 64).

The sodium chloride solution was placed in the upper container, and this flowed into the pipette at $c-b$ alongside a thermometer. The flow of this solution could be regulated by the taps $r, r', r,''$ whole on the $M-N$ bottom plate there was a place for a flask into which the solution was placed. D is a dish which is used for running the titrant into when adjusting to the zero mark on the burette; k is a sponge, used to dry the lower part of the pipette.

From the figure the method of filling and pouring out of the pipette is not very clear]74].

This piece of apparatus soon fell into disuse, probably because it was inconvenient to use. Gay-Lussac prepared his standard solutions in very large amounts, probably for use in the mint over a considerable length of time.

> Take 0·5424 kg of rock salt and 99·4573 kg water; from this 100 kg of a solution is obtained. 100 g of this solution precipitates exactly 1 g of silver [75].

In 1835 Gay-Lussac published a paper which was of great importance in the development of titrimetry. This paper dealt with the determination of hypochlorite or chlorinated lime once again, as the earlier method which he had devised proved too difficult to operate in untrained hands. In the new method the use of indigo was replaced by various reducing agents.

Standard solutions of arsenious acid, potassium hexacyanoferrate(II) and mercury(I) nitrate were used [76], and the apparatus was essentially the same as that used in earlier methods. The experimental details of the method were only published for the titration with arsenious acid, 1 l. of the standard (normal) arsenious acid solution reacting with exactly 1 l. of chlorine gas dissolved in water. He weighed accurately 4·439 g of arsenic trioxide and dissolved this in a little hot hydrochloric acid and then diluted this solution to 1 l. with water. In order to avoid loss of chlorine through volatilization he added the chlorinated lime solution from a burette into a known amount of arsenious acid, the end point of the titration being indicated by the decolourization of two drops of indigo solution.

> If the arsenious acid solution is coloured slightly with a sulphuric acid/indigo solution, and chlorine is added dropwise, the blue colour remains until there is an excess of chlorine; this only occurs when all the arsenious acid is transformed to arsenic acid.

This is the first recorded use of redox indicators in titrimetry! This indicator, however, is irreversible, and it is characteristic of the accuracy with which Gay-Lussac worked that he subtracted a half drop from the total titre in order to compensate for the excess of arsenious acid present at the end-point. He also determined the indicator blank by measuring the amount of solution required to decolourize two drops of the indicator and took this value into account for the correction.

For the determination of less concentrated chlorine water solutions he used 1/10th normal arsenious acid (normality is not used in the modern sense). Potassium hexacyanoferrate (II) was also used as a titrant, 35 g of the pure solid being dissolved in 1 l.

> This solution of potassium ferrocyanide has no effect on chlorinated lime, or only very slightly, but if acid is added previously it reacts immediately.

This, as we know today, is due to the change in the redox potential of the system on increasing the hydrogen ion concentration. The indicator for this titration was also indigo. When a standard solution of mercury(I) nitrate was used for the titration, one drop of sodium chloride was added, which caused a turbidity due to the formation of calomel, Hg_2Cl_2. When the chlorinated lime was present in excess this turbidity disappeared.

Gay-Lussac mentioned that this method had been recommended earlier by Balland de Toul in a report to the Academy in 1829, but that he had carried out the titration in the reverse manner, titrating the chlorinated lime with mercury(I) nitrate until precipitation occurred. As mentioned previously this method introduces errors caused by the volatility of the hypochlorite solution. Marozeau had also used mercurous nitrate but had not examined the experimental conditions [77]. Earlier Penot had attempted to titrate chlorinated lime with barium sulphide solution [78], and Morin [79] had investigated the use of a manganous(II) salt [80], both of these methods involving a precipitation titration. Gay-Lussac also examined the stability of his three standard solutions, and found that after six months, arsenious acid and potassium hexacyanoferrate(II) solutions were only oxidized slightly, while the change in the mercurous nitrate solution was barely perceptible.

We have seen that Gay-Lussac developed accurate and scientifically investigated methods in all the main branches of titrimetry, but did not realize the possibility of establishing a general system of titrimetric analysis. The concentration of his standard solutions had no chemical basis, apparently owing to the differences in atomic weights, and the undeveloped state of stoichiometry in those days. Therefore his standard solutions could only be used for specific analyses and for a given weight of sample, and in addition his burettes, although similar in construction to each other, had to be calibrated separately for each determination.

3. FROM GAY-LUSSAC TO MOHR

As a result of the work of Gay-Lussac, titrimetry became a convenient, rapid, and reasonably accurate method which was increasingly used in practical analysis. In its infancy the method was confined to France, but foreign students and scientists from other European countries working in France, became acquainted with the new method and took it back to their own countries. In the early part of its development titrimetry was mainly used for industrial and technical purposes, but the new analytical methods devised by chemists as celebrated as Gay-Lussac aroused the interest of the universities, so that the purely scientific possibilities of the method came to be investigated. In the next few decades it is almost true to say that every known chemical substance was tried as a standard solution for the determination of all the other known substances. The scientific journals of this period published a large number of methods, most of which did not survive beyond their publication. Some of these methods, however, are still used by analysts all over the world, for process or quality control or even in research work. For example iodometry and permanganatometry were developed during this period, but a large number of other methods were introduced and then immediately forgotten. Many of these disused methods have since been resurrected, some as long as one hundred years after their original publication, and owing to some new knowledge or different circumstances have been found to be useful. The numerous redox methods suggested in the last century were made impracticable by the lack of suitable indicators, and unnecessarily complicated methods of end-point detection were used. A very large number of titrimetric methods were tried out in the last century, but there is only space here for a consideration of a few of the most important, and a few unusual ones.

As we have seen, acidimetry, alkalimetry, and argentometry are the oldest methods of titrimetric analysis and as was mentioned previously arsenious acid, mercury(I) nitrate and potassium hexacyanoferrate(II) were among the earliest standard solutions.

The discovery of the use of iodine and iodides in titrimetric analysis was made by Houton de La Billardière [81] in 1826, soon after the discovery of the element. He boiled iodine with a solution of sodium bicarbonate, potato starch and sodium chloride, and then diluted the solution to 1 l. He used this solution for the titration of chlorinated lime. In this method sodium iodide is first formed, and this is then oxidized by hypochlorous acid to iodate, and at the end-point of the titration iodine is liberated, which imparts a blue colour to the solution [82]. This method was repeatedly used, and it often appeared as a new method. Several workers attempted to make the method more accurate by altering the experimental conditions, but were never completely successful [83].

Du Pasquier [84] was the first to use iodine as a standard solution in 1840. He titrated the hydrogen sulphide content of water with a solution of iodine in alcohol, using starch as an indicator. In this method he used a device which he called a *sulphydrometre*, which was a graduated glass tube, with a capillary attached to the bottom. The top was closed with a stopper, which on removal allowed the solution to flow out of the capillary in small drops. The concentration of the iodine solution was adjusted so that each large graduation corresponded to 10 mg of iodine, and each small graduation to 1 mg of iodine. For the basis of the calculation the author only mentions that it must be found how much hydrogen is needed for the saturation of the amount of iodine consumed in the titration. From this the amount of hydrogen sulphide can easily be calculated because the same volume of hydrogen is present in hydrogen sulphide as in hydrogen iodide [85].

Berzelius made some interesting comments on this method. He established that although Du Pasquier's method is very ingenious, it is incorrect, because hydrogen iodide is formed from iodine in alcohol, which will dissolve iodine to form H_2I_4. Thus one of the iodine atoms is not available to react with hydrogen sulphide. Thus, the only way in which accurate results can be obtained is by dissolving the iodine in potassium iodide instead of alcohol [86].

Fordos [87] and Gélis [88] described a method in 1843 for the simultaneous determination of the different sulphur acids. They first of all precipitated sulphate with barium chloride in hydrochloric acid medium, and then titrated the sulphite in the filtrate with iodine in the presence of magnesium oxide. The latter was used to bind the hydrochloric acid. The sulphate formed from this titration was then precipitated, filtered, dried and weighed. If the weight of sulphate did not agree with the sulphite titre then dithionic acid must have been present, which is oxidized to tetrathionic acid, accounting for the excess of iodine consumed. The amount of this acid could be calculated by differences [89].

Duflos [90] determined iron iodometrically in 1845. He added potassium iodide to the iron(III) solution, and titrated the liberated iodine with a standard solution of stannous chloride. 1 l. of a standard solution containing one equivalent (atomic weight) of tin (5·90 g) consumed 12·5 g of iodine [91]. This was therefore a normal solution, in the modern sense! An iodometric method for the determination of tin was developed by Gaultier de Claubry [92] in 1846, quite independently from Duflos. In this method the tin sample was dissolved in hydrochloric acid, reduced with iron or zinc and the Sn(II) titrated with an alcoholic solution of iodine, using starch as indicator. During the prior reduction arsenic, antimony, lead, mercury and copper were precipitated in the form of the metal and therefore did not interfere [93].

The most important contribution to the development of iodometry was made by Bunsen in 1853, when he utilized a general iodometric method for the determination of oxidizing substances. In this method he added potassium iodide to the solution and titrated the liberated iodine with a standard solution of sulphurous acid. In a very short paper, less than twenty pages altogether, he describes the determination of the following, iodine, bromine, chlorine, hypochlorite, chlorate, chromate, lead, manganese, nickel and cobalt oxides, cerium(IV) salts, iodate, vanadate, ozone, selenic acid, permanganate, iron(III) and arsenious acid and its salts [94].

In all the literature of titrimetric analysis this is probably the most valuable and informative paper; a modern chemist would write five or even as many as ten papers to describe this amount of work. It is interesting that he determined all oxidizing substances in the following manner: hydrochloric acid was added to the sample and the chlorine formed was passed into a solution of potassium iodide solution. It could not have occurred to him that it was possible to treat the substance directly with potassium iodide, or possibly he did try this on one or two compounds but without success. Another interesting observation that he made in connection with the Du Pasquier reaction, was that the iodine–sulphurous acid reaction can be reversed under certain conditions because as he explains

> the values of those numbers, which are called affinity coefficients, are changing according to the circumstances. For the Du Pasquier method therefore it is necessary to choose the conditions so that the reaction expressed by the equation
> $$HISO_3 = I, OH, SO_2$$

becomes negligible. If the acid content of the sulphurous acid solution is not more than 0·04—0·05 per cent the reverse reaction does not take place.

Although sulphurous acid solution is unstable, and is difficult to prepare, it was used for a long time, despite the fact that in the same year that Bunsen's paper was published, Schwarz recommended the use of thiosulphate for the titration of iodine [95]. This method, undoubtedly the most effective for this determination, has been in use for the past hundred years without being superseded. In the words of Schwarz:

> It is difficult for me to say why I consider this method for the determination of iodine to be the best; I think however, that I can recommend it with a good conscience because it is so simple and accurate [96].

During this same decade another important oxidation method was introduced, permanganatometry. Potassium permanganate was introduced by Margueritte [97] in 1846, who used it to determine iron.

In the introduction he points out the growing importance of the determination of iron, and that there is no suitable method which is both rapid and accurate. This method, however, is so simple that it could be carried out by the foreman in charge of the casting, and is based on the reaction between chamaeleon ($KMnO_4$) and the lower oxidation state of iron (Fe(II)).

> The reaction of iron protoxide and chamaeleon can be expressed by the following equation:
> Mn^2O^7, $KO = Mn^2O^2 + O^5 + KO$
> $Mn^2O^2 + O^5 + KO + 5Fe^2O^2 = Mn^2O^2 + 5Fe^2O^3 + KO$
> It can be seen that one equivalent of the permanganate oxidizes 10 equivalents of iron protoxide. It is unnecessary to mention that the solution must contain an excess of acid to prevent the formation of iron and manganese oxide precipitates...
> One of my experiments proved the validity of the equations given above: To 0·350 g iron, which I dissolved in hydrochloric acid, I added 1·98 g of crystallized potassium permanganate, i.e. the exact amounts which should neutralize each other theoretically. And truly, if I added a further 0·002 g of chamaeleon to the solution, it immediately became pink.
> On the basis of these experiments the following steps must be made:
> 1. The ore must first be dissolved in some acid, for example, hydrochloric acid.
> 2. The iron oxide formed must be converted to iron protoxide, and to accomplish this zinc, sulphurous acid or sodium sulphite can be used; in the last case the solution must then be boiled to remove sulphurous acid.
> 3. Then the normal chamaeleon solution must be added cautiously, until the solution becomes pink, and the amount consumed noted from the burette [98].

This was the first paper in which redox process is expressed by a chemical equation. Margueritte also examined the effect of any other elements which could be present with the iron, titrating a known amount of iron together with an added amount of the impurity. He found that if sulphurous acid is used for the reduction, arsenic and copper are reduced to the protoxides, and therefore these also consumed the standard solution. He therefore recommended reduction with zinc as being more selective.

Concerning the accuracy of the method, anyone can determine the amount of iron in an ore with an accuracy between 1 and 1·4 per cent while a well trained analyst can obtain an accuracy of better than 1 per cent.

The permanganate-standard solution is fairly stable, I used the same solution for one month without any perceptible change in its titre. The only precaution that must be taken is that the solution should not come into contact with any organic substances [98].

He standardized the solution against 1 g of piano wire, choosing a concentration of the solution so that 30 ml were equivalent to the 1 g sample, i.e. the solution was approximately 0·5 N.

Margueritte then described the examination of ores and alloys, and pointed out that the method can be applied to the analysis of any metal which forms a protoxide. Potassium permanganate solution was called *chamaeleon* for a very long time and even today very old chemists often use this name.

The importance of the new method was not readily recognized by contemporary chemists, and this was apparent in the later fate of this method. It would appear that it was not the content of the work, but rather the importance of the author which decided the importance of a method. Margueritte's method was reviewed in the *Jahresbericht* by Berzellius in 12 lines, while on the next page, a long forgotten method devised by Pelouze [99], the well known chemist and University Professor, which consisted of a titration of copper with sodium sulphite in ammoniacal medium until the blue colour disappeared, had a review of 92 lines [100].

A year later Bussy [101] used permanganate for the determination of arsenious acid, titrating with a very dilute (0·4 g/l.) solution in hydrochloric acid medium [102]. In the same year Pelouze used the method of Margueritte for the indirect determination of nitric acid and nitrate. He dissolved a weighed amount of pure iron in hydrochloric acid in a stoppered flask, and then added the solution of the substance being determined. The solution was boiled until it became clear, and then after dilution he titrated the excess of iron(II) using the method of Margueritte [103]. This was the beginning of the many methods using back-titration of iron(II) with permanganate; this was followed by the titration of chromate in the same way by Schwarz [104].

Hempel [105] used a standard solution of potassium permanganate for the determination of oxalic acid [106] and, as Hempel himself pointed out, this method provides the basis of several indirect procedures.

In addition to iodometry and permanganatometry several other important titrimetric methods originated in this fertile and prolific decade.

Clark [107] in 1841 worked out a method for the determination of water hardness. This method, which is still in use today, was based on the titration of the water with a standard soap solution, until a persistent foam remained after shaking the solution for 5 minutes. He observed that the amount of standard solution consumed is not directly proportional to the calcium content but that as the amount of calcium increases the volume of soap solution required decreases slightly. Clark therefore constructed a table for evaluation, giving the results in degrees of hardness. Even today in the determination of the hardness of water the results are given in degrees of hardness, reminiscent of an age when all the results of analysis were given in degrees of some kind. One degree of hardness, according to Clark, is equivalent to 1 gr of calcium carbonate dissolved in 1 gallon water, the so-called English hardness degree. He standardized his soap solution against a calcite obtained from Iceland [108], and he also patented the method. Clark differentiated between permanent and temporary hardness, determining the two from the results of two titrations. First of all he determined the total hardness and then boiled the water, and after cooling he measured the permanent hardness in a similar manner. This method was soon adopted on the Continent, the only difference being that the value of the degree was altered. Bolley [109] used a French scale, where 1° represented 0·01 g of calcium carbonate in 1 l. of water, in equivalent amounts of calcium and magnesium salts [110]. Faisst expressed the hardness of water as calcium oxide [111], and from this the German degree of hardness developed.

The determination of ammonia by a volumetric method, i.e. distilling the ammonia into a known amount of hydrochloric acid, and back-titrating the excess originates from Péligot [112], who collected the ammonia in sulphuric acid. As an indicator Péligot used iron and tannic acid, which turned violet when the alkali was present in excess. The standard solution he used was lime dissolved in a sugar solution. He ignited *marmor of Carrara*, made a pulp from it with water, and then added crystallized sugar until the solution became clear, i.e. he increased the solubility of the lime. This standard solution does not contain carbonate, as this was precipitated. This had to be filtered off, and the solution had to be standardized frequently [113].

The principle of back-titration was also used by Bineau [114], who determined the calcium carbonate content of marga. This he dissolved in a known amount of hydrochloric acid, and back titrated with a standard sodium hydroxide solution (1846). If magnesium carbonate was also present, then two determinations were made; the first being carried out as originally, while to the second a known amount of alkali was added in the presence of sugar, so that only the magnesium was precipitated. After filtration the excess of alkali was titrated [115].

The first example of the titration of alkaloids was the determination of nicotine, devised by Schloesing [116]. He extracted the tobacco with ammoniacal ether using a reflux condenser, and after boiling the ammonia out of the solution he titrated the base with standard acid [117].

The use of a potassium chromate standard solution was introduced simultaneously by two scientists, Schabus [118] and Penny [119], independently of each other, both of these workers using the method for the determination of iron. Schabus dissolved 1 equivalent (14·77 g) of potassium chromate in 1 l. of water, which could convert 1·68, i.e. 6 equivalents of iron(II) to iron(III). He therefore weighed 1·68 g of the sample, reduced it with zinc and then titrated this with the chromate solution until one drop of the solution did not give a blue colour with potassium hexacyanoferrate(III) on a spot plate. The amount of chromate solution consumed gave the percentage iron content directly [120]. Penny's method is similar to this except that the concentrations of the solutions were different [121]. Schwarz in his book comments sarcastically on this method saying that it is superfluous as the permanganate method exists and is quite adequate. He also refutes the view that permanganate solutions easily decompose, saying that if the reagent is prepared as described by Pelouze then it is adequately stable. He also states that:

> Experience shows that all such analyses where the end of the reaction must be tested by taking out drops of the solution to test on a spot plate with a reagent, tend to be inaccurate, and should only be used if no alternatives are available [122].

This opinion is quite valid.

Penny later used chromate for the determination of tin(II) in the presence of a small amount of lead acetate [123]. Schwarz had no sympathy for this method either, and reviewed it with the following words:

> This gentleman has applied his already mentioned inadvisable method of iron determination for the examination of tin also [124].

Other oxidizing and reducing standard solutions were examined during this period, but few of them proved useful, mainly owing to the lack of a suitable indicator. As was mentioned earlier some of these later became important.

Levol [125] in 1842, used a standard solution of potassium chlorate for the titration of iron [126].

Stannous chloride solution was used by Duflos for the same purpose, and although the present author was not able to see the original paper [127] which was not reviewed by Berzelius in his *Jahresbericht*, Beckurts has described the method for us. According to him it was a direct method: to the hydrochloric acid solution sodium acetate was added, and the solution titrated until the iron(III) acetate colour disappeared [128]. Schwarz also refers to a method for the determination of iron using tin(II) chloride, introduced by Duflos, but this was the method using potassium iodide referred to earlier [129].

Potassium iodate was first used by Berthet in 1846 for the determination of iodide. He established that when potassium iodate is added to a soluble iodide in sulphuric acid medium, iodine is liberated, and that for every 5 atoms of iodide, 1 atom of iodate is consumed. The standard solution used was sodium iodate. The method was rather difficult in practice, the iodine having to be boiled out of the solution many times, and each time tested with a drop of iodate solution. Berzelius mentioned that "the method seems to be better in writing than in practice" [130].

A paper by Becquerel [131] describes the detection of sugar with a solution of a copper salt (1831) and later Barreswill [132] developed this into a quantitative method, for which he was awarded 1000 francs.

He dissolved 20 g of sodium carbonate in 200 ml of water, and to this solution he added 40 g of potassium hydrogen tartrate and 40 g of potassium hydroxide. Another solution was prepared by dissolving 30 g of copper sulphate in water, and then the two solutions were mixed, filtered and the filtrate diluted to 500 ml. He also dissolved 10 g of sugar (in case of crude sugar it was first treated with sulphuric acid to bring about inversion) in 500 ml of water. 500 ml of the copper solution were measured into a porcelain dish with a pipette; this was then heated and the sugar solution added from a burette. Red copper oxide was precipitated, and the blue colour of the solution gradually decreased, the end-point of the titration being indicated by the disappearance of the blue colour. The solution was standardized against invert sugar [133].

The method was soon used by other workers and has since been modified numerous times. Fehling [134], in 1849, investigated the method and established that 1 atom of *saccharum uvae* reduces 10 atoms of copper oxide, and he also recommended the use of different copper salt solutions [135]. However his contribution to this method was not sufficient to warrant naming it after him, as has been done.

In the field of precipitation titrations the argentometric chloride determination was adapted by Duflos [136] in 1837 for the determination of cyanide. A large number of standard solutions for precipitation titrations were suggested, of which only a few proved to be useful in practice. A few examples of these are the methods of Pelouze, using a standard solution of sodium sulphide for the determinations of metal ions and the use of a uranium solution for the determination of phosphoric acid [137]. Liebig recommended the use of ferric chloride for the latter determination, but none of these three methods were of practical importance. In this latter method the end-point was indicated by a spot test with potassium hexacyanoferrate(II) [138].

A very useful indicator for precipitation titrations was introduced by Saint Venant. He added a small amount of calcium hydroxide to the chloride solution, so that the end point was indicated by the precipitation of silver oxide. He published his method in 1846, and mentioned that he had used it since 1819 [139].

Around this time the earliest methods based on complex formation appeared, the first being the cyanide determination of Liebig, which is still in use today.

If to a solution, containing cyanide, an excess of potassium hydroxide is added and then silver nitrate solution added dropwise, the precipitate which is first formed will disappear when the solution is shaken. This can be observed until sufficient silver nitrate is present in the solution, to form silver cyanide with a half of the blue acid. A CyAg + CyKa compound is formed, which is not decomposed by the excess of alkali and which is soluble in the latter. If, however, an excess of silver nitrate is added, potassium cyanide gives down one cyanide and silver cyanide is precipitated [140].

Liebig was also the first to use mercury(II) nitrate solution; his original method was for the determination of chloride. He noted that mercury(II) gives a white precipitate with urea in a neutral solution, but that this precipitate does not form if chloride ions are present. This observation was utilized in his method; chloride ions were titrated in the presence of urea with mercury(II) nitrate solution until the mixture became opalescent. For accurate determinations an indicator correction had to be applied [141]. The method could also be reversed so that mercury(II) nitrate could be titrated with sodium chloride solution, using sodium phosphate as an indicator, the precipitate of mercuric phosphate disappearing at the end-point. This indicator had already been used for argentometric chloride determinations by Levol [142].

During Gay-Lussac's time and even for a considerable period after him, the concentrations of the solutions used in titrimetry were chosen to suit a special purpose, so that for a given weight of sample the consumption of standard solution would give a direct percentage of the component sought. Even today many industrial laboratories still use this system. From about 1840 onwards the practice of making up standard solutions based on the chemical units of the elements, atomic, molecular or equivalent weights gradually increased. The interpretation of these units was by no means unanimous; there was even disagreement over the values of the atomic weights. The use of these standard solutions meant that the methods could be unified; a single standard acid solution would be used to titrate a variety of alkalis. Chemists, however, avoided the need for calculation wherever possible, so that the authors who introduced the use of these standard solutions based on chemical weight units, adjusted the weight of sample so that, as in the earlier methods, the consumption of standard solution gave a direct result. But it was also a fact that owing to the lack of uniform atomic weight, calculations based on stoichiometry often gave conflicting results.

The use of normal standard solutions originated in England, and was probably not accidental. In France, from the beginning of the last century, the weight and volume units were based on the metric system, the great advantage being that it allowed weight and volume measurements to be correlated. Thus it was a simple matter to calculate the relation between sample weight and amount of standard solution consumed, and hence the percentage of the component sought. In the other European countries the metric system was not yet used, and the old system of measures was still in use.

In the German translation of Gay-Lussac's volumetric silver determination, which appeared in 1833, Liebig added separate tables in which he converted the weight and purity into lots and grains. From the middle of the last century the use of the metric system became widespread throughout Europe. In England, however, the metric system was not accepted and the system of measures was the same as is used today, so that it was very complicated to convert these to the metric system, and even more complicated to devise a method of analysis where the consumption of reagent gave the percentage of component sought directly. This no doubt hastened the acceptance of a generalized system of standard solutions in England.

Andrew Ure was apparently the first to consider that the "atomic weight" dissolved in the unit of volume would give a convenient standard solution. Ure wrote a dictionary of science entitled: *Dictionary of arts, manufacturers and mines*, and to this he added an *Appendix to the Supplement to A. Ure's Dictionary, etc.*

and in this a chapter on *Alkalimetry*. He describes the dilution of ammonium hydroxide solution so that it should neutralize just one atomic weight of acids. Thus, as he wrote, "a universal acidimeter is available" [143]. It is interesting that his standard solution of acid was not prepared on the same basis. The standardization of the solution was made with a glass bulb which was the same specific weight as the required solution. The bulb was placed in the solution of the concentrated acid, and then water added until the bulb began to float. This is the earliest reference to the use of a normal solution, in the modern sense, that the author has been able to find. In the following years the number of papers describing the use of these solutions increased rapidly, but whether this was a result of Ure's work or whether the advantages of the use of a solution of this type was discovered separately is not known. It was not until the publication of Mohr's book, however, that the use of normal solutions became really widespread.

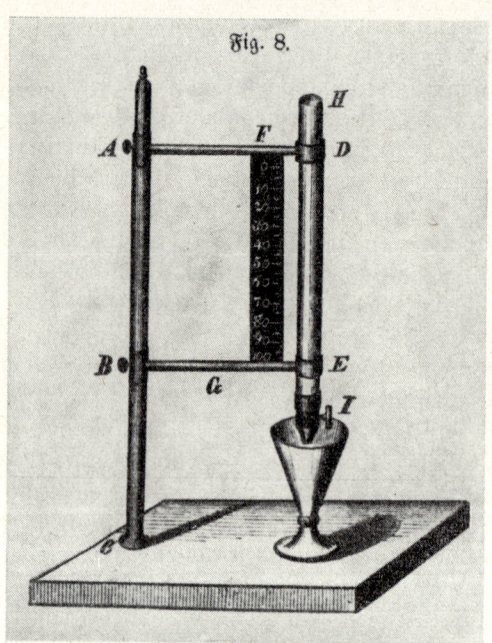

FIG. 65. The first burette with tap used by Henry in 1846. (From Schwarz: *Praktische Anleitung zur Massanalysen* [1853])

Also in this decade, the ancestor of the modern burette appeared, but did not meet with immediate success. The burette designed by Gay-Lussac was by far the most widely used in those days.

Ure describes in his *Appendix* a narrow glass tube with volume graduations, attached to the upper end of which was a tap. The reason for attaching the tap at the top is completely unknown, possibly to avoid contamination of the solution with the lubricating wax.

Andrew Ure was born in Glasgow in 1778, and studied at the university there, and in 1801 he became a Lecturer in Chemistry and Natural History at the Anderson University of Glasgow. In 1813 he edited a comprehensive book on pharmacy. He gave a great many popular lectures for the public, mainly for workers, because he was of the opinion that if the workers had some slight knowledge of science this would improve their production. The popularity of his public lectures was largely responsible for their introduction into England. In 1830 he went to live in London where he worked as an analyst and an industrial consultant. From the Board of Customs he was paid 2 guineas for each analysis, and for one of his methods which he devised for the determination of the sugar content of cane sugar he was awarded £800. As his analytical results were accurate and reliable he enjoyed a considerable reputation, and wrote several books about analytical and technological questions. He died in 1857.

A method developed by Henry [144] at that time, the precipitation titration of potassium with sodium perchlorate solution in alcoholic medium, has since been discontinued, but is of interest because it includes a description of a burette with a tap, which Henry recommends (Fig. 65) [145]. The tube was made of glass, while the stopper was made of copper. It would seem that the reason for the late development of the burette in its modern form was the inability of the glass technologists to make a tap out of glass. It would also appear that the Henry burette did not prove to be too useful in practice, because the Gay-Lussac burette still continued to be universally used almost until Mohr appeared on the stage of titrimetry.

4. FRIEDRICH MOHR

The 20 years described in the previous section was the most successful period in the development of titrimetric analysis. However, in spite of the numerous papers and methods which appeared, it would be an exaggeration to suggest that titrimetric analysis was employed in analytical chemistry to anything like the extent it is nowadays. The field of influence of this subject was not very wide, and if we examine the names of the authors of the many papers, we find the same names recurring many times. The narrow circle of chemists using these methods was originally composed of Frenchmen, but gradually chemists of other nationalities such as Liebig or Schwarz who had studied in France, appeared on the scene. Clark studied in Giessen under Liebig, and later took the method back with him to England.

The famous analytical chemists of this period were contemptuous of the new methods of titrimetric analysis. Gay-Lussac was also a brilliant chemist, but his work was always new and he pioneered many branches of science. Contributions of this nature are always esteemed more highly by posterity than by contemporaries. Gay-Lussac never dealt with the analysis of minerals, which at that time was considered the most important topic in analytical chemistry.

Berzelius, who was the leading figure in chemical society at that time, never used titrimetry, and comments about Gay-Lussac's determination of borax;

> As I hope this method will never be introduced into science, and that it will never be used where a sufficiently accurate method is already available, because with these methods at the best only approximate results can be obtained, and these results depend greatly on the practice and skill of the user [146].

In the analytical textbook by Pfaff (1821), (Chapter VI. 5), there is no mention of titrimetry. Rose's *Handbuch der analytischen Chemie* in the 2nd edition of 1831 barely mentions the titration of hypochlorite with indigo and with mercury(I) nitrate. In his book *Anleitung zur quantitativen chemischen Analyse* (1846) Fresenius gives a brief discussion of the alkalimetric and chlorometric methods, but he also writes the following in the preface:

> Sometimes solutions are measured, especially in applied industrial analysis... But it is difficult to use the apparatus required to obtain accurate results in important analyses, therefore it is preferable to use the balance rather than this method ...

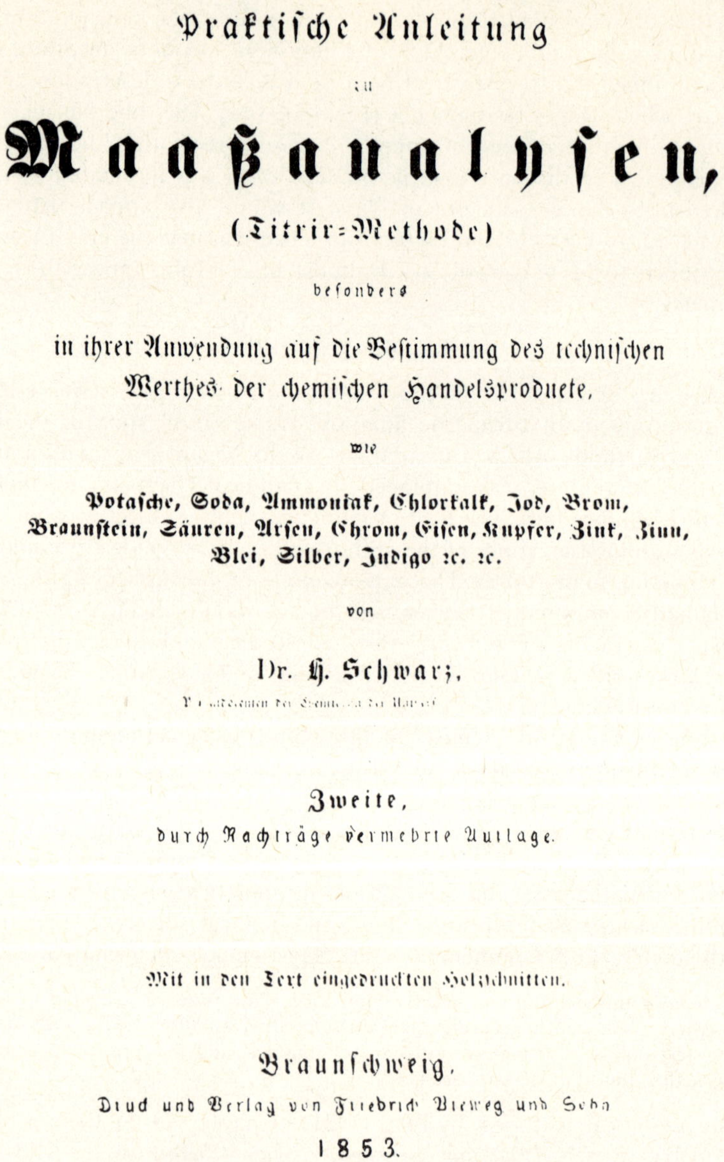

Praktische Anleitung

zu

Maaßanalysen,

(Titrir-Methode)

besonders

in ihrer Anwendung auf die Bestimmung des technischen
Werthes der chemischen Handelsproducte,

wie

Potasche, Soda, Ammoniak, Chlorkalk, Jod, Brom,
Braunstein, Säuren, Arsen, Chrom, Eisen, Kupfer, Zink, Zinn,
Blei, Silber, Indigo ꝛc. ꝛc.

von

Dr. H. Schwarz,

Zweite,
durch Nachträge vermehrte Auflage.

Mit in den Text eingedruckten Holzschnitten.

Braunschweig.
Druck und Verlag von Friedrich Vieweg und Sohn
1 8 5 3.

FIG. 66. Title page of the first book on titrimetric analysis. Schwarz: *Praktische Anleitung zur Massanalysen* (1853)

It is very likely that the main objections of these writers to titrimetry was due to their unfamiliarity with the technique. It can be noted in chemistry as well as other branches of science that new methods, even if they are very good, take a long time to become widely known if they are only published in journals. Books, on the other hand, are very effective vehicles for disseminating new methods, and it is often the case that someone who describes a method in a book has a greater influence on its general acceptance than the original discoverer.

Titrimetry only became widely used after the first books had been written on the subject, so that the existing knowledge was arranged systematically and critically examined.

The first textbook of titrimetry was written by Karl Heinrich Schwarz, and entitled, *Praktische Anleitung zu Maasanalysen (Titrir-Methode)* and was published n Braunschweig in 1850.

Karl Leonhard Heinrich Schwarz was born in 1824 in Eisleben, and studied at several universities, including the University of Paris. It was here that he worked under Pelouze, who was the successor to Gay-Lussac at the École Polytechnique, and who also dealt with titrimetric methods of analysis. Schwarz became acquainted with the methods of titrimetry during his time here, and when he returned to Germany he continued to work in this field. He became a Privat Dozent at the University of Breslau, and it was here that he published his book, and developed his methods, of which undoubtedly the use of sodium thiosulphate in iodometry is the most important. He later worked in industry, occupying several positions in Austrian heavy industry. In 1863 he returned to the University of Breslau to become Professor of Chemistry, and he later occupied the Chair of Chemical Technology at the University of Graz. In the latter part of his career his interest turned mainly to chemical technology. He died in 1890.

Schwarz's book is quite small, and in its title we find the first use of the name *Massanalyse*, which according to the preface was created by the author from the French expression *dosage à liqueurs titrées*. This word still exists in the German language, and is also used in other languages. The term "volumetric analysis" originates from this word.

Schwartz pointed out the importance of titrimetric analysis to industry.

> With the aid of titrimetric analysis, analytical chemistry could be introduced into practical life. I would be contented, if, even to a small extent, I could open the door through which science could enter into the life of industry and technology in Germany [147].

His book only mentions the burette designed by Gay-Lussac, and in the section describing the method of Henry, he refers to the tap burette which Henry used in the following way:

> Henry used instead of a burette a special measuring device ... [148].

So presumably he did not regard this as a burette. He divides the text into three parts: Saturation analysis, Oxidation and reduction analysis, and finally Precipitation analysis. The standard solutions used were mainly selected so that for a 1 g sample the consumption of solution gave the percentage composition directly. He also mentions that it is possible to use "aequivalent" amounts of substances, which when dissolved in unit volume yield "rational solutions".

Most of the methods dealt with in this book have already been described, but apart from these there are some less important methods and many that have been completely forgotten. It is interesting that Schwarz systematically uses reaction equations, and in this respect he precedes contemporary chemical books in which this treatment was lacking. For example:

$$10FeO + Mn_2O_7 = 5Fe_2O_3 + 2MnO$$
$$SO_2 + Cl + OH = SO_3 + ClH$$
$$2CrO_3 + FeO = Cr_2O_3 + 3Fe_2O_3$$
$$2NaO, S_2O_2 + I = INa + NaO, S_4O_5$$

Schwarz devised many other methods. He determined chromate by reducing it with iron(II) sulphate, and titrating the excess of ferrous iron with potassium permanganate. He later used this method for other determinations in a very complicated way, e.g. for lead and sulphate determinations, by precipitating the sulphate with a known amount of lead nitrate, filtering off the lead sulphate, and precipitating the excess of lead in the filtrate with potassium chromate. After filtration the excess chromate was determined after reduction with ferrous sulphate as described previously. His method for the determination of copper is also very complicated. The copper solution was reduced with *saccharum uvae* to the copper(I) salt, and the cuprous salt was then dissolved in an excess of iron(III) chloride, and the resulting ferrous salt titrated with potassium permanganate.

As can be seen from these examples, even in those times some scientists tended to devise *l'art pour l'art* methods which were of no practical importance. Such complicated titrimetric methods can be found even today in large numbers in the various scientific journals.

Only one edition of Schwarz's book was published (1853), and soon after this Mohr's famous book *Lehrbuch der chemisch-analytischen Titrimethode* (1855), which was far more extensive, appeared.

The conclusion to the last section stated that Mohr appeared on the stage of titrimetry, and there is a considerable amount of truth in this sentence. Mohr made an appearance in many fields of science, and after a brief allegiance to each he moved on to another. In each field he published several papers and often a book or two. Someone once said about him that "in the history of science Mohr will always play a secondary role". It would seem that Mohr, both in his personality and in his work, shows a little of the character of a dilettante. Part of the reason may be that Mohr was not a very successful man, and his accomplishments were not acknowledged either in his own time or even today. But Mohr did discover the law of the conservation of energy.

> With force we can calculate just as any other measurable quantity. It is divisible, part of it can be taken away or new amounts can be added to it, and the force is not lost, nor will its amount change. Apart from the 54 known chemical elements there exists one other agent in the world, and this is force; according to the circumstances this can exist as movement, chemical affinity, cohesion, electricity, light of heat, and all these phenomena can be transformed into one another. The same force which lifts up a hammer can cause all the others [149].

However, in those days in Germany the so-called "university professor" type came into being; this term representing the popular conception of a university professor, bearded, bespectacled, standing aloof from every-day life, immersed in his books, and usually absent-minded. If ever this type existed, then it was in Germany at the time when the study of the human sciences played such a dominant role in university life.

When we consider the people who contributed to the development of the sciences, we find that nearly all those of German nationality were university professors, whilst among the people of other nationalities, particularly in England, this was not so. In Germany scientific research was the prerogative of the universities so that the words, "scientist" and "university professor" came to mean the same, and anyone who was not a university professor was not considered a scientist.

Mohr did not have a professorship, and therefore remained beyond the fringe of official scientific society. His lack of success was partly caused by his vivid imagination, which made him dabble in various sciences, and was partly due to his irritable, critical and generally rather disagreeable nature. This is borne out in a letter to Mohr from Liebig concerning a new geological theory which Mohr had proposed. Liebig, who himself had a passionate nature, advised Mohr to be patient:

> ... and do not strike immediately—excuse me—according to your nature and habit, with a stick. People must be taught, and not forced! [150].

In another letter he wrote:

> You talk about your theories like a lover about his love, but a lover is never regarded as capable of speaking about his lover with unbiassed judgment. Please take the contradictions without anger also. But I speak to you in vain, because you are too passionate ...

and in another letter to Mohr

> you would find much less aversion if you tried to avoid always offending others [150].

Friedrich Mohr was born in 1806 in Koblenz, the son of a pharmacist, and he also studied pharmacy. He carried out his university studies in Bonn, Heidelberg and Berlin, and it was while he was at Berlin that he came under the influence of the famous analyst Heinrich Rose. After graduating he returned to his native town and took over the running of his father's chemist's shop, but although he worked hard the business was not very profitable. During his spare time he carried out experiments on a variety of scientific problems. He first dealt with physics, and in 1837 he published a paper entitled; *Ansichten über die Natur der Wärme* and in this he presented the theory of conservation of energy, from which the above extract was taken. This was five years before Robert Mayer [152] published his theory. Mohr sent his paper to Liebig, requesting that it should be published in the *Annalen der Pharmazie und Chemie*, but Liebig refused and wrote to Mohr advising him:

> I advise you not to publish this article as you may become discredited in the eyes of other physicists [151].

Mohr then sent the article to Poggendorff, who was the editor of the *Annalen für Physik*. Poggendorff, who also later refused to publish Robert Mayer's article, informed Mohr that his paper could not be published, as it did not contain any original experimental work. Finally Mohr sent the paper to Vienna where it was published in the journal *Zeitschrift für Physik* [153]. By the time the paper was published the author had lost interest in it, especially as it did not arouse much enthusiasm in the scientific world.

During this time Mohr wrote many additions to the Prussian Pharmacopoeia [154], and also edited a pharmaceutical practicum (1847), which was very successful, and in the following year was published in England and America [155]. In the course of this work he

FIG. 67. Friedrich Mohr (1806—1879). Photograph

became interested in the methods of titrimetry, and collected the isolated published papers, studied them and tested them experimentally. Many of these methods he modified, some beyond all recognition, and others he abandoned altogether and replaced them with better methods. The results of this work were published in 1855 in his *Lehrbuch der chemisch-analytischen Titrirmethode*. His interest then turned to the problems of agricultural chemistry, in which Liebig was actively concerned, in particular with the problem of chemical fertilizers. His interest became so great that he finally sold his apothecary's shop, and bought a small estate where he became a partner in a newly established chemical fertilizer plant. He investigated problems relating to enzymology and agricultural chemistry, and eventually published a book on the subject [156]. However, it needed time for Liebig's views on fertilizers to become accepted, but before this could happen the factory was ruined, and in 1863, at the age of fifty-nine, Mohr had to search for a job.

He became a Privat Dozent at the University of Bonn, and in 1867 he became the Deputy-Professor of Pharmacy. This was as far as he went in his occupation, for just as he had become established in pharmacy his interest turned to geology, and he published a book about the formation and development of the earth [157]. His geological theories aroused sharp criticism from the experts. The phenomenon of the melting of glaciers again turned his attention towards heat, and he realized the importance of the principle of conservation of energy to physics. From an English paper [158], which cited his earlier article on this subject, he discovered that his paper had actually been published. He could therefore honestly claim that he had announced this important principle thirty years earlier, and had thus preceded all the other applicants for the honour. He published his old article, in the preface of his book entitled *Die mechanische Theorie der chemischen Affinität und die neue Chemie* in 1868, and his comments are very characterstic:

> These lines I wrote thirty years before, and when I read them once more, I see that essentially I presented the basis of the mechanical theory of heat in them.

These few lines prove that he did not recognize the importance of his principle at the time of his earlier publication. He did realize however that the boundaries between physics and chemistry would slowly disappear, and wrote in the same publication:

> The distinction between physics and chemistry can hardly be made even today, and since physics is the science of forces, chemistry will be a branch of this [159.]

Mohr considered that his duty was to continue this work. Mohr's insufficient knowledge of mathematics led him, in his previously mentioned book [160], to dispute the entropy principle. Mohr died in 1879, and even before he was buried the university authorities approached his widow, requesting the return of some insignificant book that he had borrowed, evidently a sign of the dislike shown to him by his colleagues and co-workers.

Mohr's literary accomplishments were outstanding; his publications numbered more than one hundred, and dealt with meteorology, analytical chemistry, mechanics, bee-keeping, toxicology, and geology. He also published books on several of these topics. Probably his most successful works were those on the subject of analytical chemistry. His book on titrimetry was published in many editions, and as the subject developed so it was revised and enlarged, so that by the time the final edition appeared in 1914, edited by Classen, the only resemblance to the original version was in the title.

In every modern laboratory some trace of the work of Mohr is to be found; many methods and pieces of apparatus bear his name. The determination of iron and chloride, the Mohr pinch-cock, Mohr salt, Mohr balance are only a few. Also the pinch-cock burette [161], the calibrated pipette, the Liebig condenser [162], the cork-borer and many other devices are the result of his fertile brain. For many years he was on very good terms with Liebig, but later on their friendship deteriorated. All his inventions Mohr immediately communicated to Liebig, who was the first to test them. So it was with the condenser; Liebig found this to be very useful and propagated its use so that his name became associated with it. It is interesting to note that Liebig greeted Mohr's invention of the cork-borer with the following words:

> Your cork-borer is a very ingenious invention, and this will soon be used in all laboratories. You have made a great contribution to organic chemistry, which needs many fine stopper holes [163].

As an analyst Mohr is mainly famous for his great practical skill and inventiveness, as well as for his technical ability, which is proved by his long list of inventions. Although he did not introduce any new methods, he altered and improved many others which had previously been impracticable. Titrimetric analysis was not widely known until after the publication of Mohr's book, when it slowly became the most important branch of analytical chemistry.

The *Lehrbuch der chemisch-analytischen Titrirmethode* appeared in two parts, in 1855 and 1856. The various chapters are headed by the names of the discoverers of the methods which they describe. Thus, under the heading "Gay-Lussac" he deals with acidimetry and alkalimetry, under "Marguerite" (which he misspelt) he deals with permanganatometry, while under the heading "Bunsen" the titrations based on the sulphurous acid–iodine–iodide system are described. This latter chapter is very brief, the reason for this being apparent when we come to the next chapter headed "Mohr", which describes the titrations based on the arsenious acid–iodine system. Arsenious acid was first titrated with iodine by Mohr, and the method was then extended to the determination of reducing substances (by adding an excess of iodine and back-titrating the excess with arsenious acid), and also for oxidizing agents (by reducing with hydrochloric acid, similar to Bunsen, and passing the liberated chlorine into arsenious acid, and titrating the excess with iodine). It could therefore be said that he "*Mohrificated*" the methods of Bunsen. In favour of his own procedure he states that sulphurous acid is very unstable, and that it is inconvenient to use. Although this is quite true, one cannot help feeling that his treatment of Bunsen was rather harsh, and this view is borne out by the fact that in the next chapter under the heading of "Streng" [164] there is a description of the titrations involving the stannous chloride–potassium dichromate system which Mohr considered far more important than the iodometric titrations. Finally, under the heading "Liebig" precipitation titrations were described. This statement also was not very accurate because as we have seen, several other workers before Liebig used these methods.

Mohr not only described titrimetric methods, but he also reviewed them critically on the basis of his own experience. According to Liebig he sometimes carried this out in too great detail, but he comments:

> But this is not bad, because there are many persons who need an explanation for everything [150].

He also gave a list of results which he had obtained under various conditions. At the end of the book he classifies the methods according to the elements and evaluates each separately, as well as giving a recommendation on the choice of method under various circumstances. It is undoubtedly true that in many cases Mohr received the credit which was due to others. There are very few references in his book, and often one has the impression that it is Mohr's own discovery

LEHRBUCH

DER

CHEMISCH - ANALYTISCHEN

TITRIRMETHODE.

NACH

EIGENEN VERSUCHEN UND SYSTEMATISCH DARGESTELLT

VON

DR. FRIEDRICH MOHR,

Königl. Preuss. Medicinalrathe,
pharmaceutischem Mitgliede des Medicinal-Collegiums zu Coblenz, Hofapotheker
Sr. Königl. Hoheit des Prinzen von Preussen, vieler gelehrten Gesellschaften
Mitglied und Ehrenmitglied.

FÜR

CHEMIKER, ÄRZTE UND PHARMACEUTEN,
BERG- UND HÜTTENMÄNNER, FABRIKANTEN, AGRONOMEN,
METALLURGEN, MÜNZBEAMTE ETC.

IN ZWEI ABTHEILUNGEN.

ERSTE ABTHEILUNG.

MIT 104 IN DEN TEXT EINGEDRUCKTEN HOLZSCHNITTEN UND
ANGEHÄNGTEN BERECHNUNGSTABELLEN.

BRAUNSCHWEIG,
DRUCK UND VERLAG VON FRIEDRICH VIEWEG UND SOHN.
1855.

FIG. 68. Title page of Mohr's *Lehrbuch der chemisch-analytischen Titrirmethode* (1855)

that is being detailed. There are several examples of Mohr incorrectly asserting that a discovery was his own. He claimed that the discovery of the principle of back-titration was his, and also the use of sodium hydroxide as a standard solution, as well as the introduction of normal solutions, although as we have seen these were already in use. He did not refer to Schwarz's book, although it is quite certain that he knew about it as it was published by the same publishing house as his own. In the appendix, however, there is a criticism of one of Schwarz's methods [165], concerning the use of sodium thiosulphate as a standard solution in iodometry [166].

The use of normal solutions became much more widespread after Mohr's book had appeared.

> In the titrimetric method the various strengths of the standard solutions caused a great problem. While the work of the chemists was decreased by the introduction of titrimetry the number of bottles in his laboratory increased. Every discoverer of a new method prescribed the use either of an entirely arbitrary standard solution, or one which is correlated to the substance which is being determined... To avoid this I have introduced a system which forms a unit with the calculation. This system is based on one litre of the solution containing the small atomic weight of one tenth of this expressed in grams of the substance dissolved [167].

> In a later edition of his book he mentions a chemist called Griffin, who, according to the best of his knowledge, also used this type of solution, but he gave no further reference. By "small atomic weight" he meant the equivalent weights in the sense of today.

Mohr introduced the burette with a clip (Fig. 69). Although the glass-tap burette (Fig. 70) was also known in those days, it was rarely used because of the many problems it entailed, particularly leakage owing to the imperfection of the tap.

He took great care in the accurate calibration of his volumetric apparatus. After much trouble he managed to obtain a 1 kg weight of platinum, which had been standardized with the original 1 kg *etalone* in Paris. This proved to be only 0·41 mg lighter than the standard weight [168]. Mohr calibrated his own weights on this etalone several times a year.

For an acidimetric standard he used oxalic acid, for as this was a solid it was a simple matter to prepare an accurate standard solution. With the liquid acids the preparation of a standard solution was more difficult. For certain determinations such as the alkaline earth carbonates this acid is unsuitable, and in these cases nitric acid is used. Oxalic acid was used, however, for the standardization of sodium hydroxide solution. He prepared the standard solution from calcium oxide and sodium carbonate, and into the stopper of the storage vessel he inserted a tube filled with a mixture of calcium hydroxide and Glauber's salt, in order to prevent carbon dioxide being absorbed.

He also determined hydroxide and carbonate in the presence of one another, by a modification of the method of Barreswill. In this method Barreswill precipitated the carbonate with barium chloride, filtered off, and weighed. He then passed carbon dioxide into the solution until the hydroxide was completely converted to carbonate, and this was then precipitated, filtered and weighed. Mohr determined the total alkali content by titration with oxalic acid, and then precipitated the carbonate with barium chloride. He then filtered off this precipitate, dissolved it in a known quantity of nitric acid, and determined the excess of the acid alkalimetrically [169].

For the estimation of ammonia, Mohr used the following method:

The sample was boiled with a measured amount of sodium hydroxide until all the ammonia was evolved, and the excess of alkali was titrated [170].

Mohr used a litmus indicator for the titration of weak acids, and mentioned that in these cases the colour change was not so well defined, and that it was advisable to titrate until the appearance of the final blue colour [171].

For the standardization of nitric acid solution Mohr originally used barium carbonate, but he later found that sodium carbonate was more convenient to use owing to its solubility in water [172].

In the introduction to the section on redox titrations Mohr gave a list of the standard

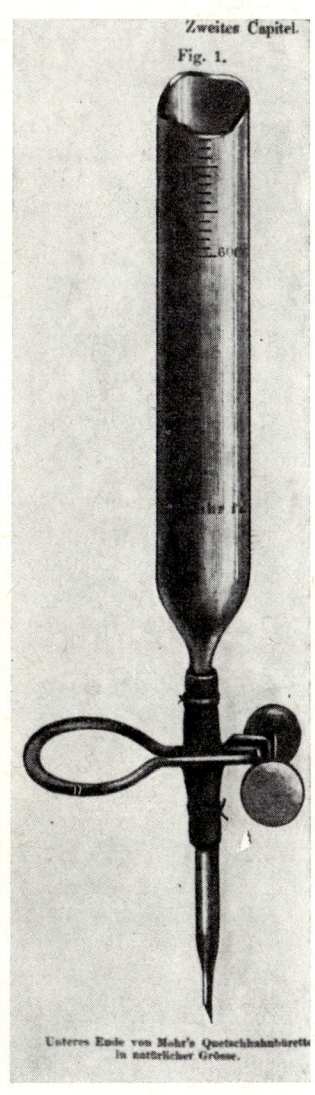

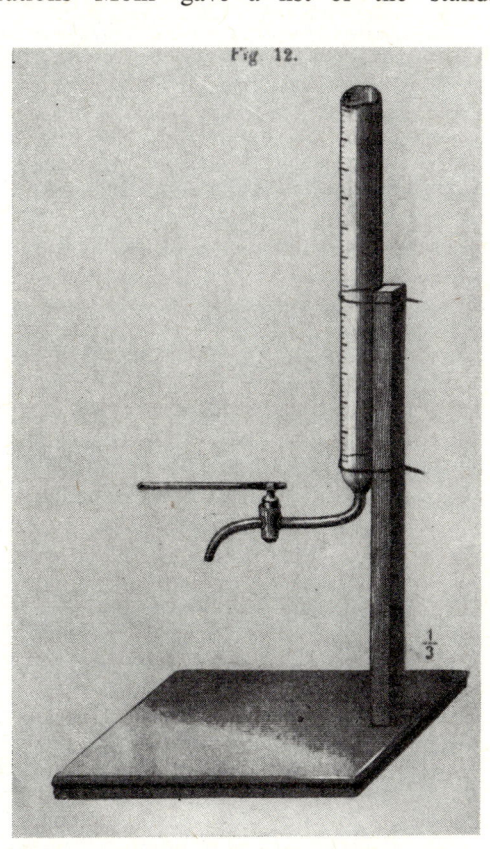

Fig. 69. Burette with a clip of Mohr. (From his book: *Lehrbuch der chemisch-analytischen Titrirmethode* [1855])

Fig. 70. Cock burette in Mohr's times. (From Mohr: *Lehrbuch der chemisch-analytischen Titrirmethode* [1855])

solutions which could be used. He gives as examples of oxidizing agents suitable for standards, potassium permanganate, chlorine water, iodine and potassium chromate, and as reducing agents sulphurous acid, iron(II), oxalic acid, arsenious acid and potassium hexacyanoferrate(II). On the basis of ease of preparation, stability and clarity of end-point detection he recommends the use of potassium chromate as a standard oxidizing agent, and arsenious acid as a reducing agent [173].

Mohr also describes the preparation of potassium permanganate, which would indicate that this salt was not commercially available in those days. He was aware that hydrochloric acid decomposes potassium permanganate, but only when it is fairly concentrated and in hot solution, and considered that the error involved in carrying out permanganate titrations in hydrochloric acid solution was negligible as the amount of reaction during the course of a titration was very small. However, if chlorine was detected by its smell then the results obtained were inaccurate [174]. For the standardization of potassium permanganate solution Mohr used either piano wire, oxalic acid, or ferrous ammonium sulphate, which he prepared specially [175].

It is interesting to note that although Mohr used the gram system for the calculation of his results, and for equivalent weights, he often used the older system of ounces and drachms in the text.

The analysis of brown stone was very important in those days because of the Weldon method of chlorine preparation and Mohr describes in this connection the method of sampling for analysis. He records that as today, every tenth shovelful must be placed on one side, and then the large pieces in this broken with a hammer until they are small. This is then heaped into a pile and divided into quarters by making a cross. Two opposite quarters are then taken and the rest of the sample rejected. This process is repeated until only 1 to 2 oz remain, this being taken to the laboratory. The manganese dioxide content was determined with oxalic acid [176].

He determined some other metals by this method, for example, lead and calcium, either by dissolving the oxalate precipitate, or by titrating the excess of oxalate in the filtrate [177]. He also describes several permanganatometric titration methods based on the titration of ferrous ions, and some of these hardly match the criteria which Mohr states in the preface to his book.

> The main advantage of volumetric analysis is a considerable reduction in time and work, whilst the results are in most cases as accurate as those of gravimetric ones, or are even more accurate. These advantages must always be taken into consideration. But no one should use unsuitable methods of titrimetric analysis, namely those that are less accurate and require more, rather than less, time to carry out than the corresponding gravimetric methods. Many methods have been based on a separation followed by dissolution of the precipitate in order to carry out a titration, where it would have been much simpler to weigh the precipitate. In cases such as these titrimetric methods are not to be recommended [178].

This danger always exists with new and fashionable methods; we can see for example with chelatometry how this method has, in many cases, been applied when it is not really suitable. However, it is only time and repeated use which will decide whether a method is of any importance.

One of the early analytical methods which did not prove to be very successful was the determination of phosphate by precipitation with iron(III) sulphate in the presence of sodium acetate. The precipitate was filtered and washed, and then dissolved in hydrochloric acid, and the iron after reduction with zinc was titrated with permanganate. Mohr himself did not have much faith in this method, and doubted whether the precipitate corresponded strictly to the formula $Fe_2O_3 +$ $+ PO_5$, and in fact Mohr left no doubt that the method in this case was not his [179].

Mohr also developed the first method for the determination of dissolved oxygen in water. He added a known amount of ferrous salt to the solution and made it alkaline, and then after allowing it to stand for a time he titrated the excess of ferrous ion with permanganate. His results did not agree with the absorption coefficients established by Bunsen: Mohr attributed the error in his own work to the presence of carbon dioxide [180]. Thirty years later Lajos Winkler proved with his experiments that the values given by Bunsen were inaccurate.

Mohr also described the first permanganatometric titration of an organic substance; this was the oxidation of uric acid to urea. This method was devised by Mohr following the observations of a military physician by the name of Scholz [181].

Mohr's own redox methods were based on the reaction of arsenious acid with iodine. This reaction was first used by Mohr who also discovered that the reaction was reversible in acid medium. Mohr first of all used sodium carbonate to buffer the medium, but later used sodium bicarbonate [182]. Although it is true that arsenious acid gives a very stable standard solution its use is rather restricted.

Mohr also described Dupré's method for the determination of iodide [183], and he amplified the sensitivity of the procedure. In this method the iodide is oxidized with chlorine to iodate, and the excess of chlorine boiled off. Potassium iodide is then added and the liberated iodine titrated in the presence of chloroform. The equations for the reactions are as follows:

$$IM + 6Cl + 5HO = ClM + 5ClH + IO_5$$

$$IO_5 + 5IK = 5KO + 6I$$

The amplification factor in this method is six, and even a concentration of 1 mg/ml of iodide could be determined [184].

In the field of precipitation titrations we do not find much that is new in his book. He describes the determination of zinc and mercury using potassium hexacyanoferrate(II) introduced by Kieffer, the end-point being indicated by a spot test. From the turbid solution he took a drop and placed it on a filter paper, alongside this he placed a drop of ferrous salt solution, and if a blue colouration was formed where the two spots impinged on one another, then the titrant was present in excess [185].

At the end of his book Mohr gives a table which shows the titres of the required substance which consume from 1 to 9 ml of the standard solution. This is the earliest example of the calculation tables in use today.

Mohr finally gave a price list for the various calibrated flasks and standard solutions, which he supplied. The standard solutions and other chemicals were produced by a factory in which Mohr had an interest, and which later ruined him. A few examples of these are given, together with their prices.

A burette with a clip stopper, 60 ml, cost 20 groschen; 50 ml, 15 groschen; 10 ml, 10 groschen; volumetric flasks with a graduation on their necks calibrated to 500 ml, 20 groschen; 100 ml, 10 groschen; measuring cylinder with a ground stopper, 1000 ml with graduations 2 thalers. 1 l. of standard N nitric acid, 10 groschen. 1 l. of potassium permanganate solution unstandardised, 15 groschen; while a 0·1 N solution of various other substances were 10—20 groschen per litre, except for 0·1 N silver nitrate, which cost 1 thaler and 15 groschen.

Chemicals could also be ordered by the pound; some prices of these are given:

potassium dichromate, oxalic acid, iron(III) chloride, nitric acid, iron(II), free potassium hexacyanoferrate(II) all cost 1 thaler per pound. Sodium chloride and barium chloride only cost 10 groschen, and sodium thiosulphate, tin(II) chloride, iron(II) ammonium sulphate cost 15 groschen. Mercuric oxide was more expensive costing 2 thalers 20 groschen. Potassium iodide was 4 thalers and 14 groschen, and iodine 6 thalers. For 1 pound of potassium hexacyanoferrate(II) Mohr charged 3 thalers; molybdic acid and paraffin both cost 6 thalers per pound, and silver nitrate 24 thalers. Urea was the most expensive costing 30 thalers [186].

These prices would be of more significance if we knew the cost of other commodities of that time for comparison. However, the only comparison is to be drawn from the annual expenditure on apparatus and chemicals, which at Liebig's Institute at Giessen was only 100 thalers. We do know, however, that Liebig considered this sum ridiculously small.

5. THE DEVELOPMENT OF TITRIMETRIC ANALYSIS BEFORE THE INTRODUCTION OF SYNTHETIC INDICATORS

Mohr's book ended the long period of the early history of titrimetric analysis. By the time of its publication the methods and apparatus were in general use, and in many cases the special volumetric apparatus, burettes, pipettes, and volumetric flasks were essentially the same as those in use today. Many substances were examined for use as standard solutions, and those that proved useful such as sodium hydroxide, hydrochloric acid, permanganate, dichromate, iodine, sodium thiosulphate, mercury(II) nitrate and silver nitrate are still the most important reagents in volumetric analysis. In the next few decades following the middle of the last century no important developments were made. The existing standard solutions were applied to a wider range of methods, and the available methods improved and refined. Application of the methods was completely empirical, for little was known of the theory of the processes occurring. Also in the fields of acidimetry and alkalimetry no important developments were made.

Péant de Saint Gilles [187] developed methods in his private laboratory for the permanganatometric determination of nitrite [188], iodide [189], and several organic compounds (formic acid, tartaric acid, apple acid, etc.) [190].

In 1862 there was great interest in the suggestion that the well known permanganometric iron determination of Margueritte did not give very accurate results when carried out in hydrochloric acid medium, and that this should be replaced with sulphuric acid [191]. The investigation of this reaction by the highly respected Fresenius proved this assertion fully [192]. Kessler [193] reduced iron(III) with tin(II) chloride, and complexed the excess of this with mercury(II) chloride [194]. He recommended this reduction for the chromatometric determination of iron, as the hydrochloric acid liberated would interfere in the determination of iron with permanganate. Later, Kessler found that in the presence of manganese(II) ions the permangatometric method gave reasonably accurate results if hydrochloric acid was present [195]. It is unfortunate, however, that he did not think of combining his own convenient method of reduction with this latter method. None of his suggestions was accepted, and Kessler's name was soon forgotten. Twenty years later, in 1881, Zimmermann [196] rediscovered the favourable effect of manganese(II) salts [197] and the method now became widely used, but the reduction with zinc was still carried out. After another five years Reinhardt combined this method with Kessler's reduction, and this method is still in use [198]. For the reduction of iron(III) many other methods were recommended, for example reduction in a tube filled with finely divided zinc powder was recommended by Jones [199] in 1889 [200], and 10 years later Shimer discovered that amalgamation increased the reducing activity of the metal [201].

An important and widely used method in metallurgical analysis is the determination of manganese by the Volhard–Wolff method. When one considers the development of this method, the first impression is that it is named after someone other than the discoverer. This is not an uncommon occurrence, for as soon as a new method is published many people try to use it, and modifications and improvements soon appear so that it is soon very difficult to decide just who has made the most important contribution to its development. In the scientific papers of the last century references were not given so frequently, so that it is often difficult to track down the source of a method with any degree of certainty. It is not always possible to decide whether one author had any knowledge of the work of another or whether he made the discovery independently. Then again other methods only became practicable after certain modifications had been made, and then it is probably justifiable to name the method after the person who was responsible for its practicability. In many cases it is impossible to decide why a method was named after a certain person, but it is probably due, in several instances at least, to the fact that most of the writers of analytical textbooks in the last century were Germans who, being mainly familiar with the German literature, reported the name of some German responsible for modification. The Volhard–Wolff method is such an example.

The principle of the method, namely the oxidation of manganese(II) ions with potassium permanganate in hot solution, was originally used by a Frenchman, Guyard, in 1863 [202]. Guyard found the method to be sufficiently accurate, but he did not give any experimental data. The method was then examined by several other scientists, some of whom found the method suitable, but there were others who did not. Finally, in 1879, Volhard made an extensive investigation of this method. He found that accurate results could only be obtained if the salt of a bivalent metal was also present in the solution [203]. Volhard precipitated and filtered iron as the hydroxide, and for the evaporation of chlorides he boiled the solution until the appearance of sulphuric acid fumes. Volhard, however, worked in a slightly acidic medium, and therefore his results were rather low. Wolff eliminated this error by using an excess of zinc oxide [204]. These are only a few of the people who contributed to the development of this method; it would be impossible to name them all.

Permanganate titration was also applied to the determination of other metals, but the majority of these procedures did not survive for very long. A few, however, did survive, for example, in 1863 Czudnovicz determined vanadium after reduction with zinc or hydrogen sulphide [205]. The reduction with hydrochloric acid or hydrogen bromide was introduced by Roscoe [206, 207]. The arsenious acid-permanganate reaction was used for the determination of manganese by Deshayes in 1878, who used lead peroxide for the oxidation of manganese [208]. (The oxidation with lead peroxide as a qualitative test for manganese had already been used by Boussingault [209].) In place of lead peroxide Smith used silver nitrate and persulphate [210]. Titanium dioxide was determined for the first time by Pisani[211] in 1864, who carried out the reduction with zinc in the absence of air [212]. The oxygen consumption of water was determined by Forchhammer [213] as early as 1849; he titrated hot water with permanganate to a permanent pink colour [214]. Schrötter [215] acidified the water [216], while Schulze [217] made it alkaline and then back-titrated the excess of permanganate with oxalic acid [218] (1868). This is the same method as is in use today.

In iodimetry the use of the sodium thiosulphate standard solution became widespread. The standardization of this solution, however, was a great problem for a long time, as the substances used today were not commercially available in those days in a sufficient state of purity. For a time the only standardization possible was the one using iodine. The purification of this was complicated and inconvenient. The use of potassium bi-iodate provided a much simpler method of standardization. The use of bi-iodate was due to Károly Than [219].

Károly Than was born in Óbecse in Hungary in 1834. At the age of 15 he became a soldier in the Hungarian War of Independence. After the collapse of the war he found his mother dead and his father ruined, so the young Than became apprenticed to pharmacists in several Hungarian towns. A pharmacist helped him to complete his secondary school studies. In 1855 he went to the University of Vienna and studied chemistry, and in 1858 he was awarded his doctorate and became the assistant lecturer to Professor Redtenbacher. Later he took a fellowship in Heidelberg and studied under Bunsen. After this he returned to the University of

FIG. 71. Károly Than (1834—1908). Oil painting

Vienna. In 1860 he became a Professor at the University of Budapest. Than played a very important part in the development of the teaching of chemistry in the Hungarian universities, and was an inspiration to many generations of chemists. He died in 1908. Than was also responsible for the introduction of potassium bicarbonate as a standard for acidimetry and alkalimetry [220], and long before Arrhenius's ionic theory, in 1865 he recommended that the results of analyses should not be presented in the form of salts, but in the percentage amount of the equivalents of each constituent. This procedure only came into general practice later, following the suggestion of Ostwald who, in the 2nd edition of his book *Wissenschaftliche Grundlagen der analytischen Chemie* acknowledges the earlier suggestion of Károly Than.

After potassium bi-iodate, iodic acid [221], and later potassium iodate [222] were also recommended for the standardization of thiosulphate. The establishment by Zulkovsky [223] in 1868, that under certain experimental conditions, dichromate reacts stoichiometrically with potassium iodide was an important discovery in iodimetry [224], for it provided a possibility for the analytical application of dichromate oxidations. A number of organic substances can be either oxidized or decomposed with dichromate, so that from the amount of dichromate consumed, the amount of the organic substance present can be found. Reischauer, in 1862, was the first to apply this method for the determination of methyl and ethyl alcohols; the excess of chromate was determined with iron(II) sulphate and permanganate [225.]

This method did not arouse much interest until considerably later when the same method was applied for the determination of glycerine. This possibility was first pointed out by Legler [226] who measured the volume of carbon dioxide given off during the reaction. The titrimetric determination was first introduced by Hehner [227], who back-titrated the excess of dichromate with iron(II) sulphate using the spot test with potassium hexacyanoferrate(II) to indicate the end-point [227]; while finally Steinfels introduced the iodometric back-titration of the excess of dichromate [228].

The scope of iodimetry was considerably increased when the bromination of organic substances could be used for analytical purposes. Landolt, in 1871, utilized this for the bromination of aniline [229]. He mentions this primarily as a qualitative test, but also pointed out the possibility of the gravimetric determination of the tribromophenol precipitate [230]. Waller adapted the reaction to a volumetric method, titrating the phenol in sulphuric acid solution, saturated with alum, directly with bromine solution until the yellow colour persisted [231]. Koppeschaar added an excess of bromine to the solution, and determined the excess by iodimetry. It is interesting to note that in the article describing this method, Koppeschaar uses the original symbols and abbreviations of Berzelius. He also refers to the use of a bromate/bromide mixture; both of these reagents he prepared himself from bromine and potassium hydroxide [232].

The determination of the iodine number of unsaturated fatty acids is similar in many respects to the previous method. The first method was devised by Hübl in 1884, and consists in treating the sample in chloroform solution with iodine in the presence of a mercury chloride catalyst, the excess of iodine being titrated with sodium thiosulphate [233]. The rate of addition of iodine is much slower than with bromine which, except for one or two isolated examples, is at present the most widely used reagent for the determination of unsaturation. Even today the results are generally expressed as iodine numbers.

The most general method of iodine-bromine number determination originates from Lajos Winkler. But it must be remarked that the determination of bromine numbers was attempted before iodine was used for the determination, but owing to the occurrence of substitutive side reactions, reproducible results were very hard to obtain. The experimental conditions were not very carefully examined so that the method was considered not very satisfactory. The first bromine number determinations were made gravimetrically, but in 1854 Knop devised a volumetric method [234]. He added a standard potassium bromate solution to the molten fats, together with hydrogen bromide, and back-titrated the excess of bromine with sodium thiosulphate until the solution was decolourized [235]. He defined the bromine number as the amount of potassium bromate in cg consumed by 1 g of the fat.

For the determination of dissolved oxygen in water Lajos Winkler developed a method which proved very useful, and which is still in use even today. Instead of the iron(II) sulphate recommended by Mohr, he used manganese chloride in an alkaline medium. He eliminated the oxidative action of air by pouring the solutions under the water in a layer, and carried out the rest of the operation in a closed bottle (1888) [236]. Winkler's results did not agree with those of Bunsen who had determined the air-absorbing capacity of water by a manometric method. Winkler, however, was so certain that his titrimetric method gave reliable results that he repeated Bunsen's experiments and found that the values obtained were not sufficiently accurate.

In the field of precipitation titrations the introduction of alkali thiocyanate standard solution was made during this period, the discovery being made by Charpentier in 1870. He determined silver with a standard thiocyanate solution [237], and also determined chloride by an indirect method using iron(III) ions as an indicator. Charpentier published his methods in rather obscure periodicals which were never read by the vast majority of chemists, with the consequence that the method was named after its second discoverer, Volhard. In addition to using this reagent for the determination of silver, and of chloride [239], Volhard attempted to apply it to the determination of a variety of other substances of which the determination of mercury is the most important [240].

Jacob Volhard was born in 1834. His father, who was a solicitor in Darmstadt, was a close friend of Liebig. Jacob wanted to be a historian, but his father forced him to become a chemist, because—as Volhard himself later wrote—"in those days all the fathers in Hesse thought that from their son a second Liebig will be developed". He pursued his studies, reluctantly, in Giessen under Liebig, but "with more diligence than competence". Thus he became, as he called himself after Molière, a *chimiste malgré lui*, a stock-made chemist. Liebig took Volhard with him as an assistant to Munich, but Volhard was not inclined to work too hard for which Liebig often had cause to reprimand him. Volhard made many friends among artists and was determined to change his occupation. His father then confided his worries about his son to another friend, A. W. Hoffmann, who took the unwilling Volhard with him to London, where as Volhard records "he ensured my attention by his tremendous hard-working enthusiasm".

One day Volhard left his work and announced to his friends that he was going back to Germany to become a writer. Hoffmann, however, went to his lodgings for him and took him back to the laboratory. As it appeared that all the greatest personalities of chemistry were combining their efforts to make Volhard into a chemist, he slowly accepted the situation. On returning to Germany he worked with Kolbe in Marburg, and it was here that he published his first paper in 1861, but his next paper did not appear until six years later. Liebig again assisted his career by employing him as a Privat Dozent at Munich where his father also installed a laboratory for him. Shortly afterwards Volhard married, so it became necessary for him to find a secure occupation. Liebig recommended him for a post as director of an agricultural research institute. At the same time Hoffmann arranged for him to be offered a professorship at the University of Torino, but Volhard declined this appointment. Volhard had published the results of his studies on the history of chemistry, and when Liebig died he was disappointed not to be appointed as his successor, the post being given to Baeyer instead. This setback proved to be a great stimulus to Volhard who ebgan to work with an increased intensity. His most important work was summarized in papers which appeared between the years 1875 and 1880 and dealt with analytical and inorganic chemistry. The results of his efforts were rewarded when he was offered a professorship at the University of Erlangen, and in 1882 he was offered a similar post at the University of Halle. Although in his later life Volhard published very little, he was the editor of Liebig's *Annalen* for 39 years and so occupied a very important position in chemical society. He died in 1910 [241].

Several substances other than those previously mentioned were tried as standard solutions during this period. For example, acids were titrated with ammoniacal copper sulphate solution, the basis of this titration being the formation of a precipitate of cupric hydroxide when the first drop of excess titrant was added, as the amount of ammonium hydroxide was insufficient for the complexation of the copper ions in solution [242]. This experiment illustrates what has been the greatest obstacle to the development of titrimetric analysis, namely the lack of suitable indicators.

The extreme usefulness of ceric salts in analysis had been known for some considerable time. Lange, in 1861, used a standard cerium(IV) solution for the determination of iron and hexacyanoferrate(II). The end-point was indicated by the yellow colour of ceric sulphate. Lange also mentioned that one of the advantages of ceric salts over permanganate was that they are more stable and less sensitive to acids [243]. Because of the lack of suitable indicator the use of ceric salts did not achieve much popularity.

Hexacyanoferrate(III) standard solution was used for the first time by Gentele in 1859 for the determination of reducing sugars [244], and for the determination of manganese, arsenic, antimony and chromium in alkaline medium by backtitration with potassium permanganate.

The search for reducing agents suitable for use as standard solutions was continued, but the susceptibility of reducing agents to atmospheric oxidation was a big problem then, at it is even today to a limited extent. Mulder [245] however, in 1858 designed an apparatus in which tin(II) chloride solution could be kept under an inert atmosphere so that oxidation was prevented [246].

In addition to the following reductometric standard solutions iron(II) sulphate, arsenious acid, tin(II) chloride, hexacyanoferrate(II) and oxalic acid, several other reducing agents were tried in this period. Dithionite was used for the determination of dissolved oxygen in water [247], mercury(I) nitrate [248] for the titration of permanganate, hydrogen peroxide [249] for the titration of permanganate, cerium(IV) salts and chromate and other oxidizing substances. None of these, however, proved to be of any practical importance.

Lead [250], and later, barium and bismuth were determined by precipitation titration with potassium chromate; in most cases silver nitrate was used as the indicator. The sulphate determination of Andrews [251] is of interest. He precipitated sulphate with barium chromate suspension, then, after neutralization of the solution, he filtered off the dichromate and titrated this iodiometrically [252]. This method is the ancestor of the precipitate-exchange methods which are still frequently used today.

The use of complex-forming standard solutions has also been known for a considerable time, the use of mercury(II) nitrate having already been referred to. This reagent only became of practical importance when Votocek [253] found a suitable indicator in the form of sodium nitroprussiate [254]. Liebig used silver nitrate in alkaline solution for the titration of cyanide [255], the end-point being indicated by the precipitation of silver cyanide. Thiocyanate and the halides do not interfere. The use of potassium iodide as an indicator was suggested by Drehschmidt [256, 257]; this gives a better end-point than the former procedure. Denigès [258] also suggested the use of iodide for this purpose, and suggested that the reaction be carried out in ammoniacal medium [259]. The complex forming action of cyanides was applied to the determination of other metals, for example mercury [260], copper [261], nickel [262], etc., and of which the determination of nickel proved the most useful and is still in use today.

It is apparent from these examples that a great many useful methods never achieved any practical importance owing to the lack of a suitable indicator. This was most marked in acid–base titrations, for in the field of redox titrations both permanganate and iodine were widely used reagents, because of their self-indicating properties. During the last decade of the 19th century synthetic indicators appeared in considerable numbers, and this led to a further rapid development in titrimetric analysis, but before describing this development a brief account of the history of indicators should prove of value [263].

6. DEVELOPMENT OF INDICATORS

It was no doubt known in very ancient times that the extracts of some plants change colour when they are treated with certain substances. However, for the alchemistic classification of their materials this phenomenon was not of importance. It was not until the age of iatrochemistry when classification of compounds into three main groups, i.e. into acids, bases and salts, was begun that these phenomena were examined, and the mechanism whereby a substance changes the colour of an indicator investigated. Robert Boyle defined acids as follows:

Acids act as solvents with varying strengths according to the nature of the substance being dissolved, they precipitate sulphur from solution in alkalis, they change the colour of a number of plant extracts to red; they replace the colour of the same plant fluids changed by alkalis, extracts which have previously been changed with alkalis, and they lose their corrosive properties by unification with alkalis [264].

This definition of acids remained unchanged for the next couple of centuries and is fairly satisfactory even today.

Boyle used plant extracts as indicators in a systematic method of qualitative tests. His work *Experiments on Colours* gives an account of these experiments. As indicators he used the extracts of cornflower (Exp. XXI), cochineal (Exp. XXIV), litmus (Exp. XXXVI), brazilian tree (Exp. XXXIX), blue-tree (Exp. XXXIX) and curcuma (Exp. XLIX). It is interesting to note that Boyle attempted to estimate the strengths of acids with indicators. He carried out this experiment with the extract of a tree called *lignum nephriticum:*

> I drop into the infusion (extract of *Lignum Nephriticum*) just as much distilled vinegar or other acid liquor as will serve to deprive it of its blueness which a few drops, if the sour liquor be strong, and phial small, will suffice to do.

Later he describes experiments with ammonium hydroxide and urine, with which the colour of the liquid can be restored. Finally, he writes:

> And therefore I allow myself to guess at the strengths of the liquors examined by this experiment, by the quantity of them which is sufficient to destroy or restore the ceruleous colour of our tincture [265].

Boyle also used indicator papers, and records:

> Take good syrup of violets, impregnated with the tincture of the flowers, drop a little of it upon a white paper (for by that means the change of colour will be more conspicuous, and the experiment may be practised in smaller quantities) and on this liquor let fall two or three drops of spirit, either of salt of vinegar, or almost any other eminently acid liquor, and upon the mixture of these you shall find the syrup immediately turned red [266].

Later he describes the effect of potash on the syrup. He expressed his results so that if the material to be tested changes the colour of the syrup to red then it shows that *acidic salts* are in excess, while if the colour turns to green it shows that the material is of opposite nature, and such materials Boyle called "alkalic salts" [266].

The use of plant extracts as indicators became widespread during the phlogiston period in qualitative analysis. F. Hoffmann stated that *spiritum minerale* (CO_2) in mineral waters is an acid, because it turns litmus solution red [267]. Boerhave describes as the main feature of alkaline substances that *Cum succo heliotropi, rosarum, violarum & similium viridescit, qui cum acidis rubebat* [268].

It is possible that Neumann [269] was the first to consider the possibility of observing the end of neutralization processes with indicators:

> Denn vermische ich das Sal alkali fixum mit einem Sale acido so lange, dass nach nachgelassener Effervescenz und abgeschiedenem Wasser, als welches nur zum Vehiculo der Solution gedienet, den Syrupum Violarum, wenn demselben hievo etwas beigemischet wird, in seiner Farbe nicht verändert, weder roth noch grün machet, so nennt man als den dieses neue Mixtum ein Sal medium, oder Sal neutrum [270].

In the discussion of the early history of titrimetry we have seen that Venel used extract of violets for the end-point detection in the semi-quantitative determination of the alkalinity of waters [15]. In the more specific field of titrimetric operations Lewis was the first to use an indicator in the determination of potash, in this case litmus [19] (Chapter VIII. 1). After this the use of indicators for acid–base titrations slowly became accepted.

It was soon observed that the various plant indicators do not react uniformly with acids, and that they require different amounts of acid to change their colour, in other words they change colour in different pH regions. Fontana, in 1775, noted that water containing carbonic acid changes litmus to red, but has no effect on the extract of violets [271]. Bergman also observed this:

> Blue plant-juices are sensitive to acids to varying degrees. Thus nitric acid makes sugar paper (this presumably refers to the blue paper used for wrapping of sugar, see Lewis) turn red, whereas vinegar does not possess this property. Litmus, but not syrup of violets, is made red by air-acid (carbon dioxide, in aqueous solution, i.e. carbonic acid). When in this way all blue plant extracts are examined with regard to their sensitivity. a suitable progression is obtained to measure the comparative strength of acids [272].

This, in fact, is the method of pH determination using indicators, but it was to be more than one hundred years before this idea could be realized.

Meyer [273] in 1783 wrote:

> Der Violensaft scheinet nur einen andern Sättigungspunkt zu erfordern, als die Lackmustinktur [274]. (Violet extract seems to have a saturation point different from that of litmus tincture.)

In his encyclopedia, Fourcroy describes the use of indicators in the following way:

> The coloured plant extracts are used primarily to distinguish between acids and bases. Acids change the colour of some plant extracts such as aqueous or sugar-containing violet syrup and litmus tincture to red. While the former turns red only in the presence of strong acids, the latter although its initial colour is not as deep a blue, turns red when treated with very weak acids. Thus these two extracts can be used for the measurement of acid strength. Alkalis usually change the blue plant extracts to green, and this is especially noticeable with the extracts of violets and a few other plants. Curcuma is very sensitive to alkalis, and can therefore be used to detect very small amounts of alkali [275].

In the field of indicators there was no change for a considerable time, and although a large number of plant extracts were tested litmus remained in most general use. Mohr mentions in his book the use of the extracts of litmus, pernambuk-tree and camphor tree as well as of curcuma [276]. He mentions that chemists often complain about the instability of solutions of litmus, referring to solutions stored in a stoppered bottle. In an unstoppered bottle litmus solution remains unchanged over a period of years.

However, none of the natural indicators was entirely satisfactory, their colour change was not very sharp, especially when large salt concentrations were present, or in the titration of weak acids or carbonates. They also tended to be easily decomposed by bacteria when stored. Mohr sought for an indicator which was coloured in alkaline medium but which became colourless when acidified.

The rapid development of organic chemistry and particularly the dye-stuff industry was soon to produce synthetic dyestuffs with similar but superior properties to the dyes occurring naturally.

Weiske, in 1875, recommended salicylic acid as an indicator. This gives a violet colour with ferric chloride, which disappears in the presence of acids, but reappears at the neutralization point. In the presence of alkalis a brownish-yellow colour is observed, possibly due to slight amounts of iron hydroxide being precipitated [277]. This method is of interest only because it is the first application of a synthetic compound as an indicator. In actual fact the colour change was far inferior to that of the available natural indicators.

Krüger, in 1876, proposed the use of fluorescein as an indicator, namely as a fluorescent indicator. This compound emits a fluorescence induced by ordinary light in alkaline medium, but the fluorescence is destroyed by one drop of free acid. Krüger stressed that carbon dioxide does not interfere with colour change of the indicator, and that it could also be used when coloured or turbid solutions were being analysed [278].

This indicator did not achieve a great success for in the following year, 1877, the first synthetic colour change indicator appeared. This was phenolphthalein, and was introduced by Luck [279], but it was rapidly followed by tropeolin [280], and by methyl orange introduced by Lunge in 1878 [281].

Georg Lunge was born in 1839 in Breslau into a merchant family. He completed his university studies in the same town and later worked in a chemical fertilizer factory in Silesia. Soon afterwards he went to England which in those days was the leading country in chemical technology in the world, where he worked in the coal tar and soda industry. He also married in England, and eventually occupied a leading position in the industry. In 1876 the Technical High School of Zurich offered him the chair of chemical technology, which he accepted, and during the 32 years he spent there he carried out research tirelessly, publishing 675 works, among which were several important books. His main achievements were in inorganic chemical technology, notably the lead chamber process for sulphuric acid and the process for hydrochloric acid manufacture.

Lunge also made many important contributions to analytical chemistry, and devoloped many methods of applied industrial analysis, particularly gas-analysis. His industrial analytical handbook, the *Lunge—Berl* which he compiled together with Professor Berl of Darmstadt, is still used as a reference book in most industrial laboratories. Lunge retired in 1908, but remained a co-worker of his institute for a long time after. He died in 1923.

In the succeeding years very many synthetic organic compounds were recommended for use as acid—base indicators. A paper published in 1893 [282] mentions 14 synthetic indicators. The colour ranges of synthetic indicators differed more markedly that those of the natural dyes. This soon led to a difference between acid-sensitive and base-sensitive indicators, and led to the introduction of differential determinations, for example, in the case of polybasic acids.

The first theoretical interpretation of the function of acid−base indicators was made, on the basis of the ionic theory, by Wilhelm Ostwald in his book *Die wissenschaftlichen Grundlagen der analytischen Chemie* (1894). In this he wrote:

> Damit ein Farbstoff als Indicator brauchbar sei, muss er entweder saurer oder basischer Natur sein und muss im nicht dissoziierten Zustande eine andere Farbe haben als im Jonenzustand. Ferner darf er keine starke Säure (oder Basis) sein, da er sonst schon im freien Zustande in seine Jonen zerfallen ware und keine Änderung seiner Farbe bei der Neutralisation zeigen wurde ... Eine schwache Säure existiert aber zum grossen Teil nicht als Jon, sondern als undissoziierte Molekel in der Lösung und erst durch die Neutralisation, d.h. durch den Übergang in ein Neutralsalz tritt die Jonenbildung ein, da die Neutralsalze auch der schwachen Säuren sehr vollständig dissoziiert sind ...

(For a dyestuff which is used as an indicator, it is essential that it should have either an acidic or alkaline nature, and it must have different colours in non-dissociated and in the ionic form. Furthermore it should not be a strong acid (or base), because in this case it would decompose to its ions even in a free state, and no colour change would be observable on neutralization ... A weak acid however does exist mainly as non-dissociated molecules and not in ionic form in the solution, and the formation of ions takes place only on neutralization, i.e. when a neutral salt is formed, since even neutral salts of weak acids are completely dissociated ... [283].)

According to Ostwald if the indicator itself is a weak acid, then other weak acids can give up their hydrogen ions so that the indicator exhibits the colour of the non-dissociated form. These are therefore acid-sensitive indicators, phenolphthalein for example, where the undissociated indicator is colourless while its ions are coloured. Methyl orange on the other hand is a moderately strong acid, whose ions are yellow, the undissociated indicator itself being red. In solution it dissociates considerably and therefore shows a mixed colour. When hydrogen ions are added the amount of dissociation decreases and the solution becomes red. A weak acid, however, contains far fewer hydrogen ions capable of changing the equilibrium when passing through the neutralization point. Methyl orange is therefore more sensitive to bases.

The explanation of Ostwald is plain, simple and understandable. It is a good example of the concise interpretation of Ostwald, which made clear and understandable a series of phenomena of chemistry, which had hitherto seemed quite incomprehensible. It is probably no exaggeration to say that in both general and analytical chemistry everything appeared to be well ordered in this Ostwaldian age, all the phenomena were understood and rationally explained.

This order, however, did not last for very long; various workers observed, here and there, phenomena which did not agree with Ostwald's theories. One of the earliest problems to appear concerned the mechanism of indicator change, and arose when it was observed that the colour change is not instantaneous in many cases, implying that some structural change, as well as the ion process, is occurring. Hantzsch [284] proposed his well known chromophoric theory of indicators, which interprets the colour change from the structural organic chemical aspect. The colour change is caused by a structural change in which the ionogenic form is formed from the pseudo one [285]. This theory was received with more favour by organic chemists than was Ostwald's theory, but from an analytical point of view it was not very favourable as it did not give an adequate interpretation of the quantitative features of the indicator change.

The two theories were reconciled by Kolthoff [286], who devised a theory which retains the most useful parts of Ostwald's theory. According to Kolthoff the colour of the indicator is determined by the equilibrium between the pseudo and ionogenic forms as well as by the dissociation equilibrium of the latter [287]. Recently theoretical organic chemists have tried to interpret the colour change of indicators as being due to mesomeric phenomena. At the end of the last century it seemed as if scientists were able to give definite unambiguous answers to most of the major questions. Nowadays, however, although our knowledge is greater, our room for presumption and idle speculation is less, so that our answers to vital questions are full of reservations, probabilities and uncertainties. This is indeed the case with the acid—base theories in general, and also with the theory of acid—base indicators. There is no suitable explanation for many observed phenomena.

According to the Ostwald theory the relation between hydrogen ion concentration and colour change became clear. The hydrogen ion concentration at which a given colour of the indicator occurs was established and conversely hydrogen ion concentrations were determined by means of indicators. Friedenthal [288] in 1904 [289] was the first to devise a colorimetric determination of hydrogen ion concentration by means of various indicators using solutions of known hydrogen ion concentration. His paper is of interest in another respect in that it contains the first reference to the use of a buffer solution. The solution contained mono- and dihydrogen phosphates and had a fairly well defined hydrogen ion concentration even near the neutral point. The use of these solutions was recommended by a co-worker of Friedenthal's Pál Szily [290], who can therefore be regarded as the inventor of buffer solutions. The subject of buffer solutions and colorimetric determination of pH is treated in greater detail in a later chapter (Chapter XII. 3, 4). Friedenthal recorded the pH range of the colour change for 15 indicators including several naturally occurring ones. In 1907 Salm investigated the colour change of 55 synthetic indicators [291], so we can see that by the beginning of this century a considerable number of synthetic indicators was available for the use of analytical chemists. For example, Soerensen, who later simplified the expression of hydrogen ion concentrations by the introduction of the pH function in 1909, had examined about a hundred indicators although he only found twenty-two of them to be completely satisfactory. He experienced for the first time errors due to proteins and to salts, and drew attention to these phenomena [292].

One of the most commonly used acid—base indicators today, methyl red, was introduced in 1908 [293]. Most of the sulphonphthalein group indicators were prepared and employed by Lubs and Clark in 1915 and by using the different sulphonphthaleins they were able to develop a colorimetric method for pH determination which is now widely used [294].

Proszt [295] established in 1929 that certain indicators either change their colour, or exhibit a second colour change at very high acidities, in effect at negative pH values. He examined the properties of crocein scarlet and neutral red from this aspect [296], but since then other indicators exhibiting similar characteristics have been found.

The introduction of the first fluorescent indicator, fluorescein, as we have already seen, was made before the dyestuff indicators. The use of fluorescent indicators, which emit in alkaline or acid medium a fluorescence when illuminated is particularly advantageous when turbid or darkly coloured solutions are being titrated. In such cases colorimetric indicators cannot be used for the end-point detection. However, there are only a very few substances which will emit a fluorescence when irradiated with visible light but, in 1910, Lehmann discovered that a much wider range of substances emitted a fluorescence when irradiated with ultra violet light. This fact is self-explanatory now that fluorescence has been shown to be due to excitement of electrons within the molecule [297]. In recent years the number of reported fluorescent indicators has become very great and their fluorescence changes cover a wide range of pH values. Recently Kenny [298] and Kurtz, as well as Erdey [299] have recommended the use of chemiluminescent acid—base indicators. These also detect the end-point by means of emission of light but their advantage over ordinary fluorescent indicators lies in the fact that the energy required to produce the luminescence is obtained from a chemical reaction and no external energy source is required.

Redox indicators constitute another important class of indicators. Most indicators in this class were developed later than the acid–base indicators, the majority after the first World War. Several isolated experiments using redox indicators were made much earlier. We have mentioned that Gay-Lussac determined hypochlorite by titration with arsenious acid with a few drops of indigo solution as indicator, the end-point being detected by the decolourization of the indigo [300]. The determination was not very accurate due partly to the reversibility of the reaction which is influenced by the acid concentration, and also because the high local concentrations of hypochlorite which were formed where the titrant entered the solution decomposed the indigo causing the indicator colour to fade. Despite these disadvantages the method became fairly widespread. The indicator is not reversible in this case.

Several redox methods utilized a spot test as an external indicator. These methods, which were mainly widespread in the middle of the 19th century, were developed because of the lack of suitable internal indicators. Potassium hexacyanoferrate(III) and potassium iodide with starch were used as external indicators. Crum [301] used the former as the indicator for the titration of hypochlorite with iron(II) sulphate [302] in 1840, while potassium iodide was used by Penot, also for the determination of hypochlorite, this time by titration with arsenious acid [303].

One rather interesting method of end-point detection for a redox reaction is suggested for the titration of iron(III) with tin(II) chloride. The flame of a bunsen burner is viewed through the solution; while Fe(III) ions are present the flame appears to have a green colour [304].

At the end of the last century Linossier determined dissolved oxygen in water by direct titration with iron(II) solution, using an alcoholic solution of phenosaphranine as an indicator. The red colour of the latter became colourless when the first drop of excess iron(II) solution was added [305].

Chromatometric methods, although they were well known from a very early period, did not become widely used owing to the lack of suitable indicators. Brandt in 1906 recommended diphenylcarbazide as an indicator for the chromatometric determination of iron(II). This indicator gives a violet colour with an excess of chromate [306]. The mechanism of the colour reaction is not easily understandable from his paper which records that the indicator became violet when the first drop of chromate was added, but that the colour disappeared when iron(II) ions were consumed, i.e. when the end-point was reached. Reynolds [307], in 1908 recommended an indicator for the chromatometric determination of tin, but he did not name the indicator, but only gave a description of its preparation. This was the first reversible redox indicator. Although the results he obtained were excellent there must have been some unforeseen difficulties because his indicator did not achieve general recognition. He prepared his indicator in the following way:

He heated azobenzene with concentrated sulphuric acid, when a vigorous reaction occurred. The molten mixture was then poured into a large volume of water, imparting to it a deep red colour. When tin(II) chloride was added to this solution the red colour disappeared, but the addition of chromate restored the colour [308].

The first definite redox indicator was diphenylamine, introduced by J. Knop [309] in 1925, and recommended for the chromatometric determination of iron(II) [310], and later for other redox titrations [311]. Knop and his wife, Olga Kubelkova, over the next few years discovered a large number of triarylmethane-type compounds which proved to be suitable for redox indicators [312]. During the last few years (and even at the present time) many investigators have been attempting to develop new redox indicators, so that a very wide and versatile range of these compounds now exists.

In addition to redox indicators which undergo a colour change, chemiluminescent redox indicators have also been recommended recently, for example, Siloxene [313], luminole and lucigenine [314]. Redox indicators which undergo a colour change usually function in acid medium, so that redox reactions which take place in alkaline medium have hitherto been seldom used for analytical purposes. Some of the more recent chemiluminescent indicators have the advantage that they can also be used in alkaline medium (Erdey and Buzás) [314].

In the field of acid–base indicators many interesting results were obtained by enzymologists working in the fermentation industry research institute at the laboratory of Carlsberg in Denmark, under the leadership of Soerensen. It is interesting to note that detailed practical and theoretical investigations of redox indicators were also made, not by analysts, but by biochemists and medical workers. The pioneers of these investigations were Clark [315] and his co-workers, who carried out their experiments with redox indicators in the laboratory of the Public Health Service of the United States. Their results were published between 1920–30 in the issues of the Public Health Reports. Even the term "redox indicator" was coined by a medico, Michaelis [316], who is also notable for his important contributions to the theoretical investigations of redox systems and indicators [317].

The third important group of indicators, adsorption indicators are a more recent innovation. The first of these was fluorescein, which was introduced in 1923 by Fajans [318] and Hassel [319] for the argentometric determination of chloride. These authors also provided a theoretical interpretation of the indicator mechanism [210] according to which at the equivalence point the dye-stuff anion which functions as an indicator, is adsorbed or desorbed instantaneously on the surface of the precipitate, owing to the change in the charge of the latter. Because of the adsorption, the electron system of the ion is deformed, and this results in a colour change. A more recent interpretation of the mechanism of adsorption indicators was developed by Schulek [321] and Pungor [322]. According to their theory no instantaneous adsorption or desorption occurs at the end point; the dye-stuff anion is adsorbed on the surface of the precipitate (silver halide) from the beginning of the titration. A colour change occurs because the adsorbed indicator removes silver ions from the excess of the titrant and the colour of the slightly soluble product is different from that of the indicator anion [323].

During recent years substances have been discovered which can act as acid–base, redox and adsorption indicators at the same time. The first of these so-called multiform indicators was *p*-ethoxychrisoidine, which was recommended by Schulek and Rózsa [324, 325].

7. THE DEVELOPMENT OF TITRIMETRIC ANALYSIS UP TO THE PRESENT DAY

The introduction of synthetic acid–base and later redox indicators resulted in great progress in titrimetry. Many new methods were developed which had hitherto been impossible because of the lack of suitable indicators.

Although I did not carry out a statistical analysis, it is probable that almost half of the analytical publications made during this century have been concerned with titrimetric analysis. At the beginning of this century titrimetric analysis was, without doubt, the most important and most frequently used method of analytical chemistry, and still today it holds this position. Among the large number of methods and modifications of methods, it is very difficult to select those methods which made some important contribution to the progress of chemistry, or which originated some new branch of analysis.

Time is the best judge of the applicability of an analytical method. But for modern times, this judge has not pronounced judgement and the author of this book has only his own judgement to guide him, and this is far inferior. It is therefore difficult to assess the importance of recent developments in analytical chemistry, and because of the amount of material available it is only possible to give a brief outline, which is far from complete.

In the field of acid–base titrations the different colour change regions of the various indicators has made possible the titration of various substances, as well as the differential titration of mixtures. The first differential method, and certainly one of the most impressive, is the method of Warder (1881) for the simultaneous determination of hydroxides and carbonates, utilizing methyl orange and phenolphthalein indicators [326]. The barium chloride method also became very useful, and C. Winkler [327] using phenolphthalein modified it so that the filtration of the barium carbonate could be omitted. For the titration of weak organic acids the acid-sensitive indicators permitted better and more accurate results. The titration of organic acids using phenolphthalein was carefully studied by Degener [328].

Initially, standard solutions of alkali hydroxide were prepared from alkali carbonates and calcium hydroxide, and because of this they were not contaminated with carbonate. The carbonate content of standard solutions is only a problem, since sodium hydroxide is produced industrially by electrolysis of brine and the standard solutions are prepared from this material. For the preparation of carbonate-free standard solutions many methods have been devised. Soerensen and, independently from him, Gowles described a very simple method in which a concentrated sodium hydroxide solution was prepared and the carbonate precipitated [329]. This method was replaced after the first World War by the procedure devised by Kolthoff [330].

A very important new method was the introduction of the determination of water-hardness with sodium hydroxide and sodium carbonate standard solutions. The method was developed by Wartha [331] and Pfeiffer [332, 333], and until the recent introduction of the EDTA method was used all over the world.

Thomson discovered that boric acid could be titrated in the presence of glycerine as a monobasic acid (1893) [334].

The determination of ammonia after displacement from its salts is a very old method. Its importance was increased when Kjeldahl applied it to the determination of the nitrogen content of organic substances. The absorption of ammonia in boric acid and the direct titration with acids originates from Lajos Winkler (1913) [335]. The reduction of nitrates to ammonia with metals was also known for a considerable time before Devarda in 1892 [336] recommended the use of the alloy which is now named after him.

The greatest development occurred in the field of redox titrations, where a large number of oxidizing and reducing agents were recommended for use as standard solutions. Potassium bromate was recommended by Győry [337] in 1893 [338] as an oxidizing standard solution, and even today is the most frequently used reagent for the determination of arsenic and antimony. For a long time there was no suitable reversible indicator for this procedure until Schulek introduced naphthoflavone or p-ethoxychrysoidine [339].

The practical application of reducing standard solutions was limited because of their sensitivity to atmospheric oxidation which causes them to decompose easily. This can be overcome by careful storing in an apparatus containing an inert atmosphere. At the beginning of this century titanium chloride (Knecht [340] and Hibbert (1903) [341]) was introduced. The solution must be kept under a carbon dioxide atmosphere. As there was no suitable indicator, the reagent could only be used for the determination of organic nitro groups, which are reduced to a colourless product during the titration. Iron(III) could be reduced to iron(II), and on this basis some other indirect determinations became possible. The indicator for these titrations was thiocyanate.

In addition to classical permanganatometry and iodimetry, the development of new redox methods began to flourish when Knop discovered in diphenylamine the first reversible redox indicator (Chapter VIII. 6). He himself introduced the indicator for chromatometric titrations, resulting in the method becoming much more convenient than before. But chromatometry was not the only method to take advantage of this indicator; a number of older methods which had previously suffered from a lack of a suitable indicator were resurrected.

One of the most important of these was cerimetry, which today is equally as important as permanganatometry of chromatometry. Willard and Young [342] applied this method in 1928 to the titration of iron(II) and oxalate, using diphenylamine as indicator [343]. They subsequently published a series of papers on the use of this standard solution for the determination of other substances. It should be mentioned that Atanasiu [344] used cerium(IV) sulphate as standard solution in potentiometric titrations rather earlier [344].

Tin(II) chloride was also reinvestigated as a standard solution. Szabó [345] and his co-worker used this reagent for determination of iron(III) and also vanadate and chromate, using diphenylamine as indicator [346].

Lang [347] and Gottlieb also used diphenylamin, and recommended ammonium vanadate for a rather complicated molybdenum determination [348]. The same combination of standard solutions was also used by Sirokomskii [349] and Stepin [350].

Sirokomskii and his co-workers used potassium periodate standard solution for the determination of iron, antimony, arsenic, tellurium, using diphenylamine indicator in each case [351].

Although not all of these standard solutions are very important in practice today, the examples show how the discovery of the first redox indicator made possible the development of a whole series of new methods!

Another important analytical application of periodic acid originates from Malaprade [352]. Periodic acid oxidizes certain linkages ($-OH$, $C=O$, $-H=O$, $\overset{|}{C}$ NH, or $-NH_2$) when any two of these are proportional to one another, to carboxyl or carbonyl-containing compounds. The periodate is reduced to iodic acid which can be determined either by iodimetry or alkalimetry [352].

Chlorine water and hypochlorite were the first oxidizing standard solutions. Their use, however, was limited because of their instability. Noll recommended the use of Chloramine T (p-toluene sulphochloramide) as a standard solution; this can be regarded as a stabilized chlorine solution [353].

Among reducing standard solutions ascorbic acid proved to be useful in many branches of analysis. This solution was recommended by Erdey [354, 355]. Its main advantage is that its redox potential is sufficiently low to be able to carry out reductometric determinations of a large number of substances, while at the same time it is fairly inert to atmospheric oxidation, and can therefore be stored without any special precautions. Ascorbic acid standard solution has been applied to the determination of iron(III), silver, chlorate, bromate, iodate and a number of other ions [356].

Mercury(I) nitrate and perchlorate standard solutions were re-examined in this century and proved to be of use for certain reductometric procedures, particularly for the determination of iron(III). It was recommended for the first time by Bradbury and Edwards [357], and later Belcher [358] and West, as well as Burriel-Marti [359], and Lucena examined it in more detail and developed several procedures [360, 361].

Many other substances were recommended as reductometric reagents, most of them having already been tried in the last century, for example: vanadium(II) [362], thallium(I) [363], molybdenum(V) [364], and copper(I) [365] salts. Their applicability is very limited.

The most negative redox potential is exhibited by the chromium(II) system, -0.41 V. It was first used by Dimroth [366] and Fister, and although it has a very vigorous reducing action, its use is rather limited because of the difficulty of its preparation and the precautions needed to prevent atmospheric oxidation during storage [367].

The determination of moisture is one of the oldest problems of analytical chemistry, and in many cases the decrease in weight after drying does not give sufficient accuracy. Many other chemical and physico-chemical methods have therefore been worked out to overcome this problem, and the most notable of these is the iodimetric method devised by Karl Fischer in 1935 [368]. This method is now universally used both with iodimetric or electrometric end-point detection.

Before 1950, volumetric methods based on complex formation did not play a very important role in analytical chemistry. The introduction of complexones, however, has completely changed the situation, and during the last ten years complexometry has probably been the most active field of analytical chemistry.

During the 1930's it was found that certain amino-polycarboxylic acids formed stable, soluble complexes with a large number of metals, notably with alkaline earths. Under the name "Trilon", the I. G. Farbenindustrie prepared one of these compounds which became widely used in industrial processes. The theoretical examination of these complexes was begun in the 1940's and as a result of these investigations Schwarzenbach [369] developed analytical methods for the determination of calcium and magnesium, and also for the hardness of water. The first method involved titration of the acid which was liberated during the complexation reactions. By far the best of the large number of complexones appears to be ethylenediaminetetra-acetic acid, or in the form in which it is normally used, the disodium salt. The introduction of metallochromic indicators led to the development of a large number of methods involving direct titration of the metal ion solution with a standard solution. The basis of these methods is that the strength of the metal — EDTA complex is greater than that of the metal indicator complex, so that at the end-point the colour of the free indicator appears. Schwarzenbach also introduced the first of the metallochromic indicators, murexide (ammonium purpurate) [370] and coined the name "Komplexon" for the aminopolycarboxylic acid chelating agents, the name that is nows used for trade purposes. Since these reactions are based on the formation of chelates they are often referred to as chelatometric titrations, the term suggested by Pribil [371]. In recent years chelatometric determinations have been developed for a wide range of metals; many new metallochromic indicators have been introduced and by suitable adjustment of pH or by the use of various auxiliary complex forming agents, selective determinations can be carried out. These procedures have become very important in industrial analysis, where rapid methods are essential. The extent to which this field has developed can be judged from a review of the analytical applications of ethylenediaminetetra-acetic acid, which records about one thousand original papers [372]. As always with "fashionable" analytical methods illogical determinations often appear and it is advisable to remember the warning given in Mohr's book [167] that methods which are more complicated than those they are supposed to replace are superfluous.

Micro methods of volumetric analysis have also been developed further, on the one hand by the use of more dilute titrants, and on the other hand by the miniaturization of the measuring device. Titrants as dilute as 0·001 N have been used, and this is almost the limit, for any further dilution would involve serious indicator errors. The other alternative is to reduce the volume of the solution, and it has recently been found possible to carry out titrations using a total volume of 0·4 ml with an accuracy of 0·0004 ml.

Mylius [373] and Fœrster [374] were the first to carry out acid — base titrations with 0·001 N solution, using eosin as indicator in 1891 [375]. The first micro burette which had a volume of 3 ml was constructed by Pilch [376, 377]. One design for a microburette utilising two parallel tubes was introduced by Bang [378], and is stil in use today.

Finally a brief review of one other new field of analysis which was not introduced until this century, and will no doubt soon become of even greater importance. During the age of alchemy chemical reactions were usually carried out in melts, and it was not until the advent of iatrochemistry that aqueous solutions came to be used. Since that time nearly all the reactions of analytical chemistry have been carried out in aqueous solution, until in recent years non-aqueous solutions came to be studied first of all from the theoretical, and later from the practical or analytical aspects. The use of non-aqueous media made many determinations feasible which had hitherto been impossible, for example, the titration of very weak acids or bases, which undergo a levelling effect in the appropriate solvent and can easily be titrated.

Non-aqueous solutions have been used from a very early period; alcohol has long been used as a solvent, and Boyle recorded the insolubility of numerous water-soluble salts in alcohol. The solubility-decreasing effect of alcohol was used for analytical separations. For example, Lowitz was able to separate calcium chloride from barium chloride using absolute alcohol (in which the latter compound is insoluble [379]. Lowitz was the first to prepare moisture-free solvents, namely absolute ether and glacial acetic acid [380]. Pelouze discovered that the iron thiocyanate reaction and the precipitation of calcium with oxalic acid, also occurs in alcoholic solution as in water [381]. Obviously there are many more scattered references in the literature to the use of non-aqueous solutions. At that time water was not regarded as a special solvent, and it was only the ionic theory that placed it in a special category, which is, however, logically unacceptable although understandable, as water plays an important part in nature.

Dissociation and ionization phenomena were also observed in several non-aqueous solvents, Cady used liquid ammonia [382], and then Walden [383] examined these phenomena in liquid sulphur dioxide [384]. The first person to titrate in non-aqueous media was Vorländer [385] in 1903; he titrated aniline with hydrochloric acid dissolved in benzene [386]. The theoretical aspects of this subject were initially considered more important than the practical possibilities, but with the advent of Foreman's method for the determination of amino acids [387] the practical importance of non-aqueous titrimetry was firmly established. This method involved the titration of amino acids in a mixed alcohol/acetone/formaldehyde medium.

The investigation of the theoretical and practical aspects of determinations in non-aqueous media were becoming increasingly important in analytical research and practice. Acid–base determinations were the most thoroughly studied. Acetic acid, piridine and dimethyl formamide are the most important solvents. Among many others, Connant, Hall, Werner, Folin, Tomiček [388], Fritz, Kolthoff, Brückenstein, Gautier [389], Pellerin [390], and Charlot and his co-workers, must be mentioned. Oxidation–reduction determinations in non-aqueous media have not been as widely used as the former. The first steps on this subject were taken by Tomiček and co-workers and by Erdey and Rády [391].

NOTES AND REFERENCES

1. Dossie, Robert was a pharmacist in London (?—1777). The title of the work in question: *The laboratory laid open or the secrets of modern chemistry*. London (1758).
2. *Poggendorffs Biographisch-litterarisches Handwörterbuch.* Leipzig (1863) **2** 673
3. Roebuck, John (1718—1794) was a physician and an industrialist and owned a sulphuric acid factory, an iron foundry and a salt-mine.
4. Métherie, Jean Claude de la (1743—1817) was a physician and was Professor of Natural Science at the Collège de France in Paris.
5. Tennant, Smithson (1761—1816) was a physician but did not practise as he had a private income. Later (1781) he became Professor of Chemistry at the University of Cambridge, and a member of the Royal Society.
6. Lampadius, W. A.: *Handbuch zur chemischen Analyse der Mineralkörper.* Altona (1801) 13
7. Koninck de, L. L.: *Bull Assoc. Chim. Belg.* (1901) **15** 28 73
8. Lüning, O.: *Apoth. Ztg.* (1911) **26** 702
9. Rancke-Madsen, E.: *The Development of Titrimetric Analysis till 1806.* Copenhagen (1958)
10. Glauber, J. R.: *Opera chymica.* Frankfurt **1** 524
11. Geoffroy, C. J.: *Mém. Acad. Paris* (1729) 68; cf. Rancke-Madsen, E.: *The Development of Titrimetric Analysis till 1806.* (1958) 25
12. Le Monnier, Louis Guillaume (1717—1799) was a physician, and Professor of Botany at the Jardin des Plantes, as well as court-physician to King Louis the XVI.
13. Le Monnier, L. G.: *Mém. Acad. Paris.* (1747) 252; cf. Rancke-Madsen, E.: *The Development of Titrimetric Analysis till 1806* (1958) 28
14. Venel, Gabriel Francois (1723—1775) was a physician. He analysed the French mineral waters on the authority of the government. Later (1759) he became Professor of Chemistry at the University of Montpellier.
15. Venel, F.: *Mém. prés. par. sav. étrang.* (1755); **2** 80; cf. Rancke-Madsen, E.: *The Development of Titrimetric Analysis till 1806* (1958) 29
16. Scheffer, Henrik (1710—1759) Professor at the University of Uppsala. His lectures were published in a book *Chemiske Föreläsningar* (1755) by Bergman. cf. Rancke-Madsen, E.: *The Development of Titrimetric Analysis till 1806.* (1958) 30
17. Home, F.: *Experiments on Bleaching.* Edinburgh (1756) 100
18. *ibid.* 299
19. Lewis, W.: *Experiments and Observations on American Potashes. With an easy Method of determining their respective Qualities.* London (1767) **4** 28; Quotation from Rancke-Madsen, E.: *The Development of Titrimetric Analysis till 1806* (1958) 51
20. Macquer, P. J.: *Dictionnaire de chymie, contenant la Théorie et la Pratique de cette Science.* 2. ed. Paris (1778) 559; Rancke-Madsen, E.: *The Development of Titrimetric Analysis till 1806* (1958) 59
21. Lavoisier, A. L.: *Oeuvres.* Paris (1854) **4** 311
22. Gionatti, Victor Amé: *Analyse des eaux minérales de S. Vincent et de Courmayeur dans le Duché d'Aoste avec une appendice sur les eaux de la Saxe etc.* Torino (1779) 371
23. Wiegleb, Johann Christian (1732—1800) was a pharmacist in Langensalza, a very dedicated chemist and the author of a large number of papers on analytical chemistry, and of several chemistry books.
24. Wiegleb, J. Ch.: *Handbuch der allgemeinen und angewandten Chemie.* Berlin—Stettin (1781) **1** 549
25. Oesterreicher, J. M.: *Untersuchung der Ofener Wasser mit vorgesetzten Untersuchungsmethode solcher Wasser, die von Herrn Professor Winterl angegeben worden ist.* Nürnberg (1783) 369; Rancke-Madsen, E.: *The Development of Titrimetric Analysis till 1806* (1958) 81
26. Bergman, T.: *Opuscula physica et chemica.* Stockholm—Uppsala—Abo (De Analysi Aquarum) (1779) **1** 132
27. Nyulas, F.: *Az erdélyországi orvosvizeknek bontásáról közönségesen* Kolozsvár **2** 179

28. GUYTON DE MORVEAU, L. B.: *Nouv. Mém. Acad. Dijon* (1782) **2** 16; quotation from RANCKE-MADSEN, E.: *The Development of Titrimetric Analysis till 1806* (1958) 91
29. CAVENDISH, H.: *Phil. Transact.* (1767) **57** 92
30. GUYTON DE MORVEAU, L. B.: *Nouv. Mém. Acad. Dijon* (1784) **1** 85
31. KIRVAN, R.: *Elements of mineralogy.* London (1784); quotation from RANCKE-MADSEN, E.: *The Development of Titrimetric Analysis till 1806* (1958) 115
32. GADOLIN, J.: *Kgl. Vet. Acad. nya Handl.* (1788) **9** 115; quotation from RANCKE-MADSEN, E.: *The Development of Titrimetric Analysis till 1806* (1958) 119
33. DUVAL CL.: *Chimie analytique.* (1951) **33** 196 228; *J. Chem. Ed.* (1951) **28** 508
34. DESCROIZILLES, A. H.: *Ann. chim.* (1806) **60** 17
35. *Poggendorffs Biographisch-litterarisches Handwörterbuch.* **1** 559
36. DUVAL, CL.: *J. Chem. Educ.* (1951) **28** 508
37. BERTHOLLET, C. L.: *Ann. chim.* (1789) **2** 151
38. LAVOISIER, A. L.: *Oeuvres.* Paris (1893) **6** 84
39. ROE, R.: *Experiments on the Oxygenated Muriatic Acid, used in Bleaching, with Remarks on the Blue Test Liquor,* etc. (1791); cf. RANCKE-MADSEN, E.: *The Development of Titrimetric Analysis till 1806* (1958) 136
40. KIRWAN, R.: *Physisch-chemische Schriften.* Berlin—Stettin (1793); cf. RANCKE-MADSEN, E.: *The Development of Titrimetric Analysis till 1806* (1958) 136
41. WEINLIG, G. C.: *Gründlicher Unterricht der so genannten Hausmannischen Bleiche,* etc. Berlin (1792); quotation from RANCKE-MADSEN, E.: *The Development of Titrimetric Analysis till 1806* (1958) 138
42. TENNER, JOHANN GOTTLOB (1748—1811) official physician and physicist in Chemnitz.
43. TENNER, J. G.: *Anleitung vermittelst der dephlogistierten Salzsaure zu jeder Jahreszeit vollkommen weiss, geschwind, sicher und wohlfeil zu bleichen.* Leipzig (1793) 85; quotation from RANCKE-MADSEN, E.: *The Development of Titrimetric Analysis till 1806.* (1958) 138
44. DESCROIZILLES, F. A. H.: *Journal des Arts et Manufact.* (1795) **1** 256
45. KLAPROTH, H. M.—WOLF, F.: *Chemisches Wörterbuch.* Berlin (1807) **2** 185
46. ACHARD, FRANZ CARL (1753—1821) was the director of the physics section of the Academy of Berlin, in succession to Marggraf and before Klaproth. He was responsible for the development of the industrial process for making sugar from sugar beet.
47. ACHARD, F. C.: *Nouv. Mém. Acad. Berlin* (1785) 101;
    RANCKE-MADSEN, E.: *The Development of Titrimetric Analysis till 1806* (1958) 199
48. LOWITZ, T.: *Crells Ann.* (1790) **1** 206
49. FORDYCE, GEORGE (1736—1802) was a practising physician in London, but later became a consultant physician. He was a member of the Royal Society, and the author of several chemical papers.
50. FORDYCE, G.: *Phil. Transact.* (1792) **82** 374; cf. RANCKE-MADSEN, E.: *The Development of Titrimetric Analysis till 1806* (1958) 163
51. BLACK, J.: *Lectures on the Elements of Chemistry delivered in the University of Edinburgh.* Edinburgh (1803) **2** 159 375
52. LAMPADIUS, A. W.: *Handbuch zur chemischen Analyse der Mineralkörper.* Freyberg (1801) 343 361
53. WELTER, JEAN JOSEPH (1763—1852) was the owner of a paint factory, and author of several chemical papers which mainly dealt with practical problems.
54. BERTHOLLET, C. L.: *Éléments de l'art de la teinture. Avec une descriptim du blanchiment par l'acide muriatique oxigéné.* 2. ed. Paris (1804) **1** 233
55. DESCROIZILLES, F. A. H.: *Ann. chim.* (1806) **60** 17
56. LAURENS: *Ann. chim.* (1808) **67** 97
57. BARTHOLDI, CHARLES (?—1849) was a chemistry teacher at a school in Colmar, the author of a few papers on analytical chemistry.
58. BARTHOLDI, C.: *Ann. chim.* (1792) **12** 74
59. BARTHOLDI, C.: *Journ. de Physique.* (1798) **47** 16

60. BECKURTS, H.: *Die Methoden der Massanalyse.* Braunschweig (1913) 935
61. *Ann. chim.* (1802) **11** 113
62. *Rep. of Arts. Man. Agric.* (1802) **1** 444
63. *Allg. Ann. der Gewerb.* (1803) **1** 165; quotation from RANCKE-MADSEN, E.: *The Development of Titrimetric Analysis till 1806.* (1958) 181
64. DALTON, H.: *Ann. phil.* **1**. 15
65. OTTO, FRIEDRICH JULIUS (1809—1870) was Professor of Chemistry at the Technical University in Braunschweig.
66. OTTO, F. J.: *Dinglers polytechn. J.* (1842) **85** 292
67. WELTER, J. J.: *Ann. chim. phys.* (1817) **7** 383
68. GAY-LUSSAC, J. L.: *Ann. chim. phys.* (1824) **26** 162
69. ARAGO, DOMINIQUE FRANÇOIS (1786—1853) was a physicist and astronomer. He was a member, and subsequently first secretary of the French Academy of Sciences.
70. GAY-LUSSAC, J. L.: *Ann. chim. phys.* (1828) **39** 337
71. GAY-LUSSAC, J. L.: *Ann. chim. phys.* (1829) **40** 398
72. GAY-LUSSAC, J. L.: *Instruction sur l'essai des matiéres d'argent par la voie humide.* Paris (1832)
73. GAY-LUSSAC, J. L.: *Vollständiger Unterricht über das Verfahren Silber auf nassem Wege zu probieren.* Braunschweig (1833) 9
74. *ibid.* 22
75. *ibid.* 9
76. GAY-LUSSAC, J. L.: *Ann. chim. phys.* (1835) **60** 225
77. MAROZEAU: *Ann. chim. phys.* (1831) **46** 400
78. PENOT: *Bull. soc. ind. Mulhouse* (1831) 285
79. MORIN, ANTOINE (1800—1879) pharmacist in Geneva, author of several chemical papers.
80. MORIN, A.: *Ann. chim. phys.* (1828) **37** 139
81. HOUTON DE LA BILLARDIÈRE, JACQUES JULIEN (1755—1834) was Professor of Chemistry at the Academy in Rouen, he was a corresponding member of the Academy of Paris.
82. HOUTON DE LA BILLARDIÈRE, J. J.: *J. Pharm.* (1826); cf. *Dinglers polytechn. J.* (1826) 263
83. LASSAIGNE: *Lieb. Ann.* (1842) **44** 356; PONTIUS: *Chem. Ztg.* (1904) 59
84. DU PASQUIER, ALPHONSE (1793—1848) was originally a practising physician, but later became Professor of Chemistry at the Medical Academy in Lyon. He wrote numerous analytical papers.
85. DU PASQUIER, A.: *Ann. chim. phys.* (1840) **73** 310
86. BERZELIUS, J. J.: *Jahresbericht.* (1842) **21** 158
87. FORDOS, MATHURIN JOSEPH (1816—1878) was a pharmacist, and the head of the laboratory at the Hospital "Charité" in Paris. He was the author of numerous analytical publications.
88. GÉLIS, AMADÉE (1815— ?) was the owner of a chemical plant near Paris.
89. GÉLIS, A.—FORDOS, M. J.: *Ann. chim. phys.* (1843) **9** 105
90. DUFLOS, ADOLPHE (1802—1889) was born in Artenais in France, and also studied there. Later he went to Germany, and finally became the Professor of Pharmacy at the University of Breslau. In 1869 he retired from teaching because of an eye-disease. He is the author of numerous books dealing chiefly with pharmacy and toxicology.
91. DUFLOS, A. F.: *Chem. Apothekerbuch.* Breslau (1845) **2** 101
92. GAULTIER DE CLAUBRY, FRANÇOIS (1792—1878) was a pharmacist. He became Professor of Chemistry, and later toxicology at the Pharmaceutical Academy in Paris.
93. GAULTIER DE CLAUBRY, H. F.: *J. chem. med.* (1846) **2** 425 473
94. BUNSEN, R. W.: *Lieb. Ann.* (1853) **86** 265
95. SCHWARZ, H.: *Praktische Anleitung zu Maasanalysen.* 2. ed. Braunschweig (1853) 117
96. *ibid.* 116
97. MARGUERITTE, FRÉDÉRIC, was an employee of the gas works in Paris. I was not able to establish either the date of his birth or of his death, or any details of his career. He wrote altogether some ten chemical papers which were published in the periodicals

Annales de chimie et physique and *Comptes rendus*.
98. MARGUERITTE, F.: *Ann. chim. phys.* (1846) **18** 244
99. PELOUZE, THÉOPHILE JULES (1807—1867) was a pharmacist, and assistant to Gay-Lussac. From 1830 he was Professor of Chemistry at the University of Lille, and a year later at the École Polytechnique, as well as being director of the mint. From 1837 he was a member of the Academy of Sciences. His main work was carried out in analysis; he determined the atomic weight of several elements with a high degree of accuracy.
100. BERZELIUS, J. J.: *Jahresbericht.* (1848) **27** 215 217
101. BUSSY, ANTOINE BRUTUS (1794—1883) was a pharmacist and Professor of Chemistry at the Pharmaceutical Academy in Paris. He later became the director of this institute, and a member of the Academy of Sciences.
102. BUSSY, A. B.: *Compt. rend.* (1847) **24** 774
103. PELOUZE, T. J.: *Ann. chim. phys.* (1847) **20** 129
104. SCHWARZ, H.: *Lieb. Ann.* (1849) **69** 209
105. I was not able to find any biographical data of the Hempel referred to, but he must not be confused with the gas analyst of the same name.
106. HEMPEL: *Mémoire sur l'emploi de l'acide oxalique dans les dosages á liqueurs titrées.* Lausanne (1853)
107. CLARK, THOMAS (1801—1867) was an employee of Tennant's bleaching plant, but later became a physician and worked at the hospital in Glasgow. From 1833 he was the Professor of Chemistry at the University of Aberdeen.
108. CLARK, T.: *Chemical Gazette* (1847) 5; cf. *Dinglers polytechn. J.* (1842) **83** 193
109. BOLLEY, ALEXANDER POMPEIUS (1812—1870) was a chemical technologist, the author of the first technological analytical handbook (*Handbuch der technisch chemischen Untersuchungen.* 1858). He was Professor of Chemical Technology at the Technical University in Zürich.
110. BOLLEY, A. P.: *Dinglers polytechn. J.* (1852) **124** 204
111. FAISST: *Dinglers polytechn. J.* (1852) **125** 34
112. PÉLIGOT, EUGÉNE MELCHIOR (1811—1890) analyst at the mint. From 1852 he was Professor of Chemistry at the Institution known as the Conservatoire des Arts et Métiers in Paris, and a member of the French Academy of Sciences. He was the author of numerous analytical papers.
113. PÉLIGOT, E. M.: *Compt. rend.* (1847) **24** 550
114. BINEAU, AMAND (1812—1861) was Professor of Chemistry at the University of Lyon.
115. BINEAU, A.: *Journ. pharm.* **12** 301
116. SCHLOESING, JEAN JACQUES THÉOPHILE (1824—1919) worked in the French tobacco industry, and later became the director of the school known as the École d'application annexée à la manufacture in Paris. He was a member of the French Academy of Sciences.
117. SCHLOESING, J. J.: *Ann. chim. phys.* (1847) **19** 230
118. SCHABUS, JAKOB (1825—1867) was Professor of Natural History at the Commercial Academy in Wien.
119. PENNY, FREDERICK (1816—1869) was a pharmacist. He became Professor of Chemistry at the University of Glasgow.
120. SCHABUS, J.: *Ber. Wiener Akad.* (1851) **6** 396
121. PENNY, F.: *Chem. Gazette* (1850) **8** 330
122. SCHWARZ, H.: *Praktische Anleitung zu Maasanalysen.* 119
123. PENNY, F.: *Quatern. Journ. Chem. Soc.* (1851) **4** 239
124. SCHWARZ, H.: *Praktische Anleitung zu Maasanalysen.* 133
125. LEVOL, ALEXANDRE, FRANÇOIS (1808—1876) was analyst at the Institute for Noble Metal Examination in France.
126. LEVOL, A. F.: *Journ. pharm. chim.* (1842) **1** 210
127. DUFLOS, A. F.: *Pharmazeutische Experimentalchemie.* Breslau (1845) 545
128. BECKURTS, H.: *Die Methoden der Maasanalyse.* 778
129. SCHWARZ, H.: *Praktische Anleitung zu Maasanalysen.* 58

130. Berthet: *Revue industr. et scientif.* **24** 394; cf. Berzelius, J. J.: *Jahresbericht.* (1848) **27** 210
131. Becquerel, A. C.: *Ann. chim. phys.* (1831) **47** 15
132. Barreswil, Louis Charles (1817—1870) was Professor of Chemistry at the Commercial Academy in Paris.
133. Barreswil, L. C.: *Journ. de pharm. et chim.* (1846) **6** 301
134. Fehling, Hermann (1812—1885) was Professor of Chemistry at the Technical University of Stuttgart.
135. Fehling, H.: *Lieb. Ann.* (1849) **72** 106
136. Duflos, A. F.: *Lieb. Ann.* (1837) **24** 310
137. Leconte: *Compt. rend.* (1849) **29** 45
138. Liebig, J.: *Lieb. Ann.* (1851) **78** 150
139. Saint Venant: *Compt. rend.* (1846) **23** 522
140. Liebig, J.: *Lieb. Ann.* (1851) **77** 102
141. Liebig, J.: *Lieb. Ann.* (1853) **85** 289
142. Levol, A. F.: *Bull. Soc. d'encouragement* (1853) 220
143. Szabadváry, F.: *Chemist and druggist* (1957) **168** 616
144. Henry, Étienne Ossian (1798—1873) was a pharmacist, and director of the laboratory of the French Academy of Medical Sciences.
145. Henry, É. O.: *Journ. de chim. et pharm.* **6** 301 (1846)
146. Berzelius, J. J.: *Jahresbericht.* (1829) **10** 157
147. Schwarz, H.: *Praktische Anleitung zu Maasanalysen.* 5
148. *ibid.* 26
149. Planck, M. *Das Prinzip der Erhaltung der Energie.* Leipzig (1887) 21
150. Hasenclever, R.: *Ber.* (1900) **33** 3835
151. Scott, J. M.: *Chymia.* (1950) **3** 191
152. Mayer, Robert Julius (1814—1878) was a physician. He worked as a ship's doctor and later became a practising physician in Heilbronn. During an operation for the letting of blood from somebody in Batavia he observed that the venous blood is much more red in tropical areas than it is in cooler climates. This observation motivated a series of investigations which culminated in his proposition of the principle of the conservation of energy, and his calculation of the heat-equivalent of mechanical energy. His views were violently attacked and as a result of this he suffered a severe nervous breakdown. However, he lived to see the acceptance of his ideas.
153. Mohr, F.: *Z. für Physik, Mathematik und verwandte Wissenschaften.* (1837) **5** 419
154. Mohr, F.: *Kommentar zu Preussischen Pharmakopoe.* Braunschweig (1847)
155. Mohr, F.: *Lehrbuch der pharmazeutischen Technik.* Braunschweig (1847); *Practical Pharmacy.* London (1848); *Practice of Pharmacy.* Philadelphia (1849)
156. Mohr, F.: *Der Weinstock und der Wein.* Koblenz (1864); *Der Weinbau und die Weinbereitungskunde.* Braunschweig (1865)
157. Mohr, F.: *Geschichte der Erde.* (1866)
158. Akin, C. K: *Phil. Mag.* (1864) **28** 474
159. Mohr, F.: *Mechanische Theorie der chemischen Affinitat und die neuere Chemie.* Braunschweig (1865) 5
160. Mohr, F.: *Allgemeine Theorie der Bewegung und der Kraft als Grundlage der Physik und Chemie.* Braunschweig (1874)
161. Mohr, F.: *Lieb. Ann.* (1853) **86** 129
162. Mohr, F.: *Lieb. Ann.* (1836) **18** 232
163. Kahlbaum, G.: *Justus von Liebig und Friedrich Mohr in ihren Briefen von 1834—1870.* Leipzig (1904) 16
164. Streng, Johann August (1830—1897) was assistant to Bunsen at Breslau, and later became Professor of Chemistry at the Mining Academy in Glausthal. He later became Professor of Chemistry at the University of Giessen, a post which had earlier been held by Liebig.

165. Mohr, F.: *Lehrbuch der chemisch-analytischen Titrirmethode.* Braunschweig (1855) **1** 377
166. *ibid.* **1** 382
167. *ibid.* **1** VII
168. *ibid.* **1** 32
169. *ibid.* **1** 55
170. *ibid.* **1** 66
171. *ibid.* **1** 95
172. *ibid.* **1** 69 349
173. *ibid.* **1** 136
174. *ibid.* **1** 139
175. *ibid.* **1** 149
176. *ibid.* **1** 161
177. *ibid.* **1** 197
178. *ibid.* **1** V
179. *ibid.* **1** 219; Raevsky: *Lieb. Ann.* (1851) **78** 150
180. *ibid.* **1** 234
181. *ibid.* **1** 365
182. *ibid.* **1** 286
183. Dupré, August (1835—1907) was a German physician who lived in England. He studied in Heidelberg and later became Professor of Chemistry at Westminster Hospital in London.
184. Mohr, F.: *Lehrbuch der chemisch-analytischen Titrirmethode.* Braunschweig (1855) **1** 377 381
185. *ibid.* **2** 69
186. *ibid.* **2** 262
187. Péant de Saint Gilles, L. (1832—1863) was a French chemist who died while still quite young. He had a private income and developed several analytical procedures in his own laboratory.
188. Péant de Saint Gilles, L.: *Ann. chim phys.* (1859) **55** 383
189. Péant de Saint Gilles, L.: *Compt. rend.* (1858) **46** 624
190. Péant de Saint Gilles, L.: *Ann. chim. phys.* (1859) **55** 388
191. Löwenthal, J.—Lensen, E.: *Z. anal. Chem.* (1862) **1** 329
192. Fresenius, R.: *Z. anal. Chem.* (1862) **1** 361
193. Kessler, Friedrich Christian (1824—1896) was Professor of Chemistry and Physics, and later director, at various German technical schools. He wrote numerous papers on analysis and optics.
194. Kessler, F. C.: *Pogg. Ann.* (1855) **95** 223
195. Kessler, F. C.: *Pogg. Ann.* (1863) **118** 41 **119** 225
196. Zimmermann, Julius Clemens (1856—1885) was the assistant to Baeyer and also to Volhard at the University of Munich.
197. Zimmermann, Cl.: *Ber.* (1881) **14** 779
198. Reinhardt, C.: *Chem. Ztg.* (1889) 323
199. Jones, Harry Clair (1865—1916) was professor at the University of Baltimore.
200. Jones, Cl.: *Chem. News.* (1889) **60** 163
201. Shimer, P. W.: *Am. Chem. Soc. Journ.* (1899) **21** 723; *Jahresbericht.* (1899) 328
202. Guyard, A.: *Bull. soc. chim.* (1863) **1** 89
203. Volhard, J.: *Lieb. Ann.* (1879) **198** 318
204. Wolff, N.: *Stahl u. Eisen.* (1884) 702
205. Czudnovicz, C.: *Pogg. Ann.* (1863) **120** 17
206. Roscoe, Henry Enfield Sir (1833—1915) was Professor of Chemistry at the University of Manchester, and then from 1898 onwards in London. His most important contribution was to the field of photochemistry, in which he worked jointly with Bunsen.
207. Roscoe, H. E.: *Lieb. Ann. Suppl.* (1868) **6** 77
208. Deshayes, M.: *Bull. soc. chim.* (1878) **29** 541
209. Boussingault, Jean Baptiste (1802—1882) was Professor of Analytical and Agricultural

Chemistry at the Conservatoire des Arts et des Métièrs in Paris. He was the author of a large number of papers on analytical chemistry.
210. SMITH, P.: *Chem. News* (1904) **90** 237
211. PISANI, FELIX (1831—1920) was a private analyst and professor at a French private school.
212. PISANI, F.: *Compt. rend.* (1864) **59** 289
213. FORCHHAMMER, JOHANN GEORG (1794—1865) was professor at the University of Copenhagen.
214. FORCHHAMMER, J. G.: *L'institut.* (1849) 383
215. SCHRÖTTER, ANTON (1802—1875) was Professor of Chemistry at the University of Wien, and Director of the Austrian Mint.
216. SCHRÖTTER, A.: *Ber. Wien. Akad.* (1859) **34** 357
217. SCHULZE, ERNST AUGUST (1840—1907) was Professor of Agricultural Chemistry at the Technical University of Zürich.
218. SCHULZE, E. A.: *Dinglers polytechn. J.* (1868) **188** 197
219. THAN, K.: *Kir. magyar természettud. társ. közlönye* (Journal of the Hungarian Royal Society of Nature Sciences) (1860) **1** 67; *Math. u. naturwiss. Berichte aus Ungarn.* (1890) **7** 298
220. THAN, K.: *Math. u. naturwiss. Berichte aus Ungarn.* (1889) **6** 127
221. RIEGLER, E.: *Z. anal. Chem.* (1896) **35** 308
222. TOPF, G.—GRÖGER, M.: *Z. angew. Chem.* (1890) 353
223. ZULKOVSKY, KARL (1833—1901) was an assistant at the University of Wien. He later became Professor of Chemical Technology at the University of Brünn.
224. ZULKOVSKY, K.: *J. prakt. Chem.* (1868) **103** 351
225. REISCHAUER, C. G.: *Dingl. polytechn. J.* (1862) **165** 451
226. LEGLER, L.: *Rep. anal. Chem.* (1885) **6** 631
227. HEHNER, O. (1853—1924) German analyst and food-chemist, who worked in industry in England. He died in South Africa. His publication referred to: *Analyst* (1887) **12** 44
228. STEINFELS: *Seifensiederztg.* (1910) **37** 793
229. LANDOLT, HANS HEINRICH (1831—1910) was Professor of Chemistry at the Agricultural Academy, and later at the University in Berlin. He discovered the famous reaction illustrating reaction-velocity which was named after him. His later work was confined to physical chemistry, and his book of physico-chemical constants is still used, in its later editions, even today.
230. LANDOLT, H. H.: *Ber.* (1871) **4** 770
231. WALLER, E.: *Chem. News.* (1873) **43** 152
232. KOPPESCHAAR, W. F.: *Z. anal. Chem.* (1876) **15** 233
233. HÜBL, A.: *Dinglers polytechn. J.* (1884) **253** 218
234. KNOP, AUGUST WILHELM (1817—1891) was Professor of Agricultural Chemistry at the University of Leipzig.
235. KNOP, W.: *Centralblatt.* (1854) 321 403 499
236. WINKLER, L.: *Ber.* **21** (1888) 2843
237. CHARPENTIER: *Rev. univ. des mines.* (1873) **32** 302
238. CHARPENTIER: *Mem. des proc. verbaux des seances de la soc. des ing. civ. de France.* (1870) 135
239. VOLHARD, J.: *J. prakt. Chem.* (1874) **117** 217
240. VOLHARD, J.: *Lieb. Ann.* (1877) **190** 57
241. VORLÄNDER: *Ber.* (1912) 45 1855
242. KIEFFER, L.: *Lieb. Ann.* (1855) **93** 386
243. LANGE, TH.: *J. prakt. Chem.* (1861) **82** 129
244. GENTELE, J. G.: *Dinglers polytechn. J.* (1859) **152** 68
245. MULDER, EDUARD (1832—1924) was Professor of Chemistry at the University of Utrecht between 1868—1902.
246. MULDER, E.: *Jahresbericht.* (1858) 587
247. SCHÜTZENBERGER, O.—GÉRARDIN, M.: *Compt. rend.* (1872) 75 879

248. LECLERC: *Compt. rend.* (1884) **10** 337
249. SCHNEIDER, L.: *Monatshefte* (1888) **9** 242
250. SCHWARZ, H.: *Dinglers polytechn. J.* (1853) **127** 55
251. ANDREWS, J. L. (1856—1931) was born in Canada but studied in Germany. From 1885 to 1906 he was Professor of Chemistry at the University of Iowa, and after this he occupied several important positions in industry.
252. ANDREWS, J.: *Am. Chem. Journ.* (1890) **2** 567
253. VOTOCEK, EMIL (1872—1950) studied at the Technical Universities of Prague and Mulhouse, and then worked in Göttingen under Tollens. In 1895 he became an assistant at the Technical University of Prague, and in 1900 he was appointed Privat Dozent, and in 1907 Professor of Inorganic and Organic Chemistry. His scientific work was mainly in the field of organic chemistry
254. VOTOCEK, E.: *Chem. Ztg.* (1918) **42** 257 271 317
255. LIEBIG, J.: *Lieb. Ann.* (1851) **77** 102
256. DREHCHSMIDT, HEINRICH (1853—1923) was an industrial chemist. He later became director of gas works in Berlin.
257. DREHSCHMIDT, H.: *J. für Gasbeleuchtung.* (1892) **35** 225
258. DENIGÈS, GEORGES (1859—1951) was born in Bordeaux, where he studied at the secondary school and at the university, where he later became an assistant lecturer and subsequently in 1892 Professor of Medical Chemistry. He retired in 1930. He was a corresponding member of the French Academy of Sciences. During his long life he wrote 680 papers covering the fields of organic chemistry, pharmacy and analytical chemistry. In this latter field his interest was mainly centred on microcrystallography, and even in the year of his death, at the age of 92, some scientific dissertations of his were published.
259. DENIGÈS, G.: *Compt. rend.* (1893) **117** 1078
260. HANNAY, J. B.: *Ber.* (1878) **11** 807
261. PARKES, J.: *Mining Journal* (1851)
262. MOORE, T.: *Chem-News* (1889) **59** 150
263. SZABADVÁRY, F.: *Acta. chim. hung.* (1959) **20** 253; SZABADVÁRY, F. *J. Chem. Ed.* (1964) **41** 286
264. KOPP, H.: *Geschichte der Chemie.* Braunschweig (1845) **2** 10
265. BOYLE, R.: *Works.* London (1744) **2** 44
266. *ibid.* 53
267. HOFFMANN, F.: *Dissertatio physico-medic.* Leyden (1708) 183
268. BOERHAVE, H.: *Elementa chemiae.* Leyden (1732) **2** 57
269. NEUMANN, CASPAR (1683—1737) was court pharmacist to the Prussian King Friedrich lst, and Professor of Chemistry at the Collegium Medicum in Berlin. He was a famous chemist of the phlogiston period, and enjoyed a great reputation during his lifetime.
270. NEUMANN, C.: *Lectiones chymicae von Salibus Alkalino-Fixis und von Camphora.* Berlin (1727) 77; cf. RANCKE-MADSEN, E.: *The Development of Titrimetric Analysis till 1806.* (1958) 69
271. FONTANA, FELICE (1730—1805) was an abbot. He was Professor of Physics at the University of Pisa, and later became director of the museum in Florence. The publication referred to *Journal de physique.* (1775) **6** 280
272. BERGMAN, T.: In the notes of book of SCHEFFER, H. T. *Chemiske Föreläsningar, etc.* Uppsala (1775) 7; cf. RANCKE-MADSEN, E.: *The Development of Titrimetric Analysis till 1806* (1958) 68
273. MEYER, JOHANN CARL FRIEDRICH (1733—1811) was a pharmacist in Stettin and the author of many analytical papers.
274. MEYER, J. C. F.: *Crells Neueste Entdeck.* (1783) **10** 67
275. FOURCROY, ANTOINE FRANÇOIS (1755—1809) was the son of a pharmacist. He studied to be a physician but, in 1784, he became Professor of Chemistry at the Jardin des Plantes. He was a member of the French Academy of Sciences, and was one of the first followers of Lavoisier's new chemistry, taking an active part in the devising of the new

chemical nomenclature. His main scientific contributions were in the field of organization and popularization of science rather than research, and he also occupied a prominent position in public life. He was also a member of the Convention, and Napoleon later made him a count. The citation is taken from FOURCROY: *Encyklopédie méthodique, chimie et minérologie.* Paris (1804) 4 66

276. MOHR, F.: *Titriermethode.* (1855) 43—45
277. WEISKE, H.: *J. prakt. Chem.* (1875) **120** 157
278. KRÜGER, F.: *Ber.* (1876) **9** 1572
279. LUCK, E.: *Z. anal. Chem.* (1877) **16** 332
280. MILLER, M.: *Ber.* (1878) **11** 460
281. LUNGE, G.: *Ber.* (1878) **11** 1944
282. TROMMSDORFF, H.: *Chem-techn. Untersuchungsmethoden.* (1893) 111
283. OSTWALD, WI.: *Wissenschaftliche Grundlagen der analytischen Chemie.* Leipzig (1894) 103
284. HANTZSCH, ARTHUR RUDOLF (1857—1935) was Professor of Chemistry at the Technical University of Zürich from 1885, and then from 1893 at the University of Würzburg, and finally from 1903 at the University of Leipzig. He carried out research in organic chemistry.
285. HANTZSCH, A.: *Ber.* (1907) **40** 307; (1908) **41** 1187
286. KOLTHOFF, IZAAC MAURITS, was born in the Netherlands in 1894, and studied at the University of Utrecht, where he also worked as a lecturer until 1927. He was then invited to work at the Department of Analytical Chemistry at the University of Minnesota, where he is now Professor of Analytical Chemistry.
287. KOLTHOFF, I. M.: *Der Gebrauch von Farbenindicatoren.* 2. ed. Berlin (1923)
288. FRIEDENTHAL, HANS WILHELM, was born in Breslau in 1870. He studied in Heidelberg, Berlin and Bonn, graduating in medicine in 1895, when he became an assistant at München and Erlangen. He later became a Privat Dozent at the University of Berlin, where from 1916 he was Professor of Physiological Chemistry. In 1933 he was forced to leave his position and died in 1943 by suicide.
289. FRIEDENTHAL, H. W.: *Z. f. Elektrochemie.* (1904) **10** 113
290. SZILY, PÁL, was born in 1878 in Budapest, studied medicine at the University of Budapest, worked at the Institute of Physiological Chemistry in Budapest and Berlin and in several hospitals. During the first World War he was an army physician; after the war he was a practising physician in a little Hungarian town, Magyaróvár where he died in 1945.
291. SALM, E.: *Z. f. physik. Chem.* (1906) **57** 471
292. SOERENSEN, S. P. L.: *Biochem. Ztschr.* (1909) **21** 131 231
293. RUPP, E.—LOOSE, R.: *Ber.* (1908) **41** 3905
294. LUBS, H. A.—CLARK, W. M.: *J. Wash. Acad. Sci.* (1915) **5** 609 (1916); **6** 481
295. PROSZT, JÁNOS, was born in Budapest in 1892. After his early training in chemistry he became an assistant lecturer, also in Budapest. In 1923 he was appointed Professor of Chemistry at the Mining and Forestry Academy in Sopron. Since 1948 he has been Director of the Institute for Inorganic Chemistry at the Technical University of Budapest.
296. PROSZT, J.: *Mitteilungen der berg- und hüttenmännischen Abt. Hochschule Berg-Forstwesen zu Sopron.* (1929); Zentralblatt (1930) **1** 1830
297. LEHMANN, H.: *Physik. Z.* (1910) **11** 1039
298. KENNY, FREDERIC, was born in 1894 in Minneapolis, and studied at various universities in America. After teaching at several schools he was appointed, in 1941, Professor of Analytical Chemistry at the Hunter College in New York. The paper referred to: *Anal. Chem.* (1951) **23** 339
299. ERDEY, L.: *Acta chim. hung.* (1953) **3** 81
300. GAY-LUSSAC, L. J.: *Ann. chim. phys.* (1835) **60** 225
301. CRUM, WALTER (1796—1867) was the owner of a textile dyeing factory. He was also a chemist and a member of the Royal Society.
302. CRUM, W.: *Dinglers polytechn. J.* (1846) **96** 40

303. PENOT: *Bull. soc. ind. Mulhouse.* (1852) **118**
304. MORGAN: *J. anal. Chem.* (1888) **2** 169
305. LINOSSIER, M. G.: *Bull. soc. chim.* (1891) **5** 63
306. BRANDT, L.: *Z. anal. Chem.* (1906) **45** 95
307. REYNOLDS, J. EMERSON (1844—1920) was Professor of Chemistry at the University of Dublin.
308. REYNOLDS, J. E.: *Chem. News.* (1908) **97** 13
309. KNOP, JOSEF, was born in 1885 in Mlada Vozice and graduated at the University of Prague, and then worked in Mulhouse and Lausanne. He later returned to the Technical University of Prague, where he worked at the Institute for Agriculture. In 1919 he became the Director of the Chemical Institute at the Agricultural Academy in Brno. He retired in 1957.
310. KNOP, J.: *Chem. Listy* (1915 1918)
311. KNOP, J.: *Z. an. Chem.* (1923) **63** 81; *J. Am. Chem. Soc.* (1924) **46** 263
312. KNOP, J.— KUBELKOVA, O.: *Chem. Listy* (1929); *Z. anal. Chem.* (1929 1931 1934 1941)
313. KENNY, F.—KURTZ, R. B.: *Anal. Chem.* (1950) **22** 693
314. ERDEY, L.—BUZÁS, I.: *Acta chim. hung.* (1955) **6** 78. Mrs Ilona Buzás is a research worker at the Institute for General Chemistry at the Technical University of Budapest.
315. CLARK, WILLIAM MANSFIALD, was born in 1884. He worked at the American Public Health Institute until 1927, when he was appointed Professor of Physiological Chemistry at the University of Baltimore.
316. MICHAELIS, LEONOR (1875—1949). According to his diploma he was a physician, but most of his later work was concerned with physics and chemistry. In 1908 he became Professor of the Medical Physico-Chemistry at the University of Berlin. Between 1922 and 1926 he was the Professor of Biochemistry at the University of Nagoya (Japan), and between 1926—1929 he was Professor at the University in Baltimore, where he worked with Clark. From 1929 he was a collaborator of the Medical Research Institute of the Rockefeller Foundation.
317. MICHAELIS, L.: *Oxydations—reduktionspotentiale.* Springer, Berlin (1916)
318. FAJANS, KASIMIR was born in 1887 in Warsaw. He studied in Warsaw, Leipzig and Heidelberg, and later worked in Zürich and in Manchester. In 1911 he became a Privat Dozent at the Technical University in Karlsruhe, and from 1917 he was Professor of Physical Chemistry at the University of Munich. In 1933 he was forced to emigrate to the United States, where from 1936 he has been Professor of Chemistry at the University of Ann Arbor, in Michigan.
319. HASSEL, ODD, was born in 1897, and after graduating from the University of Berlin he became in 1925 a lecturer at the University of Oslo, where in 1934 he was appointed Professor of Physical Chemistry.
320. FAJANS, K.—HASSEL, O.: *Z. f. Elektrochemie.* (1923) **29** 495
321. SCHULEK, ELEMÉR, was born in Késmárk in 1893 and graduated from the University of Budapest, where he later returned to work. In 1927 he went to work for the National Public Health Institute, and was later to become its Director. Since 1944 he has been the Director of the Institute for Inorganic and Analytical Chemistry at the University of Budapest. He died in 1964.
322. PUNGOR, ERNŐ, was born in 1923 in Vasszécsény. At present he is the Professor of Analytical Chemistry at the University of Veszprém (Hungary).
323. SCHULEK, E.—PUNGOR, E.: *Anal. Chim. Acta* (1950) **4** 213
324. RÓZSA, PÁL, was born in Komárom, in 1901. He is now head of a department at the National Public Health Institute, Budapest.
325. SCHULEK, E.—RÓZSA, P.: *Z. anal. Chem.* (1938) **115** 185
326. WARDER, R. B.: *Chem. News* (1881) **43** 228
327. WINKLER, CLEMENS (1838—1904) was Professor of Chemistry at the Mining Academy in Freiberg. He was the discoverer of germanium (1885), and was an outstanding analyst. His achievements in the field of gas analysis are particularly noteworthy.

328. Degener, P.: *Festschrift Techn. Hochschule Braunschweig.* (1897) 451
329. Soerensen, S. P. L.: *Biochem. Z.* (1909) **21** 168; the simultaneous procession is based on the same principle of Gowles, H. W.: *J. Am. Chem. Soc.* (1908) **30** 1192.
330. Kolthoff, I. M.: *Z. anal. Chem.* (1922) **61** 48
331. Wartha, Vince (1844—1914) was born in Fiume into a military family. He began his studies at the Technical University of Budapest, and completed them in Zürich. He then became an assistant lecturer at the Technical University of Budapest, and later went to Heidelberg where he worked with Bunsen. He became Privat Dozent at the University in Zürich, where from 1867 he was deputy Professor of Mineralogy. In 1870 he was appointed Professor of Chemical Technology at the Technical University of Budapest.
332. Pfeiffer, Ignác (1868—1941) studied at the Technical University of Budapest, where he became the assistant to Wartha, whom he later succeeded. In 1920 he retired on political grounds. After this he directed the research department of the Hungarian-Electric Co., and he also was the managing president of the Society of Hungarian Chemists.
333. Wartha, V.: *Az ivóvíz vizsgálata* (Examination of drinking-water) Budapest (1882)
    Pfeiffer, I.: *Z. angew. Chem.* (1902) **15** 198
334. Thomson, R. T.: *J. Soc. Chem. Ind.* (1893) **12** 432
335. Winkler, L. W.: *Z. angew. Chem.* (1913) **26** 231
336. Devarda, A.: *Chem. Ztg.* (1892) **16** 1952
337. Győry, István (1860—1954) studied pharmacy and became an assistant lecturer at the University of Budapest. Later he became Professor of Chemistry at the Academy of Horticulture, and afterwards worked in the Hungarian agricultural state management.
338. Győry, J.: *Z. anal. Chem.* (1893) **32** 415
339. Schulek, E.: *Z. anal. Chem.* (1934) **97** 186
    Schulek, E.—Rózsa, P.: *Z. anal. Chem.* (1939) **115** 185
340. Knecht, Edmund (1861—1926). Although he was born in England his father was Swiss and he studied and graduated from the University of Zürich. In 1883 he returned to England and worked at the Technical College in Bradford. From 1890 he was Professor at the College of Technology in Manchester until his death. He dealt mainly with problems related to the dyeing of textiles.
341. Knecht, E.—Hibbert, E.: *Ber.* (1903) **36** 1549
342. Young, Philena, was born in 1896 in Goshen (U.S.A.). She worked at several academies in America, and at present is Professor of Chemistry at the Wells College in New York.
343. Willard, H. H.—Young, P.: *J. Am. Chem. Soc.* (1928) **50** 1322 1334
344. Atanasiu, J. A., was born in 1894. He graduated at the Technical University of Bucharest, where he continued to work. Since 1945 he has been the Professor of Electrochemistry there.
    Atanasiu, I. A.: *Bull. Chim. Soc. Romane* (1927) **30** 73
345. Szabó, Zoltán, was born in Debrecen (Hungary) in 1908, and graduated at the University of Budapest. In 1940 he was appointed Professor of Analytical Chemistry at the University of Kolozsvár and since 1945 he has held a similar post at the University of Szeged. Since 1965 he has been Professor of Analytical Chemistry at the University of Budapest.
346. Szabó, Z. G.—Sugár, E.: *Anal. Chem.* (1950) **22** 361; *Anal. Chim. Acta* (1952) **6** 293
347. Lang, Rudolf, was born in 1887. He was Professor of Analytical Chemistry at the Technical Highschool in Brno. At present he lives in Germany.
348. Lang, R.—Gottlieb, S.: *Z. anal. Chem.* (1936) **104** 1
349. Sirokomski, Vitold Sigsmundovich, was born in 1892. He graduated at the Politechnicum in St. Petersburg, and from 1920 he worked at the Ural Mining and Mineralogical Institute in Sverdlovsk. In 1927 he became the head of the Institute for Chemistry and Technology at the University in Sverdlovsk, and from 1936 he was also in charge of the Institute for Analytical Chemistry of the Soviet Academy of Sciences in the Ural. He died in 1951.

350. Sirokomski, V. S.—Stepin, V. V.: *Zav. Lab.* (1936) **5** 144
351. Sirokomski, V. S.—Melamed, S. I.: *Zav. Lab.* (1950) **16** 131 273
     Sirokomski, V. S.—Knazeva, R. N.: *Zav. Lab.* (1950) **16** 1041
352. Malaprade, L. M., was born in 1903 in Dieppe. He graduated in Nancy, and afterwards worked at the university there. Since 1959 he has been Professor of Analytical Chemistry. His method mentioned was published: *Bull Soc. Chim.* (1928) **43** 683
353. Noll, A.: *Chem. Ztg.* (1924) **48** 845
354. Erdey, László, was born in 1910 in Szeged (Hungary), and graduated at the University of Budapest. He then worked at the University and at the Municipal Chemical Institute in Budapest. Since 1949 he has been the Head of the Institute for General Chemistry at the Technical University of Budapest.
355. Erdey, L.: *Magyar Kém. Folyóirat* (1950) **56** 262
356. Erdey, L.—Bodor, A.: *Anal. Chem.* (1952) **24** 418; Erdey, L.—Bodor, A.—Buzás, I.: *Z. anal. Chem.* (1951) **133** 265; **134** 22 412; Erdey, L.—Buzás, I.: *Acta chim. hung.* (1954) **4** 195
357. Bradbury, F. R.—Edwards E. G.: *J. Soc. Chem. Ind.* (1940) **59** 96
358. Belcher, Ronald, was born in Nottingham in 1909, studied at the University of Sheffield and later in Graz at the Pregl Institute where he became interested in microanalysis. He then worked in the Universities of Sheffield and Aberdeen and in 1948 became lecturer in analytical chemistry at the University of Birmingham where in 1959 he was appointed Professor of Analytical Chemistry to the first Chair of Analytical Chemistry in Britain.
359. Burriel-Marti, Fernando, was born in 1905, and studied in Madrid and Brussels. He was the Professor of Analytical Chemistry at the Technical University in Madrid from 1932, and since 1945 at the University in Madrid.
360. Belcher, R.—West, T. S.: *Anal. Chim. Acta.* (1951) **5** 260
361. Burriel-Marti, F. M.—Lucena, F. C.: *An real. soc. fis. quim.* (1951) **47B** 257
362. Engel: *Bull. soc. ind. Mulhouse.* (1895) **65** 61; Russel, A. S.: *J. Chem. Soc.* (1926) 497
363. Rao, U. V.—Muralikrishna, U.—Rao, G. G.: *Z. anal. Chem.* (1955) 145 12
364. Purgotti: *Gaz. chimital.* (1896) **26** 2 197; Tourky, A. R.—Farah, M. Y.—Elshamy, H. K.: *Analyst.* (1948) **73** 258
365. Winkler, C.: *J. prakt. Chem.* (1865) **95** 417; Belcher, R.—Gibbon, D.—West, T. S.: *Anal. Chim. Acta.* (1955) **12** 107
366. Dimroth, Otto (1872—1940) began his chemical career in industry, and then became assistant to Baeyer in München. In 1913 he became a Professor at the University of Greifswald, and in 1918 at the University of Würzburg.
367. Dimroth, O.—Fister, F.: *Ber.* (1922) **55** 3693
368. Fischer, K.: *Z. angew. Chem.* (1935) **48** 394
369. Schwarzenbach, Gerold, was born in 1904. He graduated at the Technical University in Zürich, where he worked from 1929. In 1947 he was appointed Professor of Analytical Chemistry in Zürich, and since 1955 he has been the director of the Institute for Analytical Chemistry at the Technical University in Zürich. Talanta Prize winner for Analytical chemistry 1963.
370. Schwarzenbach, G.: *Helv. Chim. Acta.* (1946) **29** 1338; Schwarzenbach, G.—Biedermann, W.—Bangerter, F.: *Helv. Chim. Acta* (1946) **29** 811
371. Pribil, Rudolf, was born in 1910. He graduated at the University of Prague. From 1930 he worked in the Analytical Institute at the Charles University in Prague until the University was closed down in 1939. For a time he worked in industry, but in 1946 he returned to the University. Since 1955 he has been directing one of the research departments of the Czechoslovakian Academy of Sciences.
372. Welcher, F. J.: *The Analytical Uses of Ethylendiamine tetraacetic Acid.* London—New York (1958)
373. Mylius, Franz (1854—1931) was a pharmacist, but sold his inherited pharmacy in order to devote his time to his scientific interests. He later worked at the University

of Freiburg as an assistant lecturer. In 1898 he was invited by Hoffmann to become head of the chemistry department of the newly established Physikalisch-Technische Reichsanstalt and he remained at this post until he retired in 1924.

374. FÖRSTER, FRITZ, (1866—1932) graduated in Berlin and then worked at the Physikalisch-Technische Reichsanstalt. In 1895 he went to the Technical University in Dresden, where he became Professor of Electrochemistry in 1908, and later in succession to Hempel he became Professor of Inorganic Chemistry and Technology.
375. MYLIUS, F.—FÖRSTER, F.: *Ber.* (1891) **24** 1482
376. PILCH, FRITZ, was born in 1887 in Graz, where he also graduated. After this he worked in industry. At present he is in charge of the library at the University of Graz.
377. PILCH, F.: *Monatshefte.* (1911) **32** 21
378. BANG, J.: *Mikromethoden zur Blutuntersuchung.* München (1922)
379. LOWITZ, T.: *Crells. Ann.* (1796) **1** 125 196 429
380. LOWITZ, T.: *Crells. Ann.* (1790) **1** 206 300
    LOWITZ, T.: *Crells. Ann.* (1793) **1** 220
381. PELOUZE, T. J.: *Ann. chim. phys.* (1832) **50** 314 434
382. CADY: *Journ. Phys. Chem.* (1897) **1** 707
383. WALDEN, PAUL (1863—1957) was born in Lithuania. He graduated in Riga at the same time as Ostwald, whom he succeeded at the Technical University. He dealt mainly with physicochemical problems. From 1910 he was head of the chemistry department at the St. Petersburg Academy of Sciences in succession to Butlerov, until in 1918 when he accepted the chair of chemistry at the University of Rostock. He retired in 1934, but gave lectures at the University of Tübingen as a visiting professor. He made important contributions also to the study of the history of chemistry.
384. WALDEN, P.: *Ber.* (1899) **32** 2862
385. VORLÄNDER, DANIEL (1867—1941) was Professor of Chemistry at the University of Halle. He dealt primarily with organic chemistry.
386. VORLÄNDER, D.: *Ber.* (1903) **36** 1485
387. FOREMAN. F. W.: *Biochem. Journ.* (1920) **14** 451
388. TOMIČEK, OLDRICH (1891—1953) Professor of Chemistry at the University of Prague
389. GANTIER, JEAN ALBERT, born 1903 at Pithiviers, since 1948 Professor of Organic Chemistry at the Faculty of Pharmacy in Paris
390. PÉLLERIN, FERNAND, born 1923 in Fécamp, has worked at the Paris University since 1964 as Professor of Chemistry at the Faculty of Pharmacy in Rouen
391. RÁDY, GYÖRGY, born 1923, dozent at the Technical University, Budapest

CHAPTER IX

ELEMENTARY ORGANIC ANALYSIS

1. FROM LAVOISIER TO LIEBIG

Organic chemistry has been distinguished from inorganic chemistry since the 17th century. Lémery divided substances occurring naturally according to whether they were of mineral or animal/plant origin [1]. The extent of chemical knowledge about the latter group, however, was very limited. Scheele was the first to carry out any systematic examination of the chemical constitution of plants, and in the course of this work he discovered malic acid, citric acid, tartaric acid, oxalic acid, glycerine, gallic acid, uric acid, and lactic acid, etc. On the basis of this work Scheele can be regarded as the founder of scientific organic chemistry. In the following decades organic compounds were regarded as rather curious substances which would not obey the current laws of chemistry. It was generally considered that they were only produced by living organisms, or more specifcly by a special vital power *(vis vitalis)*.

The foundation of the theories of the *Vitalisos* was demolished in 1828, with Wöhler's synthesis of urea from cyanic acid and ammonia [2]. Wöhler records that this was accomplished without the need of "any kidney, or animal, man or dog" [3]. A few people still believed in the existence of a vital power and explained Wöhler's synthesis on the grounds that the reacting materials cyanic acid and ammonia required the vital power for their own formation. In 1845 however, Kolbe [4] prepared acetic acid from its elements[5] this experiment being carried out with the intention of disproving this latter opinion. He implied that Wöhler's synthesis was an accident, and commented that Wöhler went away like "Saul, Ki's son, to find his monkey, and instead of this he found a kingdom". The Vitalist Theory was finally abandoned. Regarding the composition and structure of organic compounds very little was known at the beginning of the 19th century, and it was not until sufficiently accurate methods of analysis for organic compounds had been devised that any progress could be made. The maxim that without analysis there can be no synthesis is proved once more, and it was not until the methods of elementary analysis had been developed that any great developments in synthetic chemistry became possible.

The early experiments of van Helmont showed that water was formed when organic substances where subjected to combustion, a fact that he used to support his theory that the ultimate composition of all substances is water. Soon after the discovery of carbon dioxide it was discovered that the combustion of coal and other organic substances led to the evolution of this gas. The theory of combustion was thus disclosed, water is formed from the combustion of hydrogen, and carbon dioxide from carbon and so the conclusion was obvious: organic compounds are composed of carbon and hydrogen. Lavoisier considered that organic compounds are composed of carbon, hydrogen and oxygen. Berthollet [6] was

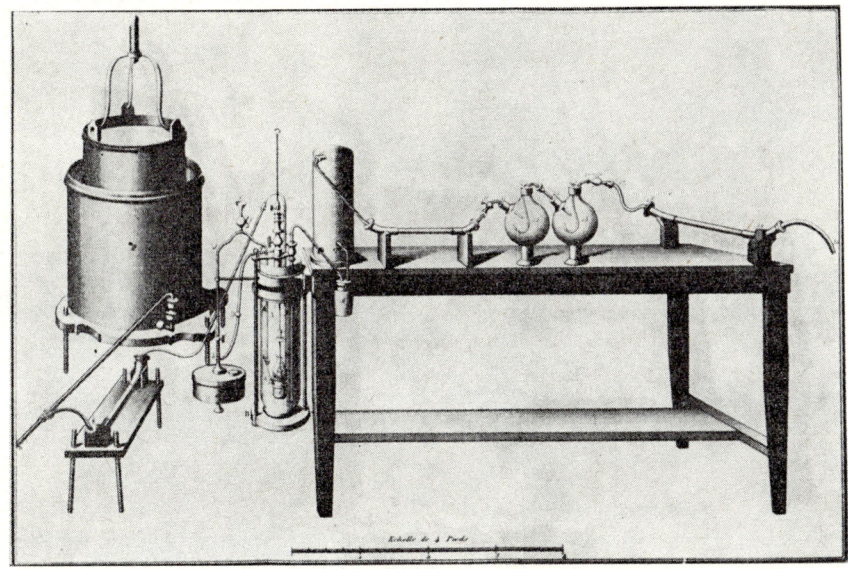

FIG. 72. Lavoisier's first apparatus for combustion analysis of organic substances. Drawing by Mme Lavoisier in the book: *Traité élémentaire de chimie* of Lavoisier (1789)

the first to discover nitrogen in organic compounds, in animal extracts, at the end of the 18th century.

The first experiment to determine the composition of organic substances was carried out by Lavoisier. He examined, first of all, various oils using a rather complicated apparatus [7]. He combusted the oil in a burner, and then led the combustion products through a cooler and then through a calcium chloride tube before absorbing in sodium hydroxide (Fig. 72). The results obtained by this method, however, were not very reliable, Lavoisier stating that:

> The difficulties encountered with this method were so great that it was impossible for me to obtain accurate results. I was able to prove that oils are transformed into water and carbon dioxide; in other words they are composed of carbon and hydrogen, but I could obtain no information about their respective ratios [8].

He did not attempt to combust alcohols in this apparatus, because of the danger of explosion, but constructed a simpler apparatus (see Fig. 73) for this purpose.

He placed a small burner under the bell *A*, which he sealed with mercury. On the wick he placed a small piece of phosphorus, which he lit with a heated bent iron rod, introduced through the mercury so that the alcohol was combusted. The bell *B* contained oxygen. He measured the gas volumes accurately before the combustion, and when the intensity of burning began to decrease, he introduced oxygen by opening the tap. The burning ceased when the amount of carbon dioxide reached considerable proportions. By reweighing the burner the amount of alcohol combusted was found. The combusted gases were passed through alkali hydroxide

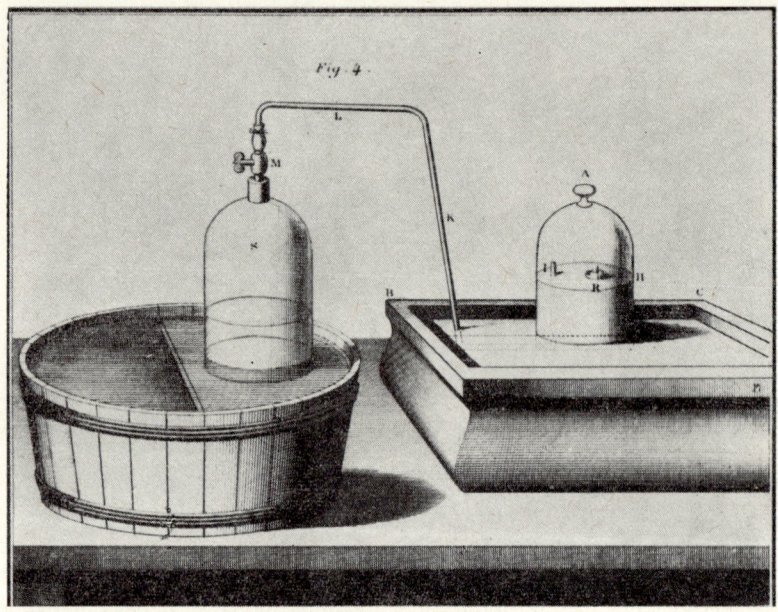

FIG. 73. Lavoisier's simple device for analysis of organic substances. Drawing by Mme Lavoisier in the book: *Traité élémentaire de chimie* of Lavoisier (1789)

(no details given) and the amount of carbon dioxide determined from the decrease in volume. Because the original volume of oxygen was known it was possible to calculate the proportion of this which was present as carbon dioxide, and from the difference in the two amounts the quantity of water formed [9].

Lavoisier mentioned that

The experiment is not successful in all cases, slight accidents often occur, and among the most frequent of these is the breaking of the bell ... From among a large number of experiments there was only one with which I was completely satisfied from all points of view [10].

According to his results the ratio of hydrogen to carbon in alcohol is 3·6 : 1, while the correct value is 4 : 1. The accuracy of this determination was quite satisfactory, especially if we consider that this was the first analysis of this type.

Lavoisier also used this method to analyse a wide variety of other substances, for example, oils, fats, waxes and ether, but it is impossible to assess the accuracy of the results as the quality of the substances analysed are not known.

Lavoisier made a further attempt to discover the composition of organic substances, using instead of oxygen or air, certain metal oxides such as mercury

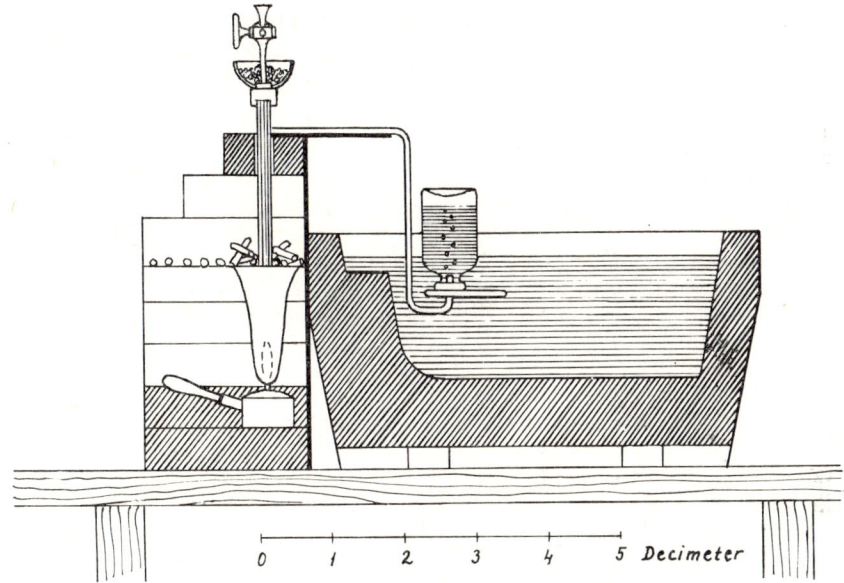

Fig. 74. Apparatus for combustion and analysis of organic substances of Gay-Lussac and Thénard. (From their *Recherches physico-chimiques* [1810])

oxide, manganese dioxide and potassium chlorate, in which the oxygen content is readily available. He attempted to determine the composition of sugar by this method, but the experiments were not successful and were never published. They were, however, recorded in his laboratory notebook which was published more than fifty years after his death, by the French authorities, in the complete works of Lavoisier [11].

Although Lavoisier only attempted to analyse a small number of organic substances, and the results he obtained are often no more than approximations to the true values, his main achievement was that he showed the correct approach to the practical methods of organic elementary analysis. Subsequent methods were essentially improvements of Lavoisier's methods. His work is also of interest when one contrasts it to that of his contemporary analysts, such as Klaproth, who attempted to analyse organic substances simply by dry distillation, followed by determination of the amounts of gases, oils and residue formed during the process.

After the death of Lavoisier only a few attempts were made to devise methods of analysis of organic substances. Berthollet distilled the substances in a retort and then passed the decomposition products through incandescent porcelain tubes, and after condensing the water in ice-cooled receivers the carbon dioxide was absorbed in alkali hydroxide solution. Any ammonia formed (for example if animal substances were combusted) was adsorbed in water [12].

Gay-Lussac and Thénard [13] in 1810 devised a technically different method, carrying out the combustion in tubes. The apparatus is shown in Fig. 74 [14].

Combustion was carried out in a vertical glass tube, 20 cm long and 8 mm in diameter, which was heated with a spirit lamp. To the upper end of the tube a tap was attached, which instead of the channel, had a small hole drilled in it. Into this hole the sample mixed with potassium chlorate was placed, and rotation of the tap caused the sample to fall into the heated part of the tube. To remove air from the apparatus a blank combustion was carried out beforehand. The combustion products were led in a side tube through mercury to a sealed vessel. Carbon dioxide was absorbed by potassium hydroxide in the mercury vessel, and the amount determined from the volume decrease of the gas. After the absorption of carbon dioxide the excess of oxygen was exploded with hydrogen, and the oxygen loss was determined. The original amount of oxygen was known; as in a previous experiment the amount of oxygen formed from potassium chlorate was determined. The decrease in the amount of oxygen was considered to be due to the combustion of the hydrogen content of the substance, and thus the hydrogen content could be determined in this indirect manner.

In principle the method can only yield accurate results provided that the original substance does not itself contain oxygen. In practice, several other sources of error existed, the most important of which was the fact that the combustion was so rigorous that particles of the sample were exploded out of the combustion zone to condense on the cooler parts of the tube. The reaction was occasionally so violent that the combustion tube shattered, but, in spite of all these difficulties, Gay-Lussac and Thénard examined about twenty organic substances and obtained reasonably accurate results.

Berzelius used a horizontal tube and heated it uniformly over its length, and so was able to combust the whole of the sample. And in order to decrease the vigorousness of the reaction he mixed sodium chloride with the potassium chlorate.

He weighed out 0.3—0.5 g of the sample, and added a 5—6 fold amount of potassium chlorate and a 9—10 fold amount of sodium chloride, and mixed these with the sample. Berzelius took every care to ensure that his samples were perfectly dry, and that they did not absorb water during the weighing and mixing process. He recommended the use of gloves, and also that the face should be averted from the mortar in which the mixing is carried out. The combustion tube was enclosed with zinc foil and iron wire to eliminate the breaking of the glass, and a 2 cm layer of potassium chlorate—sodium chloride mixture placed inside before the addition of the sample.

Another important improvement that Berzelius made was to weigh the water formed directly instead of using the indirect method of calculating from the consumption of oxygen. The apparatus is shown in Fig. 75. Most of the water condensed in the first receiver, the remainder being absorbed by a calcium chloride tube. The carbon dioxide was measured in the bell volumetrically, although it was possible to measure this gas from the increase in weight of a small vessel of potassium hydroxide which had been allowed to stand inside the bell for 24 hours [15]. Berzelius's method was much more accurate than those of his predecessors, although it still had a few faults. Liebig was of the opinion tha

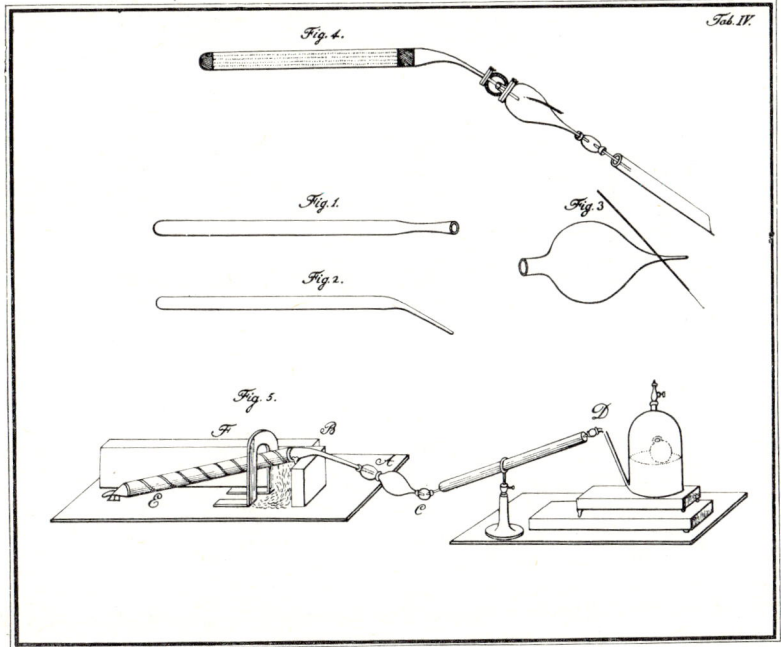

FIG. 75. Combustion apparatus of Berzelius. (From Pfaff: *Handbuch der analytischen Chemie* [1821])

the method was too slow, adding that Berzelius took eight months to carry out thirteen analyses. Berzelius was rather offended and replied that it was not the analyses that took up the time, but the preparation of the compounds in a sufficient state of purity.

Gay-Lussac later discovered that copper oxide was far more efficient than potassium chlorate [16], and in an independent investigation Döbereiner [17] reached the same conclusion. An extremely simple combustion apparatus using copper oxide was designed by Döbereiner (Fig. 76), but never came into use [18].

The method suggested by Prout is of interest, in which the combustion is carried out in oxygen in a vertical combustion tube. This method is difficult to understand, but from Liebig's description we know that the initial and final volumes of oxygen were measured, and from the difference it was possible to calculate the amounts of carbon and hydrogen. When carbon is combusted the volume of carbon dioxide formed will be equal to the decrease in the volume of oxygen so that the total volume will remain constant. When, however, hydrogen is combusted, each volume of hydrogen combines with one half volume of oxygen so that the total volume decreases. Three possibilities occur. Firstly, if the total volume remained unchanged, then the ratio of hydrogen : oxygen in the substance was the same as in water; secondly, if the volume became less then the hydrogen was in excess, while thirdly if the volume increased then the amount of hydrogen was less than in water [19]. On the basis of this argument (the reasoning seems

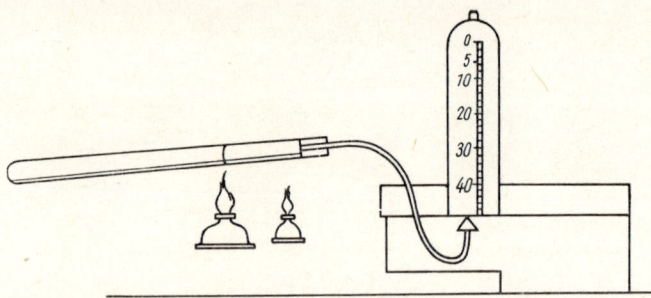

Fig. 76. Combustion apparatus of Döbereiner from 1816.
(From the original paper)

a bit obscure for a present-day reader) Prout recorded the composition of a variety of organic compounds (sugar, oxalic acid, tartaric acid, citric acid, etc.). The results he obtained were unexpectedly precise, and if it were not for the fact that Prout was the first person to analyse these compounds they would be rather suspicious. It would appear that Prout was very adept at using this method and that no one could match his skill, for a chemist by the name of Bischof [20] attempted to employ the method and angrily reported that it was impossible to use [21].

Prout's method is mentioned solely out of interest, as it was to have no influence on the further development of organic analysis which was continued on the basis of Berzelius's procedure.

A piece of apparatus which is in common use today, the U tube was designed by Bussy, who used it for the first time in 1822, filled with anhydrous calcium chloride [22].

In the next section, dealing with the period from Liebig to Pregl, we can see how the determination of carbon and hydrogen was improved by Liebig.

2. FROM LIEBIG TO PREGL

Justus Liebig as well as being one of the outstanding chemists of the last century, was also a very interesting character. He was born in 1803, and his father was a merchant who dealt in medicinal plants and herbs. He must therefore have become acquainted with pharmaceutical and chemical substances at a very early age. Liebig had ten brothers and sisters, and his family was not very rich. Liebig did not show much ability during his early educational training and his teachers had no great hopes for his future. He failed his examinations consistently so that eventually his father took him away from the school and apprenticed him to a pharmacist. We have already seen that many great chemists of those days started their careers in this way, but Liebig showed no signs of following in their footsteps. Even so he was interested in chemistry and read books on the subject, often through the whole night.

On one occasion one of his experiments exploded and he almost demolished the whole house where he was working, as a result of which he lost his job. He returned to his father's house where his father attempted to employ him in his shop. This did not appeal to Liebig who entreated his father to allow him to continue his studies of chemistry. He enrolled at the University of Bonn (no entrance examination was required in those days) and studied there, and later at Erlangen. It was here that he became acquainted with Platen, the famous poet, who recommended Liebig for a fellowship in Paris. This excursion was very useful to Liebig who realized how much more advanced was the study of chemistry in France than in Germany. The method of teaching in France was entirely different and research work flourished. In this modern world the dull German natural philosophies had no place, and only practical experimental evidence was valid. Liebig worked with Thénard, where he became acquainted with organic analysis. He worked very

Fig. 77. Justus Liebig (1803—1873). Trautschold's drawing from 1845. (From Kopp: *Geschichte der Chemie* [1847])

diligently and obtained some interesting results. Even Humboldt [23], the famous German explorer, became interested in his work. Humboldt spent most of his time in Paris and introduced Liebig to Gay-Lussac. Liebig became an assistant to the latter, but after two years in Paris he returned to Germany. Humboldt also assisted him in later life, for as a result of his recommendation the reigning prince gave him the Chair of Chemistry at the small University of Giessen, without any previous consultation with the staff. Liebig was only twenty-four years old at the time and was rather resented by the rest of the staff, but he was later to justify his appointment.

Liebig began to work, despite interferences, with great enthusiasm and introduced laboratory instruction to the University and organized the practical course. Because of lack of money he was forced to start his work in a small room in an old military barracks. This small laboratory soon became very crowded with his students who rapidly increased in numbers as the reputation of Liebig's teaching method widened.

Liebig worked tirelessly from morning till night in spite of the unfavourable circumstances. The annual grant for his institute was one hundred thalers. (We have already seen that the charge made by Mohr for one litre of 0·1 N silver nitrate was one thaler! Chapter VIII. 4.) A brief extract from his letter illustrates the conditions under which he had to work.

I shall repeat the analysis more accurately as soon as the water in my laboratory thaws.

Liebig published a great deal, but as he was a sharp but objective critic of the work of others he made many enemies. Berzelius once advised him:
Dear Liebig, please do not be a chemical executioner. You stand highly enough, so you do not need to suppress your fellows [24].

His friend, Wöhler, also advised him not to be too critical, and when Liebig attacked Mitscherlich, he wrote:

What good comes from these debates? Nothing, really nothing. You make some difficulties for Mitscherlich, the public laughs, while you yourself become more and more bitter, and your health becomes worse [25].

According to the evidence of his letters, Liebig's mood changed frequently, on some occasions he worked with feverish energy, while on others he was very depressed and his work suffered accordingly. Several extracts from letters written to Wöhler and Berzelius illustrate this aspect of his character.

About my spirits I would not like to tell you. But I am almost sick of my life and often imagine that a shot or the cutting of my throat would help me in some cases. Oh, if only my unhappy life contained a few gay moments! If I did not have a wife and three children I would be more content with a portion of hydrogen cyanide than with this life! To be truthful I hate chemistry... and this terrible writing of books, which takes me to complete despair: I will write no more books, even if they would give me mountains of diamonds!..

Wöhler attempted to reassure him:

Dear friend, you suffer from the special disease of chemists, the so-called *hysteria chemicorum*, which is caused by the hard mental work, bad laboratory air and unlimited ambition. All great chemists suffer from this disease.... You are wasting yourself and your health! Think of 1900, when we shall be once again carbon dioxide, water and ammonia... and what was once our bones will be part of the bones of a dog, which... Who will worry then whether we had lived gaily or in bad temper? Who will know about your chemical debates? Nobody. The facts which you have disclosed, however, will always be known. But how do I attempt to advise the lion to eat sugar? [25].

His friends, however, did not take Liebig's temporary hysterical outbursts very seriously, they knew that he would soon be back to normal when some new aspect of research occupied his attention.

It is probable that Liebig's irritability was increased by his poor financial position, because in spite of his reputation his income was never very great. He had a large family and therefore was always looking for new sources of income. His attempts at writing popular books was largely for this purpose, and he also founded a chemical fertilizer factory, and plants for producing infant food, baking powder, and coffee extract, but none of them was successful. Finally, in his old age one of his ideas proved to be a financial success. He had written in one of his books that cattle in South America are slaughtered only for their hides, while the meat is left to rot on the pampas. Liebig suggested that it would be possible to use up this wasted meat and to sell it in the form of meat extract. Someone followed up this idea and invited Liebig, because of his reputation, to become a partner in the project in Argentina.

The Liebig meat cube was the result of this enterprise. With time Liebig's enthusiasm for teaching lapsed.

> Working with young men, which was a great joy before, is now a terrible job for me; a question or an explanation makes me almost miserable. [25].

Liebig was offered posts at several well-known universities, and in 1852 he accepted a post at the University of München, with the stipulation that he should not be required to undertake any teaching duties, which the university accepted. He lived in München, and held frequent lectures, but his interest in research waned. In 1872 he wrote to his old friend Wöhler...

> I read scarcely any chemical literature. How can an old interest be extinguished to such an extent?

Liebig is one of the most important persons in the early history of organic chemistry. He carried out analyses on an immense number of organic substances which had never previously been examined, and the explanation of the valency

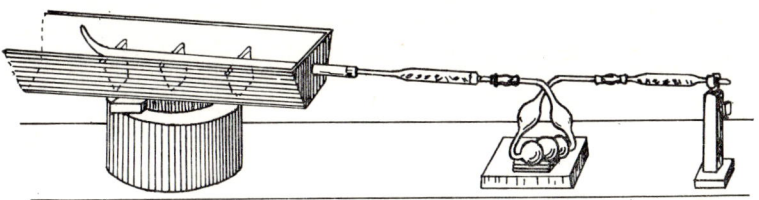

FIG. 78. Combustion apparatus of Liebig. (From his work: *Anleitung zur Analyse organischer Körper* [1837])

of acids, as well as the theory of organic radicals were also due to his work. A number of famous organic chemists studied under him. His contributions to agricultural chemistry are also notable, for example, Liebig discovered that plants obtain the carbon required for their growth from the air, and that fertile soil requires nitrogen and phosphorus. He also recognized the possibility and importance of the production of chemical fertilizers.

The determination of carbon and hydrogen in organic elementary analysis s still carried out today using Liebig's principles.

The original apparatus of Liebig is shown in Fig. 78 [26]. It has three important improvements over that of Berzelius. Heating was carried out with heated coal, but the container for the combustion-tube heater was divided into several parts so that the heating could be varied by adjusting the amount of coal. Carbon dioxide was measured by weight and not by volume. The small triangle-like glass apparatus contained potassium hydroxide, which absorbed carbon dioxide, while in the horizontal tube before the CO_2 absorber, anhydrous calcium chloride was packed to absorb the water.

An old problem of this type of procedure was the removal of the combustion products after the reaction, but in Liebig's apparatus this is overcome by breaking off the small end of the combustion tube, and sweeping air through the system. Liebig stated that the error involved from the CO_2 in the air was compensated by the loss of water vapour from the potassium hydroxide absorption vessel.

Berzelius in a later procedure passed the air through a potassium hydroxide vessel before sweeping out the combustion tube. This improvement, however, was not accepted by Liebig, and they had bitter arguments over this question, and also as to whether rubber tubing or cork stoppers were the most suitable for connecting the tubes. Liebig recommended cork stoppers, while Berzelius used rubber tubing. They both considered that the substance used to connect the tubes could not have constant weight and would therefore lead to errors in the weighing. In practice, however, rubber was found to be the most suitable. Liebig recommended the use of lead chromate for the combustion of halogen-containing samples; this was simply added to the other oxidant.

A number of suggestions for improving the Liebig method was made during the last century, but there has been no alteration of the basic principle. Winkler raised objections to the use of calcium chloride on the grounds that it becomes basic on ignition, and therefore absorbs carbon dioxide also. He recommended the use of sulphuric acid instead of calcium chloride, and constructed an apparatus for the absorption of water which was in use for a short time [27]. It was subsequently discovered that sulphuric acid causes even higher errors because it does not have a constant weight, and also that calcium chloride only becomes basic on heating if it contains any magnesium chloride.

With the widespread use of gas lighting and the invention by Bunsen of his gas burner [28] in 1857, the use of coal as a source of heat was abandoned.

Soda-lime recommended by Mulder in place of potassium hydroxide solution became widely used. This absorbent could be packed, in solid form, into U tubes. Mulder filled seven-eighths of the tube with soda lime, and the remaining one eighth with calcium chloride to retain the water which was liberated from the former [29].

With certain compounds, however, it was found that the carbon did not combust to carbon dioxide, but remained in the form of elementary carbon. In order to avoid this error, Lavoisier's original suggestion of combustion in an oxygen atmosphere was applied. Thus for the analysis of starch, Brunner [30] in 1832 attempted to use oxygen [31]. Hess [32] also used oxygen for the combustion some years later, and was the first to use a small vessel for the introduction of the sample [33], and not as previously to add the sample directly. It was Kopfer [34] who discovered the advantages of using a catalyst with the oxygen combustion. He employed platinum for this purpose, first in the form of platinum black, and later as platinized asbestos. In the determination of halogen-containing samples he also mixed silver fibres with the catalyst [35].

The accuracy of carbon determinations was limited by the accuracy with which the composition of carbon dioxide was known. According to the experiments of Lavoisier the ratio of carbon to oxygen was 800 : 311 in this compound, while Berzelius made a correction to this value, finding the ratio to be 800 : 306. This was accepted for a long time, although it was often found that in the analysis of oxygen-free compounds the sum of the carbon and hydrogen determinations exceeded 100 per cent. Dumas reinvestigated the question in 1840 and found the ratio to be 800 : 300, which corresponds to the correct value [36].

The nitrogen content of organic substances was first determined by Gay-Lussac and Thénard, using the method described previously where the gas remaining after the explosion of the excess of oxygen was considered to be nitrogen. Bussy tried to eliminate errors due to the formation of nitrogen oxide by leading the gases through a copper layer which had been formed in the apparatus of Berzelius from copper oxide [37]. After many experiments and modifications Liebig finally evolved a method for the determination of nitrogen, carbon and hydrogen using a separate sample and apparatus. The nitrogen and carbon dioxide were collected in a eudiometer over mercury (water had previously been absorbed), the carbon dioxide absorbed in potassium hydroxide and the nitrogen measured volumetrically. The method was very tedious, first because of the difficulty in removing air from the apparatus, and second because of the difficulty in ensuring the complete removal of the combustion products. In the sealed end of the combustion tube Liebig placed a vessel filled with a suspension of calcium hydroxide which he heated to form water vapour which was used to flush out the tube.

In 1831 Dumas devised a method whereby the gases were collected in a eudiometer tube filled with alkali hydroxide thus eliminating the need for separate absorption of carbon dioxide. The other important innovation in this procedure was the use of carbon dioxide, generated by heating lead carbonate, to flush out the combustion products [38]. Dumas used copper for the reduction of nitrogen oxides after first activating the copper by heating in a stream of hydrogen. This method for the determination of nitrogen is still in use.

Jean Baptiste Dumas was born in 1800, in Alais. Just as many famous chemists before him had done, Dumas started his career as apprentice to a pharmacist. He later became an assistant pharmacist in Geneva, a town where there was an enthusiastic interest in science. Dumas attracted considerable interest with some of his experiments and became quite famous in the small community. When a physicist of Geneva recommended the newly discovered iodine as a cure for goitre Dumas prepared a solution of iodine in potassium iodide as a medicine, so that his pharmacist's shop did a flourishing trade. Humboldt [23], who was quick to spot a promising young scientist, met Dumas when he was passing through Geneva, and persuaded him to go to Paris where he assisted him in his career as he had done earlier with Liebig. Dumas became assistant to Thénard at the École Polytechnique, and in 1826 he married the daughter of Brogniart, who was a well known geologist and the director of the porcelain factory at Sèvres. During this time he developed his method for vapour density measurement, as a result of which he became famous. He also discovered methyl alcohol in 1824.

At one of the receptions given by King Charles X the candles did not burn properly, giving off clouds of soot and a gas which caused the guests to choke. Dumas's father-in-law, Brogniart, who was asked to investigate the reason for this occurrence, turned to Dumas for help. Dumas discovered that hydrogen chloride gas was formed because the candle maker had used chlorine to whiten the wax for his candles so that traces remained in the finished article. As a result of this discovery Dumas examined the reactions of chlorine and bromine with other organic compounds and found that in certain cases substitution of a halogen for hydrogen takes place. This conclusion was not in accordance with Berzelius's dualistic theory, which maintained that it would be impossible to substitute a positively charged radical by a negatively charged one. Fierce argument took place between Berzelius and Dumas, but further experiments vindicated Dumas's findings. As a result of this Berzelius bore a grudge against Dumas and was always in later years prejudiced against the discoveries of the latter. It is true to say, however, that Berzelius in his type theory drew conclusions which were not supported by his experimental results, and when Dumas found that certain atomic weights calculated by Berzelius did not agree with his own findings from vapour density measurements, he dismissed all of Berzelius', atomic weights as being incorrect, and accepted the hypothesis of Prout.

Dumas's reputation during this time increased and he became Professor at the College de France and a member of the Academy, and finally he succeeded Gay-Lussac at the Sorbonne. He engaged in much polemics with Liebig. Eventually Dumas became the leading figure in French chemistry, and he then dabbled in politics, becoming for a short time Minister for Agriculture and Trade. When Dumas died in 1884 he was given a state funeral, such was his importance to France.

Fig. 79. Jean Baptiste Dumas (1800–1884). (From Bugge: *Buch der grossen Chemiker*)

Since the work of Dumas the method for the determination of nitrogen has changed very little, the only noticeable alteration being the replacement of lead carbonate by a variety of other carbonates. Zulkovszky formed the CO_2 from a mixture of sodium carbonate and potassium dichromate in a separate vessel [39]. Erdmann [40] and Marchand [41] tried to prepare carbon dioxide separately and to lead the gas into the combustion tube [42]. Although they were not successful Schiff, twenty years later, devised a similar method which was a success.

In the method of Dumas it was very difficult to fill the combustion tube and then to transfer the combusted gases over mercury into a vessel to measure the volume. Many contemporary workers record the difficulty of this method, and Dennstedt later wrote "The nitrogen determination was a punishment for everybody, until azotometers were introduced!" The first azotometer which proved to be of use was introduced by Schiff [43].

Owing to the inaccuracies and difficulties encountered in the determination of nitrogen by combustion methods, interest was turned towards the possibility of using a wet method of analysis. Methods of this type are based on the principle that when certain organic nitrogen-containing substances are ignited either alone or in the presence of other substances such as alkalis, the nitrogen is split off in the form of ammonia. The ammonia determination thus gives the nitrogen content.

Dumas was the first to suggest the use of this reaction (which had previously only been used for qualitative investigation) for quantitative purposes. In 1831 in a letter to Gay-Lussac he mentions that certain compounds, for example oxamide and urea, give off ammonia when heated with alkali and he suggests that this gas could be determined volumetrically. He did not investigate this suggestion further until ten years later when he developed a method [44], but by this time two students of Liebig, Varrentrapp [45] and Will [46] had developed a much better method [47]. The method was called after them. The procedure was based on the ignition of the compound with barium hydroxide, the ammonia which was liberated being passed into hydrochloric acid, and determined gravimetrically by precipitation as ammonium hexachloroplatinate.

As was mentioned in the earlier chapter on volumetric analysis, Péligot modified the method so that the ammonia was absorbed in a known amount of hydrochloric acid, the excess then being back-titrated [48]. For the back-titration Péligot used lime dissolved in water containing sugar.

This method soon became widely used, and almost displaced the Dumas method entirely during the last quarter of the 19th century. It was later discovered that the method was not quite as accurate as had previously been imagined, as part of the nitrogen is not converted to ammonia, but is lost in the form of nitrogen. Several modifications were made to the original method to try to overcome these difficulties, but they were eventually unnecessary as the new method introduced by Kjeldahl was far superior to all others.

Varrentrapp and Will carried out the ignition in a combustion tube, and certain alterations were made by replacing the tube with a flask; this was first done by Seegen in 1864. He also used lime water instead of barium hydroxide [49], but his method was only successful in the case of easily decomposable substances. In the succeeding years many other substances were recommended for assisting the decomposition, but none of them proved to be useful except in some isolated and special cases. Amongst many others potassium permanganate was tried in acid medium. In 1883 Kjeldahl attempted to use this method for the determination of the nitrogen content of proteins, but without any immediate success. Kjeldahl then tried sulphuric acid, which proved to be successful, and after some further studies he published his new method [50]. To assist the digestion he also used phosphoric acid and potassium permanganate added to the sulphuric acid, collecting the ammonia in a known amount of sulphuric acid and back-titrating. He also added zinc powder to the solution before distillation, and many other substances were recommended, which, however, did not alter the principle of the method. Kjeldahl did not describe the apparatus in any detail: apparently he used ordinary flasks. The method soon became very widely used and many specialized pieces of apparatus were designed for the purpose, Kjeldahl himself designing the distillation assembly which is named after him [51].

Johan Gustaf Kjeldahl was born in Denmark, in 1849, the son of a physician. He studied at the Technical University of Copenhagen, and became a lecturer at the Agricultural High School. A man by the name of Jacobsen, who was the owner of the Carlsberg beer factory

FIG. 80. Johan Gustaf Kjeldahl (1849–1900). Photograph

founded a laboratory for the purpose of research on brewing and the beer industry. This was later to become the famous Carlsberg Institute. The young chemist Kjeldahl was given charge of the chemistry department, and he eventually became director of this institute. He worked here until his early death in 1900. Most of his work was carried out on enzymological research.

The Kjeldahl method for nitrogen, however, did have a few limitations, Kjeldahl himself realizing that in the case of nitro and cyano compounds as well as alkaloids the digestion was not suitable.

Asboth [52] modified the method by omitting the use of permanganate, and he was able to digest nitro compounds by adding benzoic acid to the sulphuric acid digestion mixture [53]. Asboth's method was improved by Jodlbauer, who used phenol instead of benzoic acid, so that the time required for digestion was reduced [54]. With nitronaphthalenes, Chenel was able to effect digestion by adding iodine and phosphorus [55].

These were most important developments in this still widely used procedure.

As with the determination of nitrogen, attempts were made to replace combustion with digestion for carbon determination.

R. and M. Rogers [56] determined the carbon content of graphite in quartz sand. They heated the sample in a retort together with sulphuric acid and potassium dichromate, and the carbon dioxide formed during the digestion was passed into potassium hydroxide which was back-titrated [57]. Brunner, in 1855, used this method for organic substances, and measured the carbon dioxide volumetrically [58]. Legler used the sulphuric acid–chromate mixture for the oxidation of glycerine, and measured the volume of CO_2 formed [59]. Messinger again used a similar digestion procedure in 1888 [60].

The determination of carbon by wet procedures never achieved any real importance on the macro scale because it was necessary to carry out a combustion for the determination of hydrogen.

The detection of sulphur in organic substances was first made by Robiquet [61] and Thibierge who found sulphur present in mustral oil. The determination of sulphur was made by converting to sulphate by digestion with *aqua regia*. Zeise [62] tested for sulphur by fusing the substance with metallic potassium when hydrogen sulphide was given off from the molten mixture. Later Vohl [63] detected sulphide with sodium nitroprusside [64].

The first method for the determination of sulphur in organic compounds was devised by Henry and Plisson [65]. They mixed the compound with iron oxide and quartz sand and heated it in a stream of oxygen formed from heated potassium chlorate, in a combustion tube. The carbon and sulphur dioxide formed was absorbed in sodium tetraborate, while the sulphur dioxide was measured volumetrically [66].

Even Henry himself was not very satisfied with this method, for four years later, in 1834, he described a new one in which he digested the substance with fuming nitric acid, and then after fusing with potassium hydroxide determined the sulphate formed gravimetrically [67]. In the same year Zeise fused the substance with potassium nitrate in a crucible, following digestion with *aqua regia*, and finally determined sulphur as sulphate [68].

The principle of this method is still used in present-day methods for the determination of sulphur, only the order of the different steps as well as the purity of the oxidizing agents has changed since those times. During the last century almost all the possible oxidizing agents have been applied to the oxidation of organic compounds, but a completely satisfactory one has yet to be found. The most common feature was the incompleteness of oxidation; while increasing the reaction temperature often resulted in the decomposition of the sulphate. The most efficient oxidant was the mixture of sodium peroxide and sodium carbonate which was recommended by Asbóth [69].

Naturally occurring organic halogen compounds were first discovered in the middle of the 20th century; halogens had to be determined only in synthetic products. The determination of halogens in most cases requires no preliminary digestion. The halogen atom frequently split off easily on boiling with a suitable reagent. Kekulé [70] boiled the substitution products of organic acids with sodium amalgam and water, and split off the bromine in this way [71]. Gustavson used sodium methylate for this purpose carrying out the reaction in a sealed vessel [72]. Since then a number of other substances, especially alcohols, have been recommended for this purpose. Where this treatment is insufficient then either a combustion or a digestion must be used to break down the molecule. The earliest method of this type was devised by Erdmann who simply ignited the compound with lime [73]. Piria [74], in 1857, used a mixture of lime with soda. He placed the mixture in a platinum crucible and inverted this in a larger crucible filled with soda lime mixture, which he then heated [75]. Many other metal oxides mixed with soda were subsequently recommended, but even today the soda—lime mixture is the most frequently used.

Carius [76] devised a new method whereby both sulphur and the halogens could be determined. He first published his method in 1860, but later made several improvements and modifications. In the final method he oxidized the substance by heating it in a sealed tube (bomb) with nitric acid. Barium chloride or silver nitrate was first mixed with the substance so that after cooling and leaching out the tube, the sulphur or halogen could be determined gravimetrically as barium sulphate or silver chloride [77].

Kopfer's method, mentioned previously, using oxygen combustion with a platinum catalyst, was applied by Zulkovsky and Lepez for the determination of sulphur, halogens, carbon, hydrogen and nitrogen [78], but their method was improved by Dennstedt [79, 80]. Although Dennstedt did manage to obtain a few results by this method it was not a great success and he achieved far more fame for his apparatus for the determination of carbon and hydrogen which is still in use today. A convenient method for the complete elemental analysis of organic compounds is still being sought at the present day.

The determination of other elements in organic compounds is relatively simple because the element sought is finally converted to an ionic form.

The most common constituent of organic compounds apart from carbon and hydrogen, is oxygen. The determination of oxygen, however, is a difficult task; even at the present day this determination is avoided wherever possible, the oxygen content being found by difference after all the other elemental determinations have been completed. There have been many attempts to develop a method for the direct determination of oxygen ever since the earliest days of organic analysis, but it was only at the beginning of this century that any success was achieved.

The problem was examined for the first time by Baumhauer [81] in the 1860's [82]. The principle of this method was to carry out a combustion in the presence of a known amount of oxygen. The combustion was carried out in a nitrogen atmosphere, and the carbon dioxide and water formed, as well as the unbound oxygen, were determined. From these data the oxygen content of the original sample could be determined. In practice, however, this method was difficult to carry out. As a source of oxygen, copper oxide was first tried; later silver iodate was used, the latter being recommended by Ladenburg [83]. In Ladenburg's method the sample mixed with sulphuric acid and silver iodate is sealed in a glass bulb and placed in a vertical combustion tube. After combustion the bulb is broken, the carbon dioxide is drawn off, and the excess of iodate back-titrated. After reweighing the tube it was possible from these data to determine the oxygen content. It was impossible to understand this method, even though the paper is illustrated with a number of figures, so that for further information reference must be made to the original paper [84]. It is very unlikely that anyone has even tried to repeat this method.

In 1874 Crétier introduced a method where the combustion products were passed over incandescent magnesium. He claimed that in this procedure all the oxygen was converted to magnesium oxide [85]. Some problems must have been encountered with this method, however, as no further reports of its use are to be found.

All these procedures were in the nature of experiments, for no satisfactory method emerged from the results so that for a long time the problem was neglected.

Hempel [86] and Markert in 1904, attempted to convert the oxygen to carbon monoxide by passing the combustion products over incandescent carbon. Their attempts were unsuccessful, but later, in 1939, Schütze [87] was able to use this principle with success. He mixed the substance with carbon and heated it to a high temperature in a nitrogen atmosphere, and then led the gases through a further layer of carbon at 1000 °C. The oxygen was converted to carbon monoxide which was further oxidized by iodine pentoxide to carbon dioxide, this being measured in the usual way [88].

Another method of oxygen determination originates from Ter Meulen [89] who introduced the principle of catalytic hydrogenation into elementary analysis, using thorium and nickel as catalysts. This method can also be used for the determination of nitrogen, sulphur and the halogens, for heating to a suitable temperature in the presence of a catalyst results in the formation of water, ammonia, hydrogen sulphide and hydrogen chloride which can be conveniently estimated [90].

At the beginning of this century micro methods of organic elementary analyses were developed, and they are now the only field of chemical analysis where micro methods have displaced the corresponding macro procedures. Macro methods for organic elementary analysis are very seldom used today, the micro methods being almost universal.

Microanalysis has the advantage of being more convenient, rapid and accurate than the corresponding macro methods, and has proved invaluable for many problems which were incapable of solution by macro methods. Just as elementary macro analysis played an important part in the development of organic chemistry, so micro elementary analysis played an equally important role in the development of the newer sciences of physiology, biology and biochemistry. In these fields it is quite common to isolate very small amounts of a substance either as a reaction product or an intermediate, so that by utilizing micro methods of analysis only 2–3 mg of sample are required for a complete analysis.

Elementary analysis utilizes gravimetric methods of determination, so that the development of micro methods of analysis could not begin until a sufficiently sensitive microbalance was available. This, as we have already seen, did not come about until the beginning of this century (Chapter VII. 3). It would at first appear that microanalysis simply entailed the scaling down of the apparatus and procedures used for macroanalysis, but in actual fact it is not quite so straightforward. It required much patience, skill and original thought, three qualities which abounded in the man with whose name microanalysis is so closely linked, Fritz Pregl.

Fritz Pregl was born in Laibach (now Ljubljana) in 1869. His father was a clerk. He completed his studies in medicine at the University of Graz, where he spent most of his life. After graduating he became a lecturer at the Institute of Physiology at the same University, and was occupied with the chemical problems of this subject. He soon achieved a reputation for great skill and inventiveness and began to study chemistry more thoroughly. In 1903 he went to work with W. Ostwald for six months and then for a further half year with Emil Fischer. In 1905, after his journey he returned to Graz where he became a lecturer, and later a Reader at the Institute of Medical Chemistry at the University. His first reported scientific results originate from this period and were mainly concerned with biochemical subjects. It was in the course of this work that he realized how great was the need for suitable methods for the analysis of very small amounts of substance.

In 1910 he was appointed professor at the University of Innsbruck to teach medical chemistry, and it was here that he began to develop methods to fulfil this need. He became so engrossed in this work that he eventually became, instead of a biochemist, a microanalyst. With unlimited patience and great skill, working alone, he established the whole scheme of quantitative microanalysis. He began with the carbon and hydrogen determinations and continued his work with the determination of nitrogen, halogen, sulphur, methoxyl and carboxyl functions. In most cases the principle of the method was devised by Pregl, but in all cases the apparatus, instruments and the method itself was the work of Pregl alone. He wrote hardly any publications, but slowly improved all his methods until finally, in 1917 he published his book *Die quantitative organische Mikroanalyse*, which founded a completely new field of chemistry. He returned to Graz in 1913, where he was appointed Professor of Medical Chemistry and in 1923 he received the Nobel prize for chemistry, the first award to be made for an achievement in analytical chemistry. He continued to work in Graz until his death in 1930.

Pregl first published his methods for the microdeterminations of carbon, hydrogen and nitrogen in 1912 [91], and this was followed by a number of new microanalytical methods. Many small, apparently unimportant improvements had to be made, however, before the final method was perfected. Even if the micro method was similar in principle to the macro one, special problems arose. The absorption of water and carbon dioxide was carried out first of all with calcium chloride and soda lime respectively, while more recently the magnesium perchlorate absorbent for water, recommended by Willard and Smith, is used [92]. The universal tube filling, which was introduced by Pregl and consists of copper oxide, lead chromate, lead peroxide and silver metal, can be used for all types of organic compounds. Almost all laboratories nowadays use these methods, and although some methods have been altered in small details the principle remains the same.

FIG. 81. Fritz Pregl (1869—1930). Photograph

Both the Dumas and the Kjeldahl methods were adapted to the micro scale, but the latter is less accurate than the corresponding macro version, consequently it is not used so often. This was one case where the micro method did not entirely replace the macro one. Pilch was the first to employ it on the micro scale with the use of sulphuric acid, potassium sulphate and mercury chloride for the digestion [93]. The hydrogen iodide–phosphorus methods [94], originate from Friedrich [95] while the micro apparatus [96] was designed by Roth [97].

For the determination of sulphur many methods which are simply scaled down versions of the macro procedure are to be found. One method, the Zeise–Vohl method is only used on the macro scale for the qualitative detection of sulphur but Bürger [98] has adapted it to a quantitative microdetermination [99]. The compound is reacted with potassium in a sealed tube and the sulphide formed during the reaction is titrated iodimetrically. The first micro determination of halogens was devised by Emich [100] and Donau [101] at the beginning of this century, by adapting the Carius method to the micro scale [102.] More recently the flask method of Schöniger [103] has been introduced where the compound is combusted in oxygen [104]. The original semimicro method of Schütze for the determination of oxygen was adapted by Zimmermann [105] to the micro scale [106], and Unterzaucher [107] further modified it. He made use of an iodimetric titration after converting the iodine equivalent to carbon monoxide to iodate [108].

In organic analysis it is often necessary to determine certain functional groups in addition to the elements present. For industrial process control rapid methods for the analysis of functional groups are an important requirement and instrumental methods are often employed. In most other cases a method based on a characteristic functional group reaction is employed, usually employing a titrimetric finish.

It has been shown that quantitative organic analysis was developed before the corresponding qualitative analysis. When one considers the difficulties inherent in qualitative organic analysis this becomes quite understandable. In inorganic analysis a qualitative knowledge of the elemental composition is very often sufficient to identify the compound, whereas for an organic compound even a quantitative elemental analysis can give little information as to its structure. Whereas in inorganic chemistry there are about 100 elements which could be tested for (in practice less than half of these are of importance) which together with several radicals form some 35,000 compounds, in organic chemistry consisting of carbon, hydrogen, oxygen and occasionally nitrogen there are already more than one million recorded compounds. The reactions used in inorganic qualitative analysis number between fifty to a hundred, while naturally the reactions used in qualitative organic analysis number far more. These few facts illustrate the difficulty involved in the development of a systematic scheme of organic qualitative analysis, a process which is still being perfected even today.

At the beginning of this century Mulliken [109] began to work on this problem. He attempted to classify compounds on the basis of the physical properties and chemical reactions of their characteristic groups (1904) [110]. This has since been proved to be the only valid basis for a systematic scheme of qualitative organic analysis.

Other classification systems are also based on this principle. The work of Staudinger [111] is especially notable in this field for he contributed to the development of functional group reactions. His work, which was first published in 1923 [112], placed this subject on a very firm basis and proved very useful in many other branches of analytical chemistry.

Stig Veibel [113] and his co-workers, have approached the problem from a different aspect, using quantitative methods for qualitative analysis by preparing the derivatives of the original substance and then determining their equivalent weight. From this the original substance can be identified.

NOTES AND REFERENCES

1. Kopp, H.: *Geschichte der Chemie.* Braunschweig (1842) **4** 241
2. Wöhler, F.: *Pogg. Ann.* (1828) **12** 253
3. Wallach, O.: *Briefwechsel zwischen Berzelius and Wöhler.* Leipzig (1901) **1** 206
4. Kolbe, Hermann (1818—1884) was the assistant and successor to Bunsen at the University of Marburg. Later he became Professor of Chemistry at the University of Leipzig. His primary interest was in the field of organic chemistry. He was the editor-in-chief of the journal: *Journal für praktische Chemie* for about 20 years.
5. Kolbe, H.: *Lieb. Ann.* (1845) **54** 145 186
6. Berthollet, C. L.: *Journ. de physique* (1786) **28** 272
7. Lavoisier, A. L.: *Oeuvres.* Paris (1864) **1** 346
8. *ibid.* **1** 351
9. *ibid.* **2** 586; *Mém. Acad. Paris* (1784) 593
10. *ibid.* **2** 590
11. *ibid.* **2** 773
12. Berthollet, C. L.: *Mém. soc. Arcueil.* **3** 64
13. Thénard, Louis Jacques (1777—1857) was a simple peasant boy who learned a little mathematics and Latin from the village priest. At the age of 17 he went to Paris and became a laboratory assistant to Vauquelin, but after three years he became a co-worker of the latter. (It is an interesting analogy; Vauquelin himself started his career in a similar manner rising from laboratory assistant to co-worker with Fourcroy!) In 1789 Thénard became a lecturer at the École Polytechnique, and later at the University of Paris. In 1810 he became a member of the French Academy, while in 1831 he was made a peer of France (member of the Over House). His most important discovery was that of hydrogen peroxide.
14. Gay-Lussac, J. L.—Thénard, L. J.: *Recherches physico-chimiques.* Paris (1810) **2** 265
15. Berzelius, J. J.: *Neues Journal der Pharmazie.* (1817) **1** 130
16. Gay-Lussac, J. L.: *Ann. chim. phys.* (1814) **95** 184; (1815) **97** 53
17. Döbereiner, Johann Wolfgang (1780—1849) was a pharmacist, but later became Professor of Chemistry and Pharmacy at the University of Jena. He was responsible for many literary works, but his vivid imagination often led him astray in his ideas. He is famous because of his laws concerning the elements.
18. Döbereiner, J. W.: *Schweigg. Journ.* (1816) **17** 369; *Neues Journal der Pharmazie* (1817) **1** 356
19. Liebig, J.: *Anleitung zur Analyse organischer Körper.* Braunschweig (1837) 4
20. Bischof, Karl Gustav (1792—1870) was Professor of Chemistry and Technology at the University of Bonn.
21. Bischof, K. G.: *Schweigg. Journ.* (1824) **40** 25
22. Bussy, A. B.: *Journ. de Pharm.* (1822) 580
23. Humboldt, Alexander, baron (1769—1859) was a German mineralogist, botanist and traveller. At first he was concerned with a mine but later left this job and made several journeys of a scientific nature in America and Europe. Between 1810 and 1827 he lived in Paris. Later he taught chemistry at the University of Berlin, during which time he made several journeys in Siberia. The University of Berlin is now named in honour of him.
24. Carrière, J.: *Berzelius and Liebig. Ihre Briefe.* München (1893) 146
25. Hoffmann, A. W.: *Aus Liebigs und Wöhlers Briefwechsel.* Braunschweig (1888) **1** 224
26. Liebig, J.: *Pogg. Ann.* (1831) **31** 1; cf. Liebig, J.: *Anleitung zur Analyse organischer Körper.* Braunschweig (1837)
27. Winkler, C.: *Z. anal. Chem.* (1882) **21** 545
28. Bunsen, R. W.: *Pogg. Ann.* (1857) **100** 43
29. Mulder, G. J.: *Scheikund, Verhandelen Onderzoek.* 212; cf. *Jahresbericht.* (1858) 589

30. BRUNNER, KARL (1796—1867) was a pharmacist, and Professor of Chemistry and Pharmacy at the University of Berne.
31. BRUNNER, K.: *Pogg. Ann.* (1832) **26** 497
32. HESS, GERMAIN HENRI (1802—1850) was born in Switzerland, but lived from his childhood in Russia, where he also studied medicine. Later be became Professor of Chemistry at the University of Petrograd. He established the law of thermochemistry, which was later forgotten along with Hess himself until Ostwald drew attention to it.
33. HESS, G.: *J. prakt. Chem.* (1849) **17** 98
34. KOPFER, FERDINAND, chemist, who worked in industry (?—1893).
35. KOPFER, F.: *Ber.* (1876) **9** 1377
36. DUMAS, J. B.: *Compt. rend.* (1840) **11** 287
37. BUSSY, A. B.: *Journal de pharmacie.* (1822) 580
38. DUMAS, J. B.: *Ann. chim. phys.* (1831) **47** 198
39. ZULKOVSKY, K.: *Ber.* (1880) **13** 1096
40. ERDMANN, O.: (1804—1869) was a pharmacist. From 1837 he was Professor of Chemistry at the University of Leipzig. He became famous because of his establishment of several atomic weights.
41. MARCHAND, RONALD FELIX (1813—1850) was Professor of Chemistry first of all at the Artillery School of Berlin and later at the University of Halle. He worked in collaboration with Erdmann.
42. ERDMANN, O. L.—MARCHAND, R. F.: *J. prakt. Chem.* (1838) **14** 206
43. SCHIFF, H.: *Z. anal. Chem.* (1868) **7** 430
44. DUMAS, J. B.—STASS, J. S.: *Ann. chim. phys.* (1840) **73** 137
45. VARRENTRAPP, FRANZ (1815—1877) was a pupil of Liebig, who later became Professor of Chemistry at the Medical High School of Braunschweig.
46. WILL, HEINRICH (1812—1890) was also a student of Liebig and later became his successor at the University of Giessen as a Professor of Chemistry.
47. VARRENTRAPP, F.—WILL, H.: *Lieb. Ann.* (1841) **39** 257
48. PÉLIGOT, E. M.: *Compt. rend.* (1847) **24** 552
49. SEEGEN, L.: *Z. anal. Chem.* (1864) **3** 155
50. KJELDAHL, J. G.: *Z. anal. Chem.* (1883) **22** 366
51. KJELDAHL, J. G.: *Jahresbericht.* (1888) 2611
52. ASBÓTH, SÁNDOR, worked in Budapest as a chief assistant in the National Chemical Institute.
53. ASBÓTH, S.: *Zentralblatt.* (1886) 161
54. JODLBAUER, M.: *Zentralblatt.* (1886) 433
55. CHENEL, L.: *Bull. soc. chim.* (1892) **7** 321
56. ROGERS, ROBERT (1814— ?) was a physician, and also Professor of Chemistry at the University of Virginia (U.S.A.).
57. ROGERS, R.—ROGERS, M.: *Jahresbericht.* (1847—48) 943
58. BRUNNER, C.: *Pogg. Ann.* (1853) **95** 379
59. LEGLER, L.: *Rep. anal. Chem.* **6** 631
60. MESSINGER, J.: *Ber.* (1888) **21** 2910
61. ROBIQUET, PIERRE JEAN (1780—1840) was Professor of Chemistry at the Pharmaceutical High School of Paris, and the owner of a number of chemical factories. He was the first to analyse a number of natural organic compounds.
62. ZEISE, WILLIAM CHRISTOPHER (1789—1847) was a pharmacist and Professor of Chemistry at the Technical University of Copenhagen, and was a well known chemist in his time.
63. VOHL, EDUARD (1823— ?) was an industrial and technological chemist and worked in Köln mainly on problems concerned with gas lighting.
64. VOHL, E.: *Dinglers polytechn. J.* **168** 49
65. PLISSON, AUGUSTE (?—1832) was a pharmacist at a hospital in Paris, and the author of several papers on analysis.
66. HENRY, E. O.—PLISSON, A.: *Journal de pharmacie* (1830) 249

67. HENRY, E. O.: *Journal de pharmacie* (1834) 29
68. ZEISE, W.: *J. prakt. Chem.* (1834) **1** 458
69. ASBÓTH, A.: *Chem. Ztg.* (1895) 2040
70. KEKULÉ, AUGUST (1829—1896) studied at the University of Giessen, and later made journeys to study in Paris and London. It was while in London that he discovered the structure of benzene. He later proposed the theory of the tetravalency of carbon, and in so doing played one of the most important parts, if not the most important part, in the development of structural organic chemistry. He became a Professor at the University of Ghent (Belgium), and later at Bonn, where he taught chemistry until his death.
71. KEKULÉ, A.: *Lieb. Ann. Suppl.* (1861) **1** 337
72. GUSTAVSON, A.: *Ann. chim. phys.* (5) **2** 208
73. ERDMANN, O.: *J. prakt. Chem.* (1840) **19** 326
74. PIRIA, RAFFAELLO (1805—1865) was Professor of Chemistry at the Universlty of Turin. He worked primarily in the field of organic chemistry.
75. PIRIA, R.: *Lezioni di chim. org.* (1857) 153
76. CARIUS, GEORG LUDWIG (1829—1875) was Professor of Chemistry at the University of Marburg.
77. CARIUS, G. L.: *Lieb. Ann.* (1860) **116** 1 28; *Lieb. Ann.* (1865) **136** 129; *Ber.* (1870) **3** 697
78. ZULKOVSKY, K.—LEPEZ: *Monatshefte* **5** 537
79. DENNSTEDT, MAXIMILIAN (1852—1931) was an assistant to Hoffmann and Cannizzaro, and later from 1855 he was Professor at the Artillery High School in Berlin. From 1903 he was the head of the National Laboratories in Hamburg.
80. DENNSTEDT, M.: *Z. angew. Chem.* (1897) 462
81. BAUMHAUER, HENDRIK (1820—1885) was Professor of Chemistry and Pharmacy at the University of Amsterdam.
82. BAUMHAUER, E. H.: *Lieb. Ann.* (1854) **90** 228; *Z. anal. Chem.* (1866) **5** 141
83. LADENBURG, ALBERT (1842—1911) was Professor of Chemistry at the University of Kiel, and later of Breslau. His work was mainly concerned with organic chemistry and chemical kinetics.
84. LADENBURG, A.: *Lieb. Ann.* (1865) **135** 1
85. CRÉTIER, H.: *Z. anal. Chem.* (1874) **13** 1
86. HEMPEL, WALTER (1851—1916) was Professor of Chemical Technology at the Technical University of Dresden. He is mostly famous for his work on gas analysis, as he devised several new analytical methods in this field.
87. SCHÜTZE, MAX, was born in 1896 in Silesia. He studied in Greifswald and Tübingen. Later he worked for a short time at the universities of Tübingen and Hochenheim. Since 1927 he has been carrying out research work at the Badische Anilin and Sodafabrik.
88. SCHÜTZE, M.: *Z. anal. Chem.* (1939) **118** 245
89. TER MEULEN, HENRI (1871—1941) was a Professor at the University of Delft in the Netherlands.
90. TER MEULEN, H.: *Rec. Trav. chim. Pays Bas.* (1922) 509; cf. TER MEULEN, H.—HESLINGA, J.: *Neue Methoden der organisch-chemischen Analyse.* Leipzig (1927)
91. PREGL, F.: *Abderhaldens Handbuch biochemischer Arbeitsmethoden.* (1912) **5** 1307
92. WILLARD, H. H.—SMITH, G. H.: *J. Am. Chem. Soc.* (1922) **44** 2255
93. PILCH, F.: *Monatshefte* (1911) **32** 21
94. FRIEDRICH, A.: *Z. physiol. Chem.* (1933) **216** 68
95. FRIEDRICH, ALFRED (1896—1942) was born in Knittelfeld, Switzerland, and studied in Graz. From 1923 onwards he worked at the University of Vienna. In 1938, after the Anschluss, he lost his position. He died in Brjansk in the Second World War of petechial typhus.
96. ROTH, HUBERT, was born in 1903, and studied at the University of Graz, where he later became the co-worker of Pregl. From 1930 to 1937 he worked in Heidelberg, at the Kaiser Wilhelm Institute. Since 1937 he has been the head of an agricultural laboratory at the Badische Anilin und Sodafabrik.

97. ROTH, H.: *Mikrochemie.* (1944) **31** 287
98. BÜRGER, KARL, was born in 1911 in the Tirol. He studied in Innsbruck, Jena and Munich, and then worked with H. Fischer in Munich. Between 1949—1952 he worked as an analyst in Argentina. Since 1955 he has been head of an analytical laboratory at the chemical works at Hoechst.
99. BÜRGER, K.: *Z. angew. Chem.* (1941) **54** 479
100. EMICH, FRIEDRICH (1860—1940) was a professor for several decades at the Technical University of Graz. He was one of the pioneers of microchemistry.
101. DONAU, JULIUS FERDINAND, was born in 1877. He worked at the University of Graz, and then in industry and in 1930 returned to the University. He is Professor Extraordinarius of Microchemistry.
102. EMICH, F.—DONAU, J.: *Monatshefte* (1909) **30** 745
103. SCHÖNIGER, WOLFGANG, was born in 1920. He studied at the University of Graz, where he later worked as a lecturer. Since 1953 he has been the head of the microanalytical laboratories of the Sandoz AS Basel works.
104. SCHÖNIGER, W.: *Mikrochim. Acta* (1955) 123
105. ZIMMERMANN, WILHELM, was born in 1902, in Karlsruhe, where he also studied. From 1929 onwards he worked at the Badische Anilin und Sodafabrik, while since 1950 he has worked at the University of Melbourne.
106. ZIMMERMANN, W.: *Z. anal. Chem.* (1939) **118** 258
107. UNTERZAUCHER, JOSEF, was born in Tangern (Austria) in 1901. He studied at the University of Graz and until 1938 worked there and at the Technical University of Zurich. Since 1938 he has been the head of the microanalytical laboratories of the Bayer Works in Leverkusen.
108. UNTERZAUCHER, J.: *Ber.* (1940) **73** 391
109. MULLIKEN, SAMUEL PARSONS (1864—1934) was a professor at the Massachusetts Institute of Technology, Cambridge (U.S.A.).
110. MULLIKEN, S. P.: *Identification of Pure Organic Compounds.* New York (1904)
111. STAUDINGER, HERMANN, was born in 1902 in Worms. Since 1926 he has been Professor of Chemistry at the University of Freiburg. He was awarded the Nobel prize for Chemistry. He died in 1965.
112. STAUDINGER, H.: *Anleitung zur organischen qualitativen Analyse.* Berlin (1923)
113. VEIBEL, STIG, was born in 1898 and studied at the Technical High School of Copenhagen. Later he worked at the University and then returned to the Technical High School, where he has been Professor of Organic Chemistry since 1944.

CHAPTER X

ELECTROGRAVIMETRY

The story of Galvani and the frog's leg is very well known, and this simple experiment gave birth to all the phenomena which are known as electricity, and which is now so indispensable to our present day civilization. Although even in ancient times static electrical phenomena were observed, and also in much later times it was found that electrical current could be generated by friction machines and stored in a Leyden jar, these observations had no practical importance.

There are so many anecdotes told about Galvani's discovery that it is very difficult to arrive at the truth. Galvani, who was a physician, carried out experiments with friction machines for producing electricity, because at that time the idea that the new electrical phenomena were beneficial in medicine was very fashionable. It was in 1780 that he made his revolutionary discovery. In the course of an experiment he placed a frog's leg together with the copper hook to which it was attached on an iron plate, and when the copper touched the iron the limb began to twitch. This was the start of 10 years of research which Galvani finally published in 1790. He considered that electricity is an animal phenomenon, that it is a special type of Leyden jar generated by the marrow in the bone.

Luigi Galvani (1737—1798) was first a Professor of Anatomy and later of Obstetrics at the University of Bologna. When Napoleon occupied Italy in 1797 and founded the new Cisalpine Republic, Galvani refused to accept the new regime and was therefore dismissed from his chair. There is nobody amongst all Galvani's contemporary scientists who can compare with him in regard to moral character or the courage with which he defended his ideas. Galvani, as a result of his views became very poor and for over a year he had no income and almost nothing to eat.

Galvani's experiments caused a great sensation in Europe, and everywhere frogs were caught; adults as well as children repeated the miraculous experiment of Galvani.

Volta, who was a Professor at the University of Pavia was one of the pioneers of electricity. His views regarding its origin, however, were quite different from those of Galvani. He discovered that the frog is not an essential part of the experiment and can be omitted; all that is required is two different metals and a moist contact or solution placed between them to produce a current. He termed the metals as first class electromotors (i.e. electrical conductors), and the liquids as second class ones. This expression although it had a different meaning has survived in the term: electromotive force.

If different electromotors, that is first class or dry ones, are in contact with wet or second class electromotors, this produces electrical fluidum and gives it force. But nobody should ask me how this is done! For the present it should be enough that this phenomenon occurs! (Volta, 1794).

In 1800 Volta constructed the first galvanic cell. This was known as the Volta pile and consisted of thirty or forty (or often more) zinc and silver discs, placed alternately on top of one another and separated by a small piece of cloth impregnated with an acid or salt solution. This was the first durable and convenient source of current. Electricity could easily be generated and it was necessary only to change the number of cells to alter the voltage. Very soon enormous Volta piles were constructed and used for experimentation; some of them contained more than a thousand discs! Napoleon himself authorized the construction of a large pile which he presented to Davy. The Emperor also founded a Volta award, which was analogous in those days to the Nobel prize. The first recipient of this award after Volta was Davy.

Alessandro Volta (1745—1827) was a teacher at the gymnasium of Como. He was a well-known physicist, and many discoveries such as the electroscope are the result of his work. In 1780 he was appointed Professor of Physics at the University of Pavia. It was here that he became famous after the construction of his pile, and Napoleon, as was usual, made him a Count. Napoleon's enemy, the Emperor of Austria tried to combat this award by also giving Volta honours. Volta, however, was not swayed by this flattery and remained in Pavia. In 1819 he retired and lived on his small estate until his death.

The Volta pile was applied to a variety of scientific purposes, and some of the most amazing achievements were made in the field of chemistry. In the same year that Volta constructed his pile (1800) two English chemists, Carlisle [1] and Nicholson [2] found that electricity decomposes water [3]. A third Englishman, Cruikshanks [4] established, also in the same year, that metals are deposited on the negative pole, while acids (anions) are deposited on the positive pole. Cruikshanks also recommended electrolytic deposition of the metal as a qualitative analytical test, and described it as a test for copper [5]. This reaction is also described in Pfaff's book, where copper is deposited on zinc [6].

In 1808 Davy made his famous discoveries of the electrolytic decomposition of the alkalis and earths (which had previously been considered as elements) with the Volta pile, and isolated the strange new metals [7].

Humphrey Davy (1778—1829) was one of the greatest chemists of all time and as a personality he is one of the most interesting figures in the history of chemistry. His father died while he was quite young and he became apprenticed to a physician, where he was introduced to chemistry. Davy later worked in a hospital where various gases were used to treat the patients, and it was here that Davy made his famous discovery of nitrous oxide (laughing gas). This was soon to be used with success by an American dentist as an anaesthetic for the extraction of teeth. Lord Rumford, who married the widow of Lavoisier, had recently founded a scientific institute, the Royal Institution, the aim of which was to put scientific results to a practical use, and also to publicize scientific discovery and make science popular among the people. He invited Davy to become a lecturer. He soon became a great success and the lectures he gave were always filled to capacity. Such was his fame, and the enthusiasm for science in England at that time, that even the aristocracy and fashionable society came to listen to his lectures.

He also made his great discoveries during this period, which added to his fame. Apart from working very hard he also lived a gay life and went to endless parties. He wasted no time, however, for if he had a social engagement he would work until within minutes of his appointment, and then after the dinner or party, often well after midnight he would return to the laboratory. The next morning he was always the first to arrive at the Institute. One of his biographers writes that he was often in such a hurry that he had no time to change his clothing before going to a party and would simply put a clean shirt on over the old one. It is also said that he once wore five pairs of stockings on top of one another.

He was knighted. His marriage in 1817 to a rich and attractive widow, enabled him to join the life of high society. Berzelius, whose impression of Davy has already been mentioned (Chapter VI. 2), wrote about him in his memoirs as follows:

FIG. 82. Humphry Davy (1778—1829). Painting of Thomas Lawrence

> His marriage and the wealth he obtained by it, as well as the title he received from the king, opened new possibilities for his ambitious character; he became proud and pretended to be an aristocrat, but he did not realize that in this field only those who are brought up to it can shine. The scientist is highly esteemed in these circles, but can easily become ridiculous if it is obvious that he wants to be an aristocrat but does not succeed [8].

Davy is also famed for his discovery of the arc-light, the miner's safety lamp as well as the discovery that chlorine is an element. He travelled widely, always accompanied by a portable laboratory complete with chemicals and equipment, and carried out experiments wherever possible.

Davy was accompanied on these expeditions by his laboratory assistant who was in effect his apprentice and whose name was Michael Faraday. The latter's reputation eventually exceeded that of his employer and it is not to Davy's credit that he was jealous of Faraday's success and tried to place obstacles to his progress wherever possible. For example, when the question of electing Faraday to membership of the Royal Society was raised there was only one black ball in the voting urn and everyone knew that this was contributed by Davy. However, everyone has their little weaknesses and many other masters have not been as enthusiastic as they ought when the success of their pupil has exceeded their own. In these cases it is possible for the master to consider his pupil's success as his own, and when shortly before his death Davy was asked what he regarded as his most important discovery he replied, "Faraday".

Davy did not look after his health, and several times he had attacks of paralysis, but managed to cure them in Italy. He died in Geneva during one of his journeys at the early age of 50.

After the discoveries of Cruikshanks and Davy it should have been obvious that the phenomenon of electrolysis could be used for the determination of metals, and it is very strange that this only came about more than half a century later.

For qualitative examination, however, electrical current was already used; the first example of Cruikshank's detection of copper has already been mentioned. In 1812, Fischer detected arsenic by electrolysis [9]. In 1830, Becquerel [10] observed that on electrolysis lead and manganese are deposited at the positive pole [11]. Cozzi [12] used an electric current to identify metals in metallic salt solutions [13], and in 1815 Gaultier de Claubry recommended the use of electric current for the detection of toxic metals in toxicological analysis. He deposited the metals after digestion with nitric acid on platinum. Despretz [14] separated lead from copper by electrolysis, the latter being deposited on the anode while the former was deposited on the cathode [15]. Bloxam also used an electric current for the detection of arsenic, which formed hydrogen arsenide at the electrode and thus gave an arsenic mirror on the side of the tube. The description of the method is difficult to understand because no figures are given [16].

During this period a great deal of work was carried out on the deposition or electrolytic reduction of various metals, but the interest was only in the electrochemical aspect. In 1864 Wolcott Gibbs became the first to apply these methods to analysis by weighing the amount of deposited metal (he is not to be confused with the famous American chemist Josiah Willard Gibbs, who derived the Phase Rule). Gibbs determined copper by electrolysis. He used a platinum crucible as the cathode attached to the negative pole of a Bunsen cell. The positive pole of the cell was connected to a platinum wire which was immersed into the solution in the middle of the crucible. Gibbs passed the current through the solution until a small sample of the solution gave no precipitate with hydrogen sulphide. After the electrolysis he rinsed the crucible with distilled water, dried it in a vacuum over concentrated sulphuric acid and then reweighed it. He mentions that the coating must be smooth, because if it is spongy it is rapidly oxidized by the air, as well as being very difficult to wash free from occluded electrolyte. He carried out the determination of nickel in a similar manner but from ammoniacal medium [17].

Wolcott Gibbs (1822–1908) was the first great personality in American chemistry. He was born into a wealthy family, his father was a property owner, while his mother was the daughter of a Minister of Finance. His father was very interested in mineralogy and the mineral gibbsite is named after him. Wolcott Gibbs soon showed an interest in the natural sciences, and even the classically inclined teaching of those days did not rob him of it. He graduated as a physician in New York, and as he wanted to work in the field of science he undertook an extended European tour; he worked for a year with H. Rose, and for six months with both Liebig and Dumas. After returning to his native country he became a teacher of physics and chemistry

FIG. 83. Wolcott Gibbs (1822–1908). Photograph

at the College of New York City, and later he became Professor of Natural Sciences at Harvard University, and subsequently head of its Chemical Institute when this body became separate. He played a very important role in the scientific life of America, which was just beginning to achieve importance during his lifetime. He was the editor for many years of the *American Journal of Sciences*, and was one of the founders of the Academy of Sciences. His face is sculptured as a relief on the wall of the Capitol in Washington. His scientific career began with a study of the complexes of cobalt, and later dealt with spectroscopy and thermodynamics. His greatest achievement, however, was the discovery of electrogravimetry.

Luckow, who was a chemist with a railway company, discovered electrogravimetry independently of Gibbs. At least, he wrote in 1865 that since 1860 he had been determining copper and silver by electrolysis from potassium cyanide solutions [18].

In 1869 a German metallurgical company advertised a competition for the development of a rapid method for the determination of copper. A method that could be carried out in 5—6 hours was needed, so that one operator would be able to carry out about 18 determinations in a day. The error of the method was not to exceed a given value. Luckow entered his method for the competition, but it did not win, the first prize of 300 thalers being awarded to Steinbeck who suggested a titrimetric method using potassium cyanide. However, the examining board, realizing the importance of the electrolytic method made an award to Luckow also. It is interesting to note that the objection to Luckow's method was that the time required to carry out the determination was too long whereas the electrolytic method later came into use because of its rapidity. [19]. The reason was that in Luckow's original method electrolysis was carried out from cold sulphuric acid solution, but he later discovered that the addition of a small amount of nitric acid makes the procedure much more rapid.

As the electrolysis method for copper spread rapidly during the next few years, it was quite understandable that methods were developed for the determination of other metals in this way. Thus methods for zinc, lead [20], mercury [21], cadmium [22] and manganese [23] were developed in rapid succession. Yver also carried out separations; he separated cadmium and zinc by electrolysing first from alkaline solution and then from acetic acid medium [24].

Alexander Classen published his first paper on electrogravimetry in 1881 [25], and this was followed by a series of other similar studies. Classen's work is very important with regard to the development of electrogravimetric analysis. Before Classen's work the only investigations that had been made were in regard to the nature of the electrolysis solution. The quality of the deposit was improved by variation of the solvent, addition of electrolyte, etc. Classen was the first to examine the influence of the current and applied voltage, and he also introduced the use of measuring devices into the circuit [26]. He also used accumulators in place of galvanic cells and was the first to discover the advantages of using warm solutions, so that by this means, combined with efficient mixing of the solution he was able to develop rapid methods of electrolysis. The use of a rotating anode to ensure mixing of the electrolyte was introduced by Klobukhov [27, 28]. The first book on electrogravimetry was written by Classen entitled: *Quan-*

titative Analysis auf elektrolytischen Wege. The book became equally as important to analytical chemistry as the treatises of Fresenius or Mohr. It was published in several editions, each new edition including recent developments. Although the first edition was only 52 pages long, at the time of publication of the third edition 10 years later, it had grown to 212 pages. Ten years later the fifth edition was 336 pages in length.

Alexander Classen was born in Aachen in 1843 and after studying at the Universities of Giessen and Berlin he returned to his native town where he worked as a private chemist. In 1870 he was invited to teach analytical chemistry at the Technical University of Aachen, where he became a Professor and head of the Institute in 1883. He died in 1934. His exemplary equipped laboratory in Aachen was the centre of electrogravimetry work, and from all parts of the world chemists came to Aachen to learn this new method of analysis. Many of them came from industrial firms. Classen also contributed to other branches of analytical chemistry with success, for example, he transcribed the new editions of Mohr's books at the beginning of this century. His own titrimetric textbooks were also very popular at that time.

The original crucible or pot-shaped platinum electrodes were used for a long time without any alteration. Paweck [29] introduced the use of metal-net electrodes for the electrolysis of mercury. The electrode was disc-shaped, and was made of brass [30]. C. Winkler developed the cylinder-shaped net electrodes, which are still used today, as well as the spiral platinum wire anode [31].

The development of theoretical electrochemistry clarified many puzzling phenomena. Le Blanc who was very much concerned with the problem of decomposition potential, determined this value for a great many metal salt solutions [32], and as a result of this work Freudenberg, working in the Institute of Ostwald, attempted to separate various metals by variation of deposition potential [33].

Platinum was the most common material for the construction of electrodes, and as a result of its high price many workers examined the possibility of using other types of electrodes. Many different metals were tried unsuccessfully and it was concluded that platinum was the best; even if it is not cheap it becomes economical as it is durable and resistant to attack. Apart from platinum only the mercury cathode is of interest, because it is still used today for many separations because of the special properties of mercury. Gibbs himself, the founder of electrogravimetry was the first to attempt to deposit metals on a mercury cathode, and to determine these as amalgams [34]. Luckow separated zinc and silver with the aid of a mercury cathode from iron, nickel, cobalt and manganese, as the latter do not form amalgams [35]. The deposition at controlled potential was first suggested by Sand [36] who used an auxiliary electrode for this purpose [37].

One branch of electrogravimetry is almost as old as the original method itself; this is known as *internal electrolysis*. At one time there were great hopes of this technique, but it never achieved much importance. In internal electrolysis no external current source is used, the electromotive force being produced by the electrode used for the deposition of the metal and by another electrode immersed in the solution. It can only be used for very small amounts of metals, so that its main application is for the determination of trace impurities, especially in noble metals. Thus it was used by R. Fresenius, on a qualitative basis only for the testing of copper. As a quantitative determination Ullgren [38] used the method for the first time in 1868. He used a zinc rod, immersed into saturated sodium chloride solution, and placed the solution to be electrolysed into a platinum crucible, which was connected to the zinc rod with a wire. In this way he could determine very small amounts of copper [39]. During this century Sand has developed several methods using this principle, and has constructed a more suitable apparatus. The term "internal electrolysis" was coined originally by Sand [40].

Pregl adapted the electrogravimetric methods to the micro scale, and essentially the method is simply a scaled-down version of the macro method.

The technique known as *electrography* is also included in the field of electrogravimetry. This is a qualitative analytical method and has the advantages of being very sensitive and also that the sample is not altered in any way. The principle of the method is that the substance under examination is made to function as an anode, and is separated from the cathode by a strip of moistened filter paper, impregnated with (in addition to some neutral salt to increase conductivity) a reagent which will detect the required component. This method was discovered in 1929 by Glazunov [41] and Fritz [42] independently.

Coulometric analysis is also of interest. This is a relative method and is the direct analytical application of Faraday's Laws. The method is based on the measurement of the amount of current passed through the cell, and obviously can only be applied where the electrode process is accurately known, and when no side reactions take place. The end point is indicated by a suitable chemical reaction; for example, in the case of an acid a common acid–base indicator can be used. Coulometry was introduced by Szebellédy [43] and Somogyi [44] in 1938. They carried out their measurements with a silver coulometer, and determined the titre of hydrochloric acid, sulphuric acid, thiocyanate, hydrazine, sodium hydroxide and hydroxylamine standard solutions [45].

NOTES AND REFERENCES

1. CARLISLE, ANTHONY (1768–1840) was a physician and Head of the Westminster Hospital. He was also court doctor to the Prince of Wales, and a member of the Royal Society
2. NICHOLSON, WILLIAM (1753–1815) was at first a clerk with the East India Company and then headmaster of a school in London. He later became a private engineer and a designer of canals. Finally he became the publisher of a natural science journal.
3. CARLISLE, A.—NICHOLSON, W.: *Nicholsons Journal* (1800) **4**. 179; *Gilb. Ann.* (1800) **6**. 340
4. CRUICKSHANKS, WILLIAM (1745–1800) was a physician, and Professor of Anatomy. He was also a chemist to the artillery and a member of the Royal Society.
5. CRUICKSHANKS, W.: *Nicholsons Journal* (1800) **4** 187
6. PFAFF, C. H.: *Handbuch der analytischen Chemie.* Altona (1822) **2** 332

7. Davy, H.: *Phil. Transact.* (1807—1808)
8. Berzelius, J. J.: *Selbstbiographische Aufzeichnungen* Leipzig (1903) 57
9. Fischer, N.: *Schweigg. Journ.* (1812) 6
10. Becquerel, Antoine Cesar (1788—1878) first worked as an engineering officer, but later resigned and became Professor of Physics at the Musée d'Histoire Naturelle. He was a member of the French Academy. He was the grandfather of the discoverer of the uranium radiation.
11. Becquerel, A. C.: *Ann. chim. phys.* (1830) **43**
12. Cozzi, Andrea (?—1852) was a pharmacist in Florence.
13. Cozzi, A.: *Arch. delle scienze med. fis.* (1840) **50** Sem. **2** 208
14. Despretz, César (1792—1863) was Professor of Physics at the Sorbonne and a member of the French Academy.
15. Despretz, C.: *Compt. rend.* (1857) **45** 449
16. Bloxam, Ch. L.: *Chem. Soc. Quart. Journal* (1860) **13** 12 338; citation: *Jahresbericht.* (1860) 645
17. Gibbs, W.: *Z. anal. Chem.* (1864) **3** 334
18. Luckow, C.: *Dinglers polytechn. J.* (1865) **177** 43 296
19. Luckow, C.: *Z. anal. Chem.* (1869) **8** 1
20. Parodie, G.—Mascazzini, A.: *Z. anal. Chem.* (1877, 1879) **16** 469 587
21. Hannay, J. B.: *Journ. Chem. Soc.* (1873) **11** 565
22. Smith, Edgar F.: *Ber.* (1878) **11** 2048
23. Riche, A.: *Compt. rend.* (1877) **85** 226
24. Yver, A.: *Bull. soc. chim.* (1880) **34** 18
25. Classen, A.—Reis, M. A.: *Ber.* (1881) **14** 1622
26. Classen, A.: *Ber.* (1894) **27** 2060
27. Klobukhov, Nikolai (1860—1899) studied in Munich, and worked as an assistant lecturer at the University of Harkov.
28. Klobukhov, N.: *J. prakt. Chem.* (1866) (1889) **33** 475; **40** 121
29. Paweck, Heinrich, was born in Vienna in 1870. He was Professor of Electrochemistry at the Technical High School of Vienna from 1919. He died in 1941.
30. Paweck, H.: *Z. f. Elektrochem.* (1896) **5** 221
31. Winkler, Cl.: *Ber.* (1899) **32** 2192
32. Le Blanc, M.: *Z. phys. Chem.* (1889) **8** 299
33. Freudenberg, H.: *Z. phys. Chem.* (1891) **12** 97; *Ber.* (1892) **25** 2492
34. Gibbs, W.: *Chem. News* (1880) **42** 291
35. Luckow, C.: *Chem. Ztg.* (1885) **9** 338
36. Sand, Henry Julius Salomon (1873—1944) studied in Dresden and Zurich, and later worked with Ramsay and Frankland. From 1921 he was Head of the Department of Chemistry at the Sir John Cass Institute in London.
37. Sand, H. J. S.: *J. Chem. Soc.* (1908) **93** 1572
38. Ullgren, Clemens (1811—1868) was Professor of Chemistry at the Technical University of Stockholm, and a member of the Swedish Academy of Sciences.
39. Ullgren, Cl.: *Z. anal. Chem.* (1868) **7** 442
40. Sand, H. J. S.: *Analyst* (1930) **55** 309
41. Glazunov, Alexei Iljich, was born in Petrograd in 1888. From 1920 he was the Professor of Metallurgy at the Mining High School of Pribram (Czechoslovakia), but later worked in France and South America. He died in Chile in 1951.
42. Fritz, H.: *Z. anal. Chem.* (1929) **78** 418; Glazunov, A.: *Chim. ind.* (1929) **21** 2
43. Szebellédy, László (1901—1944) studied at the University of Budapest pharmacy, in the institute of Lajos Winkler, and was his successor in 1938 at the Chair of Inorganic and Analytical Chemistry.
44. Somogyi, Zoltán, studied at the University of Budapest and afterwards worked in industry. He died in 1945 as a result of the war.
45. Szebellédy, L.—Somogyi, Z.: *Z. anal. Chem.* (1938) **112** 313 322 338 391 395 400

CHAPTER XI

OPTICAL METHODS

1. PRELIMINARIES OF SPECTROSCOPY

Optical methods of analytical chemistry, as with other branches of instrumental analysis, are nowadays progressing in two different directions. The first of these is in the application to specific analytical procedures, and the second is in the development of instrumental methods.

The gulf between these two paths is constantly widening, and the apparatus used by the analyst becomes more and more complicated. When his instrumental devices were confined to colorimeters and accumulators, the analyst had a certain measure of security, but now that he is committed to using complex electronic instruments, that security is lost. If his instrument develops a fault he generally has to call upon the services of an electronics engineer to correct it. This can be seen to be a sign of the increasing specialization that is apparent not only in chemistry and the sciences but in all walks of life. We now find that in addition to analysts who use the instruments, there are also those who are solely concerned with the design, construction or improvement of the instruments themselves, and the latter are generally qualified as physicists or electronics engineers. This division is now so marked that no important analytical department can survive without the services of an electronics expert.

This, no doubt, is the pattern of the future. In the history of this branch of science the development of physics played a very important part, but in this brief account the analytical aspects are of most concern, and the physical side will be treated as briefly as possible.

The most important optical-analytical methods are spectroscopy, spectrography, flame photometry, spectrophotometry and colorimetry. These divisions are only approximate, as many methods are not easy to define, for example colorimetry and photometry are often used to express the same thing. When an exact definition of each method is attempted the situation becomes even more difficult. What, for example, is the difference between flame photometry and spectroscopy? The answer is that in flame photometry the excitation is achieved with a much lower energy source (flame) than in spectroscopy, where an arc or spark is used. But Bunsen and Kirchhoff in their pioneer studies used flame excitation for their spectroscopic methods. It can also be argued that in flame photometry the evaluation is made directly, while in spectrography it is achieved by photographing and then measuring the optical density of the lines.

However, in the most recent quantometers evaluation is also made directly. As another example, a modern Beckman type or similar spectrophotometer is suitable both for colorimetric measurements and also for obtaining the absorption spectrum. Thus it is very difficult to differentiate between these methods, because their basis is the same and this is shown more markedly in their historical development.

Spectroscopy is the oldest of the optical methods of analysis and originated with the discovery that white light can be dispersed into its component parts, and the subsequent discovery of spectra and the investigation of dispersion.

Most of this work is associated with the name of Isaac Newton [1]. In 1666 he observed that the image of a narrow slit on the lattice of a window became wider and was coloured if a prism was inserted between the light source and the screen. He concluded from this experiment that the white light of the sun is composed of light rays of differing refractive indexes, and that the colour of these individual rays is related to their refractive indexes. He then attempted to obtain a better dispersion by investigating the effect of different prisms, lenses and slit widths. Newton was finally able to obtain a spectrum 25 cm in length by using a 1 mm slit width [2].

In this way spectroscopy originated as a branch of physics. The first chemist to make a contribution to this field was Marggraf who noted the difference in the colour imparted to a flame by sodium and potassium [3] (Chapter IV. 3). The flame emission of the alkaline earth metals was first observed by Lowitz.

In 1800 Herschel [4], the famous astronomer, measured the temperature of the radiation in different parts of the spectrum of sunlight, and found that the temperature at the red end of the spectrum is the highest. He continued his examination of the spectrum beyond the red end and found that the temperature of the radiation in this region is even higher, and concluded therefore that there must be some invisible radiation [5].

A year after the discovery of infra red radiation Ritter [6] discovered ultra violet rays by their effect on silver chloride. He also established that the reducing effect of this radiation, in effect the energy, is much higher than the violet end of the visible spectrum [7].

In the following year Wollaston observed that the spectrum of the sun is not continuous, and that black lines are to be found. If the sunlight was passed through a narrow slit, and then through a flint glass prism, 10–20 feet away from the eye, then between the single colours of the spectrum black lines could be seen. He described this discovery very inaccurately, so that it is only now that we can understand what he was trying to describe [8]. Wollaston himself was not very interested in this phenomenon, and did not pursue the subject any further. It is very probable that these lines had already been discovered a century earlier because many scientists at about that time had investigated the spectrum of the sun, and had probably attributed these lines as being due to some flaw in the prism.

In 1802 Young [9] calculated the wavelengths of the different spectral lines, using a diffraction grating [10].

Following all these investigations Fraunhofer, in 1814, rediscovered the black lines in the spectrum of the sun.

> Josef Fraunhofer was born in 1787 in Straubing. He came from a poor working class family, being the tenth child of a glass factory worker. He had to begin work at an early age, in order to earn some money for the family. He had no schooling, and at the age of 15 he could not read or write. By this time he was an apprentice at the glass factory and one day the building in which the workshop was located collapsed and the boy was buried under the ruins. Luckily he was unhurt, and this remarkable piece of luck was reported in the press, as well as the young boy's biography. A generous merchant read the account of the young boy's miraculous escape and of his poor circumstances, took pity on him and took him into his care. He paid for him to be educated, and although his generous benefactor died soon afterwards Fraunhofer managed to find a job in an optical workshop. His talent was soon recognized and he became very useful to the owner of the workshop. He studied optics and mathematics during his leisure time and after three years he became a partner in the workshop which by now, mainly due to Fraunhofer's work, had become a factory for producing a variety of optical devices, particularly acromatic lenses which were in great demand. Fraunhofer was always experimenting to improve the quality of the lenses they produced. He also published the results of these investigations, as a result of which he was elected a member of the Bavarian Academy of Sciences. His active and strenuous life was, however, destined to be short for he died in 1826, at the early age of 39.

In 1814 Fraunhofer was seeking for a new and better method to determine the refraction and dispersion of his acromatic glasses. First of all he projected a spectrum on to a surface with a glass prism of known refractive index, with the object of measuring the degree of dispersion by the size of the spectrum. But the limits of the spectrum were not sufficiently sharp for this purpose, so he repeated the experiment in darkness using the light from a candle. In the course of this experiment he noticed that between the boundary of the red and yellow lines there is a very sharp band, and that the position of this band is always constant. He decided to use this line to measure the refraction, but when he tried to repeat the experiment the following day with sunlight, he found that this band was absent and in its place he found a dark line. In order to observe this strange phenomenon he designed a special apparatus, which consisted of a slit and prism as well as a theodolite, using the telescope of the theodolite to examine the spectrum. The slit had a 15° axis with the objective, this being three times that which is used in metallurgical analysis today. He describes the experiment as follows:

> I wanted to examine whether in the spectrum of the sun, there is to be found a similar light band as in that of candle light, but found instead innumerable vertical dark lines, some being very weak but others were so dark that they were almost black.

Fraunhofer proved that these lines are a part of the solar spectrum, and did not originate from any errors connected with the apparatus. He made a diagram of these lines and assigned capital letters to the most prominent ones. He also counted the smaller lines, for example between the B and H lines he recorded 754. By using a diffraction grating he was able to measure the position of the lines relatively accurately, and so was able to calculate their wavelengths. Fraunhofer also examined the light from stars, and found that their spectra also contained dark lines, but nowhere near so many as in the spectra of the sun. Another important discovery he made was that the dark lines in the spectra of the planets are the same as those in the spectra of the sun. He also observed the spectrum of electrical discharges,

> but the spectrum of this is quite different from that of the sun and of fire. In its spectrum there are several light lines to be found, and one of them in the green part of the spectrum is almost glittering [11].

Fraunhofer did not pursue this study any further, his business commitments making it impossible, as he records, to investigate anything which was of no practical importance, and commending the further examination of this subject to the scientists. Little did he realize that his discovery would one day be of great technical importance.

In 1822 Herschel [12], the son of the famous astronomer [5] published the results of his investigations of the spectra of various flames. He established that:

> It is very probable that these colours originate from the molecules of the coloured substances which after being transformed to the vapour state are in vigorous motion.

But he concluded, however, that at a certain temperature all flames become yellow. Presumably the yellow colour of the sodium flame which can be observed almost everywhere deceived him, and he failed to realize that the colour of the flame is characteristic of the substance and not the temperature [13].

In 1826 Talbot [14], whose name is well known in the history of photography, constructed a device for the examination of flame spectra. He dipped a wick into the substance to be examined, and then after drying it he lit it, and passed the light from the flame through a slit and a prism, and examined the emergent spectrum on a screen. This was a very primitive attempt at spectroscopy. Talbot observed that potassium salts emitted a characteristic red line, while sodium salts gave a yellow line. Talbot was the first to connect the appearance of a given line with the presence of a certain compound [15].

Later Talbot was able to distinguish strontium from lithium on the basis of their spectral lines, although both impart a red colour to a flame. The spectra of strontium, however, shows several red lines together with dark lines whereas lithium shows only one very strong red line. Talbot concluded his report with the following words:

> I do not hesitate to state that by optical analysis the smallest amounts of these two substances can be distinguished at least as well, if not better, than by any other methods [16].

In this same publication he records the result of his examination of the spectrum of a cyanogen flame, noting that he observed a strong band at 388 mμ, the characteristic cyanide band.

Brewster [17], in 1834, proposed a theory to explain the existence of Fraunhofer lines. He suggested that the sun emits a continuous radiation, and that the black lines are formed because of the absorption of some of the radiation by the high temperature gaseous envelope surrounding the sun. He proved this theory experimentally by passing light through nitric acid vapour and showing the presence of dark lines [18]. Miller [19] had carried out some similar experiments even earlier. He passed light through tubes containing bromine or iodine vapour, and showed the existence in the emergent spectrum of dark lines [20, 21].

The younger Herschel [12] was the first person to photograph spectra, in 1840. He passed the sunlight through a narrow slit on to a brominated light-sensitive paper, and found that the area of the paper affected by the ultraviolet region was much larger than expected. He also observed that the ultra-violet region of the spectrum caused far more darkening than the infra-red region [22]. A short time later Becquerel [23] also photographed the spectrum of the sun on a Daguerre-plate [24]. Draper [25] also carried out similar experiments but his photographs were not quite as good as those of Becquerel and Herschel. The apparatus used by Draper is of interest, however, as it incorporated a variable slit, constructed from two knife edges which could be adjusted with a micrometer screw. He projected from the slit through a prism and then on an achromatic lens and finally on to the objective of a telescope. Draper also projected the spectrum on to a white screen so that he could mark the lines of the sun's spectrum. In the visible violet region alone he counted over 600 lines [26]. He later designed an apparatus which incorporated a diffraction grating [27]. Miller [28] described and published in figures the spectra of copper chloride, wine-stone, strontium, barium chloride and many other substances [29].

Stokes [30] in 1852 drew the absorption lines of the ultra violet region of the sun's spectrum on a white screen, which he made with quinine sulphate solution. Stokes also mentioned that the part of the ultra violet spectrum of the sun that is missing is due to absorption by the glass, and suggested that replacement of the glass material of the optical arrangement by some other substance would allow this part of the spectrum to be studied further [31].

As a result of all this investigation it gradually became clear that the absorption lines in the sun's spectrum and the flame emission lines are identical. This was to be seen most clearly in the D line of the sun's spectrum and the yellow emission line due to sodium. We shall not consider the further experiments in the progress of this field, but restrict our attention to those matters which are of most concern to analytical chemistry.

Masson, in 1850, examined the spectrum of electrical discharges, and in order to obtain spectra as distinct as possible he tried different metal electrodes. He found that many of the lines were identical in each experiment, but that a few lines were different and these he attributed to the vapourized metal electrodes.

Masson did not realize that the lines which were present every time originate from the air, he considered that they were due to the discharge [32].

Ångström [33] published a long and detailed paper describing the major differences in the spectra of solid and gaseous substances, in which he records the fact that metals give the same spectrum as their compounds. He came very near to the discovery of the reversal of the spectral lines, but this was later to be elaborated by Kirchoff. Ångström made a diagram of the sun's spectrum alongside a spark spectrum, and commented on it.

> When looking at this, one has the impression that the first is the reverse of the second [34].

In 1854 Alter [35], who was an American, stated that:

> The spectrum emitted by an element differs from all others in its number of bands, intensity, and position, so that the element can be identified simply by observation... The colours seen in the Polar Light possibly indicate the elements being present. By the use of a prism it is possible that the elements of the stars and the Earth can also be identified.

TABLE 11

ALTER'S TABLE OF SPECTRAL LINES

Red	Orange	Yellow	Green	Blue	Indigo	Violet	
,	,	,	,				Ag
	,,	,	,,,				Cu
,	,,	,		,,,	,		Zn
	,	,			,	,	Hg
,,	,	,,	,,,	,,			Pt
,	,	,	,				Au
	,,,,	,,	,				Sb
,	,	,	,,		,	,	Bi
,	,,	,,,	,	,	.	,	Sn
,	,,	,,	,,,,		,,	,	Pb
	,		,,,,				Fe
,	,,	,	,,,	,,,	,		Brass

Alter determined the spectral lines of the individual elements in the visible region, and published this in the form of tables [36] (Table 11).

Swan [37], in 1856, discovered that the R line of the spectrum is due to sodium, and that this can be used to detect sodium even when it is only present at a concentration of one part in two and a half million [38].

Helmholtz [39] was the first to use a quartz prism to investigate the ultra violet region of the sun's spectrum [40], and Robiquet similarly was the first to use an arc for the production of spectra [41].

The year 1859 was the most important year in the history of spectroscopy. Plücker [42] published an account of his investigations of the spectra of gases, and established that the spectrum of the discharge of a gas contained in a sealed tube is characteristic for the gas. He also discovered the first three lines of the hydrogen spectrum [43]. These experiments were carried out in discharge tubes constructed by his glass technician, Geissler. Plücker and his student Hittorf [44] established, several years later, that gases produce two types of spectra, i.e. line and band spectra. They mentioned that "the nature of the gases and their chemical changes are indicated characteristically by their spectrum lines" [45]. They also measured the wavelengths of the more important spectral lines, but with only a very rough accuracy. In the same year van der Willigen [46] investigated the effect of different electrodes on gaseous spectra, as well as the effect of metal salts evaporated on the electrodes and the effect of the separation of the electrodes.

Finally, in 1859 Kirchhoff and Bunsen published the results of their experiments with the first spectroscope, and this marked the start of a new chapter in analytical chemistry.

2. KIRCHHOFF AND BUNSEN

The majority of people consider that spectroscopy was the invention of Bunsen and Kirchhoff, but as we have just seen there had already been considerable progress made in this field when Bunsen and Kirchhoff came on the scene. Although in their famous paper the only reference they make to previous investigation is to the work of Swan, we have shown that several other investigators, notably Talbot and Alter, had commented on the analytical possibilities in the use of spectra. These earlier workers had been mainly concerned with the characteristic lines of the alkali metals and their relation to the Fraunhofer lines in the sun's spectrum. Nevertheless, these were only isolated thoughts and from a practical point of view the method of spectroscopy originates from Kirchhoff and Bunsen. They converted the examination of spectra into "spectrum analysis", which is now an important and self-contained branch of analytical chemistry. Spectrum analysis immediately produced astonishing new results because even the first publication recorded the discovery of new elements found in samples which had previously been examined by chemical methods. This indicated that the method was far more sensitive than all other existing analytical methods. Since then spectroscopy has displaced wet methods of analysis in very many cases, especially in qualitative analysis. It is now accepted that the first step in the analysis of an unknown sample is to carry out a spectrographic assay.

It is typical that in this early period in the development of physico-chemical analysis Kirchhoff was the Professor of Physics while Bunsen was Professor of Chemistry at the University of Heidelberg, and that as a result of their collaboration this method was developed. Bunsen was mostly concerned with the problem of qualitative tests for the elements by their flame colourization, and examined

FIG. 84. First spectroscope of Bunsen and Kirchhoff. (From their original paper in the year 1860)

the flame through various coloured glasses and solutions, while Kirchhoff who was more familiar with the optical side, advised Bunsen on the use of prisms. The results of their joint efforts were made public in a lecture given before the Academy of Berlin 1859 (this was published in 1860) [48].

Kirchhoff and Bunsen first examined the alkali and alkaline earth metals. They purified the samples very carefully, by recrystallizing eight to ten times. In addition to the coal-gas flame they also used many other flames, for example by burning sulphur, hydrogen, hydrogen-oxygen, etc. They established

> ... variation in the compounds of an element, the multitude of the chemical reactions taking place in the flame, as well as differences in flame temperature or different types of flame, have no influence on the position of the characteristic spectral lines of the single elements.

In a later publication they amended this by saying that although different compounds always exhibit the characteristic lines of the metals, this does not mean that compounds necessarily always have indentical lines with their elements. Their first spectroscope is shown in Fig. 84.

They examined the single elements separately; the following extract from their paper illustrates this:

Sodium. Of all the spectral reactions that of sodium is the most sensitive. The yellow sodium alpha line is the only one which can be seen in the spectrum of sodium, and this is the same as the D line of Fraunhofer. This is famous for its well defined broadening and its high light intensity. If the temperature is very high and the amount of substance in the flame is great, there is a continuous spectrum to be seen around this line. Lines of other elements which are near to the sodium line are much less intense and in many cases can only be observed when the sodium reaction disappears ...

The next experiment confirms the claim of the sodium flame colouration to be the most sensitive chemical reaction. In the opposite corner of the room (which is 60 m^2 in area) to where the spectroscope is situated 3 mg of sodium chlorate mixed with lactic acid was exploded and at the same time we observed the slit of the spectroscope. It did not show the sodium line immediately, but after a few minutes, however, a strong sodium line appeared and this persisted for 10 minutes. From the amount of sodium salt taken and the volume of the room it was calculated, that for one weight of the air in the room, less than one twenty millionths of this weight of sodium smoke was present. Since the reaction can easily be observed in one second, during which time only 50 ml, i.e. 0·0647 g. of air goes through the burner at the usual flow rate, then this means that the eye can detect even less, i.e. one thirty millionth of a milligram of sodium salt can be detected ...

Lithium. The glowing vapour of a lithium compound shows two well defined lines, a yellow Li beta, and a light red Li alpha line. As regards reliability and sensitivity this reaction is also better than any other chemical reaction of lithium.

We must also mention that for mixtures of volatile lithium and sodium salts the lithium flame reaction is only slightly less sensitive and distinctive than that of sodium. If a mixture containing one-thousandth part of lithium is placed in the flame the lithium colour is observed first, owing to the greater volatility of lithium salts, and then the persistent yellow sodium flame appears [48].

Kirchhoff and Bunsen clearly understood the importance of their method; in a letter Bunsen wrote to Roscoe before the publication of their method he states:

Now I am working with Kirchhoff which hardly gives us time for sleep! Kirchhoff has made a miraculous discovery, he has discovered the cause of the black lines in the sun's spectrum. What is more he can magnify them and also produce lines in a colourless flame spectrum which correspond exactly to the Fraunhofer lines. You will understand that this now makes it possible for us to determine the composition of the sun and stars just as accurately as we can identify chloride or sulphate in the laboratory. With the same degree of accuracy we can identify the individual elements on earth also. For example, we were able to detect lithium in 20 g of sea water! For the identification of certain substances this method is far more sensitive than any other. If you have a mixture consisting of lithium, sodium, potassium, barium, strontium and calcium, you need only give me one milligram for with my apparatus, looking at it through a telescope, without touching the substance at all, I can tell you which elements are present [49].

In their paper they also referred to the possibility of the discovery of new elements through spectrum analysis. Any element which occurred in nature, in an amount which was too small to separate and detect by conventional analytical chemical techniques, would probably be detected by its flame spectra. Their forecast was proved in the same year with the discovery of rubidium and caesium, which brilliantly illustrated the applicability of their method [50].

They also mentioned that in addition to the examination of the substances found on earth the possibility of chemical investigations will be opened up for previously unattainable areas which may range far over the earth, even to the boundaries of the solar system. Since it is sufficient only to observe the light from the glowing substance to apply this method, then it is obvious that it can also be used for the investigation of the atmospheres of the sun and of the light-emitting stars [48].

Kirchhoff was the first to report the phenomenon of reversed spectra [51], by which he meant that a coloured line disappears from the spectrum and a black

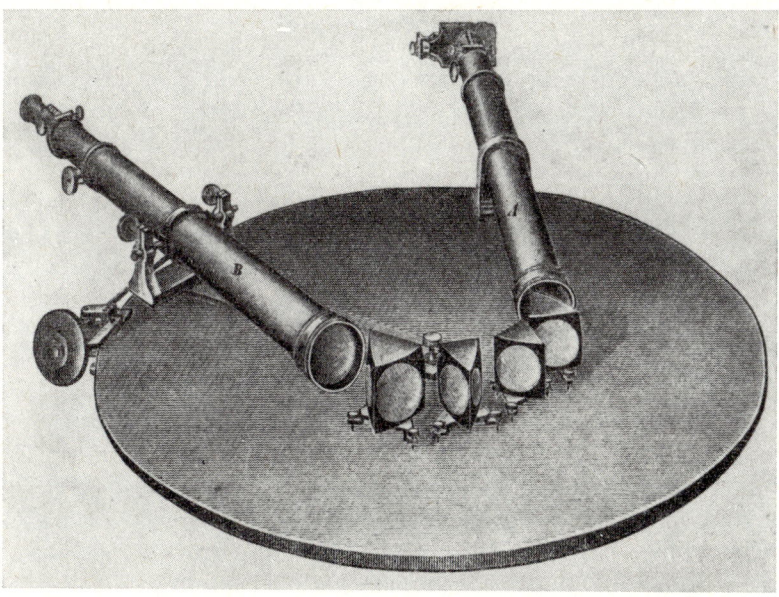

Fig. 85. Kirchhoff's spectral device, 1862. (From his original paper)

line appears in its place, if a light source which gives a continuous spectrum is passed through a gas.

This experiment formed the basis of Kirchhoff's explanation of the origin of the dark lines of the sun's spectrum, that they are caused by absorption of radiation by the gases which surround the sun's glowing nucleus. This theory was proved experimentally several years later when Janssen [52] found some lines of unknown origin in the sun's spectrum [53]. Lockyer [54] repeated these experiments and explained the results by postulating the existence of an unknown element, which he called helium [55].

> The elements—continue Kirchhoff and Bunsen—have lines of varying appearance. Just as analytical precipitates are different—they may be powder-like, gelatinous or crystalline— so the spectral lines also have their characteristics; some of them are sharply defined, others are diffuse on one or both of their sides, some are broad and others are thin. Again just as in analytical chemistry we only use those precipitates which are also formed in dilute

solution, so in spectrum analysis we only use those spectrum lines which are formed from small amounts of substances and at low temperatures. In this respect the two methods are similar. Spectrum analysis, however, has a great advantage over wet methods of analysis, in that the spectral lines are independent of any interferences, as the formation of a spectrum is a chemical property of the element which is just as specific as the atomic weight and consequently can be determined to a high degree of accuracy [48].

In the following year Bunsen and Kirchhoff published a second paper on spectroscopy, in which they reported the discovery of rubidium and caesium. For these studies they used an improved spectroscope, where the spectra of two light sources could be examined simultaneously, and compared with a scale [50]. In the next year Kirchhoff replaced the single glass prism with four prisms in a semi-circle, and this resulted in a lengthening of the spectrum as well as an improvement in the resolution.

In 1861 Kirchhoff reported on his investigations of the sun's spectrum, in which he indicated the elements represented by the different single lines in the spectrum. He gave a diagram showing the elements corresponding to each line, but this was not calibrated in length values but only contained an arbitrary scale. In his first paper Kirchhoff only recorded the sun's spectrum between the D and F lines, because his eyes became so tired with the work that he found it impossible to continue [56]. With the help of his co-workers he eventually finished the work in 1863 [57].

The work of Kirchhoff and Bunsen as well as attracting great attention also aroused some violent opposition. The main objection came from English scientists who contested the originality of Kirchhoff's theory of the origin of the absorption lines. Notable scientists such as Kelvin, Brewster, and Crookes entered into this controversy, maintaining that Stokes and Talbot, or even Wheatstone was the discoverer of this phenomenon, and they accused Kirchhoff of plagiarization. Wheatstone [58] in fact made some very important observations about the origin of the lines, as well as on their position, and also established that all the elements can be identified from their spectral lines. It would have been very difficult for Kirchhoff and Bunsen to know of this, however, for the lecture at which Wheatstone made public the results of his investigations was held in 1835, and was not published until 1861 by Crookes. In 1835 only a summarized extract was published [59]. Ångström [60] was only one of several others who also claimed priority in this discovery over Kirchhoff.

Kirchhoff subsequently published a paper in which he discussed the previous history of spectrum analysis, as well as detailing the failures experienced by his predecessors in contrast to his own original discoveries. Nevertheless if he had included this in his first paper, in 1860, then all these disputes could have been spared [61]. It is obvious that, as in all other branches of science, spectrum analysis did not originate from nothing, there was a great deal of preliminary investigation upon which Bunsen and Kirchhoff were dependent. But it is an indisputable fact that it was only as a result of their work that spectrum analysis was developed into a usable method. That this method is still associated with their names is very appropriate, for even in the 1860's the method was named after them. In a catalogue of laboratory equipment published by the English firm, Griffin (who produced the squat beaker which is named after them) in 1866, there is an entry which records that a Kirchhoff–Bunsen type spectroscope can be purchased for 5 guineas [62].

FIG. 86. Gustav Kirchhoff (1824–1887). (Engraving from the journal *Berichte* [1887])

Gustav Kirchhoff was born in 1824 in Königsberg and subsequently studied natural science at the university there. After graduation he chose the scientific life, and remained at the university. He later became a Privat Dozent in Berlin, until the University of Breslau appointed him Professor of Physics. Shortly afterwards Bunsen went to Breslau to become Professor of Chemistry, and a great friendship developed between the two. Kirchhoff at that time was mainly interested in electricity, and his name is commemorated by several laws which he introduced and for which he carried out most of the basic experiments whilst still a student, in 1845. In 1852 Bunsen was offered the chair of chemistry at the more illustrious University of Heidelberg, which he accepted, and soon contrived that Kirchhoff was appointed there also. It was here that they collaborated in their research on spectroscopy and developed their method of spectrum analysis. After they had developed the method, Bunsen's interest turned to the chemical applications, whilst Kirchhoff was more interested in the spectra of stars which he examined.

In 1861 Kirchhoff gave an account of the elements present in the sun, based on an examination of the sun's spectrum. This opened an entirely new field in astronomy. In 1879 he was offered the chair of physics at the University of Berlin, which he accepted. However, ill health overtook him and after several illnesses he died in 1887.

Robert Wilhelm Bunsen is one of the greatest personalities in analytical chemistry. His name is associated with a diverse collection of discoveries which range from spectroscopy, iodimetry, and gas analysis to the Bunsen burner, which is probably the most widely used piece of laboratory apparatus today. Bunsen also introduced the laboratory stand and the water pump, both of which commemorate his name, and for many years the Bunsen battery was used in electrochemistry. This contained carbon and zinc electrodes and had the advantage of being cheap as well as producing a uniform current. In a slightly modified form the Leclanché and dry batteries still use this principle. The Bunsen photometer which is simply a piece of paper with a wax spot, is still used in secondary schools to illustrate experiments on light intensities. He also introduced a special type of burette-tap for use with alkaline solutions. Bunsen can be considered as the Edison of analytical chemistry, for during his long lifetime he worked in very many fields to each of which he made some very important practical contribution.

Robert Wilhelm Bunsen was born in 1811 in Göttingen, a famous German university town. He grew up in the university where his father was a librarian, and he later entered the university to study of his own choice physics, chemistry and geology. His main interest was in chemistry, and at that time the Professor of Chemistry was Strohmeyer, a well known chemist in those days, and the discoverer of cadmium, who was primarily an analytical chemist. After graduating, Bunsen undertook a travelling fellowship and later returned to Göttingen as a Privat Dozent. In 1834 he became a teacher at the technological school in Kassel, in succession to Wöhler, and in 1839 he was offered a Professorship at the University of Marburg from where in 1851 he went to the University of Breslau. A year later he was given the chair of Chemistry at the University of Heidelberg, the chair having previously been occupied by Gmelin. He was a professor here for thirty-eight years, during which time he lectured in addition to carrying out his extensive research work. In 1889 he retired and lived very quietly in a street which had been named after him. He died in 1899.

Bunsen was essentially a practical scientist, he never involved himself in the theoretical disputes which raged during this period, always remaining disinterestedly apart. In his old age this disinterestedness developed into an antipathy towards all problems of a theoretical nature... "A hypothesis, which only leads to further possibilities has no value." "A single determination of one fact is more valuable than the most beautifully constructed theory" — illustrate his views.

Bunsen was a very great teacher and his lectures were interesting and well attended by the students. His lectures were never reported or published, and Bunsen made no changes in them throughout a quarter of a century. Thus, they eventually became rather out of date. His laboratory teaching, however, remained famous during his whole lifetime, for he lived and worked among the students right up to his retirement. He himself took charge of the laboratory classes throughout his whole working life, and his practical skill was a living legend. Bunsen expected a very high degree of accuracy from his students, but it must

Fig. 87. Robert Wilhelm Bunsen (1811—1899). Photograph

be stated that it was comparatively easy to learn analysis from his example. Students from all over the world came to work with Bunsen, and after spending one or two years alongside him they left for other universities to take their doctors' degrees. The reason for this was that Bunsen never promoted anyone, he even objected to his students using books saying that they only contained useless theories! Among his students many were later to become famous; they included Carius, Roscoe, Beilstein, Pauli, Volhard, Baeyer, Graebe, Bunte, Kolbe, Curtius, Nessler, Friedländer, C. Winkler, Treadwell, etc.

Bunsen was a very modest man, both in behaviour and in his requirements. He never married, saying that he never had time for marrying, but in his later years he often complained of loneliness and his isolation. He did not attribute much importance to the many honours and awards which he received and when someone once congratulated him on receiving a famous award he replied, "Oh my God! The only good thing about these honours was that they made my mother happy, but now she, poor soul, is dead."

He had a good sense of humour, and was as absent-minded as all professors are supposed to be. Many stories are told about him, how for example, when one of his colleagues visited him, and after talking for an hour and a half the visitor was about to depart Bunsen said, "You can imagine how poor my memory is, but when I first saw you I thought that you were Kekulé". The visitor looked astonished and said: "But I am Kekulé!"

The first scientific works of the young Bunsen were concerned with organic chemistry. He examined the cacodyl compounds, determined their composition, and established that part of these compounds, namely the cacodyl, can pass from one compound to another unchanged, i.e. it is a radical. His investigations provided invaluable proof of the new organic chemical theory of radicals. It is strange that Bunsen did not pursue his interest in organic chemistry, even more so when one remembers that in Germany in that century chemistry in effect meant organic chemistry and all chemists were organic chemists. Indeed in Germany in those days it was very difficult to find anyone to fill the chairs of inorganic chemistry.

In 1838 Bunsen investigated the combustion processes occurring in iron furnaces. He carried out gas analysis first of all, and found that about 50 per cent of all the fuel was lost. The results of his work caused great interest in the iron industry and he was invited to England to examine the furnaces there which were worked with coal. Here he found that the situation was even worse, 81 per cent of the fuel being wasted. These investigations of Bunsen were very important from an economic point of view, and with the introduction of his methods of gas analysis it became possible to investigate the effect of constructional changes in the furnace on the economy of the combustion. Bunsen developed the methods of gas analysis to a high degree of accuracy, by using absorption and volumetric procedures. These methods were published in 1867 in his book entitled *Gasometrische Methoden* in 1857. This book was the beginning of modern industrial gas analysis. The first chapter of the book deals with sampling, and subsequently analysis of gas mixtures and determinations of individual gases are discussed. There is also a detailed description of gas density measurements as well as a chapter on the absorption of gases in various liquids, and finally one on gas diffusion and gas combustion. The thoroughness with which Bunsen approached every subject can be seen from his introduction where he mentions that he had investigated all the subjects discussed in his book. This comment could equally well be applied to his other investigations.

In 1846–7 he went on a trip to Iceland, at the invitation of the Danish government. Here he carried out geological observations and investigated the geysers. Bunsen also made important contributions in the field of petrography, but geology was really only his hobby and it was not until his later years that he was fully occupied by it. He also worked with Roscoe, on photochemistry, examining the flame reactions of individual elements for qualitative analytical purposes. The system he devised was published in 1866 under the title *Flammenreaktionen*. In these studies his Hungarian co-worker, Vince Wartha was his assistant. In 1867 Wartha who was then a Privat Dozent in Zurich published a book in which he describes a complete analytical system based on the flame reactions of Bunsen [63]. It was from these investigations that Bunsen came to develop the spectroscopic method already referred to. In 1857 he constructed the Bunsen burner, which produced a non-smoky flame, and could also be easily regulated [64]. It was about this time that gas lighting was introduced. After working for 10 years on spectroscopy he turned his attention to the vapour and ice calorimeter. The discovery of the water pump was made in 1868 [65].

3. FURTHER DEVELOPMENT OF SPECTRUM ANALYSIS

The great practical importance of spectroscopic analysis was proved in the next few years by the discovery of several new elements. After the initial discovery of caesium and rubidium, thallium, indium and gallium were also discovered by spectrum analysis.

The calculation of the wavelengths of the spectral lines was, however, very inaccurate. Mascart [66] recalculated the wavelengths in 1863/64, first of all of the ultraviolet lines, using a diffraction grating. This grating contained 400 lines per millimetre, but the results obtained were rather inaccurate [67]. This problem was investigated several times during this period and many scientists made measurements of the spectral lines [68]. Finally Ångström made a thorough investigation and measured the absolute wavelengths of a large number of Fraunhofer-lines with a diffraction grating, and compared these results with the spectral lines formed by the elements [69]. Ångström carried out his measurements so carefully, and with instruments of such precision that the question was considered completely solved. Ten years after the death of Ångström, Thalén [70], one of his co-workers, reported that an error in the measurement of the grating width had made all the results inaccurate [71]. He also stated that Ångström himself was aware of this error in 1872. It is probable that Ångström could not bring himself to publish this error, before his death in 1874. Thalén also gave a correction according to which 13/100 000 of the value of each wavelength must be added to it. Subsequently many new measurements were made, but Rowland [72] had in the meantime constructed a grating marking device which could inscribe 1720 lines per millimetre. He later constructed concave gratings which resulted in an increase in the dispersion and sharpness of the spectral lines. Using this grating several workers, including Rowland himself, again measured the wavelengths of the sun's spectrum, Rowland measured about 1100 lines of the sun's spectrum as well as arc spectra, with an estimated error of less than 0·01 Å. The limits of wavelength over which he made his measurements was 2152·91 Å to 7714·68 Å units [73]. In the following years he recorded all the visible lines of the sun's spectrum and of arc spectra of the elements on a plate. As a result of Rowland's work the visible and ultraviolet regions of the spectrum were completely charted.

The investigation of the infrared spectrum was also commenced during this period, when Langley [74] introduced the bolometer for the measurement of the radiation in this part of the sun's spectrum [75].

The development of the techniques of photography during this time resulted in plates being available for the recording of spectra, so that from 1880 onwards spectrographers used photographic plates to record their spectra.

Quantitative spectrum analysis has only been developed in the present century, but preliminary examination had been carried out before this. Lockyer [54], the discoverer of helium, as well as being one of the pioneers of the introduction of photography into spectroscopy, observed that some spectral lines are dependent on the method of excitement. Some lines occur in all spectra, others only when a certain form of excitation is used. Lockyer carried out a series of experiments and concluded that while qualitative spectrum analysis is based on the position of the lines, quantitative analysis could only be based on the length, strength and number of lines [76].

This idea was later proved correct. Quantitative spectra analysis, however, is based even today on purely empirical grounds. The basic physical and chemical processes are not completely understood even though the method is used with increasing frequency. It is, however, more suitable for informative rather than for accurate quantitative analysis. The development of quantitative spectrum analysis will only be treated briefly here.

Pioneering work in this field was carried out by Hartley [77] and his co-workers. They established that in the spectrum of all of the elements there are lines which cannot be detected under a given concentration limit. Thus, with decrease in concentration the number of lines decreases, and those lines which remain in the spectrum at the lowest practical concentrations are called persistent lines [78]. On the basis of this conclusion Hartley himself, as well as Leonard and Pollok [79], and de Gramont [80] attempted to develop a system of quantitative analytical spectroscopy at the beginning of this century [81]. All their methods were based on the fact that the amount of a given element can be found from an examination of the number of lines appearing in the spectrum. The choice of suitable lines, and a study of their concentration dependence was made by using reference standards. This method proved rather inaccurate in practice because other factors influenced the disappearence of the lines, particularly the quality of the plate, the amount of electrical excitement, and the time and method of exposure of the plate.

Work on quantitative spectrum analysis lapsed for some time and in 1910 Kayser [82] the author of the most comprehensive spectroscopical handbook, wrote the following: "Summarizing the results of all the experiments carried out I concluded that quantitative spectrum analysis is impracticable" [83].

This forecast proved to be wrong, for in 1924 Gerlach [84] solved the problem, and paved the way for quantitative spectrum analysis to develop [85]. The method devised by Gerlach involved the use of homologous line-pairs. In the next few years this method was refined and further developed by Schweitzer [86, 87]. The method is based on the following principle: If several samples in which the amount of the base element is virtually constant, but which contain varying amounts of the element to be determined are examined, then whereas the spectral lines of the former occur with constant intensity, provided of course that the experimental conditions are constant, the spectral lines of the latter will vary according to the amount of the element present. It is then only necessary to find a pair of lines, one of the base and one from the element being determined, which exhibit similar intensity (blackness). Thus once two lines of equal intensity have been found in the spectrum, then the concentration can be found from a calibration curve or table. If there are not enough homologous line-pairs in the spectrum then a suitable emergency spectrum can be chosen for comparison.

Scheibe [88] and Neuhäusser introduced the use of the logarithmic sector. This is a rotating disk which rotates in front of the slit and which is cut in the shape of a logarithmic curve. The light intensity falling on the slit therefore was not uniform; the lowest part receiving most of the light, while the light intensity decreases logarithmically towards the upper part of the slit. This resulted in the spectral lines of less intensive radiation being shorter than those of the more intensive radiation, and hence from the relative length of two homologous lines their relative concentration could be estimated. This was an important improvement over the homologous line-pair method, because instead of the subjective estimation of line strengths the measurement of line lengths could be made, this being a much easier task than the former [89]. The measurement of line intensities by photoelectric methods was introduced by Lundegårdh [90] in 1929 [91]. This is now the most widely used method of evaluation. More recently quantometric (spectrometric) methods have been introduced; these are based on the scanning of the spectrum lines with a photomultiplier unit. The first application of photomultipliers for this purpose originates from Dieke and Crosswhite (1945).

The use of X-rays had already been known for some time; in analytical chemistry, however, it had no applications. It was observed then that if a substance were irradiated by X-rays a secondary emission took place. This fluorescent light could be transferred through an analysator crystal, where its differences in wavelength and strength could be used for a qualitative test and quantitative determination of the metals with atomic numbers higher than 12. X-ray fluorescence analysis has the advantage over emission spectrographic analysis that it can also be used for components present in high concentrations. Although Hadding (1923) as well as Hevesy and Alexander (1932) were interested in these phenomena in the twenties, the method has been developed mainly in the last decade, when the technical improvement of radiochemical instruments has made the work in this field easier.

Lundegårdh was also the founder of flame photometry, for although the history of flame photometry is difficult to separate from that of spectroscopy, it has several features which merit separate consideration.

The possibility of carrying out flame photometric measurements (as we know it today) was first realized by Janssen who noted it in a paper published in 1870.

> If we presume that all the metal molecules formed from metal compounds by excitation undergo the same process of ignition during the excitation, and if in so doing they produce equal amounts of light, then it can be concluded that the total amount of emitted sodium light — from the beginning of the decomposition of the sodium salt until the cessation of any colour in the flame — will be proportional to the amount of sodium present in the salt. Therefore all the processes which give out illumination can be used for the quantitative determination of the amount of metal concerned. By this means the analysis can be simplified to the measurem entof the amount of light for all cases in which the metal produces a characteristic emission in the flame [92].

Janssen's theory was proved experimentally three years later by Champion, Pellet [93] and Grenier. They determined sodium with an apparatus which consisted of a spectroscope and an instrument in which the extinction of light from a known amount of sodium, using a blue glass was compared with the required extinction for the substance to be determined [94]. Truchot, in 1874, determined lithium in the following way: He immersed a platinum spiral in the solution to be determined, and then placed this in the flame of a Bunsen burner, just in front of the slit of a spectroscope. He then compared the intensity of the Li-α line with the intensity of lines obtained with solutions of known concentration. The comparison was made simply by observation of one sample after another [95]. Ballmann re-examined this method shortly after its publication and found that it was not even approximately accurate. He tried viewing the two spectra simultaneously, but this also was not suitable, because the two pictures in the spectroscope did not show the maximum light intensities at exactly the same moment. Ballman also found that the measurement of time taken for the disappearence of the line from the spectrum was of no value. Finally he devised a dilution method, whereby the solution to be determined was diluted until the Li-α line disappeared. This was compared with solutions containing known amounts of lithium [96].

The next important event in the development of flame photometry was the introduction by De Gramont, in 1923, of the oxygen-acetylene flame for use in spectrography [97]. Lundegårdh, in 1928, developed a flame photometer which consisted of an atomizer, pressure control unit and an acetylene-air flame [98]. He first of all used a spectroscope, but later employed a single monochromator for the dispersion of the light measuring the light intensity with a photocell. Schuhknecht [99], in 1937, replaced the monochromator with simple coloured filters [100] and in this form the flame photometer is still used today.

4. COLORIMETRY

Long before it was summarized as an exact law it had been observed that there was a correlation between the colour of a solution of a coloured substance and the concentration of the solution. This obviously had been noted in numerous every-day observations.

I found the first description of a method based on colour comparison in an article dating from 1845.* Its author Heine [101], determined the bromine content of mineral waters [102]. The method was based on the known reaction of bromine, namely that it dissolves in ether to give a reddish brown colour ...

> First I prepared a series of solutions, containing known amounts of bromine. To 25 g of water I added 5 to 50 mg of bromine in small increments. This was achieved by adding 5, 10, 15, 20, 25, 30, 35, 40, 45 and 50 mg of potassium bromide, and dissolving in water. To these solutions equal amounts of ether sulphurious was added and the bottles immediately sealed.
>
> Using the same vessel which I used for the measurement of the ether, I also added the chlorine water. Preliminary tests had shown how much of the latter was required because even the strongest bromine solution showed no further darkening in colour after sufficient chlorine water had been added. The addition of chlorine water was made very rapidly by two people, and the solutions vigorously shaken. After allowing the solutions to stand for one minute the ether layer separated from the water and showed a very regular colour scale from yellow to brown, indicating that my supposition that these solutions are suitable for comparison was true. For samples greater than 50 mg the comparison is inaccurate as slight changes in the dark solution could not be detected [102].

At the end of his paper Heine mentioned:

> For the determination of small amounts of bromine my method based on the colour scale is more accurate than the analytical determination according to Rose ... [102].

He also mentioned that iodine gives a colour reaction with starch, but as none of the waters that he examined contained iodine, he made no quantitative determinations based on this reaction.

Jacquelain [103], in 1846, devised a method for the determination of copper, based on the comparison of the colours of copper–ammonia complexes. The comparison was made with a reference solution which contained 0·5 g copper in 1 l. From this he took 5 ml in a test tube and added two drops of ammonia. The colorimeter which he describes consisted of three glass tubes of similar diameters. One of them had a cm^3 scale, with a white background, and was examined through an opaque, blue glass plate with a diaphragm. He diluted

---

* Recently Snelders refers (*Chemie and Techniek* **17** 498 [1962]) to an earlier colorimetric determination by W. A. Lampadius [*J. prakt. Chem.* **13** 385 (1838)] who estimated the iron and nickel content of a cobalt-ore by comparing the colour of the corresponding filtrate with the colours of solutions with known metal contents between 5—50 per cent Fe or Ni and 50—95 per cent Co in cylindrical tubes.

the solution under examination until its colour corresponded with that of the reference solution. For example he dissolved 2 g of brass in nitric acid, added ammonia, and diluted to 200 ml. He poured 50 ml of this solution into the test tube and added water in small portions until its colour was the same as that of a standard reference solution. If for example 25 cm³ of water was needed, then

the colours became similar and therefore the solution of 30 cm³ contains six times the amount of copper present in 5 cm³ of reference solution, thus

$$5 \text{ cm}^3 : 30 \text{ cm}^3 = 0{\cdot}0025 \text{ g} : x; \quad x = 0{\cdot}015 \text{ g copper,}$$

and

$$5 \text{ cm}^3 : 200 \text{ cm}^3 = 0{\cdot}015 : x; \quad x = 0{\cdot}6 \text{ g}$$

which is equal to 30 per cent copper [104].

Herapath [105] determined iron in 1852 by a colorimetric procedure using thiocyanate. He also used reference solutions which contained iron in amounts from 1 to 250 mg. He compared the colour of the solutions using a white paper screen [106].

In 1853 Müller [107] constructed an apparatus which he claimed made the evaluation more accurate than could be achieved with the eye. He called his apparatus "a complementary colorimeter" (Fig. 88). The principle of its operation is as follows:

The coloured solution is placed in the vertical cylinder A. This is made of glass, and at the bottom (e) a colourless biplan glass plate is used to seal it. On the side ($d.d.$) there is a scale graduated in millimetres. To the top of this cylinder a stopper (cc) is fitted and this contains an adjustable glass tube through its centre, this tube also being sealed by a colourless glass plate. This telescopic tube system stands on the wooden box. The mirror (i) can be adjusted by the knobs (kk), and allows sunlight to pass through the diaphragm and through the glass disk g (made of the complementary colour to the colour of the solution under investigation). Thus, the coloured light passes through the solution placed in A, and can be observed at a.

If the solution in A consists of diluted iron thiocyanate solution then a kingfisher-blue plate (g) is used. The tube A is closed with an opaque shield, and if the small tube is adjusted then when it is in the highest position the bottom plate appears to be reddish-yellow, but as the telescopic tube is gradually lowered the colour of this plate gradually changes, passing through white until finally it appears to be kingfisher-blue.

If one has some sense of colour, even in the case of slightly coloured solutions, one can determine the position where the plate appears to be white with an accuracy of some ten millimetres, and I base my colorimetric method on this fact.

We find the zero value of a dilute solution of known concentration, and note the position on the scale, this could be, for example n. We mark this distance on the coloured glass plate used, i.e. the height of the boundary liquid column. From the latter we can calculate the amounts of coloured substances using this value and the value obtained from the measurement; from the separation of the a and e bottom plates the concentrations can be calculated by using inverse proportions [108].

In a subsequent paper Müller describes the modification of the instrument as well as further uses for his method. He recommended here that the evaluation with the "standard liquid" should be made in all cases. This means that the one final "calibration" of the glass plate was not sufficient; this also being apparent from the fact that he had also recommended that the standard solution used for comparison should be equal not only in colour, but also in chemical composition, with the solution to be determined [109]. With his apparatus he investigated the behaviour of iron thiocyanate, and chromate solutions as well as the copper/ammonia system. He found that the iron thiocyanate method is

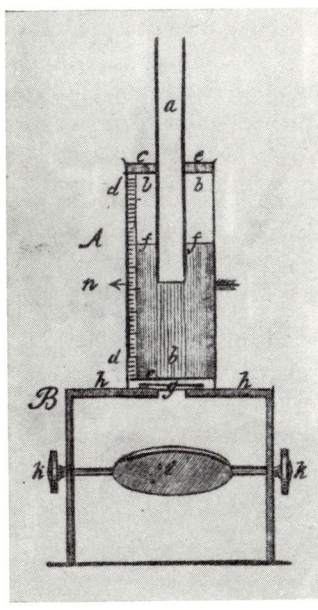

Fig. 88. Müller's "Complementär colorimeter", 1853.
(From his original paper)

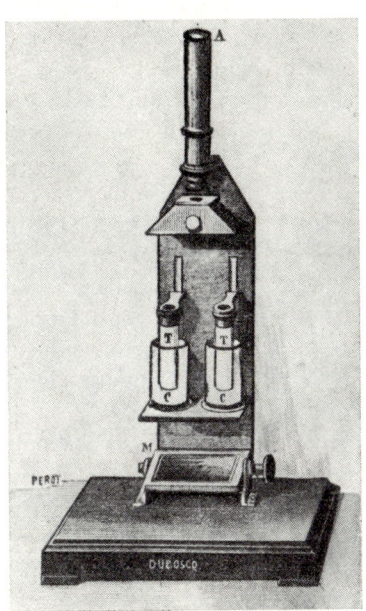

Fig. 89. Colorimeter of Duboscq.
(From *Chemical News*, 1870)

suitable for the determination of 0·00075 – 0·003 g iron/100 ml, and that the colour depends on the free acid and thiocyanate concentrations. He also claimed that by this method as little as 0·0001 mg of iron can be determined. He also examined the other two solutions in a similar manner.

The weakest point of this procedure is the method of illumination, which produces a completely uncontrollable light.

In 1864 Dehm constructed a new type of colorimeter for the determination of copper. This colorimeter was produced by the firm of Siemens-Halske. In this apparatus illumination was again made with sunlight and a mirror. In this case, however, there were two cylinders side by side, containing two closed tubes, which could be adjusted, inside them. These two tubes could be examined simultaneously by means of a lens. In the first of the tubes there was water, while in the second the ammoniacal copper solution was placed. The height of the water column was altered until the colours became equal. The calculation of concentration was also made on the basis of results obtained from solutions of known concentrations [110].

The latter method appeared to be more convenient, although Müller accepted it with reluctance, saying that it was similar to someone determining the concentration of acetic acid by diluting the solution with water until it gave the same taste as an acetic acid solution of known concentration [111].

Dehm's method was followed by Duboscq [112] who, in 1870, measured the colour density of the solution to be determined directly with a comparison solution. He passed the light through two glass prisms so that one half of a circular plate was illuminated by the light passing through the sample solution, while the other half was illuminated by light passing through the comparison solution. The instrument is shown in Fig. 89 [113]. In principle this device is very similar to the colorimeters in use today so that with the introduction of the Duboscq colorimeter this field had reached its peak, and we can pass on to a consideration of another branch, photometry.

Photometry, however, which is nowadays often referred to as colorimetry, differs from the latter in the method of measurement. While in colorimetry the optical part of the instrument is adjusted until equal colour strengths are obtained, in photometry monochromatic light is used mainly and the intensity of this light is adjusted accordingly.

Absorption spectroscopy is based on the law of light absorption. This is known as the Beer–Lambert law. These two workers did not collaborate, for a hundred years separated them. We could equally well call this law the Bougouer–Bernard law as these two Frenchmen established it independently of their German colleagues. Moreover, Bougouer long preceeded Lambert, while Bernard was only a few months behind Beer.

Bougouer's [114] work was published in 1829 in Paris under the title: *Essai optique sur la gradation de la lumière*. In this we can read:

> To examine the truth or the falsity of this idea, I passed a light of 32-candle power perpendicularly through two sheets of glass, after which I found its intensity to have been halved, for it was then only 16-candle power. Now if another thickness of two sheets of glass had caused an equal diminution, all the rays would have been intercepted. Nevertheless, the addition of two more sheets certainly did not form an absolutely opaque body. The light was still very bright, and when I passed it through ten sheets, it was still as intense as that from one candle.
>
> But doubtless the reader already sees quite clearly that, for a second thickness to intercept exactly the same number of rays, then exactly the same number of rays must be incident upon it. But if only one-third or one-quarter of the total number of rays are incident upon the second layer, it is certain that this layer must intercept only one-third or one-quarter as many rays as the first. Thus the equal layers should absorb not equal quantities, but rather proportional quantities. That is, if a certain thickness intercepts half of the light, the thickness that follows the first, and which is equal to it, will not intercept the entire second half, but only the half of this half and will consequently reduce it to one-quarter. Since all of the layers absorb like portions, it is obvious that the light will always be diminished in geometric progression.

In Lambert's work, however, far more mathematical exactness is to be found, although he already knew and referred to the work of Bougouer.

Lambert himself was not a chemist, but a physicist and mathematician. He was born in 1728, in Mühlhausen, into a poor family. He was completely self-taught and when he was later employed as a teacher by various aristocratic families he studied very hard during his spare time. Finally he settled in Berlin and carried out research there, as a result of which he was made a member of the Academy. Lambert was very conceited and there are many anecdotes recording this. One of these records how he was very worried because the king was very slow in making his appointment to the membership of the Academy. His friend Achard tried to cheer him by saying that he was certain that the king would make the appointment very soon.

> I am not a bit impatient,—replied Lambert—because this is a matter for his own glory. It would be a discredit to his reign in the eyes of posterity if he did not appoint me.

His appointment was eventually made and King Frederick the Great, speaking to him at a reception asked him which of the various branches of science he was most expert at, to which Lambert replied shyly "With all of them". "Thus, you are an eminent mathematician also?" asked the king. "Yes, Sir". "Who was your master in this science?" "I myself, Sir". "That means, that you are a second Pascal?" "At least, Sir", replied Lambert. After Lambert had left, the king commented that it seemed he had appointed a great fool to the Academy. The king was not quite right in his assessment of Lambert, for Lambert indeed was a brilliant mathematician, and his work in this field as well as in optics played a very important role in the history of sciences. Lambert died in Berlin in 1777.

His book, dealing with optical measurements, entitled *Photometria* was published in 1760. In it the following observation can be found:

> If a light beam with an intensity I passes through a layer of width l (in this case through glass), its intensity will decrease to $I\frac{1}{n}$-th of its original intensity while if it passes through a further similar layer its intensity will decrease to $I\frac{1}{n} \cdot \frac{1}{n}$ of its original value. If the width of the layer is m fold of the first, the intensity of the exit light will be

$$I' = \frac{I}{n^m}$$

In this work we can also find the following very interesting sentence:

> The amount of captured light is the greater, the greater the number of particles within a given volume and the larger the surface area of a single particle.

It is not certain whether by this he meant concentration, but it can be assumed that this is what he is referring to.

Later, in 1852, Beer [115] pointed out that Lambert's law is also valid for solutions, and that in certain cases concentration plays a similar role to the layer width [116]. Beer also defined the absorption coefficient, as did Bernard in the same year [117].

The concept of the absorption coefficient was introduced into analytical chemistry by Bunsen and Roscoe during their photochemical investigations.

Their definition of this function is the reciprocal of the layer width at which the original light intensity is decreased to 1/10th of its original value [118]. The absorption and concentration are proportional to one another, and this furnishes a convenient relationship for the determination of concentration. This, however, was only used for analysis by Vierordt.

Bahr [119] and Bunsen were the first to use absorption spectroscopy for quantitative analysis. In their paper dealing with erbium and yttrium we can read the following:

> For the rough determination of neodymium oxide in gadolinite in the presence of lanthanum we used a spectrum analytical titrimetric method. This is based on the comparison of the absorption bands of a neodymium solution of known concentration with another of an unknown one, and by diluting the latter until the same band intensities can be seen. The experiment can best be carried out as follows: Using two similar, calibrated glass tubes one (I) is filled with the solution to be determined, and placed in front of the free slit of the spectroscope, while the other (II) is filled with the standard neodymium solution, and placed before the prism slit. In this way two absorption spectra are obtained, one over the other. If the tube I contains T_1 cm^3 of solution, while the tube II contains T_2 cm$_3$ of solution, the latter containing $a\,g$ neodymium oxide per cm^3, to tube II, t cm^3 of water has to be added in order to obtain equal absorption strengths. In this case the weight of neodymium oxide present in the I solution is
>
> $$d = T_1 \frac{T_2}{T_2 + t} a$$
>
> The observation of the exact position where the two absorption bands show equal intensities is rather difficult [120].

Bunsen did not realize the analytical possibilities of the absorption coefficient. The comparison and evaluation was made on the basis of concentration-changes and not by the regulation of the light intensity. This latter possibility was first used in spectroscopy by Govi [121] and Vierordt [122]. In the spectroscope of Govi two comparison prisms were placed one on top of the other in front of the slit. This projected the light obtained from two sources on to a half-opaque plate. A narrow slit was used for examination of the chosen parts of the spectra separately. Govi obtained equal light intensities by adjusting the distance of the light sources [123] so that in effect what he was doing was making light intensity measurements, although later he claimed priority over Vierordt [124]. This, however, was an unjustified claim for even if his apparatus was constructed before that of Vierordt, the latter was the first to apply this method to analysis.

The founder of photometry as we know it today was Vierordt. His apparatus and the principle of his method was more perfect than that of Govi, and he also investigated the theory of photometric analysis. He used the laws of Lambert and Beer, and discovered that the use of the absorption coefficient of Bunsen and Roscoe gives a simple method for the evaluation of photometric measurements.

> The slit of Vierordt's apparatus was divided into two parts, and the width of both the upper and lower parts could be adjusted with a micrometer screw. He placed a glass cell in front of the slit, and filled it half full with the solution to be determined. The light from the petroleum

lamp passed partly through the solution and the lower slit, and partly over the solution and through the upper slit. The light from the two slits was reflected into the spectroscope one over the other. In the case of similar slit widths the spectrum of the light beam passing through the solution was naturally weaker than that passing over the solution. Thus by making the upper slit narrower with the micrometer the two light intensities gradually became equal. The screw used for the adjustment of the upper slit had a scale calibrated from 0—100, which was a measure of the transmission of the solution. Vierordt published the optical density values in a table [125]. Later, in order to reduce the width of the light-band Vierordt used two adjustable diaphragms in the ocular telescope. Using this method he was able to determine permanganate, chromate, copper-ammonia, fuchsine, blood stains and other substances.

The error of Vierordt's method was that the monolateral alteration of the slit width caused changes not only in the light intensity, but also in the quality of the light. In the case of coloured solutions the light falling on the two slits was not quite the same. This error was overcome by Glan [126] and Hüfner [127], independently of each other, by decreasing the light intensity, not by altering the slit width, but by polarization. Glan polarized both light beams [128], while Hüfner only polarized one [129]. The determination of the intensity ratios was made from the rotation angles of the Nicol-prisms.

These instruments were, however, far too complicated for use in everyday routine analysis. A simplification was needed, and this need for simplification was emphasized by G. Krüss [130] in his book published at the end of the last century. This book can be regarded as the first comprehensive colorimetric monograph [131]. Krüss himself attempted to devise a simpler instrument.

Krüss was of the opinion that absorption spectroscopy should be combined with the instruments and principles of colorimetry. He investigated several problems connected with this and constructed several instruments. This part of his work was greatly facilitated by the fact that his father owned an optical factory. He made several attempts at producing homogeneous light not only with a spectroscope but also with simple glass filters, while for the equalization of light intensities he suggested several ideas such as the variation of layer-widths, and the use of grey glasses, etc. He called his instruments spectrocolorimeters.

Thus we can see that two types of photometers were developed, namely those based on slit-width variation and those on the polarization of light, and both these types were subsequently defined and modified. A large number of commercial instruments are now available, and although most of them have some new innovation, they are all essentially similar. The polarization type of photometer was initially the more common, but recently the slit-variation types of instrument are the more popular.

As we have seen the original spectrophotometers are older than the slit-variation types, the latter being constructed for laboratory routine work. Photometers incorporating photocells have only been developed in this century. According to Willard [132], Berg was the first to produce an instrument of this type in 1911. This utilized a selenium element. A year later Pfund described the conditions for the use of selenium cells in photometry [133], but in practice, however, this type of instrument only came into general use after 1925.

NOTES AND REFERENCES

1. NEWTON, ISAAC (1642—1726) was the son of a smallholder. He was one of the greatest, if not the greatest personality of physics. From 1669 to 7701 he was Professor of Mathematics at Cambridge University, and he later became Master of the Royal Mint in London. In 1705 he became an earl. He was a President of the Royal Society. Newton's most notable achievements were the introduction of differential and integral calculus, as well as the derivation of the basic laws of mechanics and optics. He is buried in Westminster Abbey.
2. NEWTON, I.: *Philosophiae naturalis principia mathematica.* London (1687); *Optics or a treatise of the reflection, refractions, inflections and colours of light.* London (1704); New theory of light and colours. *Philos. Transact.* (1672)
3. MARGGRAF A. S.: *Opuscules chimiques.* Paris (1762) **2** 338 374
4. HERSCHEL, FRIEDRICH WILHELM (1738—1822) was born in Hanover. The princes of Hanover at that time were also the Kings of England. Herschel came to London as a member of an orchestra of the guards. After leaving the army he remained in London as an organist. He studied astronomy as a hobby and in 1781 he discovered the planet Uranus. As a result of this he became famous and from then on devoted all his time to astronomy. The King appointed him Astronomer Royal. He was a member of the Royal Society.
5. HERSCHEL, F. W.: *Philos. Transact.* (1800) 255 284
6. RITTER, JOHANN WILHELM (1776—1810) was a physician, who practised in Jena and Gotha. He also carried out scientific investigations, and made important discoveries regarding the chemical effect of electricity (Volta pile). The importance of Ritter's work is still very much underrated.
7. RITTER, J. W.: *Gilb. Ann.* (1801) **7** 527; (1803) **12** 409
8. WOLLASTON, W. H.: *Philos. Transact.* (1802) **2** 365
9. YOUNG, THOMAS (1773—1829) was first of all a practising physician, but from 1800 he was Professor of Physics at the Royal Institution. He later became the chief physician at a London Hospital. He was also a member of the Royal Society.
10. YOUNG, TH.: *Philos. Transact.* (1802) **2** 387
11. FRAUNHOFER, J.: *Denkschr. Münch. Akad.* (1817) **5** 193; *Gilb. Ann.* (1817) **56** 264
12. HERSCHEL, JOHN FREDERICK WILLIAM (1792—1871) son of the astronomer Herschel; he became an astronomer, and a member of the Royal Society.
13. HERSCHEL, J. F. W.: *Transact. Edinb.* (1823) **9** 445
14. TALBOT, WILLIAM HENRY FOX (1800—1877) was a wealthy Englishman who devoted his life to scientific research. In 1839 he made the first photographs on paper (Talbotypia). He was a member of the Royal Society.
15. TALBOT, W. H. F.: *Brewsters Journ. Sci.* (1825) **5** 77
16. TALBOT, W. H. F.: *Phil. Mag.* (1834) **4** 112
17. BREWSTER, DAVID (1781—1868) was first a pharmacist, and then an advocate in Edinburgh, and finally became Professor of Physics at the University of St. Andrews. He published many papers, dealing mainly with optics. He was a member of the Royal Society.
18. BREWSTER, D.: *Transact. Edinb.* (1834) **12** 519
19. MILLER, WILLIAM HALLOWS (1801—1892) was Professor of Mineralogy at Cambridge University.
20. MILLER, W. H.: *Phil. Mag.* (1839) **15**
21. MILLER, W. H.: *Phil. Mag.* (1833) **2** 381
22. HERSCHEL, J. F. W.: *Philos. Transact.* (1840) **1** 1
23. BECQUEREL, ALEXANDRE EDME (1820—1891) was the son of the Antoine Caesar Becquerel who worked in the field of electrochemistry, and was the father of the discoverer of uranium radiation (Antoine Henry); he was a Professor at the Musée d'Histoire Naturelle.

24. Becquerel, A. E.: *Bibl. Univ. Genève* (1842) **40** 341
25. Draper, John William (1811—1882) was a physician and Professor of Chemistry at the University of New York.
26. Draper, J. W.: *Phil. Mag.* (1842) **21** 348
27. Draper, J. W.: *Phil. Mag.* (1845) **26** 465
28. Miller, William Allen (1817—1870) was Professor of Chemistry at Kings College in London.
29. Miller, W. A.: *Phil. Mag.* (1845) **27** 81
30. Stokes, George Gabriel (1819—1903) was Professor of Mathematics at Cambridge University. He was a member of the Royal Society, and was also the discoverer of fluorescence.
31. Stokes, G. G.: *Philos. Transact.* (1852) **2** 463
32. Masson, A.: *Ann. chim. phys.* (1851) **31** 295
33. Ångström, Anders Jöns (1814—1874) was an astronomer, and head of the Observatory of Uppsala. He later became Professor of Physics at the University of Uppsala, and a member of the Swedish Academy of Sciences.
34. Ångström, A. J.: *Vetensk. Akad. Handl.* (1852) 327
35. Alter, David (1807—1881) a physician in Pennsylvania.
36. Alter, D.: *American J.* (1854) **18** 55; (1855) **19** 213
37. Swan, William (1818—1894) was Professor of Physics at the University of St. Andrews from 1859 to 1880.
38. Swan, W.: *Pogg. Ann.* (1857) **100** 306
39. Helmholtz, Hermann (1821—1894) was a physician. He worked first as an army doctor, but later became Professor of Physiology at the University of Königsberg, and then at the University of Bonn, and finally at Heidelberg. From 1871 onwards he was Professor of Physics at the University of Berlin. He is famous for his important contributions to all branches of physics.
40. Helmholtz, H.: *Pogg. Ann.* (1855) **94** 1
41. Robiquet, E.: *Compt. rend.* (1859) **49** 606
42. Plücker, Julius (1801—1868) was Professor of Mathematics at the University of Halle. He later became Professor of Physics at the University of Bonn.
43. Plücker, J.: *Pogg. Ann.* (1858) **103** 88; (1859) **107** 497 638
44. Hittorf, Johann Wilhelm (1824—1914) was Professor of Physics and Chemistry at a High School in Münster. His work was mainly concerned with electrochemistry (transference numbers).
45. Hittorf, J. W.: *Pogg. Ann.* (1869) **136** 1 197
46. Willigen, Volkert Simon, van der (1822—1879) was Professor of Mathematics at the High School of Deventer (Netherlands).
47. Willigen, V. S. van der: *Verslagen en Mededeeling d. Kon. Acad. v. Wet.* **7** 209 266; **8** 37 189 308; **9** 300
48. Kirchhoff, G.—Bunsen, R.: *Pog. Ann.* (1860) **110** 160
49. Roscoe, H.: *Gesammelte Abhandlungen von Bunsen.* Leipzig (1904) **1** 34
50. Kirchhoff, G.—Bunsen, R.: *Pogg. Ann.* (1861) **113** 337
51. Kirchhoff, G.: *Pogg. Ann.* (1859) **109** 275
52. Janssen, Jules Pierre (1824—1907) was a French astronomer, who is famous for his discovery of the atmosphere of the planet Mars. He was a member of the French Academy, the Royal Society and many other scientific societies.
53. Janssen, J. P.: *Compt. rend.* (1868) **67** 838
54. Lockyer, Joseph Norman (1836—1920) was director of the Astrophysical Institute at Kensington, and a member of the Royal Society.
55. Lockyer, J. N.: *Proc. Roy. Soc.* (1870) **18** 354
56. Kirchhoff, G.: *Abhandl. Berlin. Akad.* (1861) 93
57. Kirchhoff, G.: *Abhandl. Berlin. Akad.* (1862) 227; (1863) 225
58. Wheatstone, Charles (1802—1875) was a musical instrument manufacturer. He also

worked in physics apart from this and eventually became Professor of Physics at Kings College in London, until he retired and lived from the royalties of his many patents.

59. WHEATSTONE, CH.: *Rep. Brit. Assoc.* (1835) 11; WHEATSTONE, Ch.: *Chem. News* (1861) **3** 198
60. ÅNGSTRÖM, A. J.: *Pogg. Ann.* (1862) **117** 290
61. KIRCHHOFF, G.: *Pogg. Ann.* (1863) **118** 94
62. *Classified and Descriptive Catalogue of Chemical Apparatus.* John Griffin and Son. London (1866) 377
63. WARTHA, W.: *Die qualitative Analyse mit Anwendung der Bunsenschen Flammenreaktionen.* Zürich (1867)
64. BUNSEN, R.: *Pogg. Ann.* (1857) **100** 43
65. BUNSEN, R.: *Lieb. Ann.* (1868) **148** 269
66. MASCART, NICOLAS (1837—1908) was Professor of Physics at the Collège de France, and a member of the French Academy.
67. MASCART, N.: *Compt. rend.* (1863) **57** 789; (1864) **58** 1111
68. BERNARD, F.: *Compt. rend.* (1864) **58** 1153;
    DITSCHMAIR, L.: *Ber. Wien. Akad.* (1865) **52** 289;
    STEFAN, J.: *Ber. Wien. Akad.* (1866) **53** 521
69. ÅNGSTRÖM, A. J.: *Recherches sur le spectre normal du soleil.* Upsala (1868)
70. THALÉN, TOBIAS ROBERT (1827—1905) was a co-worker of Ångström at the University of Uppsala, and later succeeded him as Professor of Physics there. He was a member of the Swedish Academy of Sciences.
71. THALÉN, T. R.: *Nova acta reg. soc. Ups.* (1884) **12** 1
72. ROWLAND, HENRY (1848—1901) was Professor of Physics at the Johns Hopkins University in Baltimore.
73. ROWLAND, H.: *Astrophys.* (1893) **12** 321
74. LANGLEY, SAMUEL (1831—1906) was director of the observatory at Alleghan (U.S.A.).
75. LANGLEY, S.: *Amer. J. Sci.* (1881) **21** 187
76. LOCKYER, J. N.: *Philos. Transact.* (1874) **164** 805
77. HARTLEY, WALTER NOEL (1843—1913) was a physician. From 1879 he was Professor of Chemistry at the University of Dublin.
78. HARTLEY, W. N.: *Philos. Transact.* (1884) **175** 50 257
79. POLLOK, JAMES H.: (1868—1916) was senior lecturer in physical chemistry at the University of Dublin.
80. GRAMONT, ARNAUD DE (1861—1923) was a French chemist and minerologist, and a member of the French Academy of Sciences.
81. POLLOK, J. H.—LEONARD, A. G.: *Roy. Dublin Soc. Proc.* (1907);
    GRAMONT, A. DE: *Compt. rend.* **159** 171 1106.
82. KAYSER, HEINRICH (1853—1940) was Professor of Physics at the University of Bonn. He retired in 1920.
83. KAYSER, H.: *Handbuch der Spektroskopie.* Leipzig **5** 27
84. GERLACH, WALTER, was born in 1889. He studied at the University of Tübingen. On the completion of his studies he became a lecturer there and also became a Privat Dozent. He afterwards worked for a while in industry, until in 1921 he was appointed Professor Extraordinarius in Frankfurt. In 1924 he became a Professor of Physics in Tübingen, and later, in 1929, he was appointed Professor of Physics at Munich. In 1957 he retired.
85. GERLACH, W.: *Z. anorg. Chem.* (1925) **142** 383
86. SCHWEITZER, EUGEN (1905—1934) was a physicist, who studied at the Technical University of Stuttgart, and later at the University of Tübingen. He then worked in industry, and although his early scientific work gave the promise of a great future he unfortunately died at the age of 29.
87. SCHWEITZER, E.: *Z. anorg. Chem.* (1927) **164** 127
88. SCHEIBE, GÜNTHER, was born in Munich in 1893. He studied in Munich and Erlangen, and later became Professor of Physical Chemistry first of all at the University of Erlangen, and then at the Technical High School of Münich.

89. SCHEIBE, G.—NEUHÄUSSER, A.: *Z. angew. Chem.* (1928) **41** 1218
90. LUNDEGÅRDH, HENRIK GUNNAR, was born in 1888. From 1918 he was Professor of Plant Physiology at the University of Uppsala.
91. LUNDEGÅRDH, H.: *Quantitative Spektralanalyse*, Jena (1929)
92. JANSSEN, J.: *Compt. rend.* (1870) **71** 626
93. PELLET, HENRY (1848— ?) was a chemist with various industrial and mining companies.
94. CHAMPION, P.—PELLET, H.—GRENIER, M.: *Compt. rend.* (1873) **76** 707
95. TRUCHOT, P.: *Compt. rend.* (1874) **78** 1022
96. BALLMANN, H.: *Z anal. Chem.* (1875) **14** 297
97. GRAMONT, A. DE: *Compt. rend.* (1923) **176** 1104
98. LUNDEGÅRDH, H.: *Ark. Kemi Mineral. Geol.* (1928) **10** 26
99. SCHUHKNECHT, WOLFGANG, was born in Leipzig in 1908. He worked in industry, and is now the director of the laboratories of Saargruben A. G.
100. SCHUHKNECHT, W.: *Z. angew. Chem.* (1937) **50** 299
101. HEINE, CARL (1808— ?) was an official analyst at ore laboratories in Eisleben, and later director of a Czech mine and iron foundry. He was the author of several analytical papers and books.
102. HEINE, C.: *J. prakt. Chem.* (1845) **36** 181
103. JACQUELAIN, AUGUSTIN (1804—1885) was born in Italy. He was a chemist, and worked in industry, in France.
104. JACQUELAIN, A.: *Compt. rend.* (1846) **22** 945
105. HERAPATH, THORNTON JOHN (1830—1858) was an industrial chemist in Bristol, but later worked at a metallurgical plant in South America. While returning to England his ship was wrecked and he was lost.
106. HERAPATH, T. J.: *Quarter. Journ. Chem.* (1852) **5** 27; *J. prakt. chem.* (1852) **56** 255
107. MÜLLER, ALEXANDER (1828—1906) was Professor of Agricultural Chemistry at the industrial school of Chemnitz, and later at the University of Stockholm.
108. MÜLLER, A.: *J. prakt. Chem.* (1853) **60** 474
109. MÜLLER, A.: *J. prakt. Chem.* (1855) **66** 193
110. DEHM. F.: *Z. anal. Chem.* (1863) **2** 143
111. MÜLLER, A.: *J. prakt. Chem.* (1865) **95** 36
112. DUBOSCQ, JULES (1817—1886) was an optician and the owner of an optical workshop. He lived in Paris, and wrote several scientific papers.
113. DUBOSCQ, J.: *Chem. News.* (1870) **21** 31
114. BOUGOUER, PIERRE (1698—1758) was a French mathematician and astronomer. He made astronomical observations for several years in South America. BOUGOUER, P.: *Essai optique sur la gradation de la lumière.* Paris (1729). Referred in: MALININ, R.—YOE, J. H.: *Journ. Chem. Ed.* (1961) **38** 129
115. BEER, AUGUST (1825—1863) was Professor of Mathematics at the University of Bonn.
116. BEER, A.: *Pogg. Ann.* (1852) **86** 78
117. BERNARD, F.: *Ann. chim. phys.* (1853) **35** 385
118. BUNSEN, R.—ROSCOE, H.: *Pogg. Ann.* (1857) **101** 235
119. BAHR, JONS FRIDRIK (1815—1875) was a lecturer at the University of Uppsala, and a member of the Swedish Academy of Sciences.
120. BAHR, J.—BUNSEN, R.: *Lieb. Ann.* (1866) **137** 1
121. GOVI, GILBERTO (1826—1889) was a physicist. He was Professor of Physics at several Italian Universities and high schools including Mantova, Florence, Turin and Naples.
122. VIERORDT, CARL (1818—1884) was a physician. He first had a private practice in Karlsruhe, and also carried out research during this time. In 1849 he became Professor of Physiology at the University of Tübingen.
123. GOVI, G.: *Compt. rend.* (1860) **50** 156
124. GOVI, G.: *Compt. rend.* (1877) **86** 1044 1100
125. VIERORDT, C.: *Pogg. Ann.* (1870) **140** 172; *Die Anwendung des Spektralapparates zur Photometrie der Absorptionsspektren und zur quantitativen Analyse.* Tübingen (1873)

126. GLAN, PAUL (1846—1898) was a professor at the University of Berlin.
127. HÜFNER, CARL GUSTAV (1840—1908) was a physician, and Professor of Organic and Physiological Chemistry at the University of Tübingen.
128. GLAN, P.: *Wied. Ann.* (1877) **1** 351
129. HÜFNER, G.: *J. prakt. Chem.* (1877) **16** 290
130. KRÜSS, GERHARD (1859—1895) studied in Munich, where he later became a Privat Dozent, and subsequently Professor Extraordinarius in the institute of Baeyer. His early death spoiled a very promising scientific career. He founded the journal *Zeitschrift für anorganische Chemie*.
131. KRÜSS, G.—KRÜSS, H.: *Kolorimetrie und quantitative Spektralanalyse.* Hamburg—Leipzig (1891)
132. WILLARD, H. H.: *Anal. Chem.* (1951) **23** 1728
133. PFUND, A.: *Physik. Z.* **13** 567

CHAPTER XII

THE DEVELOPMENT OF THE THEORY OF ANALYTICAL CHEMISTRY

1. THE DEVELOPMENT OF PHYSICAL CHEMISTRY

It has already been mentioned that Immanuel Kant, the great German philosopher of the 18th century, stated that

> In the various branches of the natural sciences only that which can be expressed by mathematics is true science.

Ever since Kant gave this definition chemists have suffered from a feeling of inferiority. Although, as a result of the work of Lavoisier and the pioneers of stoichiometry, chemistry obtained a quantitative basis the mathematics involved was merely the use of direct proportions. This was trivial compared with the highly complex mathematical reasoning encountered in physics or astronomy, where the use of differential equations had been commonplace a hundred years earlier. It was in physical chemistry that any mathematics more complicated than the calculation of direct ratios first appeared, and this involved the use of fashionable differential equations. Wilhelmy [1] was the first to establish that in the inversion of cane sugar the rate of disappearance of the sugar dz/dt is proportional to its concentration (1850). Ostwald later commented cheerfully that this equation for the reaction velocity was the first possibility of expressing the course of a chemical reaction in the language of mathematics.

Physical chemistry originated about the middle of the last century, but Lomonosov had indicated the nature and scope of this science a hundred years earlier:

> Physical chemistry is that science which, using the laws of physics, explains how compounds are formed in chemical processes.

There were two methods of approach to this problem, the first originated from the old concept of affinity. Ever since the time of Albertus Magnus, this phenomenon of affinity held great interest for scientists and many famous names are associated with it, such as Glauber, Geoffroy and Bergman.

At the beginning of the 19th century Berthollet proposed a physical explanation for this concept. His ideas on the subject originated mainly from speculation;

he considered that many factors such as cohesion and elasticity, as well as the mass of substances taking part in the chemical reaction and also the gravitational force should be taken into account. Berthollet, however, did provide several correct establishments in this explanation. Guldberg [2] and his brother-in-law, Waage [3] began their investigations on the basis of Berthollet's speculations, but they replaced the mass of the reactions by their concentration. Thus, in 1867 they derived the Law of Mass Action. The name of this law recalls Berthollet's original ideas. Affinity also can be used to explain the reaction velocity. This, as has already been mentioned, was pointed out by Wilhelmy in 1850 for the inversion of cane sugar, while the research works of Berthelot [4] and Menshutkin [5] proved this law for other reactions.

The great importance of heat in chemical and physical processes had already been realized, but for a long time the nature of heat was a mystery to chemists who mainly regarded it as a corpuscular phenomenon. Berzelius included heat among his list of "unmeasurable substances". The idea which originates from the 17th century, that heat is caused by the unobservable movement of the tiny particles of matter was still accepted.

The general subject of heat, and the problems which it introduced became the basis of the new science of thermodynamics.

This field of science originated with the research work of Carnot [6], who in 1824 examined the course of the energy transformations in steam engines. His theory, in which he introduced the second law of thermodynamics, did not cause any great interest even after Clapeyron [6] clarified it with a mathematical interpretation. In 1841 Mayer published the law of conservation of energy, while Joule [7] determined the mechanical equivalent of heat in 1843. In the next ten years as a result of these discoveries Helmholtz, Thomson (Lord Kelvin) [8] and Clausius [9] developed the theory of classical thermodynamics. This work was carried out in physics at a time when most of the scientific interest of chemists was concerned with organic chemistry, and it seemed very unlikely at that time that these two sciences would ever meet.

Horstmann [10] applied thermodynamics to the problems of chemistry for the first time, in 1870, to calculate the natural dissociation of gases. On the basis of the entropy-law he was able to deduce the law of mass action. A very important work was the 300 page study of Willard Gibbs [11], entitled *Thermodynamical principles determining chemical equilibria* which was published from 1871 onwards in instalments in a little known journal of the Connecticut Academy. In this work he applied the principles of thermodynamics to chemical processes, and also introduced the phase rule as well as the idea of chemical potential. "With this thesis chemistry reached that degree of versatility and precision, which had been attained by mathematical physics a century ago." This was Ostwald's comment on this work.

In 1877 a botanist, Pfeffer [12] constructed a semipermeable membrane. The idea behind this probably came more naturally to a botanist who often observed these membranes in living cells than it would to a chemist. His apparatus was successful; by immersing the membrane-vessel attached to a manometer filled with sugar solution into water he was able to observe the height of the liquid in the manometer rise in an amount proportional to the concentration of the sugar solution. Pfeffer reported the result of this experiment to his friend Clausius, who did not believe it, but who, in any case, was not really interested in the subject. In 1883 Raoult [13] established that for a given solvent equimolecular amounts of different solutes cause the freezing point to be depressed and the boiling point elevated by equal amounts. The young van't Hoff [14] concluded from Pfeffer's results that the gas laws are also valid for solutions. His theory proposed in 1886 proved correct and gave an explanation for the observations of Raoult, at least for organic substances. When inorganic solutes were investigated values two or three times the expected value were obtained.

This anomaly was solved by the electrolytic dissociation theory of Arrhenius. At this point the development of physical chemistry originating from the idea of affinity crossed the other original approach, the phenomena of galvanism.

Grotthus [15], in 1819, suggested the first theory to explain the chemical reactions induced by the Volta column. According to this theory molecules are so arranged that if an electrical current is applied to them their negative side moves to the positive electrode, while their positive side moves to the negative electrode. This effectively cuts the molecule into two pieces, those that are in the neighbourhood of the electrodes are deposited, while the others reunite with different partners. Faraday [16], in 1833, discovered the laws of electrolysis, which are named after him. Daniell [17] established that salts are decomposed by electricity to give metals and acid radicals, whereas according to the dualistic theory of Berzelius they would be decomposed to give metal oxides and anhydro acids. He also observed that in the neighbourhood of the electrodes the components of the solution are either concentrated or diluted.

The question whether the cation and anion play equal parts in the transference of the current, in other words whether their velocities are equal or not, was raised by Hittorf. By a series of ingenious experiments in 1853, he determined the ion mobilities and introduced the concept of transference numbers. His results did not attract any attention for a considerable time. In 1867 Kohlrausch [18] developed a new method for the measurement of conductance, and this method is still used today in a slightly more refined form. He examined Hittorf's transference numbers and discovered that the velocity of an ion is always constant, and is independent of the nature of any other ions in solution or of the partner with which it forms a molecule. This conclusion aroused great controversy as it contradicted the existing theories which supported the idea that different compounds are bound with differing affinities. It would therefore be unreasonable to expect potassium to have equal velocities in the presence of acetate or chloride where the affinities of the two molecules are vastly different. It was also discovered at about this time that electrolytes will conduct a current even if very low voltages are applied, whereas on the basis of the dualistic theory it would have been expected that a considerable voltage would be required for the decomposition of the molecules. As a result of this Clausius proposed that the decomposition of the molecules is not caused by the current, but by the fact that the molecules themselves are in collision and that the decomposition products are responsible for conducting the current. If this was in fact the case then it would follow that

FIG. 90. Wilhelm Ostwald (1853—1932). Drawing by Klamrath

in dilute solution the conductivity should decrease, as the frequency of collision is lowered. The experimental evidence, however, showed the reverse of this situation to be true, and it was not until 1887 when Arrhenius [19] proposed his theory of dissociation into ions that the problem was solved. Arrhenius's theory also explained the phenomena of galvanism, as well as clarifying the anomalies observed in connection with the van't Hoff theory.

Thus the basic principles of physical chemistry had been developed, and what was most needed at this stage was someone who could collect and correlate the isolated facts. This had already been achieved, in part, by Arrhenius, but the major contribution to this problem was made by Wilhelm Ostwald. It was he who related Arrhenius's theory to the law of mass action, and thus applied the equilibrium constant to processes involving dissociation. In 1893 he determined the ionic product of water, and followed this with his new theory of acids and bases, and also stated the dilution law. All this was described in a continuous, logical argument, applied to well known phenomena of chemistry in his book *Lehrbuch der allgemeinen Chemie*, which was published in several editions, as well as being published in many articles in the journal *Zeitschrift für physikalische Chemie*, which was founded by Ostwald himself.

This is a very brief review of the early development of physical chemistry. Many new developments in physical chemistry have been used to explain phenomena encountered in analytical chemistry. At the end of the century the time came to develop the theory. This was done by Wilhelm Ostwald, too.

2. WILHELM OSTWALD

Wilhelm Ostwald was born in 1853 in Riga, Latvia, into a wealthy German family. He was educated in Riga, and proceeded from there to the University of Dorpat. When he first went up to the university he did not do very much work, but after a stern warning from his father he reformed somewhat and from then on his examination results were very good. After completing his studies he remained at the university as a lecturer. In 1882 he was offered the chair of chemistry at the Technical University of Riga. Five years later, in 1887, he went to the University of Leipzig, where his famous scientific career began. Ostwald married, and his three sons all had names beginning with W, and they all became chemists, the most famous being Wolfgang Ostwald, who was the well known colloid chemist. In 1906 Wilhelm Ostwald retired and he died in 1932.

Ostwald began his intensive research works in Riga. From measurements of reaction velocity he was able to calculate affinity constants and thus arrange the various acids in order of increasing strengths. During this time he heard of the work of Arrhenius who had just submitted a doctoral thesis at the University of Uppsala. Arrhenius's thesis, which expounded the ionic dissociation theozy, was accepted rather dubiously by the University. Ostwald immediately realized the importance of this theory and travelled to Sweden to meet Arrhenius. The interest of the famous professor from Riga was very encouraging to the young Arrhenius, whose theory had met with doubts on all sides. Ostwald later told of his meeting with Cleve, who was Professor of Chemistry at Uppsala. Cleve took Ostwald into his laboratory, showed him a flask containing sodium chloride solution and asked him, "Do you really think that the sodium atoms are swimming in this glass by themselves?" When Ostwald replied "Yes", Cleve was very astonished.

In 1887 Ostwald was offered the chair of chemistry at the University of Leipzig, and the next decade was to be his scientific "golden age". It is true to say that he "organized" physical chemistry during this time, for he developed the dilution law, introduced the concept of the dissociation constants of acids and bases, calculated the ionic product of water, as well as investigated the process of neutralization, devised the theory of indicators and also examined the problem of the heat of neutralization, where he revived the long forgotten law of Hess, and developed a theory of catalytic action. In addition he wrote several books, *Allgemeine Chemie, Geschichte der Elektrochemie* as well as the work entitled *Die wissenschaftlichen Grundlagen der analytischen Chemie*, with which we are mainly concerned in this chapter. His Institute in Leipzig became the world centre for physical chemistry and it soon became a famous scientific forum. To be able to achieve this reputation in so short a time was remarkable because at that time Germany was the Eldorado of synthetic organic chemistry, and this was regarded as the only chemistry worth studying, the "great old chemists", Hoffmann, Baeyer, Wislicenus, Kekulé, etc., regarding the activities of this young man as well as his strange ideas with good natured amusement. Ostwald was also a very brilliant teacher, and he had soon founded a new school from his students; it was in Ostwald's Institute that Nernst developed his famous theory on the osmotic pressure of galvanic cells, and Beckmann his method for the molecule-weight determination by ebullioscopic measurement, and Bodenstein his investigations on catalysts. Other famous names among his assistants were Böttger, Bredig, Tamman, Mittasch, Freundlich, etc., and of his students sixty-three were later to become professors and thus influence the development of this new science.

Ostwald had an amazing capacity for hard work; he carried out research, published, and organized tirelessly. He was also interested in the history of chemistry, and this science is much indebted to him for his series of monographs entitled *Ostwalds Klassiker der exakten Naturwissenschaften*, which contained the most important papers on the natural sciences in a new edition. Ostwald was the founder and leader of a journal and a society. In the *Zeitschrift für physikalische Chemie* he published several thousand reviews of books and papers. He made journeys all over the world to give lectures, for not only did he carry out most of the research into this new science but he publicized it whenever he could. For these efforts he was nicknamed the "Little German Lavoisier". This is rather an apt description, for just as Lavoisier was the founder of the quantitative basis of chemistry, so Ostwald introduced the physico-chemical nature of chemical reactions. If, as sometimes occurs, there are now some chemists who consider that only physical chemistry is a true science, then this can also be attributed to the influence of Ostwald.

Physical chemistry began essentially from the concept of the conservation of energy. Ostwald was of the opinion that thermodynamics can be used to explain every phenomenon of chemistry. In 1907 he stated, "There are no problems which cannot be solved by thermodynamics and by the equations of Gibbs", but already the new discoveries in the field of radioactivity seemed to refute this statement.

Ostwald was, as he freely admitted, a fierce opponent of natural philosophy. He believed that the natural philosophy of the 18th century in Germany had been the cause of a complete lapse for half a century in the progress of the German natural sciences. He opposed natural philosophy so rigorously, and so long that he himself eventually created a new natural philosophy, which he called energetics, and in his later years he became far more concerned with his new "philosophy" than with chemistry.

Ostwald also opposed the theory of atomism, because he considered the atom as a remnant of the ancient, mystical philosophy. That Ostwald held this view was very unfortunate especially because chemists, with very few exceptions, did not investigate any of the phenomena of radioactivity. This is to some extent the reason why the subject of the atom, its disintegrations and transformations, which was for so long one of the mysterious aspects of chemistry, is nowadays included in the science of atomic physics.

Ostwald attempted to introduce thermodynamics into other branches of science; he tried to explain the phenomena and laws of physiology, and even to apply it to the human sciences and the arts, philosophy, psychology, sociology, as well as social life. He was also an artist, a talented painter and musician, so that his interests often extended in opposite directions. As a result of this he played the role of philosopher, but his philosophical views were only accepted courteously and secretly ridiculed. In 1906 he resigned his chair, partly because of his new interests. His colleagues were probably not sorry to see him go for Ostwald was not a very agreeable man; he had passionate views which he defended vigorously, and he was very shrewd and bitter in both criticism and in his editing, and was rather conceited about his work. His versatile talents, however, are indisputable; he painted well, was a talented musician, and his writing was excellent. His 27 books would be considered an accomplishment even by a novelist! His chemical textbooks are famous for their clear and understandable text, for Ostwald was a master at the explanation of the most complicated subjects in a logical, easily understandable, manner. His style is excellent as can be clearly seen in his philosophical and autobiographical works.

Wilhelm Ostwald was awarded the Nobel Prize for Chemistry in 1908. This honour was in recognition of his earlier accomplishments, which had been made mainly in the preceding century, for the work he was doing at the time of the award had no bearing on chemistry at all. The titles of the books that he wrote in this latter period of his work indicate this change: *Energetical imperativus* or *The biology of genius, Energetic fundamentals of cultural history*, etc. To show more clearly the misguided direction in which this former great scientist had turned one can refer to his famous "happiness-equation". Ostwald considered that the human senses can be explained in terms of thermodynamic equations, and this led him to propose a happiness-equation, as follows:

$$G = k(E + W)(E - W) = k(E^2 - W^2)$$

where G is the amount of happiness, E is the sum of energies directed by the will, W is the energy used up for experiences opposing the will, i.e. the sum of the energies used for overcoming obstacles, while k is a factor for energetical-physical transitions which is dependent on the individual. A more detailed description can be found in the original text [20]. Ostwald was very attached to this subject, and even twenty years later, in his last work which is a very interesting autobiography, he devoted a chapter to this equation of happiness.

The theoretical explanation of analytical phenomena Ostwald described in his book *Die wissenschaftlichen Grundlagen der analytischen Chemie*, i.e. Theoretical Background to Analytical Chemistry. This volume was published in 1894, but it was quite small, altogether only 187 pages. The object of the work is explained by the author in the preface [21].

> Analytical chemistry, which is the art of testing substances and their constituents, plays a very important role among the many applications of scientific chemistry, because those questions which it answers arise whenever chemical processes are used for scientific or technical purposes. It is a measure of its importance that it has been used since the earliest times, and has thus collected almost all the scientific observations of quantitative chemistry. It is deplorable, however, that while the technique of analytical chemistry stands on a very high level, its scientific treatment is almost completely neglected. In this field even the best textbooks, at the most review the various reaction equations, which are only valid in ideal cases. In fact these processes do not occur according to this ideal scheme, for there is no completely insoluble precipitate, and no absolutely accurate separation and determination. I am afraid, however, that these points are not made quite clear, not only to the student, but also to the practising analyst, to the detriment of the exact justification of analytical methods and results.
>
> Thus we can see that analytical chemistry plays the role of a servant to other sciences, being at the same time subordinate but also indispensable. While in all other branches of chemistry there is a vigorous activity in the theoretical evaluation of the scientific results, these problems often being of more interest to the scientists than mere experimental problems, analytical chemistry readily takes up the theoretical old suit which has been discarded by other sciences, and which is fifty years out of fashion. For example, it is still permissible, even today to record the components of potassium sulphate according to the dualistic theory of the 1820's as K_2O and SO_3.

Ostwald tried to overcome this handicap.

> This aim was not able to be achieved before, because scientific chemistry did not have adequate knowledge. We have now, however, reached the stage where the phenomena of analytical chemistry can be examined in a scientific manner.

His book begins with the identification and separation of different substances. He deals with precipitation in great detail, with the methods for increasing the crystal size of the precipitate, and with filtration and washing. In this section there is no theoretical treatment, and only one mathematical formula is deduced for the washing of precipitates. Here he deduces the relation between the decrease in the concentration of any contaminants, and the number of washings of the precipitate [22]. However, he mentions that this relationship is not valid in practice, for in the calculation the phenomenon of adsorption is not taken into account. According to his calculation, after ten washings the contaminant should be completely removed, whereas practical experience has shown that this is not

DIE
WISSENSCHAFTLICHEN GRUNDLAGEN
DER
ANALYTISCHEN CHEMIE.

ELEMENTAR DARGESTELLT

VON

W. OSTWALD.

LEIPZIG
VERLAG VON WILHELM ENGELMANN
1894.

FIG. 91. Title page of Ostwald's book *Die wissenschaftlichen Grundlagen der analytischen Chemie*

true. Unfortunately, Ostwald notes, we know hardly anything about the laws of adsorption, it being probable that it is proportional to the surface area as well as also being dependent on the nature of the solid and dissolved substance, and on the concentration of the latter. The result of a more recent calculation is that a factor k must be applied, and the previous value must be multiplied so that in a process where adsorption occurs the process of washing proceeds similarly, but only the k-th part of the washing solution is active during the process. Thus the concentration of contaminant decreases in a geometric series with the number of washings.

In the next chapter he examines the problem of increasing the particle size of precipitates. He explains the fact that allowing the solution to stand assists the formation of crystals of a large particle size, by the fact that the solubility of small crystals is greater than the solubility of larger ones. The surface tension between the liquid and solid phases therefore causes the decrease in surface area, i.e. the formation of larger crystals. The formation of larger crystals is also more beneficial because the amount of adsorption only occurs to a smaller extent. If the crystals are too large, however, occlusion of the mother liquor may occur.

The fourth chapter which has the title "Chemical Separation" is the most important in the whole book. After a presentation of the laws of solutions, in which he discusses the theory of electrolytes, Ostwald considers the law of mass action and applies this to the case of electrolytic dissociation. In this he introduces the concept of dissociation constants. He also describes the dissociation of multibasic acids:

> We can readily assume that in the case of a H_2A type acid the dissociation occurs according to the following scheme:
>
> $$H_2A \geq 2H + A''$$
>
> when the equilibrium would be as $ab^2 = kc$. This however is not so, as multibasic acids dissociate according to the scheme
>
> $$H_2A \geq H\cdot + HA'$$
>
> the univalent ion formed dissociates further according to the equilibrium
>
> $$HA \geq H\cdot + A''$$
>
> The dissociation constant of the second process is always much smaller than that of the first [23].

Ostwald then deals with the interactions of electrolytes, and establishes that in every reaction it is always the least dissociable substance that is formed. Thus the formation of water is the explanation of all neutralization processes. He then explains displacement reactions:

> If we add a strong acid to the salt of a weak acid, then as we have already mentioned it does not matter whether the alkali or acid corresponding to the salt is weak or strong, the salts are almost completely dissociated. The solution of a salt of weak acid therefore contains essentially only free ions. Now if to this a strong acid is added, the anions of the salt are united with the hydrogen ions, and form the undissociated weak acid. Thus, in the solution, apart from the cation of the weak acid, the anion of the strong acid will be present, i.e. the salt formed in the displacement is present [24].

On this basis he also explains the phenomenon, often used in analytical chemistry, that to decrease the acidity of the solution the salt of a weak acid must be added.

If, for example, an excess of sodium acetate is added to hydrochloric acid, not only is the weak acid acetic acid formed, but the excess of sodium acetate strongly represses the dissociation of the acid. Thus a solution is obtained which although slightly acidic behaves almost as a neutral salt. Thus, if a strong acid is added to the solution it will immediately be converted into the slightly dissociated weak acid, resulting in only a minute increase in the hydrogen ion concentration [25].

After a treatment of the roles of hydrolysis, heterogeneous equilibria, reaction velocity and temperature Ostwald turns to precipitation processes. He introduces here the concept of solubility product.

The object of the analyst is to develop such circumstances in the solution so that the solubility of the precipitate is decreased as much as possible... This can be reached by low temperature or by the addition of solvents which decreases the solubility. The solubility of a precipitate which is an electrolyte can be considerably decreased by the addition of a soluble electrolyte which has a common ion with the precipitate.

In the saturated aqueous solution of an electrolyte a complicated equilibrium is set up. The solid substance is in equilibrium with the undissociated substance in the solution, and this in turn is in equilibrium with the dissociated parts, i.e. with the ions. The first equilibrium is governed by the relative concentrations, mentioned previously, and as the concentration of the solid substance is constant then the concentration of the undissociated part in the solution must also be fairly constant. In the second equilibrium for the most common case, i.e. dissociation to give monovalent ions, then if the concentration of the ions are a and b, while that of the undissociated part is c, then:

$$ab = kc$$

As c is constant at a given temperature then the product kc and also ab are constant. Therefore between the solid substance and the solution above it an equilibrium exists so that the product of the concentrations of the two ions has a definite value. This product can be referred to as the solubility product... In certain cases where the solubility product of a solid substance in solution is exceeded then super-saturation results. In cases where the solubility product is not reached, then the solution will dissolve part of the solid material. This is very briefly the basis of the theory of precipitation...

The aim of the analyst is always to achieve the precipitation of a given ion. Barium sulphate is precipitated either for the determination of SO_4'' or $Ba^{..}$ ions... If we consider the precipitation of SO_4'' ions, when an exactly equivalent amount of barium salt is added an amount of the SO_4'' ions remain in solution, this amount being equal to the $Ba^{..}$ ions formed from the solubility product equilibrium. If further amounts of barium salt are added then one factor in the solubility product equation will increase, resulting in a decrease in the SO_4'' concentration, so that more barium sulphate will be precipitated. A further addition of barium salt will cause a similar effect; the concentration of SO_4'' ions however will never decrease to zero as the concentration of $Ba^{..}$ ions can never reach infinity.

Thus the well known rule that in a precipitation the precipitant must be present in excess is easily explained. The greater the solubility of the precipitate, then the greater is the excess of precipitant required.

In actual fact most precipitations only require a slight excess of precipitant, as for analytical purposes only the most insoluble precipitate can be used... [26]

Ostwald later points out that the same argument is valid for the washing of precipitates. It is therefore better to wash lead sulphate with a dilute sulphuric acid solution than with water. Similarly it is preferable to wash a mercurous chromate precipitate with mercurous nitrate solution rather than with water.

Manipulations with salts are relatively simple, but in case of acids and bases — because of their different dissociation — it is rather more complicated. For example calcium salts can be precipitated by carbonates but not by free carbonic acid.

> This is due to the fact that carbonates are normally dissociated so that if calcium ions are added to a carbonate solution then the product of carbonic acid and calcium ions exceeds the solubility product of calcium carbonate, and thus precipitation occurs. An aqueous solution of carbonic acid however contains only a very small amount of carbonic acid ions, owing to the slight dissociation of carbonic acid. Therefore even though a large concentration of calcium ions may be present the solubility product will not be reached and therefore no precipitation will occur [27].

Thus Ostwald used the solubility product principle together with the dissociation constant to explain phenomena which had been observed for a very long time.

The fifth chapter deals with the methods of weighing, and of making volume and other measurements. In the second part of the book Ostwald describes the reactions of the various ions, and discusses and explains them in detail.

He begins by treating the hydrogen and hydroxyl ions and in this section he illustrates his theory of indicators, which has previously been referred to. He also noted the interference of carbon dioxide experienced with acid—base titrations, as well as phenomena occurring in the titration of polybasic acids. Ostwald explains how to titrate phosphoric acid with various indicators. It must be mentioned, however, that Ostwald in his discussions of hydrogen ion concentrations, dissociation constants and solubility products did not use a quantitative treatment, and there are very few numerical values for the deduced constants. Among the very few numerical data that he gave was the dissociation constant of ammonia, which he found to be $k = 0.000023$, and mentions that in 0.1 N solution only 1·5 per cent of the ammonia dissociates [28].

Here are a few extracts from this part of his book:

> Among the salts of monovalent copper only the halogen compounds are known... the iodide is sufficiently insoluble to be used for the determination of copper. If to the copper(II) salt, potassium iodide is added, then cuprous iodide and iodine is liberated:
>
> $$Cu' + 2I' \geq CuI + I$$
>
> The reaction does not proceed to completion as the reverse reactions may also take place. To shift the position of equilibrium to the right it is necessary for one of the reaction products to be removed. Therefore the addition of sulphurous acid to the solution removes the iodine as it is liberated. This decrease in the concentration of the substances on the right hand side of the equation causes the reaction to go to completion ... [29].
>
> A solution of mercury nitrate gives a precipitate with urea, but a solution of mercury chloride gives no precipitation. In mercury nitrate solution hydrolysis results in sufficient

mercury(II) ions being present to exceed the solubility product of the urea complex, wherea in mercury chloride solution the number of mercury ions is very small and no precipitation occurs. Therefore if mercury nitrate is added to a solution containing chloride ions and urea, no precipitation will occur until sufficient mercury nitrate has been added to combine with all the chloride ions. The first excess above this amount results in the precipitation of the urea complex [30].

The very slight solubility (of barium sulphate) is responsible for the formation of a very finely divided solid, which absorbs strongly. This feature results in considerable errors in quantitative determinations. This error can be avoided by the formation of a coarse precipitate, i.e. by precipitating from hot, acidic solution... [31]

The last chapter of Ostwald's book deals with the calculation of the results of analyses. He first states that in the analysis of a complex substance (a mineral or mixture of salts) it is impossible to conclude from the results of the analysis how the various ions were originally combined. It is more exact, therefore, to present the results of analyses in the form of the amount of each ion present. This method of presentation of analytical results had been suggested much earlier (1865) by the Hungarian chemist Károly Than in his analyses of mineral waters. Than suggested this method even before the introduction of the ionic theory, and in the fourth edition of his book Ostwald acknowledges his priority [32].

These extracts from Ostwald's book illustrate very clearly the importance of his ideas, and no comment is needed to emphasize this.

3. DEVELOPMENT OF THE CONCEPT OF pH

Ostwald, as we have seen, explained the theoretical basis of analytical chemistry with great clarity, and by his definitions many phenomena which had previously been either unexplainable or only incompletely understood became apparent. In the author's opinion this was a period in chemistry when everything seemed to fit into a logical order, perhaps because Ostwald was so expert at explaining things. However, it is true to say that in some cases his explanations only just touched the surface, and on a deeper and more thorough investigation many anomalies in the theories of Ostwald and Arrhenius were found.

There is one very important correlation, however, which concerns many aspects of analytical chemistry, and this is known as the Nernst equation and, as we are aware, it gives a quantitative interpretation of redox processes. Ostwald did not realize the great importance of this principle to analytical chemistry, because he did not mention it in his book, and even the editions which appeared as late as 1921 contain no reference to it. This is very difficult to understand for even at that time redox processes were important in analysis. It is true, however, that Ostwald made only very few alterations to the subsequent editions of his books, although they were published for nearly 30 years, and during this time much progress was made. It is even more interesting that he did not mention the works of Nernst [33], as Nernst developed his theory in Ostwald's institute in 1889 [34]. The young Nernst based his arguments on the thermodynamic aspects of the osmotic pressure of cells, and from this he was able to calculate the electromotive force of galvanic cells.

The Nernst equation can also be used for the calculation of the potential difference between solutions of different concentrations. This was also established by Nernst.

The ionic theory provides a special relationship between the hydrogen and hydroxyl ion concentrations, and Nernst's formula opened up the possibility for concentration determination. Someone was required who could use this theory reversibly, i.e. determine the concentrations from the basis of the e.m.f. measurements of cells. A discovery by Le Blanc made this a practical possibility. He found that if a stream of hydrogen is allowed to flow over the surface of a platinum electrode coated with platinum (platinised platinum) then this behaves as a hydrogen electrode . . ."

> In the case of palladium or platinum the hydrogen has no resistance to be overcome and this metal, filled with hydrogen of a given pressure, behaves in an acid solution like a metal electrode immersed in a metal ion solution [35].

In the case of the interaction of two solutions the diffusion potential influences the results. Nernst tried to eliminate this error by calculation, but this was only possible in a very few cases. Tower [36] discovered that if potassium chloride solution is used to effect the electrical contact between the two solutions, then the diffusion potential is decreased to such an extent that it becomes almost negligible [37]. Böttger, in 1897, was the first to use a hydrogen electrode in the titration of an acid.

Several scientists about this time calculated the dissociation constant of water on the basis of various experimental results. First of all Arrhenius, in 1887, calculated this from the degree of hydrolysis of sodium acetate [38], and then in 1893 Ostwald used a concentration cell [39] for the same purpose. In the same year Wijs determined this constant from the saponification rate of methyl acetate with water, with hydroxyl ions as a catalyst [40]. Finally Kohlrausch and Heydweiller [41] calculated it from conductivity measurements [42]. All these measurements were in agreement and gave values for the ionic product between $1·41 \times 10^{-14}$ and $1·1 \times 10^{-14}$.

It had earlier been observed that different indicators change colour at various acidities, and the reason for this was given by Ostwald (his theory was discussed in the history of indicators) (Chapter VIII. 6). The mere qualitative establishment as to whether an indicator was sensitive to acids or bases, gradually became insufficiently accurate, and a more suitable method for measuring the transition points of indicators was sought. The first step towards the solution of this problem was made by Friedenthal [43], who prepared a series of solutions of known hydrogen ion concentrations, and used these solutions for examining the colour changes of fourteen indicators. His paper is important from several aspects, especially as it records the first method for the colorimetric determination of hydrogen ion concentration. This will be discussed in more detail later. Friedenthal found that after a certain degree of dilution it was impossible to prepare solutions of well defined hydrogen or hydroxyl ion concentrations from pure acids or alkalies. The very small H^+ or OH^- concentration was affected by the merest trace of any contaminant. One of his students, Pál Szily, discovered that it was possible to prepare solutions of small but constant hydrogen ion concentrations, and one of these solutions used was a mixture of mono- and dihydrogen phosphates (alkali salts). Szily had stated earlier in regard to his investigation of blood serum that it behaves with acids and alkalis like a bicarbonate solution which is saturated with carbon dioxide. This is the first study and application of buffer solutions [44]. Friedenthal finally recommended that the reaction of a solution should be characterized by its hydrogen ion concentration, because according to the equation

$$H_2O \rightleftharpoons H^+ + OH^-; \quad C_{H+} = C_{OH+} = 10^{-7}$$

the hydrogen ion concentration also determines the hydroxyl ion concentration, because 10^{-14} divided by the hydrogen ion concentration gives the hydroxyl ion concentration.

In the same year Salessky measured the hydrogen ion concentration at the transition point of several indicators. The system he used for this determination consisted of two hydrogen electrodes and two separate solutions, the first solution being 0·1 N with respect to hydrochloric acid, while the other solution contained the indicator to which acid or base was added until the colour changed. For a measurement of the e.m.f. at the transition point Salessky was able to calculate the corresponding hydrogen ion concentration [45].

Shortly after the work of Friedenthal, Fels carried out an investigation of the dependence of the transition points of several indicators on the hydrogen ion concentration. He calculated the hydrogen ion concentrations of his solutions from the results of conductivity measurements. In the neighbourhood of the equivalence point he used combinations of ammonium hydroxide—ammonium chloride and acetic acid — sodium acetate solutions [45], thus Fels also used buffer systems. This expression had already been used previously by Fernbach in 1900 at a congress on beer processing in Paris. He pointed out that a malt extract gives an acidic reaction with phenolphthalein, while with methyl orange it reacts as a base. Fernbach also observed that the enzyme action of a malt extract is only influenced by acids or bases if a certain critical amount of the acid or base is exceeded. The extract is therefore protected against acids and bases to a certain degree, and this protection is due to the presence of mono and dihydrogen phosphates in the extract. Fernbach stated that "the mixture behaves like the buffer disc of a waggon (tampon), it lessens the effects of acids and bases" [46]. The English word buffer derived from this.

It is particularly interesting that the earliest work on pH measurement, buffer solutions, as well as on redox potentials was carried out by biochemists and physiologists. But this is not unreasonable to expect because buffer solutions are an essential part of living organisms.

The pH sign was introduced by Sørensen [47], who attempted, in this way, to simplify the expression of hydrogen ion concentration used by Friedenthal He replaced the whole number expression by the exponent (i.e. the negative logarithm) which he termed the pH.

The title of Sørensen's original paper in which he describes the pH and its colorimetric determination is *Enzymstudien*, in which he examines the dependence of enzymatic effect on the acidity [48].

It is well known that the activity of enzymes is increased by the addition of acid until it reaches a maximum, but then decreases rapidly after reaching this characteristic acid concentration. It was also experienced, however, that this maximum depends not only on the composition of the enzyme, but also in the case of the same enzyme on the nature of the acid used. Sørensen considered that the effect was not due to the acid itself, but only to the hydrogen ions, and therefore he needed a simple method of expressing the hydrogen ion concentration. For the same reason he had to devise a reliable method for the measurement of hydrogen ion concentration, and also to prepare solutions of accurately known hydrogen ion concentrations. The Sørensen type buffer solutions are the result of this work. These, as a matter of fact, are based on silico- phosphate buffer solutions.

The concept of pH eventually became one of the most important and widely used factors in the whole of theoretical and practical chemistry. But in Sørensen's time, and even later, it was of interest only to a very small group of scientists, with very few chemists among them.

It was first of all the biochemists and enzymologists who realized the practical importance of pH. The co-workers of the Carlsberg Laboratory published the results of a succession of experiments on the effects of pH in enzymology. A series of similar papers were published in rapid succession. The first monograph on this subject was published in 1914 by Michaelis, who was a physician-biochemist, entitled: *Die Wasserstoffionkonzentration*. Its subtitle was: *Its importance for biology*. In the preface Michaelis complains that the concept of acidity and alkalinity is not treated with sufficient importance by the majority of physiological chemists, with the result that there are many phenomena which are incompletely understood. He gives several examples to illustrate this point; one of these is of the author who found that the efficiency of invertase is diminished by the addition of blood serum. The author of this paper concluded that the blood serum must contain some of the anti-substance of the invertase. This is a mistaken conclusion, however, for this decrease in activity is due simply to the pH of the blood serum, and in fact any solution of this pH would have a similar effect on invertase. Michaelis pointed out the importance of hydrogen-ion concentration to the physician, chemist and bacteriologist. Finally, he mentioned two people who were already using this new concept, the Hungarian chemists Liebermann [49] and Bugarszky [50].

In 1922, in the second edition of his book Michaelis mentioned that whereas in the first edition of his book the subject had only been of interest to a very small band of scientists, in the past eight years the material of this subject had increased so much that it was impossible to summarize in one volume. Owing to this Michaelis intended the second edition of his book to be of use mainly to biochemists. In 1920 Clark, a physiologist, wrote a book on hydrogen ion concentration, entitled *The Determination of Hydrogen Ions*, and this renewed the interest in this subject. The first book devoted to the analytical aspects of this subject was Kolthoff's: *Der Gebrauch von Farbindicatoren*, published in 1921.

4. THEORY OF TITRATION

The development of theories is a far more difficult problem to trace than is the development of practical achievements. It is even more difficult to establish who was the discoverer of a certain concept. Certain ideas were picked up by someone who developed it further before the idea passed on to someone else. The theory of titrations developed in this way.

It was well known that for the titration of weak acids indicators different from those used in the titration of strong acids were required. It had also been established that the titration of certain polybasic acids gave entirely different results with different indicators. In his indicator theory Ostwald indicated the cause of this phenomenon, but he did not investigate the subject very deeply.

Ostwald pointed out, for example, that polybasic acids dissociate in several steps. If the dissociation of the second step only occurs to a small extent then the transition is not very sharp, and in all probability the titration is impossible to accomplish because of the lack of a suitable indicator.

> Phosphoric acid behaves against methyl orange as a monobasic acid, that is, that only the first hydrogen atom of the phosphoric acid is sufficiently dissociated to ensure that sufficient hydrogen ions are available for the formation of the red, non-dissociated methyl orange molecule. Against phenolphthalein, which is a much weaker acid, phosphoric acid can be titrated as a dibasic acid, because this indicator needs far fewer hydrogen ions for the formation of the colourless, non-dissociated molecule. The third hydrogen of phosphoric acid is such a weak acid, that its corresponding alkali salt dissociates in water to such an extent that it cannot be titrated [51].

As the number of synthetic indicators available increased rapidly, it became increasingly difficult for chemists to decide which indicator was suitable for a given titration. At the beginning of this century, therefore, an international committee was appointed to study the indicators available for use in acidimetry and alkalimetry. It has not been possible to find any account of the work of this committee, so possibly it did not reach any conclusions. At that time the dissociation constants of only a very few indicators were known, for in 1907 only that of *p*-nitro phenol [52] and phenolphthalein [53] had been recorded.

Salm, in 1907, determined the dissociation constants of a number of indicators by a colorimetric method. He made the important discovery that the dissociation constant of an indicator corresponds to the hydrogen ion concentration at which the indicator is dissociated to an extent of 50 per cent. This is naturally the transition point of the indicator also. From a knowledge of the dissociation constant the indicator, for a given titration, can easily be selected. The titration must be continued until the colour of the indicator corresponds to the hydrogen ion concentration at the end point.

To explain this more fully it would be best to quote Salm, who explained plainly and simply the difference between equivalence-point and neutral-point:

> If a strong acid is titrated by a strong base, in the presence of equivalent amounts of the acid and base the solution will show a neutral reaction. This is not so, however, if one of the components of the "neutral" salt formed is weak. Thus, if to a solution of ammonia an equivalent amount of hydrochloric acid is added, the residual solution is acidic, while if to boric acid an equivalent amount of sodium hydroxide is added, an alkaline solution is formed. The indicator must show that point where exactly equivalent amounts of acid have been added to the base. This point we shall call the equivalence point, and this only becomes identical with the neutral point in the case of strong electrolytes.
>
> If therefore such an indicator were chosen for the titration of a weak acid with a strong base, which shows a colour change just at the neutral point (litmus, rosolic acid, alizarin), the results would be too low, while if strong acids were titrated with weak bases, they would be too high. The titration of weak electrolytes therefore can be made after considering the hydrolysis grade of the "neutral salt solution" formed at the end-point.
>
> The hydrolysis grade of the salt solution can be calculated from the dissociation constants of the weak acid or base ... For example, we want to titrate aniline with hydrochloric acid. The concentration of the solution should be 1/10 N. In this case an indicator must be used which shows a colour change just at the hydrogen ion concentration of an N/10 aniline hydrochloride solution.
>
> The hydrolysis grade of such a solution can be obtained from the following formula:
>
> $$x = \sqrt{\frac{1}{c} \cdot \frac{K_w}{K_b}}$$
>
> where K_b is the dissociation constant of aniline $= 4.9 \times 10^{-10}$
>
> $c = 1/10$ $\qquad K_W = 1 \times 10^{-14}$
>
> $x = 1.4 \times 10^{-2}$ $\qquad C_H = xc = 1.4 \times 10^{-3}$
>
> At 1×10^{-3} N hydrogen ion concentration the following indicators change their colours, as can be seen from tables: Methyl violet, tropeoline 000, dimethylaminoazobenzene, etc ... [54]

I have quoted a very large extract from this publication because I found that the principles of the theory of acid—base titrations are to be seen here for the first time, and are clearly and simply explained. This theory later became the basis for all other types of titration.

Sørensen himself established that it was impossible to determine the true acidity simply by titration because the presence of any buffering substances would consume some of the titrant [48].

These problems were investigated in greater detail by Noyes [55, 56], and later by Bjerrum [57], who summarized the theory of acid–base titrations in his monograph: *Die Theorie der alkalimetrischen und acidimetrischen Titrationen* in 1915.

The hydrogen ion concentration which Bjerrum referred to as the indicator exponent [58] was called by Noyes the apparent ionization constant, this is the pH value where one half of the indicator has been transformed. It had already been noted that the colour change range of most indicators was about 2·5 pH units and Bjerrum, using the indicator exponent, was able to arrive at this result theoretically. Noyes also investigated the proportion of the various indicators which must be transformed before there is any visible colour change. He found that in the case of single colour indicators it was necessary for 25 per cent of the indicator to change, whereas for two colour indicators the amount varied between 5 and 20, i.e. 95 and 80 per cent. The hydrogen ion present at the transition point of a given indicator Noyes called the indicator function [56]. Bjerrum expressed this value in pH units and referred to it as the titration exponent; he also established that the titration exponent must be lower by at least 0·5 pH unit than the indicator exponent [59].

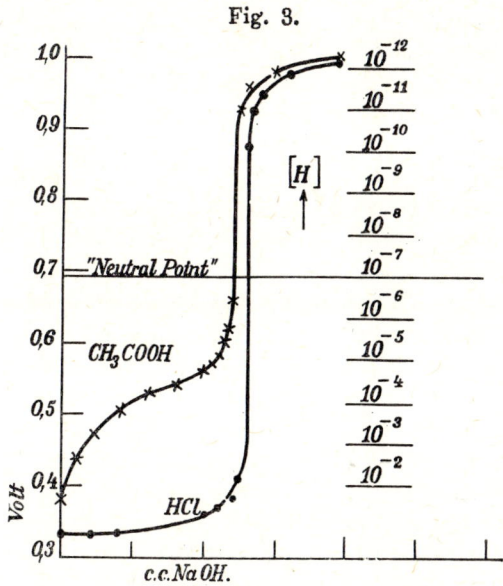

Fig. 92. Theoretical titration curves of Hildebrand (1913). (From the original paper)

During this time potentiometric titrations had been investigated using a hydrogen electrode.

Hildebrand [60] was especially interested in this subject, and it was he who first carried out a theoretical investigation of the titration curves. He recorded the hydrogen ion concentration on the ordinate, corresponding to the potential, and the inflexion point and potential break of the curves gave a clear indication of the equivalence point, as well as its correlations with the pH and the concentration and nature of the solution being titrated [61]. One of the experimental curves obtained by Hildebrand is shown in Fig. 92. As a result of the investigation of these curves many problems in the theory of titrimetric analysis were clarified. It was soon possible to select a suitable indicator for a given titration by this method.

Bjerrum published a table which recorded the pH limits for the titration of various acids and bases with different dissociation constants in order to obtain a reasonable degree of accuracy [62]. Noyes gave formulae for the calculation of the indicator function, i.e. the equivalence-point pH for the titration of weak acids and weak bases. The formulae for calculating the degree of hydrolysis had been known earlier but Salm was the first to use them for analytical purposes [63]. Bjerrum considered that there were far too many formulae in use, and attempted to find a uniform solution. In the course of this work he proposed his theory regarding the titration error [64]. According to Bjerrum three factors contribute to this error, namely the hydrogen error, the hydroxyl error and the acid or base error. On this basis he was able to calculate the pH at the equivalence point; moreover, he was able to give the pH values corresponding to the separate steps in the dissociation of polybasic acids. Bjerrum also calculated that the limiting value for the titrability of acids is the dissociation constant of 10^{-10} and for bases it is 10^{-4}. With this method of calculation the titration curves could be calculated theoretically, even for weak electrolytes. These curves and the method of calculation are today an essential part of all analytical textbooks. The first example of these is to be found in Kolthoff's book [65]. The development of the theory of acid—base titrimetry was essentially completed by the start of the first World War.

The ionic theory of Arrhenius exhibited certain anomalies; these were mostly associated with the behaviour of strong electrolytes. Bjerrum was the first to explain the reason for these deviations successfully. Bjerrum assumed that solutions of strong electrolytes are completely dissociated, irrespective of their concentrations, and that the interaction between the ions is the cause of their anomalous behaviour. He attempted to compensate for these deviations by introducing three coefficients, and of these the so-called activity coefficient was to become the most important (1916) [66]. The concept of activity was used for the first time by G. N. Lewis (1875—1945), who tried to replace the existing gas laws by one law which is valid for all pressure and concentration values [67].

Noyes introduced the idea of the activity coefficient as the ratio of activity and concentration [68], but it is unnecessary to consider the further development of the activity concept as this did not influence analytical chemistry. As dilute solutions are generally used in analytical chemistry, concentrations rather than activities are generally used. But in the concept of pH this did play a certain role. Sørensen determined the hydrogen ion concentrations of his solutions from conductivity measurements. For this he used a solution of 0·01 N HCl and 0·09 N KCl which had a hydrogen ion concentration of 18°C corresponding to 2·038 pH units. He used this as the basis for the establishment of the potential of the 0·1 N calomel electrode, referred to the potential of the normal hydrogen electrode. This value he found to be +0·338 V, and he subsequently used this to make numerous pH measurements. Sørensen himself recalculated this potential in 1924, according to the most recent theories and found it to be +0·3357 V. This value has only been modified by a negligible amount since that time. As the electrode potential is defined not by the hydrogen ion concentration, but by the activity, then the pH measurement—based on e.m.f. measurements—cannot give an accurate value for the concentration. Thus the definition of Sørensen is fundamentally incorrect, so that the meaning of the term pH is not always quite clear.

In analytical practice, however, the original definition of Sørensen can be used without involving any significant error.

In 1923 Brönsted [70] introduced a new theory of acids and bases [71]. The most important advantage of this theory was that it was not confined to aqueous systems, but could be used to explain acid—base phenomena in non-aqueous media. Briefly the theory states that acids are those substances which are capable of donating a proton while bases are those substances which are capable of accepting protons.

$$\text{acid} \rightleftharpoons \text{base} + H^+$$

Bases, therefore, always possess one negative charge in excess of acids. Ampholytes are those substances which are able both to accept and donate protons. Hence the ancient well known formula

$$\text{acid} + \text{base} = \text{salt} + \text{water}$$

according to the Brönsted theory becomes

$$\text{acid}_1 + \text{base}_2 = \text{acid}_2 + \text{base}_1$$

There are many advantages in this theory, but there are also some disadvantages; thus, the concept of a salt is rather neglected in this system. The acid and base dissociation constants are replaced by the acid exponents, but these can also be obtained from the law of mass action. By means of this theory it is possible to calculate the pH of solutions, and also the pH at the equivalence point of a titration. According to Brönsted's theory it is possible to develop the theory of acid–base titrations in a similar manner to the classical ionic theory. The first analytical textbook to treat acid—base phenomena according to the basis of the Brönsted theory was published by Charlot [72] in 1942.

Bjerrum also realized the analogy between the theory of redox and precipitation titrations, and acid–base titrimetry. He commented on this several times in the preface to his book, and gave suggestions as to how it should be accomplished:

> In this we have only considered the theory of acidimetric and alkalimetric titrations. These principles, however, can also be applied to other titrimetric methods.
>
> The theory of precipitation titrations can be similarly explained. In the precipitation of chloride, bromide or iodide with silver nitrate for example, the concentration of silver ions plays exactly the same role as does the hydrogen ion concentration in alkalimetry. The various silver ion indicators (chromate, phosphate) change colour at well defined silver ion concentrations, and from the solubilities of the various silver salts we can choose a suitable indicator for the titration. From the sensitivity of the indicator as well as a knowledge of the solubility of the corresponding silver salt the titration error can be calculated. Similarly just as the hydrogen electrode can be applied in acidimetric titrations, so the silver electrode can be used in argentimetric titrations.
>
> The theory of oxidation-reduction titrations can be built up on the same principles. In place of the hydrogen ion concentration, however, we must introduce the concept of

oxidation potential (oxygen pressure). Crotogino and Hildebrand have recommended a platinum electrode as an indicator in this type of titration. The potential of this is a measure of the oxidation potential of the solution ... [73]

The information required to develop the theory of redox titrations was readily available in the accumulated knowledge of physical chemistry. The basis of this, naturally, was the theory of Nernst which related ionic concentration with electrode potential. Ostwald later pointed out the electrode potential of a system is a measure of its oxidizing or reducing power. He made this comment in a lecture on a "chemometer", which was a device for measuring whether the energies of a system are in equilibrium or not. This was rather more a philosophical than a scientific discussion. Measurement of the potential of a solution was also mentioned briefly in this lecture [74]. The application of the Nernst equation in the case of redox systems was examined by Peters working in Ostwald's institute, who made measurements of potential on the iron(III)–iron(II) systems [74]. His experiments lead to the establishment of the following equation:

$$\pi = A + \frac{R \cdot T}{F} \ln \frac{c_{Fe\cdots}}{c_{Fe\cdot\cdot}}$$

where the $R \cdot T$ factor, which originates from the conversion of osmotic to electrical energy, is constant. The value at 17° of this factor, using logarithms with base ten, is 0·0575. The equation therefore becomes

$$\pi = A + 0\cdot0575 \log \frac{c_{Fe\cdots}}{c_{Fe\cdot\cdot}}$$

in other words the electromotive force of a cell consisting of a standard electrode and a mixture of ferrous and ferric ions, is equal to a constant A plus an additional term which consists of the logarithm of the concentration ratio multiplied by 0·0575. If the ratio of ferric to ferrous ions is unity then the concentration dependent term will be equal to unity and then

$$\pi = A$$

The constant A therefore is the e.m.f obtained when the standard electrode is combined with another electrode containing equimolecular amounts of ferric and ferrous ions [75].

This, therefore, is the definition of the standard redox potential.

In 1900 Crotogino carried out potentiometric titrations using potassium permanganate, and in the introduction to his paper [76] describing this work, he gives the oxidation for permanganate as follows:

$$MnO_4 \geqq Mn\cdot\cdot + 4\,O'' + 5F\cdot$$

$F\cdot$ here refers to the charge on the ions. He pointed out that it must be taken into consideration that the oxygen ions react with the hydrogen ions of water to a certain extent, yielding hydroxyl ions, that is

$$MnO_4' + 4H\cdot \geqq Mn\cdot\cdot + 4OH' + 5F\cdot$$

and therefore

$$\pi = \frac{RT}{5F} \ln K \frac{(MnO_4')(H\cdot)^4}{(Mn\cdot\cdot)(OH')^4}$$

is gained. Crotogino replaced the OH' ions with H· ions using the ionic product of water, and he thus obtained an equation which takes into account the effect of hydrogen ions on the redox potential. It was therefore possible, using this equation, to calculate the e.m.f. of the system at any point during the course of the titration.

The importance of this equation became apparent when redox indicators were introduced into redox titrations. The choice of a suitable indicator, having a transition potential corresponding to the equivalence point potential of the titration required that this latter value was capable of being accurately calculated.

The theory of redox titrations was developed from an investigation of redox indicators, similar to the way in which the theory of acid–base titrimetry was developed from investigation of acid–base indicators. These studies were also carried out by the same workers who made the pioneer investigations into acid–base processes, Clark and Michaelis.

The theory of redox titrations, based on these fundamental studies, was developed by Kolthoff. He examined mathematically the variation in electrode potential as a function of the amount of a standard solution added for various electrode processes, as well as calculating the equivalence-point potential in each case. This is also now a major part of all analytical textbooks [77].

NOTES AND REFERENCES

1. WILHELMY, LUDWIG FERDINAND (1812–1864) owned a pharmacists shop, but later sold it and with the proceeds went to live in Berlin, where he devoted himself to research.
2. GULDBERG, CATO MAXIMILLIAN (1836–1902) was Professor of Mathematics at the Military Academy of Oslo, and later was Professor of Technology at the University of Oslo.
3. WAAGE, PETER (1833–1900) was Professor of Chemistry at the University of Oslo.
4. BERTHELOT, MARCELIN (1827–1907) was Professor of Chemistry at the Collège de France, and a member of the French Academy of Sciences as well as a number of other academies. He was the author of a large number of books and published papers, which mostly dealt with thermochemistry. He was also very active in the political field, and partly as a result of this he was very influential in French scientific life.
5. MENSHUTKIN, NIKOLAI ALEXANDROVICH (1842–1907) who was born into a merchant family, completed his university studies at St. Petersburg, and then worked with Strecker, Kolbe and Wurtz in Germany and in Paris. When he returned he worked in Mendeleev's department. In 1869 he was appointed Professor of Analytical Chemistry at the University of St. Petersburg. His analytical chemical textbook (1871) was one of the most famous books of this period, and it was translated into German and English. In 1885 he succeeded Butlerov as Professor of Organic Chemistry, and in 1903 he was appointed director of the chemical institute at the new technical high school.

6. CARNOT, SADI (1796—1832) was the son of Nicolas Carnot, who played an active prt in the French revolution. After completing his studies at the École Polytechnique he served with the army engineering corps. He died of cholera, and as a result many of his manuscripts were burnt, as they were considered infected material.
CLAPEYRON, BENOIT PIERRE (1799—1864) was at first an army engineer, and served in Russia. Ha later became a professor at the École des Ponts et Chaussées, in Paris, and taught mechanics.
7. JOULE, JAMES PRESCOTT (1818—1889) was a brewer, who carried out research as a hobby. He was a member of the Royal Society.
8. THOMSON, WILLIAM, later LORD KELVIN (1824—1907) studied at Glasgow and in Cambridge, and later became Professor of Physics at the University of Glasgow. He was one of the most important personalities in the whole of physics. He was a member of a great many scientific academies and societies.
9. CLAUSIUS, RUDOLPH (1822—1888) was Professor of Physics at the University of Würzburg, and later in Bonn. He was a member of many academies.
10. HORSTMANN, AUGUST (1843—1929) studied at the University of Heidelberg, and later became a professor there. He became blind while quite young as a result of a disease.
11. GIBBS, JOSIAH WILLARD (1839—1903) was Professor of Theoretical Physics at Yale University.
12. PFEFFER, WILHELM (1845—1920) was Professor of Botany at the Universities of Bonn, Basle and later at Tübingen.
13. RAOULT, FRANÇOIS (1830—1910) was Professor of Chemistry at the University of Grenoble.
14. VAN'T HOFF, JACOBUS HENRICUS (1852—1911) was born in Rotterdam. He completed his studies in Delft, Leyden, Bonn and Paris. He began his career at the Veterinary High School at Utrecht, and it was here that he developed his theories which became the foundation of stereochemistry. In 1878 he became Professor of Chemistry at the University of Amsterdam, and here he worked out his new laws of solutions. In 1896 he went to Berlin at the invitation of the Academy of Prussia. He was awarded the first Nobel Prize of Chemistry in 1901.
15. GROTTHUS, THEODOR baron (1785—1822) was a landowner in Lithuania, who carried out research for a hobby.
16. FARADAY, MICHAEL (1791—1867) was born into a very poor family. At the age of thirteen he was apprenticed to a printer but later became a laboratory assistant to Davy, or more accurately, the servant of Davy. He soon became well known through his scientific work, and he eventually became a member of the Royal Society, and the director of the Royal Institution. Many discoveries commemorate his name in physics and chemistry, probably the most important being his discovery of electromagnetic induction.
17. DANIELL, JOHN FREDERIC (1790—1845) was Professor of Chemistry at Kings College. He was a member and secretary of the Royal Society.
18. KOHLRAUSCH, FRIEDRICH (1840—1910) completed his studies at Erlagen, and then became a Privat Dozent in Frankfurt and Göttingen, and finally Professor of Physics at several universities; first at the Technical University of Zürich, and later at the Universities of Würzburg, Strasbourg and Berlin.
19. ARRHENIUS, SVANTE (1859—1927) studied at the University of Uppsala. His views were first made known in his doctoral thesis. He began his career in the physics department at the university, but in 1891 the Technical University of Stockholm appointed him Professor of Physics. In 1903 he was awarded the Nobel Prize, while in 1905 he became director of the Nobel Research Institute of the Swedish Academy.
20. OSTWALD, W: *Ann. der Naturphilosophie.* (1905) **4** 459
21. OSTWALD, W: *Die wissenschaftlichen Grundlagen der analytischen Chemie.* Leipzig (1894) 5
22. *ibid.* 18
23. *ibid.* 57
24. *ibid.* 59

25. *ibid.* 62
26. *ibid.* 71
27. *ibid.* 75
28. *ibid.* 113
29. *ibid.* 140
30. *ibid.* 146
31. *ibid.* 170
32. *ibid.* 4th ed. (1904) 202
33. NERNST, WALTHER (1864—1941) was born in Thorn, and carried out his university studies in Zürich, Würzburg and Graz. In 1887 he became an assistant to Ostwald in Leipzig. It was here that he published his first book, entitled *Theoretische Chemie*. In contrast to Ostwald he regarded the atomic theory as being very important. He had a very good mathematical brain and was expert in expressing the laws of physical chemistry in mathematical terms. In 1891 he went to Göttingen where in 1894 he became the Professor of the first department entirely devoted to physical chemistry. In 1904 he was invited by the University of Berlin to become the successor to Landolt. In 1922 he became the Director of the Physikalische Technische Reichsanstalt. In 1933 he retired, and lived on his estate until his death. He was awarded the Nobel Prize in 1920.
34. NERNST, W.: *Z. physik. Chem.* (1889) **4** 129
35. LE BLANC, M.: *Z. physik. Chem.* (1893) **12** 133
36. TOWER, OLIN (1872— ?) from 1907 he was the Professor of Chemistry at the University of Cleveland.
37. TOWER, O.: *Z. physik. Chem.* (1895) **18** 17
38. ARRHENIUS, S.: *Z. physik. Chem.* (1887) **1** 631
39. OSTWALD, W.: *Z. physik. Chem.* (1893) **1** 521
40. WIJS, J. J.: *Z. physik. Chem.* (1893) **11** 492; **12** 253
41. HEYDWEILLER, ADOLF (1856—1926) studied in Berlin, while from 1908—1926 he was the Professor of Physics at the University of Rostock.
42. KOHLRAUSCH, F.—HEYDWEILLER, A.: *Z. physik. Chem.* (1894) **14** 317
43. FRIEDENTHAL, H.: *Z. Elektrochem.* (1904) **10** 113
44. SZILY, P.: *Orvosi hetilap* (1903) 509
45. FELS, B.: *Z. Elektrochem.* (1904) **10** 208
    SALESSKY, W.: *Z. Elektrochem.* (1904) **10** 204
46. FERNBACH, A.—HUBEN, L.: *Compt. rend.* (1900) **131** 293
47. SØRENSEN, SØREN PETER LAURITZ (1868—1939) completed his studies in Copenhagen, and then worked at the Carlsberg Laboratories where he was mainly concerned with enzymological research. After the death of Kjeldahl he became the director of the laboratory.
48. SØRENSEN, S. P. L.: *Biochem. Z.* (1909) **21** 131 201 **22** 359
49. LIEBERMANN, LEO (1852—1926) first Professor at the Chemistry of Veterinary School, Budapest, later Professor of Public Health at the University, Budapest
50. BUGARSZKY, ISTVÁN (1868—1941) was Professor of Chemistry at the University in Budapest.
51. OSTWALD, Wi.: *Die wissenschaftlichen Grundlagen der analytischen Chemie.* 108
52. BADER: *Z. physik. Chem.* (1890) **6** 297
53. MC COY: *Am. Chem. J.* (1904) **31** 503
54. SALM, E.: *Z. physik. Chem.* (1907) **57** 471
55. NOYES, ARTHUR AMOS (1866—1936) finished his studies at the Massachusetts Institute of Technology, and then came to Europe where he worked with Wislicenus. Later owing to the influence of Ostwald, he became interested in physical chemistry, and worked with him. In 1890 he returned to his own country and lectured in physical and analytical chemistry at the Massachussetts Institute of Technology. He later founded a physicochemical research laboratory here. After the first World War he went to the California Institute of Technology.

56. NOYES, A. A.: *J. Am. Chem. Soc.* (1910) **32** 815
57. BJERRUM, NIELS JANNIKSEN (1879—1958) completed his university studies in Copenhagen. In 1913 he became Professor of Chemistry at the Veterinary and Agricultural High School in Copenhagen, while in 1939 he became the director of this Institution. In 1946 he made important contributions to both analytical and physical chemistry. He was a member of the Danish, Finnish and American Academies, as well as of the Academy of Göttingen.
58. BJERRUM, N. J.: *Die Theorie der alkalimetrischen und azidimetrischen Titrierungen.* Stuttgart (1915) 28
59. *ibid.* 51
60. HILDEBRAND, JOEL H. was born in 1881, and completed his studies at the University of Pennsylvania. After graduating he made an extended journey in Europe, and worked at the University of Berlin. After returning to his home he worked at the University of California, where he became a professor in 1917. In 1952 he retired.
61. HILDEBRAND, J. H.: *J. Am. Chem. Soc.* (1913) **35** 847
62. BJERRUM, N. J.: *Die Theorie der alkalimetrischen und azidimetrischen Titrierungen.* Stuttgart (1915) 122
63. ABEGG: *Die Theorie der elektrolitischen Dissociation.* Stuttgart (1903)
64. BJERRUM, N. J.: *Die Theorie der alkalimetrischen und azidimetrischen Titrierungen.* Stuttgart (1915) 69
65. KOLTHOFF, I. M.: *Der Gebrauch von Farbenindicatoren.* Berlin (1923) 30
66. BJERRUM, N. J.: *Forh. Skand. Naturforsk. Mde.* **1916**. Osto (1918) 226; cf. *Z. Elektrochem.* (1918) **24** 321
67. LEWIS, G. N.: *Proc. Amer. Acad. Arts. Sci.* (1907) **43** 257
68. NOYES, A. A.—BRAY, W. C.: *J. Amer. Chem. Soc.* (1911) **33** 1643
69. SØRENSEN, S. P. L.—LINDERSTRÖM LANG, K.: *Compt. rend. trav. lab. Carlsberg.* (1924) **15** 6
70. BRÖNSTED, JOHANNES NICOLAUS (1879—1947) was the Professor of Physical Chemistry from 1912 onwards at the University of Copenhagen. A similar theory to his was presented also by the English scientist Lowry (1847—1937).
71. BRÖNSTED, J. N.: *Rec. trav. chim. Pays-Bas.* (1923) **42** 718
72. CHARLOT, GASTON, was born in 1904. He studied at the École de Physique et de Chimie in Paris, and later worked here as a lecturer and from 1945 as Professor of Analytical Chemistry. Since 1950 he has been Professor of Analytical Chemistry at Paris University. His work referred to: *Théorie et Methode nouvelle d'analyse qualitative.*
73. BJERRUM, N. J. (1915): *Die Theorie der alkalimetrischen und azidimetrischen Titrierungen.* Stuttgart (1915) 127
74. OSTWALD, Wi.: *Z. physik. Chem.* (1894) **15** 398
75. PETERS, R.: *Z. physik. Chem.* (1898) **26** 193
76. CROTOGINO, F.: *Z. anorg. Chem.* (1900) **24** 225
77. KOLTHOFF, I. M.: *Chem. Weekblad.* (1919) **16** 408
    KOLTHOFF, I. M.—HOWELL, N.: *Potentiometric Titrations.* London—New York (1926)

CHAPTER XIII

ELECTROMETRIC ANALYSIS

1. THE MEASUREMENT OF pH

At the present day the measurement of pH is one of the most frequent analytical measurements made in the whole of chemistry, applied chemistry and the related sciences.

The earliest method for the measurement of pH, and also the one which is used most frequently today is based on e.m.f. measurement. In this system two hydrogen electrodes are used in a concentration cell, where the concentration of one compartment is known and the potential is measured by the Poggendorf-type compensation. The hydrogen electrode, as has already been pointed out (Chapter XII. 3) was discovered by Le Blanc [1] in 1893. The form in which it is used today was devised by Wilson and Kern [2]. Salessky and Salm also measured pH with hydrogen electrodes, and Sørensen combined a hydrogen electrode with a calomel one. The calomel electrode had been used by Kohlrausch even earlier.

Sørensen used a calomel electrode with 0·1 N potassium chloride as the positive pole, together with a hydrogen electrode. A pressure of moist hydrogen gas of 1 atm, and a temperature of 18°C was used in all cases. Sørensen found that the voltage when 1 N acid was used was 0·338 V.

pH measurement with a hydrogen electrode is the most accurate method, and in alkaline medium it is the only method available, but it is not very convenient, hence attempts were made to replace it by a more suitable one.

The first electrode to replace the hydrogen electrode for pH measurements was the glass electrode, although it is only during the last few decades that this electrode has come into widespread use.

Cremer (who also was a biologist) carried out a series of experiments on the electrical properties of tissues in 1906. In his experiment he separated sodium chloride solutions with a glass membrane 0·02 mm thick, and he added sulphuric acid to one of the solutions. He observed that a potential difference of about 0·23 V was set up between the two solutions. Cremer considered that the membrane was responsible for this potential difference [3].

In 1909 Haber [4] and Klemensiewicz [5] investigated this phenomenon more extensively and found that the potential difference between the solutions changes as the hydrogen ion concentration changes similar to the change in electrode potential of a hydrogen electrode in solutions of varying pH. These workers measured the potential difference with a calomel electrode (Fig. 93), and they noted the possibility of replacing the hydrogen electrode with a glass electrode of Thüringen [6] glass. Their expectations were not realized for a long time. It was discovered that not all types of glass were suitable for use in the construction of glass electrodes; with most glasses the potential was influenced by other types of ions in addition to hydrogen. The high internal resistance of glass membranes was also a limitation to their use. In this case the compensation method could not be used, for apart from the very inconvenient quadrant electrometer, there was no galvanometer with sufficient sensitivity. Thus, it was not until the 1920's that interest again turned towards the use of glass electrodes, and this was partly the result of Mc Innes [7] and Dole [8] discovering a suitable type of glass for the construction of the membrane [9], and partly as a result of the introduction of vacuum tube voltmeters which made very accurate measurements possible.

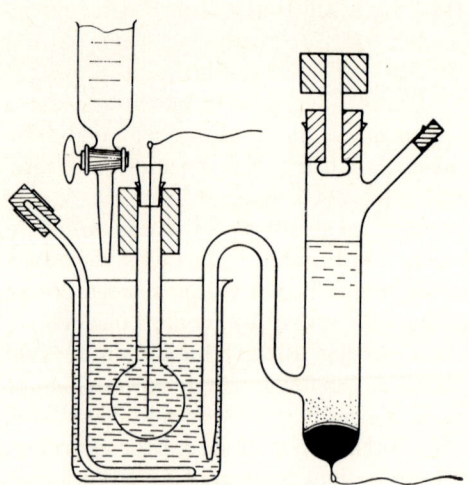

Fig. 93. Glass electrode of Haber and Klemensiewicz, 1909. (From their original paper)

The quinhydrone electrode was first recommended for pH measurements by Biilmann [10] in 1920. He found that a shiny platinum electrode immersed in a saturated solution of quinhydrone gives a potential which is proportional to the pH of the solution. This gives a very simple and convenient method for the measurement of hydrogen ion concentration [11].

The antimony electrode was recommended by Uhl [12] and Kestranek in 1923, originally for the measurement of the pH of soil. At first they added antimony oxide to the solution, but they later discovered that there was always sufficient oxide on the electrode surface to enable it to function [13].

The first attempts to devise a colorimetric method for the determination of pH were made by Pál Szily. He attempted to devise a method to test the reaction of blood serum. Szily determined the hydrogen and hydroxyl ion concentrations of the transition point of several indicators (Poirier-blue, α-naphtholphthalein, phenolphthalein, litmus, rosolic acid and alizarin). He found that the transition point is dependent only on the hydroxyl ion concentration, and is independent of the nature of the residual solution. "By the use of this we can create a measure by which the reaction of salt solutions can be determined". Finally he established that these indicators could be used to detect down to 1×10^{-4} of hydrogen ions and to 5×10^{-4} for hydroxyl ions, and that the reaction of blood serum is within these limits [14]. In the first decades of this century colorimetric methods for pH measurement were used much more frequently than electrometric methods because of their simplicity. However, now that the introduction of the vacuum tube voltmeter has made electrometric methods much more rapid and convenient, colorimetric methods are not frequently used. For very rough pH measurements, however, the so-called universal pH indicators are the most convenient.

The first exact colorimetric pH measurement was improved, as has already been mentioned, by Friedenthal [15] (1904). He prepared fifteen solutions with hydrogen ion concentrations ranging from 1 to 10^{-14} in decreasing exponents. In the acid region he used hydrochloric acid for hydrogen ion concentrations till 10^{-5}. Between 10^{-5} and 10^{-6} he used solutions of 0·00001 and 0·000001 N hydrochloric acid together with boric acid. For the 10^{-7} concentration he first of all used water, but later on at the suggestion of Szily, who at that time was working with him, he used a mono- and dihydrogen phosphate buffer. Hydrogen ion concentrations of 10^{-8} and 10^{-9} were prepared from potassium hydroxide and aniline, while the remainder were prepared from potassium hydroxide. Friedenthal examined fourteen indicators, tropaeoline, neutral red, methyl violet, methyl orange, congo red, lackmoid, litmus, galleine, rosolic acid, p-nitrophenol, sodium alizatin sulphonate, phenolphthalein, naphtholbenzoin, and poirriers-blue. He found that these indicators exhibited colour changes at different pH regions, and he was able to find an indicator which changed colour in all the hydrogen ion standard solutions that he had prepared. Friedenthal investigated the colour of successive indicators in 10 cm³ of an unknown sample and in an equal volume of the standard hydrogen ion solutions. His work was continued by Salm, who examined the hydrogen ion concentration of the buffer solutions with a hydrogen electrode, and also investigated the colour changes of several other indicators.

A more accurate method for the colorimetric measurement of pH was developed by Sørensen in 1909. With this method the pH value could be determined to within one tenth of a pH unit. The principle was identical to Friedenthal's method, except that Sørensen used a larger number of indicators and buffers. He used methyl violet, mauveine, benzene sulphinic acid, azobenzene analine, benzene-azodimethyl aniline, methyl orange, naphthol red, p-benzenesulphonic acid-azonaphthylamine and p-nitrophenol, neutral red, rosolic acid, α-naphtholphthalein, phenolphthalein, thymolphthalein, alizarin red, and the buffers consisted of hydrochloric acid, sodium hydroxide, glycine, potassium mono- and di-hydrogen phosphates, sodium dihydrogen phosphate, sodium citrate, and boric acid [16].

Several other workers later devised a series of buffer solutions, and probably the most important are those devised by Clark and Lubs [17]. They used potassium chloride, potassium hydrogen phthalate, boric acid, sodium hydroxide, hydrochloric acid, as buffer solutions and introduced the sulphonphthalein series of indicators. The sulphonphthaleins were very effective for this purpose because of their regular gradation in colour [18].

Gillespie [19] devised a method for the colorimetric determination of pH which did not require the use of buffer solutions. This method is known as the drop ratio method. Two series of nine test tubes, the first series containing an acid solution, and the second an alkaline solution are prepared so that they contain from one to nine drops of the indicator. To the unknown solution ten drops of the indicator are added and the colour of this solution compared with the colour of a combination of two tubes, one from the acid series and one from the alkaline placed one behind the other. The two tubes are chosen so that the sum of the number of drops of the indicator is equal to ten. When a correct colour match is obtained the pH of the solution can be calculated from the indicator exponent and the ratio of drops [19]. There have since appeared many modifications of this method, but in all cases the principle has remained unchanged.

Michaelis and Gyémánt developed a rather similar method to Gillespie, but their method only involved the use of single colour indicators. A single comparison solution is used together with a titration from the result of which the pH of the solution can be determined [20].

2. POTENTIOMETRIC TITRATION

This subject has already been referred to in the chapter dealing with the development of the theory of analytical chemistry; the development of electrodes was briefly described in the chapter dealing with pH measurement.

The first potentiometric titration was carried out by Behrend [21], in 1893, at Ostwald's Institute in Leipzig. The title of his paper was: *Elektrometrische Analyse* [22].

He titrated mercurous nitrate solution with potassium chloride, potassium bromide and potassium iodide, and vice versa, using a mercury electrode together with a mercury/mercurous nitrate reference electrode.

In a cell, composed of mercury N/10 mercurous nitrate — N/10 mercurous nitrate mercury, the voltage is initially zero. If potassium chloride is added to one of these solutions, mercury(I) chloride is precipitated, thus the osmotic pressure of the mercury(I) ions decreases on this side and a potential difference is set up, so that the mercury will be more negative in the solution to which potassium chloride was added relative to the unchanged mercury electrode. If further amounts of potassium chloride are added the potential difference increases, and if equal increments of potassium chloride are added, the potential difference increases slowly at first, but later at an increasing rate. The potential difference is measured by the correlation $\pi = 0.058 \cdot \log \frac{p_2}{p_1}$, provided that the small e.m.f. at the junction of the solutions is ignored. If p_2 is constant, the value of π will double if p_1 decreases to one hundredth of its original concentration. To achieve this 990 cm³ of N/10 KCl must be added to 1000 cm³ of N/10 HgNO₃, assuming that both of these salts are completely dissociated. For a further one hundred fold decrease only 9.9 cm³ of KCl solution are required, and for the next decrease of this order only 0.099 cm³ are needed. The addition of potassium chloride causes the most marked effect when all the mercurous nitrate is precipitated, a sudden increase in the potential difference then being observed. If additional increments of potassium chloride are made there is only a very slight increase in the potential difference, this of course being controlled by the law of mass action. The product of chloride and mercurous ions is constant, provided the mercury(I) chloride is present in excess, and therefore the osmotic pressure of mercury(I) ions decreases with the increasing amount of chloride ions.

Thus Behrend, according to the current views, explained why the potential break occurred at the equivalence point. He also recorded the titrations graphically (Fig. 95). The volume of mercurous nitrate is recorded along the abscissa, and the electrode potential is shown in "Leclanché units" on the ordinate. Behrend noted that it was not necessary to measure the potential difference accurately, because the drops of the titrating solution cause a small change in the mercury column height in the capillary electrometer; this must be adjusted with a variable resistance and an opposing cell until the zero point is attained. The end-point of the reaction is indicated by the sudden increase in the height of the mercury column in the electrometer. As the titration of iodide with mercury(I) nitrate did not give satisfactory results Behrend changed the electrode system to a silver plate as the indicator electrode and a silver—silver nitrate reference electrode, and titrated with silver nitrate. Finally, Behrend pointed out that the electrometric end point indication is of use, not only in the titrations referred to, but also for a number of redox and precipitation titrations. One of the great advantages of the method is:

FIG. 94. Robert Behrend (1856—1926). Photograph

> that its use is independent of the source of light, and can be used equally well in daylight or by lamp light.

This, however, was a very modest appraisal of the advantages of potentiometric titrations!

The next potentiometric method was described four years later. In 1897 Böttger [23] published a paper which dealt with the potentiometric titration of acids and bases using a hydrogen electrode. The hydrogen electrode was made of gold plated with palladium, and the voltage was measured by a compensation method. The reference system was another hydrogen electrode immersed into a solution of an acid or a base. He titrated 14 different acids and bases by this method, and also added indicators to the solutions and observed if the colour change corresponded to the biggest potential change. The titration curves he

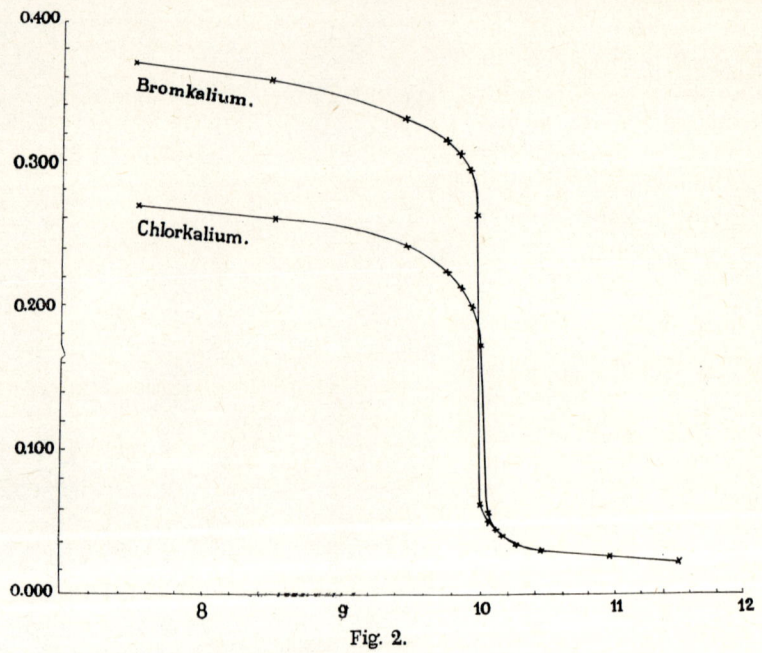

FIG. 95. Behrend's potentiometric titration curves, 1893.
(From the original paper)

obtained illustrate very clearly the difference between strong and weak acids, as well as the behaviour of polybasic acids. Böttger also calculated the potential of several of the end-points, and he also mentioned further advantages of the potentiometric method; for example certain acids and bases which cannot be titrated in conjunction with a colorimetric indicator can be titrated using the potentiometric method of end-point detection, although the accuracy is not very great. Even boric acid and carbonic acid can be titrated but, especially in the case of the latter, the titration curves are difficult to evaluate. He also pointed out that these curves give a good indication of the removal of hydrogen ions and can therefore be of use in certain cases for the investigation of some theoretical problems. They can be used, for example, to give valuable information regarding the stability of aqueous solutions of salts. Böttger also claimed that the shape of titration curves is related to the dissociation constant of the substance in question [24].

The third paper describing a potentiometric titration was the one by Crotogino, referred to previously [25]. He used the method following precipitation and acid—base methods, for redox titrations. He titrated halide ions with potassium permanganate using a shiny platinum electrode and a calomel electrode. According to Crotogino:

> The principle of the method is as follows: An inert (platinum) electrode, which is immersed into the solution to be titrated shows against a standard electrode the oxidation potential of the dissolved substance. If an oxidizing agent is added to a reducing solution the equilibrium between the reducing substance and the reaction product is shifted towards the latter. The potential, accordingly, changes only slowly until the amount of the reducing substance becomes very small. A small addition of the titrating solution will then cause a large change in the potential, as the remaining reducing substance is removed and the potential then corresponds to the oxidizing agent. The stepwise change of the potential can be used as an indication of the end point of the reaction.

With the halides he only obtained a satisfactory result (i.e. an appreciable potential change at the end point) for iodide.

The three methods just described were the first potentiometric procedures, but in the following years they were considered curiosities and were rarely used. The potentiometric method itself was the subject of investigation, and no methods based on this procedure appeared until the 1920's when as a result of the research carried out it became a little more widely known. However, the basic principle of these early methods is still used today, i.e. an indicator electrode is immersed in the solution to be titrated, and the potential change is measured against a reference electrode. This is done by compensation, or, more recently, by means of tube voltmeters. Several other potentiometric procedures were devised, and a brief reference to them is of interest.

The first of these was the recommendation of Dutoit [26] and Weisse (1911) who used polarized electrodes [27]. For a number of metal ions there are no reversible metal electrodes available, and in order to overcome this a platinum electrode is used, and if the current is kept very low a constantly renewed surface of the metal is plated on to the electrode. This method is not very often used in practice.

Another modification was the titration to zero potential, where the solution to be determined and the indicator electrode is combined with another electrode system such that the potential of this electrode corresponds to the potential of the indicator electrode at the end-point of the titration. The e.m.f. of this cell is then measured during the course of the titration, and the potential of the system decreases until zero potential is obtained. Thus, it is not necessary to plot the e.m.f. of the system during the titration, but only to find the point at which zero current is flowing in the system, i.e. with a galvanometer. The method has the advantage of being rapid, but the serious disadvantage that a different indicator system must be chosen for each titration. In redox titrations this meant that the comparison solution has to be adjusted to the potential of the endpoint of the solution, i.e. it is necessary to carry out a preliminary titration before the main determination! This method was recommended simultaneously by Pinkhof, Treadwell [28] and Weiss in 1919; Pinkhof used a capillary electrometer [29], while the others used a galvanometer [30].

Bimetallic electrodes are based on the observation that the various inert electrodes behave differently in a solution. Some of them do not aquire the potential of the solution, others only reach this potential very slowly. Thus, an inert electrode can replace a reference electrode and when combined with a sensitive indicator electrode can be used to follow the course of a potentiometric titration. Hostetter [31] and Roberts titrated iron(II) ions with dichromate, and observed that the metallic palladium electrode shows no potential break when combined with a calomel electrode at the equivalence point of the titration. On the basis of this observation they titrated iron using a platinum indicator electrode, and palladium reference electrode [32].

Willard and Fenwick devised a bimetallic electrode pair system consisting of two platinum electrodes, one of which was slightly polarized [33].

The principle of differential titration was recommended by Cox in 1925 [34]. In this method the solution to be titrated is divided into two parts, and these are connected by means of a salt bridge. Both portions of the solution are titrated in an identical manner except that one titration is maintained at a small amount (0·2 ml) in front of the other. As the titration proceeds the slight initial potential difference gradually increases until the first solution reaches the end-point when the potential difference reaches a maximum. If the results are plotted graphically then the end-point is indicated by the maxima in the titration curve.

The so-called "dead stop" end-point detection is based on the polarization of the electrodes. In this method a small voltage is applied to the polarized platinum electrodes, so that the e.m.f. of the polarization is just compensated, and no current is flowing in the system. If the standard solution used for the titration is able to depolarize the electrodes, then the first excess of this will be indicated by the flow of current through the circuit. This can be easily observed by means of a galvanometer. The method can be used in all cases provided that at least one of the electrodes is polarized. If for example arsenite ions are titrated with iodine solution, using 10–15 mV applied voltage the cathode remains polarized until an excess of iodine is present. The elementary iodine is then reduced on the cathode, and this acts as a depolarizor; correspondingly, iodide ions are oxidized at the anode. Therefore the first slight excess of iodine causes current to flow in the circuit. This method of end point detection can also be used in the reverse manner, i.e. during the titration the electrodes are depolarized, but at the end one of them becomes polarized. In such cases the cessation of current flow can be used to detect the end point. The method was described by Foulk [35] and Bawden [36] (1926) [37].

To be strictly accurate it must be mentioned that Salomon published a method in 1897, which is based on a similar principle [38]. He wrote:

> The proportionality between current and metal ion concentration provides a possibility for the application of the theory of the residual current. From the formula presented

here, it can be seen that current can pass through the electrolyte only if the solution contains ions corresponding to the metal of the electrodes, and also provided that the applied voltage is lower than the decomposition voltage of the electrolyte. Thus, between silver electrodes potassium chloride acts almost as an insulator, but if, however, silver nitrate is added to the solution then the current intensity at first remains zero until the reaction

$$AgNO_3 + KCl = AgCl + KNO_3$$

reaches completion. (Ignoring the small amounts of silver ions which are formed because of the slight solubility of silver chloride.) When all the chloride ions are combined with the silver to give silver chloride, then further addition of silver nitrate will cause a rapid increase in the current. It is to be seen therefore, that a titration can be carried out, and because of its simplicity and rapidity is to be recommended. It is also of advantage in those cases where the available methods of volumetric analysis are unsuitable.

In the next part of his paper Salomon gives a description of this type of method together with the results obtained.

He mentions that his method is rather similar to that devised by Behrend who also used the change of potential to follow the change in the ionic concentration, and used the break in the potential to detect the end point in the titration. Salomon considered that his method was considerably simpler, in practice, than Behrend's. This opinion has since been proved wrong as potentiometric titrations have achieved great importance.

The first monograph on potentiometric titrations was written by E. Müller [39] in 1923 and entitled: *Die elektrometrische Massanalyse*.

Potentiometric titrations can also be carried out on the micro and submicro scales, as with visual indicator methods. There is now a wide range of miniature electrodes, one of the first being constructed by Alimarin.

3. CONDUCTOMETRIC TITRATIONS

Conductometric titrations are based on the change of conductivity of the solution which occurs during the course of the titration. For example, if an acid is titrated with an alkali, then the hydrogen ions with very high mobilities are removed, and in their place metal ions of much lower mobilities appear. The conductivity of the solution therefore decreases steadily until the end point is reached and then starts to increase slightly owing to the presence of free hydroxyl ions. The conductivity of the solution is therefore a minimum at the end-point. In the titration of potassium bromide with silver nitrate the concentration of potassium ions remains unchanged throughout the titration, but the bromide ions are replaced in solution by nitrate ions which have a slightly lower mobility. The conductivity, therefore, decreases slightly during the course of the titration until the end-point is reached. Beyond the end-point the silver ions contribute to the conductivity resulting in a slight increase. Thus the end-point is indicated by the minimum in the conductivity curve.

The investigation of conductivity also helped to solve many physico-chemical problems. The use of conductivity measurements for analytical purposes was first suggested by Küster [40] and Grüters [41].

Their paper is very short, and does not include a graphical representation of their titration results. There is also no reference to the method of measurement used; presumably they used the bridge system recommended by Kohlrausch [42], with a set of head-phones to detect the null-point. This arrangement was also suggested by Ostwald [43] for conductivity measurements. It was attempted to replace the head-phones by some instrument which would give a visual indication of the conductivity, i.e. by measuring the scale deflection on a galvanometer. As there was no a.c. galvanometer available, a d.c. instrument had to be used together with a detector or thermocouple, and Jander [44, 45], who made many contributions to conductometric analysis, was responsible for the development of this. The first monograph on this subject was published by Kolthoff in 1923 and entitled: *Konduktometrische Analyse*.

High-frequency titration is also based on the change of conductivity of a solution during the course of the titration. The main advantages of this technique are that it can be used for very low concentrations and also that no internal electrodes are required. Nevertheless, the apparatus is rather complicated and often expensive, and it is impossible to describe the theoretical background in a few lines. High-frequency titration was introduced in 1946, simultaneously, by Foreman and Crisp and by Jensen [46] and Parrack [47]. It should be noted that Blake had used this principle for concentration measurement earlier.

4. POLAROGRAPHY

One of the most important contributions to the progress of analytical chemistry during the present century has been the discovery of polarography. It is now only about thirty years since the date of the original discovery, but during this time the scope and the application of the method have become widespread. Nearly all laboratories now contain, or have access to, a polarograph, and the instrument has been applied to basic research as well as to the solution of analytical problems.

Polarography is simply electrolysis with a dropping mercury electrode. Mercury is very suitable for use as an electrode, because the overvoltage of hydrogen on this metal is very high and as it is a liquid its surface is constantly being renewed. Lippmann [48], in 1873, used mercury for the first time, in his capillary electrometer, while the anode also consisted of a large pool of mercury. When a current was passed through the cell, the mercury electrode with the small surface was immediately polarized, while the electrode with the large surface area was practically unpolarized because of the low current density. Lippmann used this device for the determination of surface tension of polarized mercury [49].

Kucera [50], in 1903, modified this device by elevating the mercury reservoir to the capillary until the pressure caused the mercury to drop out of the capillary. He determined the surface tension of polarized mercury by weighing the mercury droplets [51]. When a graph was drawn with the weight of the mercury droplets on the ordinate, and the polarizing voltage on the abscissa, Kucera obtained the electrocapillary curves and found some secondary maxima, which had not been observed on the curves obtained by Lippmann. Kucera noticed these anomalies, but did not suggest a reason for them. In 1918 Heyrovsky [52] reinvestigated these phenomena and established that they are caused by the oxygen in the air. On further investigations it became apparent to him that the uniformly dropping mercury cathode was very suitable for the investigation of electrolytic processes. Its constantly renewed surface, on which the overvoltage of hydrogen is very high, could be easily polarized. The results were very reproducible and in addition only a very small fraction of the solution under investigation is decomposed.

FIG. 96. The first polarograph of Heyrovsky and Shikata in the year 1925. (Prof. Heyrovsky's original photograph)

Heyrovsky pointed out that the current on the surface is a much more characteristic measure of the electro-chemical reactions taking place on the dropping surface than is the surface tension. These investigations led him to make a survey of the voltage–current curves and thus to the development of polarography. From the shape of these curves the depolarizing substances in the solution could be evaluated qualitatively and quantitatively. The solution was electrolysed between a dropping mercury cathode and a constant mercury pool electrode as reference electrode.

Heyrovsky first reported on the electrolysis with the dropping mercury electrode in 1922 [53], and in the following years he published a whole series of papers on the recording of the current and the plotting of the current–voltage curve. Heyrovsky, together with Shikata [54], invented the polarograph [55] in 1925. The construction of the first polarograph is shown in Fig. 96. Their paper records the polarograms of several substances, including lead, zinc, cadmium and nitrobenzene. Since then polarography has become much more highly developed and sophisticated largely owing to the efforts of Heyrovsky and his co-workers.

The evaluation of polarographic curves is made much easier if it can be differentiated. The derivative curves are very advantageous, especially where the steps are not sufficiently well defined, or where two substances whose half waves are close together need to be determined. Several methods for the production of derivative curves have been developed during the last decade, for example, a

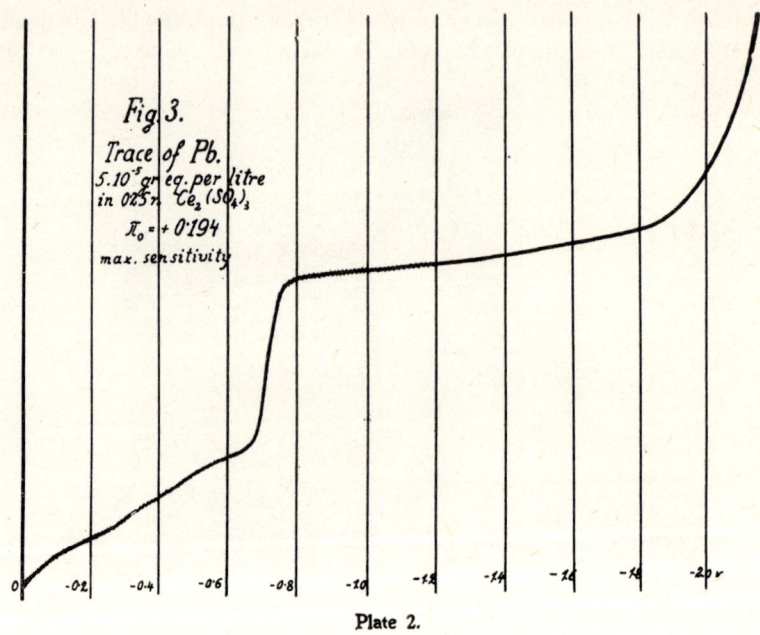

FIG. 97. Polarograms from the original paper of Heyrovsky and Shikata.

retarding electrode [56], a condenser [57], or, more conveniently, a transformer [58] method can be used.

One of the refinements of the polarographic method is oscillopolarography. In this technique the electrolytic processes occurring at the surface of the polarized electrode are observed with a cathode ray oscillograph, namely the correlations of current, time and voltage. In solutions which do not contain a depolarizing agent the dV/dt curves appear as pure ellipses, the upper part showing the cathodic and the lower part the anodic process. In the presence of a depolarizor, breaks in the curve are observed, and the height of these are dependent on the concentration. Nowadays this method, originally devised by Heyrovsky [59], is used mainly for theoretical investigations.

For the determination of very small amounts of ions Kemula [60] and Kublik [60] developed a method which uses a hanging mercury drop electrode. This is yet another direction of the development of polarographic analyses. In Kemula and Kublik's method an electrolysis is carried out prior to the polarographic analysis so that an amalgam is formed on the mercury droplet. The oxidation current of this electrode is then recorded. The method is fairly sensitive, and in suitable instances can be utilized for the determination of substances in concentrations as low as 10^{-8} mole/l. [61].

The polarographic diffusion current is suitable, not only for the direct determination of the concentrations of substances, but also for the end-point detection in a titration. According to the nature of the reaction taking place, the change in the diffusion current either at the commencement or cessation of the current flow in the system, can be used to detect the end-point. In certain cases the diffusion current reaches a minimum or changes its direction. This method which is called amperometry, or according to Majer, polarometry, was also discovered by Heyrovsky in 1927 [62]. Majer [63] was the first to investigate this method in any detail [64].

NOTES AND REFERENCES

1. LE BLANC, MAX (1865—1943) studied in Leipzig, and later became the head of the Physico-Chemical Institute of the Technical University of Karlsruhe, while from 1906 he occupied a similar position at the University of Leipzig. He retired in 1934.
2. WILSON, J. A.—KERN, E. J.: *Ind. Eng. Chem.* (1925) **17** 74
3. CREMER, M.: *Z. Biol.* (1906) **29** 562
4. HABER, FRITZ (1868—1934) studied in Berlin. After graduating he became an assistant lecturer at the Technical University of Karlsruhe with Bunte, who was a famous gas-chemist. He later succeeded Bunte, and continuing to work in this field he developed his famous ammonia synthesis here. In 1912 he was appointed head of the new Kaiser Wilhelm Institute für Chemie, where he worked until 1933, when he was forced to leave. He emigrated first of all to England where he lived in London, and then to Switzerland, but he did not have any success in either venture and eventually he became penniless. As a result of this hardship he died in 1934. He was awarded the Nobel Prize for Chemistry in 1916.
5. KLEMENSIEWICZ, ZYGMUNT (1886—1963) was born in Cracow, studied in Lwow and Karlsruhe, worked at the Technical High School in Lwow (Lemberg) and in the Radium Institute, Paris. From 1920 till 1939 he was Professor of Physics at the Technical High School in Lwow, in the second World War he went to England and tanght at the emigrant Polish Polytechnical School. In 1956 he returned to Poland and was Professor of Nuclear Physics at the Technical University in Gliwice.
6. HABER, F.—KLEMENSIEWICZ, Z.: *Z. physik. Chem.* (1909) **67** 385
7. MC INNES, DUNCAN ARTHUR was born at Salt Lake City in 1885, studied at different American universities. From 1917 till 1926 he was associate professor at the Massachusetts Institute of Technology and later worked at the Rockefeller Institute of Medical Research. He retired in 1950.
8. DOLE, MALCOM, was born in 1903 in Massachusetts, studied at the Harvard University, worked at the Rockefeller Institute of Medical Research and later at the Northwestern University in Evanston, Illinois. During the second World War he was Director of the National Defense Research Laboratory. At present he is Professor of Chemistry and Chairman of the Materials Research Center at the Northwestern University, and Professor of Physical Chemistry at the Northwestern University (U.S.A.).

9. Mc Innes, D. A.—Dole, M.: *J. Am. Chem. Soc.* (1930) **52** 29; *Ind. Eng. Chem. An. Ed.* (1929) **1** 57
10. Biilmann, Einar (1873—1946) studied in Copenhagen, and then worked with Jörgenssen. He later became a professor at the University of Copenhagen.
11. Biilmann, E.: *Festschrift Universität Koppenhagen.* (1920); *Ann. chim.* (1921) **15** 109
12. Uhl, Alfred, was born in Wienerneustadt in 1889. In 1912 he began his work at the Austrian agricultural research institute, and in 1942 he became the head of this institute. He retired in 1947.
13. Uhl, A.—Kestranek, W.: *Monatshefte.* (1923) **44** 29
14. Szily, P.: *Orvosi hetilap* (1903) 509
15. Friedenthal, H.: *Z. Elektrochem.* (1904) **10** 113
16. Sørensen, S. P. L.: *Biochem. Z.* (1909) **21** 131 201; **22** 352
17. Lubs, Herbert August, was born in Savannah in 1891. Between 1914—1919 he worked for the Public Health Service in America. He has since then been employed in industry.
18. Clark, W. M.—Lubs, H. A.: *J. Wash. Acad. Sci.* (1916) **6** 483
19. Gillespie, Louis John (1886—1941) was Professor of Physical Chemistry at the Massachusetts Institute of Technology (Cambridge, U.S.A.).
    Gillespie, L. J.: *J. Am. Chem. Soc.* (1920) **42** 742; (1923) **45** 930
20. Michaelis, L.—Gyémánt, A.: *Biochem. Z.* (1923) **109** 165
21. Behrend, Robert (1856—1926) studied in Leipzig, and then worked there at the university. He later left the study of physical chemistry and turned to organic chemistry. In 1897 he became Professor of Organic Chemistry at the University of Hanover. He retired in 1924.
22. Behrend, R.: *Z. physik. Chem.* (1893) **11** 466
23. Böttger, Wilhelm (1871—1949) was a pharmacist. He studied at several universities in Germany and Switzerland, and later worked with Ostwald in Leipzig. In 1910 he became Professor of Analytical Chemistry at the University of Leipzig. He worked until 1938, when he retired.
24. Böttger, W.: *Z. physik. Chem.* (1897) **24** 251
25. Crotogino, F.: *Z. anorg. Chem.* (1900) **24** 225
26. Dutoit, Paul (1873—1944) was born in Lausanne, and studied in Geneva. In 1898 he became Professor of Physical Chemistry at the University in Lausanne, while in 1918 he became Professor of Inorganic Chemistry at the same university
27. Dutoit, P.—Weisse, G.: *J. chim. phys.* (1911) **9** 578
28. Treadwell, William Dupré (1885—1959) was Professor of Analytical Chemistry at the Technical High School of Zürich. He retired in 1955.
29. Pinkhof, J.: *Over die Toepassing der Elektrometrische Titraties.* Amsterdam (1919)
30. Treadwell, W. D.—Weiss, L.: *Helv. Chim. Acta* (1919) **2** 680
31. Hostetter, John Clyde was born in 1886. He worked at the American Bureau of Standards, and later in industry.
32. Hostetter, J. C.—Roberts, H. S.: *J. Am. Chem. Soc.* (1919) **41** 1337
33. Willard, H. H.—Fenwick: *J. Am. Chem. Soc.* (1922) **44** 2504
34. Cox, D. C.: *J. Am. Chem. Soc.* (1925) **47** 2138
35. Foulk, Charles William (1869—1958) was Professor of Analytical Chemistry from 1898 to 1939 at the Ohio State University.
36. Bawden, Arthur Talbot, was born in 1897. He taught chemistry at various schools, and at present is at Stanton College.
37. Foulk, C. W.—Bawden, A. T.: *J. Am. Chem. Soc.* (1926) **48** 2044
38. Salomon, E.: *Z. physik. Chem.* (1897) **24** 55
39. Müller, Erich (1870—1938) studied in Berlin. After spending considerable time on his travels he took over the running of his father's silk factory. His interest however lay in science and he returned to the Technical University of Dresden as an assistant lecturer. He later became Professor of Physical Chemistry, first at Braunschweig, and then at Stuttgart and finally in Dresden. He retired in 1925.

40. KÜSTER, FRIEDRICH WILHELM (1861—1917) studied in Berlin and Münich. He worked with Nernst in Marburg and Göttingen. In 1899 he became a Professor at the Mining Academy in Clausthal. In 1904 he retired and dealt with fruit producing. He died as a result of a diving accident whilst swimming.
41. KÜSTER, F. W.—GRÜTERS, M.: *Z. anorg. Chem.* (1903) **35** 454
42. KOHLRAUSCH, F.: *Wied. Ann.* (1880) **11** 653
43. OSTWALD, Wi.: *Z. physik. Chem.* (1888) **2** 561
44. JANDER, GERHART (1892—1961) was born in Alr-Döbern, and studied in Münich and Berlin. After completing his studies he went to Göttingen, where he became a Professor Extraordinarius in 1925. Later he worked at the Kaiser Wilhelm Institute. From 1935 he was head of the inorganic chemistry department of the University of Greifswald, and since 1951 he has occupied a similar position at the Technical University of Berlin.
45. JANDER, G.—MANEGOLD, E.: *Z. anorg. Chem.* (1924) **134** 283
    JANDER, G.—PFUNDT, O.: *Z. anorg. Chem.* (1926) **153** 219
46. JENSEN, FREDERIK WILLIAM, was born in 1894. He worked at various colleges, and from 1925 he was a professor at the Agricultural and Mechanical College (Texas). He retired in 1959.
47. FOREMAN, J.—CRISP, D. J.: *Trans. Farad. Soc.* (1946) **24A** 186
    JENSEN, F. W.—PARRACK, A. L.: *Ind. Eng. Chem. An. Ed.* (1946) **18** 595
48. LIPPMANN, GABRIEL (1845—1921) was Professor of Theoretical Physics at the Sorbonne in Paris.
49. LIPPMANN, G.: *Pogg. Ann.* (1873) **149** 547
50. KUCERA, BOHUMIL (1874—1921) was Professor of Physics at the University of Prague.
51. KUCERA, B.: *Drud. Ann.* (1903) **11** 529 698
52. HEYROVSKY, JAROSLAV, was born in Prague in 1890. He began his research works at the University of Prague, and completed them in London. He became an assistant lecturer at the University of Prague in the Institute for Analytical Chemistry, and in 1926 he was made Professor of Physical Chemistry there. Since 1950 he has been the director of the Polarographic Institute of the Czechoslovak Academy. He was awarded the Nobel Prize for chemistry in 1959.
53. HEYROVSKY, J.: *Chem. Listy* (1922) **16** 256
    HEYROVSKY, J.: *Phil. Mag.* (1923) **45** 303
54. SHIKATA, MASURO, was born in 1895 in Tokyo. He studied there and later travelled to Europe, where he worked in Prague. In 1924 he became a professor at the Kioto University and in 1942 he was made vice-president of the Mandshurian Academy. He was a prisoner of war and was not released until 1953. He was a Professor at the University of Nagosha until 1959, when he retired.
55. HEYROVSKY, J.—SHIKATA, M.: *Rec. trav. Pays Bas.* (1925) **44** 496
56. HEYROVSKY, J.: *Chem. Listy* (1946) **40** 222
57. VOGEL, J.—RIHA, J.: *J. Chim. phys. Physico-Chim. biol.* (1950) **47** 5
58. PAULIK, J.—PROSZT, J.: *Acta chim. hung.* (1956) **9** 161
59. HEYROVSKY, J.: *Chem. Listy* (1941) **35** 155
60. KEMULA, WIKTOR, was born in 1902. In 1935 he became Professor of Physical Chemistry at the University of Lemberg. From 1945 he has been Professor of Inorganic Chemistry at the University of Warsaw.
    KUBLIK, ZENON, was born 1922 in Warsaw, and is working at the University of Warsaw.
61. KEMULA, W.—KUBLIK, Z.: *Anal. chim. acta* (1957) **18** 104
62. HEYROVSKY, J.: *Bull. soc. chim.* (1927) **41** 1224
63. MAJER, VLADIMIR, was born in 1903 in Prague, and studied at the Technical University there. Later he worked in several industrial research departments and then in the Physico-Chemical Institute of the University of Prague. He is now head of the Nuclear Chemistry Institute of the University of Prague.
64. MAJER, V.: *Z. Elektrochem.* (1936) **42** 120

CHAPTER XIV

OTHER METHODS OF ANALYSIS

1. RADIOCHEMICAL ANALYSIS

The investigation of the phenomena of radioactivity started in the present century, and at a very early stage in the development of this subject the analytical potentialities were realized. The only requirement was the production of measuring instruments of sufficient sensitivity. This problem is very much connected with physics and, therefore, here it will only be dealt with very briefly.

Measurement of radioactivity is nowadays mainly carried out with counting devices or by scintillation methods. The development of the counting device is entirely due to the work of Geiger [1]. Together with Rutherford [2] he prepared the first counting tube in 1908. This consisted of a cylindrical metal tube with an axial wire, and an electric field situated between the cylinder wall and the wire. The rays on entering the tube ionize the gas which is at low pressure, and these ions are attracted to the wire, and are then registered on an electroscope [3]. Geiger later improved the design of the tube with the help of his co-workers [4], and mainly owing to the work of Müller it had evolved by 1928 to the form in which it is used today [5].

The principle of the scintillation counter is the phenomenon exhibited by zinc sulphide, which when irradiated with α rays re-emits a part of the absorbed energy as visible light. This phenomenon was first observed by Crookes [6] and Elster and Geitel [6] in 1903. In 1908 Crookes and Regener designed a measuring instrument based on this priciple called the spinthariscope. Initially this instrument was of great importance, but the need for optical counting was a serious disadvantage and it was not until the introduction of the photomultiplier tube that the device became of real practical importance. Curran and Baker were the first to combine a scintillation screen and photomultiplier, and were able to measure the strength of α-radiation in this way [7]. Coltman and Marshall as well as Broser and Kallmann were the first to use this device for radioactive counting [8]. It is now possible to count β and γ particles in this way.

There are many different methods of analysis which involve radioactivity in some way. One of the most important of these is the use of isotopically labelled atoms. To the substance to be determined a radioactive isotope is added, and this eventually becomes intimately mixed with the sample. After the completion of any subsequent operations the radioactivity of a known amount of the product is measured.

The basic principle of all these methods was discovered by Hevesy [9] and Paneth [10], who published a paper in 1913 in which they described the determination of the solubility of lead sulphate and lead chromate. The following is an extract from the introduction to their paper:

> The fourth decomposition product of radium emanation, Radium D, as it is known, shows all the chemical reactions of lead; if RaD is mixed together with lead or lead salts, it cannot be separated from lead by any physical or chemical methods. When once we have mixed the two substances, then their concentration ratios will be the same even in the smallest amounts of the sample. Because of the radioactivity of RaD it is possible to determine very much smaller quantities than in the case of lead, so that it can serve for the qualitative and quantitative test for the latter. The RaD therefore acts as an "indicator" [11].

Into this group falls the so-called indicator analysis, where the distribution of a substance, for example, in two or more phases is measured by the distribution of the radiation. This method is used mainly for testing the accuracy of analytical methods. Erbacher and Philipp were the first to apply this technique, in the examination of the separation of gold and platinum, using labelled Au 198 as a tracer [12].

The pioneer of analyses using radioactive reagents was Ehrenberg [13]. Here the reaction is carried out with a labelled reagent of known concentration and activity, for example, chromate is precipitated by an excess of labelled lead, and from the activity of the precipitate, the lead, and conversely the chromate content, can be determined. Naturally, while the only radioactive substances available were those obtained from natural sources, the method was of little practical importance.

The methods using radioactive reagents became vastly more important, however, when artificial radioactive isotopes became readily available, and as a result of this a much greater range of reagents could be labelled. Since the 1940's nearly all the methods of classical analysis have been attempted using radioactive reagents.

One of the most important modifications to follow the introduction of artificial radioactive isotopes was the technique of radiometric titrations. Precipitation titrations lend themselves most readily to this type of procedure, where either the solution, or the titrant is radioactive. The end-point of the titration is indicated by a change in the activity of the solution. This method was originated by Langer [14].

The most important of the radiochemical methods of analysis is undoubtedly isotope dilution analysis. In the simple form of this method a radioactive isotope of the element to be determined is added to the sample. The specific activity of this isotope must be known, and after a separation, which need not necessarily be quantitative, a pure sample is taken and the activity measured. The ratio of this activity to the initial activity (if necessary this must be corrected for decay) together with the weight of the sample will give the amount of inactive element present. The method is very useful for determining one element in the presence of one or more similar elements. This procedure was introduced by Hevesy, who, working with Hobbie developed the method by an electrolytic determination of lead in 1932 [15]. For the analysis of trace elements the radioactive precipitate-exchange method of Bányai [16], Szabadváry [17] and Erdey is of considerable value. In this method the element to be determined is mixed with an isotopically labelled precipitate, one component of which forms a soluble complex with the element. The activity of the solution will therefore be proportional to the amount of the element present [18].

Another branch of radiochemical analysis, again introduced by Hevesy, is "activation analysis". The principle of this method is entirely different from the preceding methods, the sample in this case being irradiated with a stream of atomic particles, usually slow neutrons, which react with the nuclei of the element to be determined and convert one of its isotopes into a radioactive species. With certain precautions the measurement of the radioactivity corresponds to the amount of the element present.

Hevesy and Levi carried out the first neutron activation analysis on the rare earth elements. As a neutron source they used radium-beryllium, coated with paraffin. They examined the individual elements and investigated the isotopes which could be formed from them. They also recorded "The use of neutrons applied to analytical chemistry" in a separate chapter [19].

> The usual chemical methods of analysis fail, as is well know, for most of the rare earth elements and have to be replaced by spectroscopic, X-ray and magnetic methods. The latter methods can now be supplemented by the application of neutrons to analytical problems by making use both of artificial radioactivity and of the great absorbing power of some of the rare earth elements for slow neutrons.
> Qualitative analysis with the aid of artifical radioactivity is based on the determination of periods of decay. The method of artificial radioactivity has been used to termine the dysprosium content of yttrium preparations.

Hevesy and his co-worker have used neutron activation analysis to determine the dysprosium content of yttrium preparations. The yttrium sample to be investigated was activated under exactly the same conditions as a standard sample, and a comparison of the dysprosium activities obtained gave a value of 1 per cent for the dysprosium content of the yttrium sample. The neutron source was radium—beryllium.

Some brief reference must be made to the methods based on β-ray reflection. The techniques are rapidly increasing in importance. When β-rays are allowed to impinge on a sample they are reflected, the amount of reflection being proportional to the atomic number. The angle of reflection is in some cases as large as 180°. The great advantage of this method is that it is completely nondestructive. The method was discovered by R. H. Müller [20].

Finally γ-ray spectrometry must be mentioned. The development of this method was made possible by the fact that in addition to zinc sulphide several other substances are also capable of absorbing radiation and emitting part of the energy in the form of visible light. This takes place with β- and γ-rays as well as α-rays. The first attempts to employ this method took place after the second World War. Certain substances are very responsive to γ-rays crystal counters, and of these the NaI crystal activated with thallium is the most important. This crystal was first employed by Hofstadter [21]. As a result of this observation qualitative and quantitative measurement of substances emitting γ-radiations with varying energies became possible.

2. CHROMATOGRAPHY

Chromatography is one of the most important methods to emerge in the present century. As well as being an important tool for analytical chemistry chromatography is now invaluable to organic chemistry. The chemical industry also uses this technique. Biology, botany, biochemistry and the related medical sciences also use the techniques of chromatography to an ever increasing extent, and it is interesting to note that Tsvett, the founder of modern chromatography, was himself a botanist. As with other new methods of analysis, earlier workers had laid the important foundations. Capillary analysis (previously referred to in connection with spot tests (Chapter VIII. 1)) was an important forerunner to modern chromatography.

The history of chromatography begins with the work of Runge [22], who was a physician. Runge observed that certain coloured substances when spotted on to a filter paper spread out into concentric rings. He recorded these with the practised eye of a painter in his book *Zur Farbenchemie: Musterbilder für Freunde des Schönen und zum Gebrauch von Maler* (1850) and *Der Bildungstrieb der Stoffe* (1855). The immediate predecessor to chromatography was capillary analysis. Schönbein [23] in 1861, observed during the qualitative test for ozone that if an aqueous solution is spotted on a filter paper, the water precedes the dissolved substances, and also that different substances are drawn up the paper to varying degrees [24]. Schönbein commented on this observation:

> This can be of valuable service to the analytical chemist in cases where the usual reagents are not applicable, for example in the case of organic dyestuff solutions.

This work was continued by his student, Goppelsröder [25], who developed the final method of capillary analysis. Capillary analysis shows many similarities to paper chromatography. While in the latter method one drop of the solution is placed on the filter paper, in capillary analysis the edge of the filter paper is dipped into the solution itself. In paper chromatography the samples are moved up the paper by the flow of a solvent system, whereas in capillary analysis conclusions are based on the height to which the various components are sucked up the filter paper. Thus the sensitivity of the latter could never approach that of paper chromatography [26]. Goppelsröder drew conclusions from the suction heights of the various components on the paper so that in this respect his method was similar in principle to frontal analysis. Goppelsröder published the results of his first experiments on this subject in 1861 [27], and continued to work in this field throughout the whole of his life. He published a large number of papers, and finally in 1901 a monograph on the subject [27]. His method was used for the analysis of dyes, alkaloids, oils, drinking waters, wine, milk, salt mixtures, etc. Capillary analysis, although it did not become very widespread, was occasionally used in pharmaceutical analysis. Tsvett did not regard it, however, as being in any way connected with paper chromatography, but this view is open to debate.

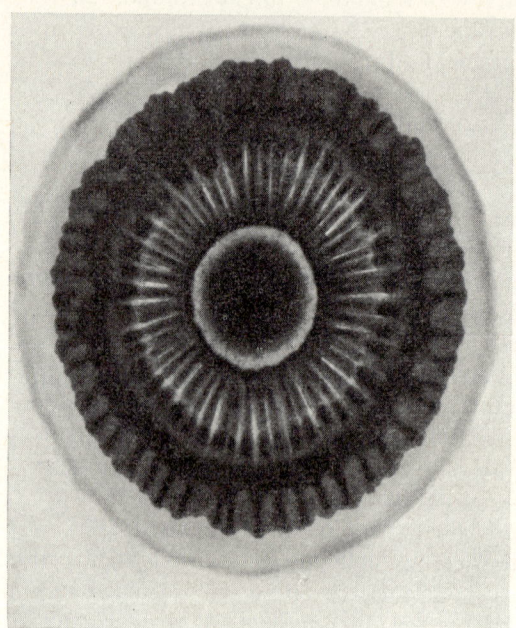

Fig. 98. Paper chromatographic picture. (From Runge: *Der Bildungstrieb der Stoffe* [1855])

One other important preliminary to chromatography was the observations which were made in connection with the ion exchange properties of the earths, and which was to result in the development of column chromatography. These will be referred to in the next section Many such investigations were made in the last century, for example Fischer [28] and Schmidmer, in 1893, separated inorganic ions using a compressed paper column, through which the solution was allowed to flow [28]. Many other experiments carried out at about this time could also be considered to have some bearing on the development of chromatography.

But it is indisputable that chromatography as we know it today was devised by Tsvett, and it is to him that all the credit is due.

Mihail Tsvett was born in 1872 in Russia. His father was Russian and his mother was an Italian, who had lived in Turkey. While he was a child his family moved to Switzerland, and it was here, in Geneva and Lausanne, that Tsvett studied. He read botany, physics and chemistry, and received his doctorate in botany in 1894. For the work that he did for this degree he received the Davy award. In 1896 he went to Russia, where he lectured on plant anatomy and physiology. In 1901 he became a Privat Dozent in Warsaw, while in 1907 he was appointed Professor of Botany and Anatomy at the University there in the Veterinary Faculty. A year later he was appointed to a similar position at the Technical University of Warsaw. It was during his years here that he carried out his most important work. During the first World War he was forced to flee, and went to live in Moscow and Nizhnii Novgorod.

By this time he was suffering from tuberculosis, and in 1918 when he was working at the University of Dorpat, the Germans occupied the town, and he was again forced to flee, this time to Voronyes. It was here that he died of tuberculosis in 1920, while he was still not yet fifty.

FIG. 99. Michail Tsvett (1872—1920). Photograph

At the beginning of this century Tsvett was working with plant pigments, but his interesting work in this field is beyond the scope of this book. He investigated the problem of the adsorption of pigments on proteins, and as a result of this work he became interested in the phenomena of adsorption in general. In 1903 he gave an account of the results of his investigations, which in fact marked the introduction of chromatography. In one experiment he passed a plant extract down a column packed with inulin. He wrote about this experiment [29]:

> The adsorption phenomena observed during the filtration through the powders are very interesting. The solution flowing down through the lower part of the tube is at first completely colourless, but then becomes yellow (carotine), until at the top of the inulin column a green band is formed, and beneath this a yellow layer slowly develops. If pure ligroin is filtered through the column, both stripes begin to expand and move.

In the following years he developed the method which he published in 1906 [30]. It is best described in his own words:

> If a chlorophyl solution dissolved in petroleum ether is filtered through an adsorbent column (I use mainly calcium carbonate suspension which I compact into a glass tube) the pigments are separated from the top to the bottom of the column, according to the order of their adsorption ability, and various coloured zones are formed. The most strongly adsorbed substances force the less strongly adsorbed ones further and further down the column. This separation is made almost complete, if, after the pigment solution is passed down the column, a solvent is passed through the column after it. Just as the light rays in the spectrum are separated, so the various components of the pigment mixture are separated on the calcium carbonate column, and therefore they can be determined qualitatively and quantitatively. This preparation is called a chromatogram, and the method is called chromatography. It is unnecessary to add that the method is not only suitable for pigment mixtures, for as can be expected all coloured and colourless chemical compounds are governed by this law.

In the same year Tsvett published a further paper, in which he gives a diagram of his apparatus together with a description of the experimental procedure [30]. The following is an extract from this paper.

> For an adsorbent any powders which are insoluble in the solvent can be used. In practice, however, as many substances have some effect on the adsorbed substances, the choice of adsorbent will be limited to those substances which are chemically neutral and at the same time capable of being finely powdered. Substances which adsorb very strongly should also be avoided because in order to achieve a separation a large amount of pigment sample is required. It is essential that the adsorbing substance must be in a very finely divided state, because coarsely divided adsorbent does not yield a well defined chromatogram. This is due to the fact that in the wide capillaries adsorption and diffusion are opposing each other... For special purposes a chemically reactive adsorbent can also be used, for instance hydrolysing, oxidizing or reducing adsorbents...

Tsvett's apparatus is shown in Fig. 100. He named the method chromatography because he wanted to indicate that it gives a picture of the composition of coloured substances. This name has persisted even though the vast majority of chromatographic samples nowadays are colourless.

Tsvett's work also yielded some very important results. For example he found that a leaf extract contained two chlorophyls, four xanthophyls and carotene. This clearly indicated the usefulness of his new method [31]. Although he published the results of his work in a monograph in 1910, it aroused little enthusiasm and his results were treated with some scepticism by his contemporaries. Karrer recently evaluated Tsvett's work as follows:

> There is no other single discovery which has had so great an influence on the field of research in organic chemistry as has the method of chromatographic analysis devised by Tsvett. Vitamins, hormones, carotenoids and many other naturally occurring substances could not have been examined so rapidly without its aid. This method illustrates the variations in the properties of a series of closely related naturally occurring compounds.

That Tsvett's method did not become widespread for a considerable time is astonishing, considering the very good results which Tsvett himself obtained with the method. Twenty five years later Kuhn [32], Wintersteiner and Lederer [33] revived this method for the investigation of carotenoids [34]. This marked the beginning of the rapid development of chromatography.

Partition chromatography discovered by Martin [35] and Synge [36], in 1941 [37], utilizes the principle of counter current distribution, the stationary

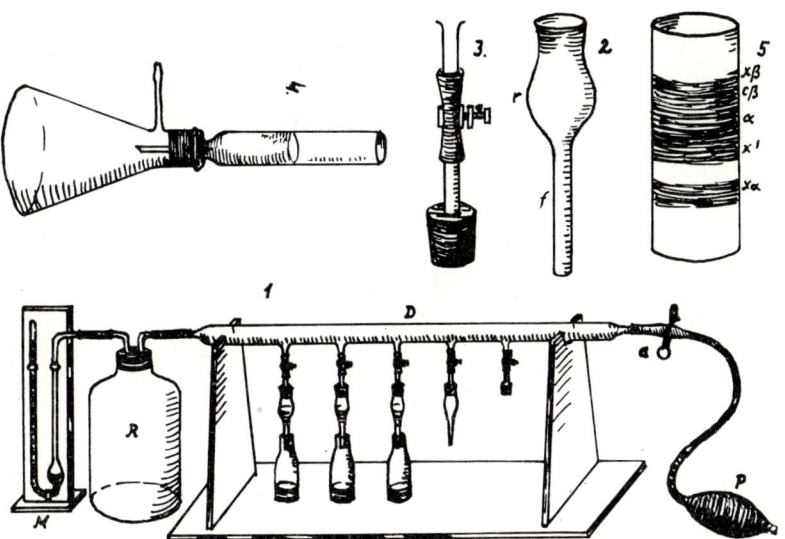

Fig. 100. Tsvett's chromatographic apparatus. (From the original paper)

phase being supported on a carrier. Consden [38], Gordon [38], Martin and Synge later used filter paper as a carrier and thus paper chromatography originated [39]. The fact that the principle of partition chromatography could be applied to gases led James [40] and Martin to develop gas chromatography [40].

3. ION EXCHANGE

Apart from the miracle of Moses who made bitter salt water drinkable by the addition of tree branches which some people may consider an application of ion-exchangers, and from Aristotle's reference to the use of clay filters for the same purpose, the introduction of ion exchange methods took place about the middle of the 19th century.

Thompson and Way, in 1850, observed that if an earth is treated with ammonium sulphate or ammonium carbonate solution, it adsorbs ammonia and at the same time calcium passes into the solution [41]. Eichhorn [42], in 1858, established that the adsorption of elements from earth waters to clay is a reversible process [43]. The ion exchange properties of earths were later examined from several points of view, but this is beyond the scope of this volume. In 1896 Harm obtained a patent for his method of removing sodium and potassium from sugar-be e liquids, treating with a naturally occurring silicate. The use of zeolites and several other naturally occurring silicates thus commenced in this way in industry as ion-exchangers. Gans [44], in 1905, prepared an artificial cation exchanger from clay, sand and soda [45].

The analytical use of ion exchangers only began in this century. The pioneering role was played, as in many other branches of chemistry, by biologists. At the end of the first World War no blood carbon was available and this was needed to bind the creatine for the determination of the ammonia content of urine. Folin [46] and Bell in 1917 investigated the use of the synthetic aluminium silicate of Gans for this purpose, the principle of the method being that the zeolite bound the ammonia, while the creatine remained in solution. Ammonia was subsequently liberated by treatment with alkali and was determined with Nessler's reagent [47].

Bahrdt, in 1927, used a zeolite column for the determination of the sulphate content of water. The calcium and magnesium were removed on passage through the column and to the resulting solution a measured excess of barium chloride was added and the excess back-titrated with a palmitate solution [48].

Nevertheless while ion exchange methods were limited to the use of natural zeolites they did not achieve very much practical importance. This was mainly due to the fact that the natural zeolites can only be used in a narrow pH range, because they are soluble in strong acids, and are peptized in alkaline medium.

The situation became vastly different with the introduction of synthetic organic resins with ion exchange properties. The ion exchange properties of certain organic substances, such as cellulose, had been known for some time, and Kullgren [49] examined the sulphite cellulose pulps and found that these were capable of replacing copper from aqueous solutions by hydrogen, and that elution with mineral acids reversed this process [50]. The first ion exchange resin of any practical value, however, was produced from phenol and formaldehyde by Adams and Holmes, in 1935 [51].

Since then, both the quality and quantity of commercially available ion exchangers has improved rapidly, and a very wide field of application has been found for them apart from water treatment. The first monograph in this field was written by Samuelson [52].

Ion exchange methods are simply aids to analytical chemistry; they must therefore be combined with some other complementary measuring technique. One of the more important developments of this subject is ion exchange chromatography. This technique was developed during the second World War in the U.S.A. for the separation of atomic fission products; it was perhaps first used by Russel and co-workers in 1944 [53].

Ion exchange papers were first prepared by Wieland and Berg [54], although they did not prove very successful; the first successful work on this subject was carried out by M. Lederer [55].

Redox reactions can also be made to occur on an ion exchange resin. This type of method was recommended for the first time by Sansoni [56], while the first actual determinations were carried out by Inczédy [57].

Resins are now available which are markedly redox in their characteristics, but these hardly come within the province of ion exchange. The pioneer work on this subject was carried out by Cassidy [58].

NOTES AND REFERENCES

1. GEIGER, HANS (1882—1945) worked with Rutherford and later returned to Germany where he was the head of the Radiochemistry Laboratory of the Physikalisch-Technische Reichsanstalt.
2. RUTHERFORD, ERNEST (1871—1937) was born in New Zealand. He was Professor of Physics at several universities in Canada, and later at Manchester, and finally at Cambridge. His laboratory there, the Cavendish Laboratory, was the birth-place of the atomic age. He made a great many important discoveries, among them the creation of the first atomic model and the first transformation of elements. He was awarded the Nobel Prize for Chemistry in 1908.
3. RUTHERFORD, E.—GEIGER, H.: *Proc. Roy. Soc. London* (1908) *A***81** 141
4. GEIGER, H.—RUTHERFORD, E.: *Phil. Mag.* (1912) **24** 618
   GEIGER, H.: *Verh. Phys. Ges.* (1913) **15** 534
   GEIGER, H.—KLEMPERER, O.: *Z. f. Phys.* (1928) **49** 753
5. GEIGER, H.—MÜLLER, O.: *Phys. Z.* (1928) **29** 839
6. CROOKES, WILLIAM (1832—1919) worked first of all with A. W. Hoffmann, but later continued his research work at his own private laboratory. He was one of the discoverers of thallium.
   ELSTER, JULIUS (1854—1920) and GEITEL, HANS (1855—1923) were both secondary school teachers in Wolfenbüttel.
7. CURRAN, S. C.—BAKER, W. R.: *Rev. Sci. Instr.* (1948) **19** 116
8. COLTMAN, J. W.—MARSHALL, F. W.: *Phys. Rev.* (1947) **72** 528
   BROSER, J.—KALLMANN, H.: *Z. f. Naturforsch.* (1947) **2A** 439
9. HEVESY, GYÖRGY, was born in 1885 in Budapest. He studied in Budapest and Zürich. Later he worked as an assistant in Zürich and Karlsruhe, and also at the Radium Institute in Vienna, and at the University of Budapest (1918—19), as well as at Copenhagen where he worked with Bohr (1919—1926). It was here that he discovered the element hafnium. From 1925 to 1935 he was a professor at the University of Freiburg, and after that at the University of Stockholm. He was awarded the Nobel Prize for Chemistry in 1944.
10. PANETH, FRIEDRICH (1887—1958) was born in Vienna, and studied in Vienna and München. He also worked in the Radium Institute in Vienna. In 1922 he became an extraordinary professor of chemistry at the University of Berlin. In 1929 he was appointed a professor at Königsberg, while in 1933 he was forced to leave, and came to England, where he worked on atomic research. Until 1953 he was Professor of Chemistry at the University of Durham, when he returned to Germany in 1953, to become Head of the Max Planck Institut für Chemie in Mainz until his death.
11. HEVESY, GY.—PANETH, F.: *Z. anorg. Chem.* (1913) **82** 323
12. ERBACHER, O.—PHILIPP, K. Z.: *Z. angew. Chem.* (1935) **48** 409
13. EHRENBERG, RUDOLF, was born in 1884 in Rostock. He was Dozent, later Professor of Physiology at the University of Göttingen. He retired in 1952.
    EHRENBERG, R.: *Biochem. Z.* (1925) **164** 183
14. LANGER, A.: *Phys. Chem* (1941) **45** 639; *Anal. Chem.* (1950) 22 1288
    LANGER, ALOIS, was born in 1909 in Paseky (Bohemia), studied in Brno, worked in different

Czechoslovakian institutes and hospitals, went in 1938 to the United States, was lecturer at the University of Pittsburgh and has been working since 1944 in the Research Center of Westinghouse Electric Corp. in Pittsburgh, as an advisory scientist.
15. HEVESY, GY.—HOBBIE, R.: *Nature* (1931) **128** 1038; *Z. anal. Chem.* (1932) **88** 1
16. BÁNYAI, ÉVA, born in 1923, is a reader at the Technical University of Budapest.
17. The author of this book.
18. BÁNYAI, É.—SZABADVÁRY, F.—ERDEY, L.: *Mikrochim. Acta* (1962) 427
19. HEVESY, GY.—LEVI, H.: *Kgl. Danske Vid. Slesk. Math. fys. Medd.* (1936) **14** 24
20. MÜLLER, R. H.: *Anal. Chem.* (1957) **29** 969
21. HOFSTADTER, R.: *Phys. Rev.* (1948) **74** 100; (1949) **75** 796
22. RUNGE, FRIEDRICH (1795—1867) was a physician, and for a short time was a lecturer in technology at the University of Breslau. He later worked in Prussia as a Secretary of State.
23. SCHÖNBEIN, CHRISTIAN FRIEDRICH (1799—1868) was Professor of Chemistry at the University of Basle. He was the discoverer of ozone and of wood-wool.
24. SCHÖNBEIN, C. F.: *Pogg. An.* (1961) **114** 275
25. GOPPELSRÖDER, FRIEDRICH (1837—1919) from 1869 until 1872 was Professor of Chemistry in Basle, and from then until 1880 he was the director of the chemistry department at Mühlhausen. After this he occupied himself with private research.
26. GRÜNE, A.: *Österr. Chem. Ztg.* (1959) **60** 301
27. GOPPELSRÖDER, F.: *Verh. naturforsch. Ges. Basel* (1861) **3** 268
28. FISCHER, EMIL (1857—1919) was Professor of Chemistry at the Universities of Würzburg, Erlangen and Berlin. His major achievements were made in the field of organic chemistry, especially notable is his research on the amino acids. He was awarded the second Nobel Prize for Chemistry in 1902. Fischer, E — Schmidmer, E: *Lieb. Ann.* (1893) **722** 156
29. TSVETT, M.: *Trudy warsavsk. obst. Jestesvoispitat, Otd. Biol.* (1903) **14** 20
30. TSVETT, M.: *Ber. deutsch. bot. Ges.* (1906) **24** 316
31. TSVETT, M.: *Ber. deutsch. bot. Ges.* (1906) **24** 384
32. KUHN, RICHARD, was born in 1902. He was a professor at the Universities of Zürich and Heidelberg. He is now the director of the Max Planck Institut in Heidelberg. He was awarded the Nobel Prize for Chemistry in 1938.
33. LEDERER, EDGAR, was born in 1908 in Vienna, studied in Vienna, worked 1930—1933 at the Kaiser Wilhelm Institut für Medizinische Forschung Heidelberg, emigrated 1933, worked in U.S.S.R. and France. Since 1956 he has been Professor of Biochemistry at the University of Paris and director of "Institut de Chimie des substances naturelles at Gif sur Yvette".
34. KUHN, R.—WINTERSTEINER, A.—LEDERER, E.: *Z. physiol. Chem.* (1931) **197** 141
35. MARTIN, ARCHER JOHN, was born in London in 1910, and studied chemistry at Cambridge and then worked in several textile research establishments. Since 1948 he has been the head of the National Institute for Medical Research. He was awarded the Nobel Prize for Chemistry in 1951.
36. SYNGE, RICHARD LAWRENCE, was born in Liverpool in 1914. He studied chemistry at Cambridge, and later worked with Martin, and received the award of the Nobel Prize for Chemistry with him, in 1951.
37. MARTIN, A. J.—SYNGE, R. L.: *Biochem. J.* (1941) **35** 91 1358
38. CONSDEN, R. was born in London in 1911, graduated at Battersea Polytechnic, was appointed (1938) to the Wool Industries Research Association, Leeds. Since 1949 he has been a research biochemist at the Medical Research Council.
   GORDON, A. H. was born in London in 1911, he was educated at Trinity College, Cambridge. From 1938 he worked at the Wool Industries Research Association, Leeds. Since 1950 he has worked at the Medical Research Council.
39. CONSDEN, R.—GORDON, A. H.—MARTIN, A. J. P.: *Biochem. J.* (1944) **38** 224
   GORDON, A. H.— MARTIN, A. J. P.—SYNGE, R. L. M.: *Biochem. J.* (1943) **37** 13

40. JAMES, A. T. was born in Cardiff in 1922, studied at the University of London, worked in different institutions (in 1952 at the National Institute for Medical Research). Since 1962 he has worked at Unilever Research Laboratory, Bedford.
    JAMES, A. T.—MARTIN, A. J. P.: *Biochem. J.* (1952) **50** 679
41. THOMPSON, H. S. J.: *Roy. Agr. Soc. England* (1850) **11** 68
    WAY, J. T.: *ibid.* (1850) **11** 313
42. EICHHORN, CARL (1816— ?) was Professor of Chemistry at the Agricultural College in Berlin.
43. EICHHORN, C.: *Pogg. Ann.* (1858) **105** 126
44. GANS, RICHARD (1880—1954) was a professor at the following universities: Königsberg, München and finally in Buenos Aires.
45. GANS, R.: *Jahrb. preuss. geol. Landesanstalt* (1905) **26** 179
46. FOLIN, OTTO (1867—1934) was born in Sweden, but went to live in the United States while still a child. He studied in Chicago; after this he made a study-tour in Europe. In 1898 he became Professor of Analytical Chemistry at the University of Virginia, and later the head of a biochemical research institute. In 1908 he was appointed Professor of Biochemistry at Harvard University.
47. FOLIN, O.—BELL, D. R.: *J. Biol. Chem.* (1917) **29** 329
48. BAHRDT, A.: *Z. anal. Chem.* (1927) **70** 109
49. KULLGREN, CARL FRIDRIK (1873—1955) was born in Uddevalla, and studied in Uppsala and Stockholm. He was Professor of Chemical Technology at the Technical High School in Stockholm between 1915 and 1938.
50. KULLGREN, C.: *Svensk. Kem. Tid.* (1931) **43** 99
51. ADAMS, B. A.—HOLMES, E. L.: *J. Soc. Chem. Ind.* (1935) **54** 11
52. SAMUELSON, OLOF, was born in 1914 in Lidingö. He studied at the Technical High School in Stockholm, and then worked in the cellulose industry. Since 1949 he has been Professor of Chemical Technology at the Technical High School in Göteborg.
    SAMUELSON, O.: *Ion Exchangers in Analytical Chemistry.* (1952) Stockholm—New York
53. SVARTOUT, A. J.—RUSSEL, E.—HUME, D. N.—KETELLE, B. H.: *Project. Work*, May 1944; cit. *J. Am. Chem. Soc.* (1947) **69** 2769
54. WIELAND, T.—BERG, A.: *Angew. Chem.* (1952) **64** 418
55. LEDERER, M.: *Anal. Chim. Acta* (1955) **12** 142
56. SANSONI, B.: *Naturwiss.* (1952) **39** 281
57. INCZÉDY, JÁNOS (born in 1923) is a dozent at the Technical University of Budapest.
    INCZÉDY, J.: *M. Kém. Lapja* (1959) **14** 409
58. CASSIDY, H. C.: *J. Am. Chem. Soc.* (1949) **71** 402 407

AUTHOR INDEX

NOTE. The figures in italics refer to notes and references.

A

Abu Bekr Al Rasi 13
Achard, F. C. 53, 117, 212, *272*
Adams, B. A. 398
Afzelius 128, 129
Agatharchides 6
Agricola, G. 26, *27*, 28
Ajtai, M. 190, *195*
Albertus Magnus 12, 16, *17*, 18, 25
Albiruni 14
Alchazini 14
Alexander 335
Alimarin, J. P. 192, *196*, 383
Alter, D. 323, *345*
Anaxagoras 90
Andrews, J. Z. 257, *278*
Angelus Sala 24
Ångström, A. J. 323, 328, 333, *345*
Archimedes 2, 7, 13
Arfvedson, J. A. 136, *157*
Aristotle 4, 10, 11, 36, 90, 397
Arrhenius, S. 351, 353, 362, *372*
Asbóth, S. 299, 300, *306*
Atanasiu, J. A. 267, *281*
Austen, T. P. 181, *194*
Averroes 10

B

Bacon, F. 21, 90
Bacon, R. 12
Badry, H. El 192, *196*
Baeyer 174
Bahr, J. F. 342, *347*
Bahrdt, A. 398
Baker, W. R. 390
Balard, A. J. 163, *192*
Balland de Toul 227
Ballmann, H. 336
Bang, J. 269
Bányai, É. 392, *400*
Barreswill, L. C. 233, 246, *275*
Bartholdi, C. 216, *272*
Basilius Valentinus 28, 29
Baudisch, O. 182, *194*
Baumhauer, H. 301, *307*
Bawden, A. T. 382, *388*
Bayen, P. 95, *112*
Bayley, T. 191, *195*
Beccaria, G. B. 95, *112*
Becher, J. *46*
Beckmann 354
Beckurts 155, 216, 233
Becquerel, A. C. 233, 312, *317*
Becquerel, A. E. 322, *344*
Beer, A. 340, 341, 342, *347*
Behrend, R. 378, 379, 380, 383, *388*
Behrens, T. 188, *189*
Belcher, R. 268, *282*
Bell, D. R. 398
Benedetti-Pichler, A. 191, *196*
Berg 343
Berg, A. 398
Bergmann, T. 43, 52, 53, 54, 64, 68, 70, 71, 72—81, 86, 87, 88, 89, 128, 162, 205, 259, 349
Berlin, N. J. 136, *158*
Bernard, F. 340, 341
Bernoulli, J. 63, *84*
Berthelot, M. 10, 350, *371*
Berthet 233
Berthollet, C. Z. 105, *106*, 107, 125, 137, 162, 198, 208, 209, 210, 211, 213, 218, 220, 285, 288, 349, 350

Berzelius, J. J. 54, 55, 98, 111, 114, 125–139, *126*, 140, 141, 142, 143, 144, 145, 146, 147, 148, 149, 150, 152, 153, 161, 165, 175, 177, 190, 218, 228, 231, 233, 237, 288, 289, 294, 295, 296, 350, 351
Bilmann, E. 376, *388*
Bineau, A. 232, *274*
Biot 218
Biringuccio 46
Bishof, K. G. 290, *305*
Bjerrum, N. J. 366, 367, 368, 369, *374*
Black, J. *60*, 61, 62, 63, 64, 86, 89, 212, 213
Blake 384
Bloxam, C. L. 312, *317*
Bodenstein 354
Boerhaue, H. 97, *112*, 150, 258
Bolley, A. P. 232, *274*
Boricky, E. *187*
Böttger, W. 362, 379, 380, *388*
Bougouer, P. 340, *347*
Boussingault, J. B. 252, *276*
Boyle, R. 14, 24, 25, 30, 31, 35, *36*, 37–40, 47, 62, 63, 92, 110, 161, 190, 257, 258, 270
Bradberry, F. R. 268
Brandt, G. 59, *83*
Brandt L. 264
Brewster, D. 322, 328, *344*
Brønsted, J. N. 369, *374*
Broser, J. 390
Brunck, O. 182, *194*
Brunner, K. 294, 299, *306*
Bruno, G. 90
Bugarszky, I. 364, *373*
Bunsen, R. W. 139, 174, 229, 230, 249, 255, 318, 324–333, *330*, 341, 342
Bürger, K. 303, *307*
Burriel-Marti, F. M. 268, *282*
Bussy, A. B. 231, *274*, 290, 295
Buzás, I. 265, *280*

C

Cady 270
Cannizzaro, S. 144, *159*
Canotanto, Paolo de 17
Carius, G. L. 300, *307*
Carlisle, A. 310, *316*
Carnot, S. 350, *372*
Cassidy, H. C. 399
Cavendish, H. *64*, 65, 66, 89, 97, 207
Champion, P. 336
Chancel, G. 184, *194*
Charles I., King of Hungary 16

Charlot, G. 369, *374*
Charpentier 255
Chenel, L. 299
Chugaev, L. A. 172, *193*
Clapeyron, B. P. 350, *372*
Clark, T. 231, 232, 237, *274*
Clark, W. M. 263, 265, *280*, 365, 377
Classen, A. 314, *315*
Clausius, R. 350, 351, 352, *372*
Coltman, J. W. 390
Consden, R. 397, *400*
Cozzi, A. 312, *317*
Cox, D. C. 382
Cramer, J. 51
Cremer, M. 375
Crétier, H. 301
Crisp, D. J. 384
Cronstedt, A. F. 53, *82*
Crookes, W. 328, 390, *399*
Crosswhite 335
Crotogino, F. 370, 381
Cruikshanks, W. 310, 312, *316*
Crum, W. 263, *279*
Curran, S. C. 390
Curtois, B. *163*
Czudnovicz, C. 252

D

Dalton, J. *108*–110, 114, 137, 140, 217, 218, 219
Daniell, J. F. 133, 351, *372*
Darby, A. 43
Davy, H. 118, 125, 130, 131, 137, 138, *310*, 311, 312
Degener, P. 266
Dehm, F. 339, 340
Democritus 24, 90
Denigés, G. 257, *278*
Dennstedt, M. 296, 300, *307*
Descartes 110
Descroizilles, F. A. H. 199, *208*–212, 213, 214, 215, 217, 222
Deshayes, M. 252
Despretz, C. 312, *317*
Devarda, A. 266
Dieke 335
Diesbach 56
Dimroth, O. 268, *282*
Diocletian 3
Diodorus of Sicily 2
Döbereiner, J. W. 289, *305*
Dole, M. 376, *387*
Donau, J. F. 303, *308*

Dossie, R. 198, *271*
Draper, J. W. 322, *345*
Drehschmidt, H. 257, *278*
Duboscq, J. 340, *347*
Duflos, A. 229, 233, 234, *273*
Duhamel de Monceau, H. L. 59, *83*
Dulong, P. L. 142, *158*
Dumas, J. B. 133, 134, 177, 294, *295*, 296, 297
Dungi, King of Ur 5
Dupré, A. 249, *276*
Dutoit, P. 381, *388*
Duval, C. 184, 185, *195*

E

Edwards, E. G. 268
Ehrenberg, R. 391, *399*
Eichhorn, C. 397, *401*
Eimbke, G. 148, *159*
Ekeberg, A. G. 54, 128, *156*
d'Elhuyar, F. and J. 45, 73, *84*
Elster, J. 390, *399*
Emich, F. 303, *308*
Engeström, G. 53, *82*
Erbacher, O. 391
Erdey, L. 175, 263, 268, *282*, 392
Erdmann, O. 177, 296, 300, *306*
Erlenmeyer, E. 168, *193*

F

Fabricius 152
Fajans, K. 265, *280*
Faraday, M. 134, 311, 351, *372*
Fehling, H. 234, *275*
Feigl, F. 191, *195*
Fels, B. 363
Fenwick, 382
Fernbach, A. 363
Firmicus, Julius Maternus 3
Fischer, E. 394, *400*
Fischer, E. G. 105, 106, *112*
Fischer, K. 268
Fischer, N. 312
Fister, F. 268, *282*
Foerster, F. 269, *283*
Folin, O. 398, *401*
Fontana, F. 259, *278*
Forckhammer, J. G. 252, *277*
Fordos, M. J. 229, *273*
Fordyce, G. 212, *272*
Foreman, J. 384
Foremann, F. W. 270

Foulk, C. W. 382, *388*
Foureroy, A. F. 45, 124, 125, 162, 163, 259, *278*
Fraunhofer, J. *320*, 321
Frederick the Great 341
Fresenius, C. R. 155, 166—171, *168*, 175, 176, 177, 178, 179, 180, 237, 251, 316
Freudenberg, H. 315
Friedenthal, H. W. 262, *279*, 362, 363, 364, 377
Friedrich, A. 303, *307*
Fritz, H. 316
Fuchs, J. N. 161, *192*

G

Gadolin, J. 73, *84*, 137, 207
Gahn, J. G. 54, 68, 73, *82*, 131
Galenus 13
Galvani, L. *309*
Gans, R. 398, *401*
Gassendi, P. 24
Gaultier de Claubry, F. 229, *273*, 312
Gautier, J. A. 270, *283*
Gay-Lussac, J. L. 110, 162, 163, 197, 199, 200, 208, 213, 217, *218*—227, 235, 237, 239, 263, 288, 289, 291, 295
Geber 11, 12, 13, 15
Gehlen 130
Geiger, H. 390, *399*
Geissler 324
Geitel, H. 390, *399*
Gélis, A. 229, *273*
Gentele, J. G. 256
Geoffroy, C. J. *200*, 201, 208, 349
Gerhardt, C. 144, *159*, 181
Gerlach, W. 335, *346*
Gibbs, J. W. 350, *372*
Gibbs, W. *312*, 313, 314, 315
Gillespie, L. J. 377, 378, *388*
Gioanetti, V. A. 204
Glan, P. 343, *348*
Glauber, R. 25, 26, *34*, 200, 349
Glazunov A. J. 316, *317*
Gmelin, C. G. 136, *157*
Gockel, E. 34
Goethe, J. W. 132, 150
Gooch, F. A. 181, *194*
Goppelsröder, F. 393, 394, *400*
Gordon, A. H. 397, *400*
Gordon, L. 184, *194*
Gottlieb, S. 267
Göttling, J. F. 150, *159*

Govi, G. 342, *347*
Gowles, H. W. 266
Gramont, A. de 334, 336, *346*
Grandcourt 209
Gratidianus, M. 6
Grenier, M. 336
Griess, P. *172*, 174
Griffin 246
Griffin, J. J. 162, *192*, 329
Grotthus, T. 351, *372*
Grüters, M. 383
Gudea, King of Babylon 5
Gudenus 29
Guldberg, C. M. 350, *371*
Gustavson, A. 300
Gutzeit 164
Guyard A. 190, 252
Guyton de Morveau, L. B. 48, 49, 125, 205—207, 208, 216
Gyémánt, A. 378
Győry, I. 267, *281*

H

Haber, F. 375, 376, *387*
Hadding 335
Hahn, F. L. 184, *194*
Hahnemann, C. S. 162, *192*
Halcs, St. 63, *84*
Hantzsch, A. R. 262, *279*
Harkort, E. 55, *82*
Harm 397
Hartley, W. N. 334, *346*
Hassel, O. 265, *280*
Haushofer, K. 188, *195*
Haüy, R. J. 148, *159*
Hehner, O. 254, *277*
Heine, C. 337, *347*
Helmholtz, H. 324, *345*, 350
Helmont, van J. B. 23, 25, 26, 32, *33*, 62, 63, 97, 284
Hempel, W. 231, 274, 301, *307*
Henry, É. O. 236, 237, 239, *275*, 299
Herapath, T. J. 338, *347*
Hermbstaedt, S. F. 152, *159*
Herschel, F. W. 131, 319, *344*
Herschel, J. F. W. 321, 322, *344*
Hess, G. H. 294, *306*
Hevesy, Gy. 335, 391, 392, *399*
Heydweiler, A. 362, *373*
Heyrovsky, J. 385, 386, 387, *389*
Hibbert, E. 267
Hildebrand, J. H. 367, 370, *374*

Hippocrates 8, 13
Hisinger 129, 130, 135
Hittorf, J. W. 324, *345*, 351
Hobbi, R. 392
Hoff J. H. van't 351, *372*
Hoffmann, A. W. 174, *193*, 255, 256
Hoffmann, F. *31*, 32, 33, 61, 161, 258
Hofstadter, R. 393, *400*
Holmes, E. L. 398
Homberg, G. 97, *112*
Home, F. 200, *201*, 202, 216
Honda, K. 184, *195*
Hooke, R. 91, *112*
Horstmann, A. 350, *372*
Hostetter, J. C. 382, *388*
Houton de la Billardière 228, *273*
Hübl, A. 254
Hüfner, C. G. 343, *348*
Humboldt, A. 219, 291, 295, *305*

I

Ibn Rusd 10
Ilinski, M. 182
Ilosvay, L. 172, *174*
Inczédy, J. 398, *401*

J

Jabir Ibn Hayyân 11, 12; see also Geber
Jacquelain, A. 337, *347*
Jacquin, N. J. 45, 62, *81*, 150
James, A. T. 397, *401*
Jander, G. 384, *389*
Janssen, J. P. 327, 336, *345*
Jensen, F. W. 384, *389*
Jodlbauer, M. 299
John, J. F. 152, *159*
Jones, H. C. 251, *276*
Joule, J. P. 350, *372*

K

Kallmann, H. 390
Kant, I. 99, 349
Kayser, H. 334, *346*
Kekulé, A. 300, *307*
Kelvin, Lord 328, 350, *372*
Kemula, W. 386, *389*
Kenny, F. 263, *279*
Kern, E. J. 375
Kessler, F. C. 251, *276*

Kestranek, W. 376
Kieffer 249
Kipp, P. J. 162
Kirchhoff, G. 318, 324, *329*
Kirwan, R. *114*—117, 151, 207, 210
Kjeldahl, J. G. 266, 297, *298*, 299
Klaproth, M. H. *117*—124, 125, 132, 150, 161, 165, 175, 211
Klemensiewicz, Z. 375, 376, *387*
Klobukhov, N. 314, *317*
Knecht, E. 267, *281*
Knop, A. W. 255, *277*
Knop, J. 264, *280*
Knorre, G. 182, *194*
Kohlrausch, F. 351, 362, *372*, 375, 384
Kolbe, H. 284, *305*
Kolthoff, I. M. 184, 190, 262, 266, *279*, 365, 368, 384
Koninck, L. L. de 199
Kopfer, F. 294, *306*
Kopp, E. 174, *193*
Kopp, H. 46
Kopp, J. H. 152, *159*
Koppeschaar, W. F. 254
Krüger, F. 260
Krüss, G. 343, *348*
Kubelkova, O. 264
Kublik, Z. 386, *389*
Kucera, B. 384, 385, *389*
Kuhlmann 189
Kuhn, R. 396, *400*
Kullgren, C. F. 398, *401*
Kunckel, J. *51*, 85
Kurtz, R. B. 263
Küster, F. W. 383, *389*

L

Ladenburg, A. 301, *307*
Lagerhjelm, P. 136, *158*
Lambert 340, *341*, 342
Lampadius, W. A. 150, 152, *159*, 199, 213, 337
Landolt, H. H. 254, *277*
Lang, R. 267, *281*
Lange, T. 256
Langer, A. 391, *399*
Langley, S. 334, *346*
Laplace, P. S. 132, 219
Laurens 215
Lavoisier, A. L. 53, 62, 64, 67, 85, *93*—97, 114, 118, 125, 204, 205, 208, 210, 285—287, 294, 349

Le Blanc, M. 315, 362, 375, *387*
Leblanc, N. *198*
Lederer, E. 396, *400*
Lederer, M. 398
Legler, L. 254, 299
Lehmann, H. 263
Lémery, N. 25, *85*, 150, 284
Leonard, A. G. 334
Leonardo da Vinci 21
Lepez 300
Leuvenhoek, A. van *185*
Levi, H. 392
Levol, A. F. 233, 235, *274*
Lewis, G. N. 368
Lewis, W. *202*, 203, 259
Libavius, A. 29, *30*
Liebermann, L. 364, *373*
Liebig, J. 126, 133, 134, 136, 137, 138, 139, 168, 178, 234, 235, 237, 241, 242, 243, 250, 255, 256, 257, 289, 290—293, *291*, 294, 295
Linossier, M. G. 264
Lippmann, G. 384, *389*
Lloyd, J. M. 191
Lockyer, J. N. 327, 334, *345*
Loewenstern baron, see Kunckel, J.
Lomonosov, M. V. *91*, 92, 93, 349
Lowitz, T. *186*, 212, 270, 319
Lubs, H. A. 263, 377, *388*
Lucena, F. C. 268
Lucianos of Samosta 90
Luck, E. 260
Luckow, C. 314
Lundegardh, H. G. 335, 336, *347*
Lunge, G. *260*
Lüning, O. 200

M

Mac Innes, D. A. 376, *387*
Macquer, P. J. 203
Magnus, H. G. 135, 136, 137, *158*
Majer, V. 387, *389*
Malaprade, L. M. 268, *282*
Malherbe 198
Marchand, R. F. 177, 296, 306
Marcus Aurelius 90
Marggraf, S. A. 51, *55*, 56, 57, 58, 59, 86, 161, 165, 186, 319
Marguart 168
Margueritte, F. 230, 231, *273*
Marignac, J. C. 177
Mariotte, E. 62, 91, *112*

Marjanovic, V. 184
Markert 301
Marozeau 227
Marsh, J. 163, 164, *192*
Marshall, F. W. 390
Martin, A. J. 397, *400*
Mary the Jewess 11
Mascart, N. 333, *346*
Masson, A. 322
Mayer, R. J. 241, *275*, 350
Mayow, J. 63, *83*
Menshutkin, N. A. 350, *371*
Mersenne, M. 91, *112*
Messinger, J. 299
Métherie, J. C. de la 198, *271*
Meusnier, J. 53
Meyer, J. C. F. 259, *278*
Michaelis, L. 265, *280*, 364, 365, 378
Miller, W. A. 322, *345*
Miller, W. H. 322, *344*
Mitscherlich, E. 135, 136, 137, *157*
Mohr, F. 155, 199, 240, *241*−250, 259, 260
Le Monnier, L. G. 201, *271*
Morin, A. 227, *273*
Mosander, K. G. 136, *157*
Moser, L. 184, *194*
Moses 397
Mulder, E. 256, *277*
Mulder, G. J. 294
Müller, A. 338, 340, *347*
Müller, E. 383, *388*
Müller, F. 45, 119, *156*
Müller, O. 390
Müller, R. H. 392
Mullikan, S. P. 304, *308*
Murray 154
Mylius, F. 269, *282*

N

Napoleon Bonaparte 106, 206, 310
Nernst, W. 184, 189, 354, 361, 362, *373*
Neuhäusser, A. 335
Neumann, C. 258, *278*
Newton, I. 319, *344*
Nicholson, W. 310, *316*
Njegovan, V. 184, *194*
Noll, A. 268
Nordenskjöld, N. G. 147, *159*
Norton 25
Noyes, A. A. 366, 367, 368, *373*
Nyulas, F. 205

O

Olympiodoros 10, 11
Osmond, F. 190
Österreicher, J. 48, *49*, 70, 146, 181, 205
Ostwald, Wi. 261, 262, 349, 350, *353*−361, 362, 365, 370, 384
Otto, F. J. 217, *273*

P

Palissy, B. *22*, 24
Paneth, F. 391, *399*
Paracelsus, T. B. 19, *23*, 28, 29, 62
Parrack, A. L. 384
Du Pasquier, A. 228, 229, *273*
Paulik, F. 185, *195*
Paulik, J. 185, *195*
Paweck, H. 315, *317*
Péan de Saint Gilles, L. 251, *276*
Pelagios 10
Péligot, E. M. 232, *274*, 297
Péllerin, F. 270, *283*
Pellet, H. 336, *347*
Pelouze, T. J. 231, 233, 234, 239, 270, *274*
Penny, F. 232, 233, *274*
Penot 227, 264
Pepys, W. 148, *159*
Peters, R. 370
Petit, A. T. 142, *158*
Petrikova 192
Pfaff, C. H. 148, 152, 153, 155, *159*, 161, 163, 164, 165, 237
Pfeffer, W. 350, 351, *372*
Pfeiffer, I. 266, *281*
Philip VI., King of France 16
Philipp, K. Z. 391
Pilch, F. 269, *283*, 303
Pinkhof, J. 382
Piria, R. 300, *307*
Pisani, F. 252, *277*
Plattner, K. F. 55, *82*
Pliny 6, 7, 8, 62, 172
Plisson, A. 299, *306*
Plücker, J. 324, *345*
Poggendorf, J. C. 176, *193*, 241
Pollok, J. 334, *346*
Pope John the XXII. 12
Porret, R. 49
Pott, J. H. *52*, 60
Pregl, F. *302*, 303, 316
Pribil, R. 269, *282*
Priestley, J. 64, *67*, 68, 95
Proszt, J. 263, *279*

AUTHOR INDEX

Proust, J. L. 107, *108*, 137
Prout, W. 141, *158*, 289, 290
Pseudo-Democritus 10
Pungor, E. 265, *280*

R

Rády, Gy. 270, *283*
Rancke-Madsen, E. 200, 212
Raspail, F. V. 186, *187*
Rault, F. 351, *372*
Rayleigh 66
Raymund Lull 19
Regener 390
Reinhardt, C. 251
Reinisch, E. H. 188, *195*
Reischauer, C. G. 254
Reynolds, J. E. 264, *280*
Rhases 13
Richter, J. B. 99—105, 110, 114, 139, 140
Riesenfeld, E. H. 184
Rinman, S. 53, *81*
Rio, M. del 45
Ritter, J. W. 319, *344*
Roberts, H. S. 382
Robiquet, E. 324
Robiquet, P. J. 299, *306*
Roe, R. 210
Roebuck, J. 198, *271*
Rogers, M. 299
Rogers, R. 299, *306*
Roland 208, 209
Roscoe, H. E. 252, *276*, 333, 341
Rose, G. 136, *157*, 165
Rose, H. 135, 137, 138, 139, 161, 164, *165*, 166, 175, 176, 237
Rose, V. 117
Roth, H. 303, *307*
Rothoff, E. 177
Rouelle, G. F. 162, *192*
Rowland, H. 333, 334, *346*
Rózsa, P. 265, *280*
Runge, F. 393, *400*
Ruprecht, A. 45, 118
Russel, E. 398
Rutherford, D. 65, *84*
Rutherford, E. 390, *399*

S

Saint Venaut 234
Salessky, W. 363, 387
Salm, E. 262, 365, 366, 368, 375, 377
Salomon, E. 382, 383
Samuelson, O. 398, *401*
Sand, H. J. S. 315, 316, *317*
Sandell, E. B. 190, 195
Sansoni, B. 398
Saussure, H. B. 55, *82*
Schabus, J. 232, *274*
Scheele, C. W. 59, 64, 68, *69*, 70, 137, 162, 198, 209, 284
Scheffer, H. 201, *271*
Scheibe, G. 335, *346*
Schiff, H. 191, *195*, 296
Schleicher-Schüll 181
Schloesing, J. J. T. 232, *274*
Schmidmer, E. 394, *400*
Scholz 249
Schönbein, C. F. 191, *195*, 393
Schöniger, W. 303, *308*
Schott 181
Schrötter, A. 252
Schubknecht, W. 336, *347*
Schulek, E. 265, 267, *280*
Schulze, E. A. 252, *277*
Schütze, M. 301, 304, *307*
Schwarz, K. L. H. 230, 231, 233, *239*, 240
Schwarzenbach, G. 269, *282*
Schweitzer, E. 335, *346*
Scopoli, J. 45
Seegen, L. 297
Sefström, N. G. 136, *158*
Shikata, M. 385, *389*
Shimer, P. W. 251
Sirokomski, V. S. 267, *281*
Smith, G. H. 303
Smith, P. 252
Soleas, N. 29
Somogyi, Z. 316, *317*
Sorby, H. 187, *195*
Sørensen, S. P. L. 262, 265, 266, 364, 366, 368, *373*, 375, 377
Stahl, G. E. *46*, 51, 55, 59
Staudinger, H. 304, *308*
Steinbeck 314
Steinfels 254
Stepin, V. V. 267
Stokes, G. G. 322, *345*
Streng, J. A. 244, *275*
Strohmeyer 330
Svanberg, L. 164, *193*
Svetonius 6
Swab, A. 52, *81*
Swan, W. 324, *345*
Szabadváry, F. 392

Szabó, Z. 267, *281*
Szebellédy, L. 190, 316, *317*
Szily, P. 262, *279*, 363, 376, 377
Synesios 10, 13
Synge, R. L. 397, *400*

T

Tachenius, O. 24, 33, *34*
Talbot, W. H. F. 321, 328, *344*
Tananaiev, N. A. 191, *196*
Telesio 90
Tennant, S. 131, 198, *271*
Tenner, J. G. 210, *272*
Ter Meulen, H. 301, *307*
Thalén, T. R. 333, *346*
Than, K. 182, *252*, 253, 361
Thénard, L. J. 152, 163, 288, 295, *305*
Thibierge 299
Thomas of Aquinas 10
Thompson 164
Thompson, H. S. J. 397, *401*
Thomson, T. 110, *113*, 116, 141
Thomson, W. *see* Kelvin, Lord
Thurneysser, L. 29
Tomiček, O. 270, *283*
Toricelli, E. 62, *83*
Tower, O. 362, *373*
Treadwell, W. D. 155, 382, *388*
Trey, H. P. 191, *195*
Trommsdorf 105
Truchot, P. 336
Tswett, M. *394–396*

U

Uhl, A. 376, *387*
Ullgren, C. 316, *317*
Unterzaucher, J. 304, *308*
Ure, A. 235, *236*

V

Varrentrapp, F. 297, *306*
Vauquelin, L. N. *124*, 125, 150, 208
Veibel, S. 304, *308*
Venel, G. F. 48, 201, 259, *271*
Vierordt, C. 342, 343, *347*
Vitruvius 8
Vohl, E. 299, *306*
Volhard, J. 252, *255*, 256
Volta, A. 309, *310*
Vorländer, D. 270, *283*
Votoček, E. 257, *278*

W

Waage, P. 350, *371*
Wachtmeister 135
Wackenroder, H. W. 164, *192*
Walden, P. 98, 270, *283*
Waller, E. 254
Wallerius 128
Warder, R. B. 266
Wartha, V. 266, *281*, 333
Watt, J. 43, 131
Way, J. T. 397, *401*
Weidacker, M. 45
Weinlig, G. C. 210, *272*
Weiske, H. 260
Weiss, L. 382
Weisse, G. 381, *388*
Weisz, H. 191, *196*
Welter, J. J. 213, 217, 218, 220, *272*
Wenzel, C. F. 59, *88*, 89, 97, 98, 99, 206
Werner 129, 130
Wheatstone, Ch. 328, *345*
Wiegleb, J. C. 205, *271*
Wieland, T. 398
Wijs, J. J. 362
Wilhelmy, L. F. 349, 350, *371*
Will, H. 297, *306*
Willard, H. 184, *194*, 267, 303, 343, 382
Willigen, V. S. van der 324, *345*
Wilson, C. 86, 192, *196*
Wilson, J. A. 375
Winkler, C. 266, *280*, 294, 315
Winkler, L. *182*, 183, 249, 254, 255, 266
Winterl, J. *48*, 49, 50, 59, 70, 146, 162
Wintersteiner, A. 396
Witz, G. 190
Wöhler, F. 134, 135, 136, 137, 138, *158*, 284, 292
Wolf, C. 91, *112*
Wolff, N. 252
Wollaston, W. H. 109, *113*, 131, 319
Woodward 57
Wren, C. 63, *83*

Y

Young, P. 267, *281*
Young, T. 320, *344*
Yver, A. 314

Z

Zeise, W. C. 299, *306*
Zimmermann, J. C. 251, *276*
Zimmermann, W. 304, *308*
Zosimos 10, 11
Zulkovsky, K. 254, 277, 296, 300

SUBJECT INDEX

A

Absorption coefficient 341
— spectroscopy 340, 342
Accuracy of analytical determination 149
— — gravimetric methods 182
Acetic acid, determination of 200
— — as standard solution 204
— —, titration of 212
— —, the use of 3
Acid, definition of 257
— exponent 369
Acid–base indicators, theory of 261
— titration 203, 205, 212, 213, 216
— — in non-aqueous solution 270
Acids 25
— as solvents for metals 79
Activation analysis 392
Activity coefficient 368
Adsorption indicator 265
Affinity 25, 349
Air, composition of 65, 218
Alchemy, origin of 3, 4
Alcohol as reagent 75
Alkali fusion 121
— hydroxide, carbonate-free solution 266
— — standard solution 266
— thiocyanate, standard solution 255
Alkalimetry 214
Alkaline earth metals, separation of 186
Alkaloids, titration of 232
Ammonia, determination 266
— estimation 246
—, volumetric determination 232
Ammonium molibdate reagent 164
— sulphide as reagent 124, 154, 163
— vanadate as standard solution 167
Amperometry 387
Analysis with radioactive reagents 391
Analytical balance 148

Analytical chemistry, origin of 71
— grade reagents 151
— journals 168, 170
Anions, form of gravimetric determination of 178
—, iodometric determination of 229
Antimony, determination of 78
— electrode 376
Aqua regia, discovery of 11
Arc spectrum 324
Argentometry 216
Aristotle's theory of elements 4
Arsenate, precipitation as lead arsenate 124
Arsenic, detection of 31, 37, 163
—, reactions of 59
Arsenious acid, permanganometric titration 231
— — as standard solution 218, 226, 249
— titration with iodine 244
Ascorbic acid as standard solution 268
Ash-free filter paper 181
Atomic weights 143, 144
— —, determination of 134
Atomist theory 24
Automatic gravimetric analysis 185
Azotometer 296

B

Back-titration 213, 232, 245
Balance 5, 16, 177, 189
Barium chloride as reagent 74
— — — standard solution 224
—, determination of 177
— sulphide as standard solution 227
Beaker 147
Beer–Lambert law 340
Bimetallic electrodes 382
"Blue Test" 210
Borax, determination of 224

Boric acid, as reagent 115
— titration 266
Boron, detection of 60
Bromate, determination by ascorbic acid 268
Bromination of organic substances 254
Bromine, determination by colorimetry 337
— number 255
Brønsted theory 369
Buffer solution 262, 363
Buffer-principle 163
Bunsen battery 330
— burner 333
— photometer 330
Burette 207, 211, 214, 220, 236, 246, 247

C

Cadmium, detection of 155
—, determination of 155
— separation from zinc by electrolysis 314
Caesium, discovery of 326
Calcium, detection of 32
— nitrate as standard solution 216
Calcium–magnesium determination 269
— separation 178
Calculation in analytical chemistry 176
— of the results of analysis 361
Calomel electrode 375
Capillary analysis 393
Carbon, determination of 293
— — by digestion method 299
— dioxide 61, 62
— —, composition of 294
— —, detection of 64
— — — —, in carbonates 70
— —, determination of, in water 207
— monoxide 116
Carius's bomb method 300
Catalytic analysis 189, 190
Catalysis 189
Cations, form of gravimetric determination of 177, 178
—, iodometric determination of 229
Cementation 7, 31
Cerimetry 267
Cerium(IV)standard solution 256
Chamaeleon 231
Chelatometry 248, 269
Chemical reagents 151, 152
— —, testing for purity 151
— symbol 142
Chemiluminescent acid-base indicator 263
— redox indicator 264

Chemistry, the origin of the word 4
Chloramine T 268
Chlorate, determination by ascorbic acid 268
Chloride, determination, Mohr method 243
—, indirect determination of 255
— ion, detection of 39
Chlorinated lime, titration of 228
Chlorine, determination of in chlorine water 211
—, discovery of 70
— water as reagent 154
Chromate, determination of 154
— — with ferrous iron 240
—, permanganometric determination 231
— as standard solution 232
Chromatometry 267
Chromium, determination as chromate 124
Chromium(II) as standard solution 268
Chromophor theory of indicators 262
Chronometric method 210
Clay-crucible 145
Cobalt–nickel separation 178
Cobalt–pearl test 54
Coins, testing of 6
Colorimetric determination of hydrogen ion concentration 362
Combustion apparatus 287, 288, 290, 293
— with catalysts 294
—, Dumas method 295, 303
— of halogen containing substances 294
— — metals 45, 47
— — organic substances 285
— in oxygen atmosphere 294
Complementary colorimeter 338
Complex formation method 234
Complex-forming standard solution 257
— — volumetric methods 268
Complexones 268
Conductometric titration 383
Conservation of energy 91
Constant proportions, idea of 102
— weight-proportions 107
Controlled potential, deposition at 315
Cooler 19
Copper, detection of in filter-paper 54
—, determination by colorimetry 337
—, — — electrolysis 312
—, electrolytic deposition as test for 310
Copper oxide in organic analysis 289
Copper–mercury separation 179
Cork-borer 243
Coulometric analysis 316
Counting device 390

Cupellation 6, 15, 16
Cupferron 182
Curcuma 259
— as indicator 206, 207
Cyanide, argentometric determination of 234
— as reagent 49
—, titration of 257

D

Dalton's atomic theory 110
"Dead stop" endpoint detection 382
Density of gases 65
Deposition potential, separation by variation of 315
Derivative thermogravimetry 185
Derivatograph 185
Differential titration 382
Diffusion potential 362
Dimethylglyoxime as reagent 172, 182
Diphenylamine as reagent 174, 267
Diphenylcarbaside as indicator 264
Displacement reactions 358
Dissociation constant 358
— — of water 362
— equilibrium constant 353
Dissolved oxygen, determination of 249
Distillation apparatus 10, 11, 13
Double decomposition reaction 98
Drop ratio method 378
Dropping mercury electrode 384
Drying 177
— of precipitates 145
Dualistic electro-chemical theory 130, 134

E

Earths, reductibility of 118
Electrochemical theory 133
Electrography 316
Electrogravimetric micro-methods 316
Electrolysis, rapid methods of 314
Electrolytic dissociation 351
Electromotive force 309
Element and compound, distinction between 95
Endpoint of neutralization process 258
Energetics 355
Equivalence-point 366
—, calculation of 368
p-Ethoxychrisoidine 265, 267
Ethyl alcohol, determination of 254

Eudiometer 65
External indicator 263

F

Faraday laws 351
Ferric chloride as standard solution 234
Ferrous ammonium sulphate as permanganometric standard 248
Filter crucible 181
— funnel 146
— paper 145, 177, 181
Filtration 146, 147, 177, 181
Flame colourization 319, 325
— photometry 336
— reaction 333
— test 58
Fluorescein as indicator 260, 265
Fluorescent indicators 260, 263
Formula of a compound, determination of 180
Fraunhofer lines 320, 326
French system of measures 149
Fresenius's scheme of cation analysis 166, 170
Functional group analysis 304
Funnel 146
Fusing agents 54
Fusion 59
—, Freiberg method 124
— mixtures 53
— with alkali hydroxide 161
— — hydrogen fluoride 161
— — potassium hydrogen sulphate 161
— — potassium hydroxide 120
— — soda 80

G

Gas adsorption 186
— analysis 332
— chromatography 397
Gases, detection of 70
—, discovery of 26
—, recognition of 62
Gasometer 63, 148
Geiger–Müller counter 390
Glass electrode 375
— filter 343
Gold, determination of 15
— — — by fire-assay 6
— examination 7
— –silver, separation of, by wet method 17, 28

Gold scratch test 7, 17
—, testing of 16
Gooch crucible 181
Griess-Ilosvay reagent 172
Gutzeit test 164

H

Halogens, determination by the flask method 303
— — in organic compounds 300, 303
Hanging mercury drop electrode 386
"Happiness-equation" 355
Hardness of water 75
— — —, determination of 202, 231, 232
— — —, degrees of 232
High-frequency titration 384
Homologous line–pair method 335
Hydrochloric acid as standard solution 202
Hydrogen, determination of 293
— electrode 362, 375
— fluoride as reagent 54
— — reaction of 124
— ion concentration, determination by indicators 262
— peroxide standard solution 256
— sulphide, preparation of 162
— — as reagent 30, 154, 161
— —, properties of 76
— —, titration of 228
Hydrostatic balance 30, 40
Hydroxide–carbonate, difference between 61
—, simultaneous determination of 266
Hydroxile ion, determination in presence of carbonate 246
Hypochlorite, determination of 210, 217, 220, 226

I

Ignition 177
— of precipitates 145, 154
Indicator analysis 391
— blank, determination of 226
— correction 224
— error 213
— exponent 367
— paper 202, 206
Indicators, acid-base 25
—, colour changes in 223, 362
—, difference in the endpoint 259
— transition point 363, 366
Indigo as indicator 263
— — redox indicator 218

Indigo solution 210, 217
Indirect methods of analysis 180
Inert gases, discovery of 66
Infrared rays 319
Interference of carbon dioxide by titration 213
Internal electrolysis 315
Iodate, determination by ascorbic acid 268
Iodide, determination of 249, 254
—, permanganometric determination of 251
Iodine as reagent 154
— number 254
—, reactions of 163
Iodine–bromine number 254
Iodine–starch reaction 163
Iodometry 228
Ion exchange chromatography 398
— — papers 398
Ionic product of water 353
Iron, chromatometric determination 232
—, detection of 7, 57
— determination by ascorbic acid 268
— — — colorimetry 337, 338
— — — precipitation titration 207
— — with Mohr method 243
— — — tin(II)-chloride 267
— —, Zimmermann–Reinhardt method of 251
—, iodometric determination of 229
—, permanganometric determination of 230
— separation from magnesium 59
— — — manganese 175
— hydroxide, reaction with alkali 55
— ores, analysis of 77
— sulphate as reagent 75
Iron(II) sulphate as standard solution 217
Isomorphism 141
Isotope dilution analysis 391
Isotopically labelled atoms, analytical use 390

J

Jones reductor 251

K

Karl Fischer method 268
Kjeldahl method 297, 299, 303
Komplexon 269

L

Laboratory stand 330
Law of atomic heat 141

Law of conservation of energy 240, 241, 350
— — constant proportions 98, 118
— — Dulong and Petit 141
— — mass action 106, 350
— — multiple proportions 108, 109, 130
— — reacting gases 140
— — the volume of reacting gases 110
Laws of solutions 351
Lead acetate 75
— chloride, reactions of 35
—, detection of 34, 162
— nitrate as reagent 116
— — — standard solution 206
— separation from copper by electrolysis 312
— ores, analysis of 77
Leblanc soda-process 198
Levelling effect 270
Liebig condenser 243
Lime water as reagent 74
Limiting value for titrability 368
Litmus 259
— as indicator 204, 207, 212, 258
Logarithmic sector 335
Lucigenine 264
Luminol 264

M

Magnesium, determination of 87, 175, 177
Manganese, determination of 80
—, — by Volhard—Wolff method 251
Manganous(II) solution for precipitation titration 227
Marsh-test 163
Masking of ions 59
Mechanical equivalent of heat 350
Mercury cathode 315
— chloride as reagent 33, 39
— nitrate as reagent 75
—, precipitation titration of 255
Mercury(I) nitrate 154
— — as standard solution 226, 227, 256, 268
Mercury(I) perchlorate as standard solution 268
Mercury(II) nitrate as standard solution 234, 257
Metal-bath 145
Metallochromic indicators 269
Metal-net electrodes 315
Metals, reactions with alkalis 55, 56
Methyl alcohol, determination of 254
— orange 260
— red 263

Metric system 149
Micro methods in organic elementary analysis 301—308
Microbalance 189
Microburette 269
Microchemical reactions 188
Micromanipulator 191
Microscope 185
—, application of 59
Microscopic analysis 185, 186
Minerals, analysis of 77, 119, 122
—, identification of, with microscope 187
Mining Academy in Selmecbánya 118
Mohr balance 243
— pinch-cock 243
Moisture, determination of 268
Molybdenum-blue reaction 124
Murexide 269

N

a-Naphthylamine-sulphanilic acid as reagent 172
Nernst equation 361
Neutrality law 98, 99, 103
Neutralization 25, 97
— process, explanation of 358
— titration 200, 203, 204
Neutral-point 366
Nickel, determination by colorimetry 337
Nitrate, detection of 172
—, indirect determination of 231
Nitric acid, discovery of 11, 12
— preparation 12, 13
Nitrite, detection of 172, 174
—, permanganometric determination of 251
Nitrogen, determination of 295, 296
— — —, by wet method 296
— discovery of 65
a-Nitroso-β-naphtol 182
Non-aqueous solution, titration in 270
Normal solution 229
— standard solution 235, 245

O

Oath of the analysts 19
Oil-bath 145
Oleum 13
Organic reagent 172
Oscillopolarography 386
Oxalic acid, as acidimetric standard 246

Oxalic acid, permanganometric determination of 231
— — as permanganometric standard 248
— — — reagent 59, 69, 73
Oxidation potential 370
Oxidizing substances, iodometric determination of 229
Oxygen consumption of water 252
— determination in organic compounds 300, 301, 304
—, discovery of 67
— dissolved in water, determination of 255
— flame 53

P

Paper chromatography 393, 397
Particle size of precipitates 358
Partition chromatography 397
Pelop 165
Periodic acid 268
Periodicals 44
Permanganometry 230
Persistent lines 334
Phase rule 350
Phenolphthalein 260, 365
Phlogiston, determination of 50
Phosphate, determination of 249
— precipitation as lead phosphate 124
Phosphoric acid, as reagent 51, 58
— — production 58
— —, volumetric determination of 234
Photoelectric measurement of line intensities 335
Photometer with slit width variation 343
Photometry 340
pH determination 259
— — by colorimetric method 263, 376, 377
— — by indicators 259
pH-sign 364
Pinch-cock burette 243
Pipette 211, 212, 220, 225
—, calibrated 243
—, device for filling 225
Plant extracts as indicators 258
Platinum chloride as reagent 154
— crucible 145
— metal, analysis of 149
— vessel 113, 121
Polarization photometer 343
Polarized electrodes 381
Polarogram 386
Polarograph 385

Polarography 384
Polarometry 387
Porcelain crucible 135, 145
Potash, examination of 201, 216, 222
Potassium, determination of 177
—, reactions of 121
— separation from sodium 59
— and sodium, difference between 58
— bicarbonate as acidimetric standard 253
— bi-iodate as iodometric standard 252
— bromate as standard solution 267
— chlorate as standard solution 233
— hexacyanoferrate(II) as reagent 56, 73
— — — standard solution 207, 226, 249
— hexacyanoferrate(III) as standard solution 256
— iodate as standard solution 233
— — — standard substance 254
— periodate as standard solution 267
— sulphate, composition of 86
Potentiometric microtitration 383
— titration 378
Precipitants 79
Precipitation 28
—, definition of 36
— from extremely concentrated solution 184
— — — dilute solution 184
— — — homogeneous solution 184
— titration 202, 206, 216, 217, 224, 227, 234, 249, 255
— — with potassium chromate 257
— — theory 369
Prout's hypothesis 141
Prussian blue 56, 57
Purity of reagents 151
Purple of Cassius 166

Q

Qualitative analysis, systematic scheme 165, 166
— organic analysis 304
— and quantitative analysis 54, 117
Quantitative determinations by blow-pipe 55
Quantometric method 335
Quartz prism 324
Quinhydron electrode 376

R

Radioactive precipitate exchange 392
Radioactivity, measurement of 390

SUBJECT INDEX

Radiometric titration 391
β-Ray deflection methods 392
γ-Ray spectrometry 392
β-Rays crystal counter 393
Reaction velocity 349
Reagents, definition 73
Redox indicator 226, 263
— —, theoretical investigation 265
— potential 370
— — measurement 370
— reactions on ion exchange resin 398
— titration 208, 217, 226
— — in non-aqueous solution 270
— — theory 370
Reductometric reagents 268
Reversible redox indicator 264
Ring-oven technique 191
Rinman-green test 53
Rock-salt, composition of 87
Rotating anode 314
Rubber tube 149
Rubidium, discovery of 326

S

Salicylic acid as indicator 260
Salts, composition of 86, 89, 122
Sampling, method of 248
Sand-bath 145
Scheme of qualitative analysis 149
Schöniger method 303
Scientific Academies 22, 40
— Societies 44
Scintillation counter 390
Separating funnel 145
Separation by hydrolysis 178
—, gold—silver 7
Sensitivity of detection 39
— — reactions 154
Shaking device 223
Silicate ores, examination of 123
Silicic acid, behaviour of 33
Siloxene 264
Silver, determination, by ascorbic acid 268
— — in form of silver chloride 85, 86
— nitrate as reagent 31, 75
— — as standard solution 207
— precipitation titration 224, 255
Sintered glass crucible 181
Soap as reagent 75
— — standard solution 231
Sodium ammonium phosphate as reagent 73
— carbonate as acidimetric standard 247

Sodium chloride as standard solution 225
— hydroxide as standard solution 245
— nitroprussiate indicator 257
— succinate, reagent for iron 124
— sulphate 154
— sulphide as standard solution 234
Solubility 177
— product 359
Solution process 25
Specific weight, determination of 13
Spectral lines 320, 324
Spectrocolorimeter 343
Spectrometric method 335
Spectroscope 325
Spectrum 319
— analysis 324
— —, quantitative 334
— of the sun 319
—, photography of 322
—, reversed 327
Spinthariscope 390
Spot test 190
— — as indicator 263
Standard redox potential 370
— solution 199, 200, 214
— — in inert atmosphere 256
Standardization 203, 246, 247, 248, 254
— of potassium permanganate 248
Stannous chloride, as standard solution 229, 233
Steel, analysis of 80
Stoichiometry, origin of the word 99, 102
Sugar, detection of 233
—, determination of 234
Sulphate, determination of 216
—, — —, Andrew's method 257
Sulphide, determination of 123
Sulphonphthalein group of indicators 263
Sulphur, detection in organic substances 299
—, determination in organic substances 299, 303
Sulphuric acid, discovery of 11
— —, manufacture of 197
— —, preparation of 13
— — as reagent 32, 34, 73
— — — standard solution 201
Sulphurous acid as standard solution 229
Synthetic indicators 260
Synthetic organic ion exchangers 398

T

Tannic acid, examination of 33
— — as reagent 7, 30, 73

Tartaric acid as reagent 121
Temperature correction of volumes 225
Test of salts 151
Test-tube 145
Thermal examinations 52
— investigation 177
Thermobalance 184
Thermodynamics 350
Thermogravimetry 184
Thiocyanic acid 49
Thiosulphate as standard solution 230, 246
Tin chloride as reagent 154
— examination of 7
—, separation from antimony 180
Tin(II) chloride as standard solution 267
Tin(II) chromatometric determination 233
Titanium chloride as standard solution 267
— dioxide, determination of 252
Titration of acids 200
— curve 367, 368
— error 368
— exponent 367
—, origin of 197
— to zero potential 381
Transference number 351
Transition point of indicators 363, 366
Tropeolin 260

U

Ultramicroanalysis 191
Ultraviolet rays 319
Universal tube filling 303
Universities, Arabian 9
—, European 10
Uranium as standard solution 234
Uranyl acetate as reagent 188
U-tube 290

V

Vanadium, determination of 252
Violet extract as indicator 201, 214
— syrup as indicator 258
Vis vitalis 284
Volumetric flask 216
— micro methods 269

W

Warder's method 266
Washing of precipitates 147, 148, 177, 356, 360
Water, analysis of 122, 154, 201, 204, 212
— bath 11, 145
—, composition of 66
—, decomposition of 310
—, determination of 268
—, examination of 8, 29—33, 57, 73, 76, 115
— hardness 266, 269
— pump 330, 333
Wavelengths of spectral lines calculation 333
Weak organic acids, titration of 266
Weights 5

X

X-ray fluorescence analysis 335

Z

Zeolite 398
Zinc, separation from silver by electrolysis 315
— sulphate, composition of 87
Zinc-reductor 251

OTHER TITLES IN THE SERIES IN ANALYTICAL CHEMISTRY

Vol. 1. WEISZ—*Microanalysis by the Ring Oven Technique*
Vol. 2. CROUTHAMEL (Ed.)—*Applied Gamma-ray Spectrometry*
Vol. 3. VICKERY—*The Analytical Chemistry of the Rare Earths*
Vol. 4. HEADRIDGE—*Photometric Titrations*
Vol. 5. BUSEV—*The Analytical Chemistry of Indium*
Vol. 6. ELWELL AND GIDLEY—*Atomic Absorption Spectrophotometry*
Vol. 7. ERDEY—*Gravimetric Analysis, Parts I–III*
Vol. 8. CRITCHFIELD—*Organic Functional Group Analysis*
Vol. 9. MOSES—*Analytical Chemistry of the Actinide Elements*
Vol. 10. RYABCHIKOV AND GOL'BRAIKH—*The Analytical Chemistry of Thorium*
Vol. 11. CALI—*Trace Analysis of Semiconductor Materials*
Vol. 12. ZUMAN—*Organic Polarographic Analysis*
Vol. 13. RECHNITZ—*Controlled-Potential Analysis*
Vol. 14. MILNER—*Analysis of Petroleum for Trace Elements*
Vol. 15. ALIMARIN AND PETRIKOVA—*Inorganic Ultramicroanalysis*
Vol. 16. MOSHIER—*Analytical Chemistry of Niobium and Tantalum*
Vol. 17. JEFFERY AND KIPPING—*Gas Analysis by Gas Chromatography*
Vol. 18. NIELSEN—*Kinetics of Precipitation*
Vol. 19. CALEY—*Analysis of Ancient Metals*
Vol. 20. MOSES—*Nuclear Techniques in Analytical Chemistry*
Vol. 21. PUNGOR—*Oscillometry and Conductometry*
Vol. 22. ZYKA—*Newer Redox Titrants*
Vol. 23. MOSHIER AND SIEVERS—*Gas Chromatography of Metal Chelates*
Vol. 24. BEAMISH—*The Analytical Chemistry of the Noble Metals*
Vol. 25. YATSIMIRSKII—*Kinetic Methods of Analysis*

MADE IN GREAT BRITAIN

1-11-67